Bioprocesses and Biotechnology for Functional Foods and Nutraceuticals

NUTRACEUTICAL SCIENCE AND TECHNOLOGY

Series Editor

FEREIDOON SHAHIDI, PH.D., FACS, FCIC, FCIFST, FRSC

University Research Professor
Department of Biochemistry
Memorial University of Newfoundland
St. John's, Newfoundland, Canada

1. Phytosterols as Functional Food Components and Nutraceuticals, *edited by Paresh C. Dutta*
2. Bioprocesses and Biotechnology for Functional Foods and Nutraceuticals, *edited by Jean-Richard Neeser and Bruce J. German*

ADDITIONAL VOLUMES IN PREPARATION

Bioprocesses and Biotechnology for Functional Foods and Nutraceuticals

edited by

Jean-Richard Neeser

Nestlé Research Center
Lausanne and
University of Geneva
Geneva, Switzerland

J. Bruce German

Nestlé Research Center
Lausanne, Switzerland, and
University of California, Davis
Davis, California, U.S.A.

MARCEL DEKKER, INC. NEW YORK • BASEL

Although great care has been taken to provide accurate and current information, neither the author(s) nor the publisher, nor anyone else associated with this publication, shall be liable for any loss, damage, or liability directly or indirectly caused or alleged to be caused by this book. The material contained herein is not intended to provide specific advice or recommendations for any specific situation.

Library of Congress Cataloging-in-Publication Data
A catalog record for this book is available from the Library of Congress.

ISBN: 0-8247-4722-4

This book is printed on acid-free paper.

Headquarters
Marcel Dekker, Inc., 270 Madison Avenue, New York, NY 10016, U.S.A.
tel: 212-696-9000; fax: 212-685-4540

Distribution and Customer Service
Marcel Dekker, Inc., Cimarron Road, Monticello, New York 12701, U.S.A.
tel: 800-228-1160; fax: 845-796-1772

Eastern Hemisphere Distribution
Marcel Dekker AG, Hutgasse 4, Postfach 812, CH-4001 Basel, Switzerland
tel: 41-61-260-6300; fax: 41-61-260-6333

World Wide Web
http://www.dekker.com

The publisher offers discounts on this book when ordered in bulk quantities. For more information, write to Special Sales/Professional Marketing at the headquarters address above.

Current printing (last digit):

10 9 8 7 6 5 4 3 2 1

PRINTED IN THE UNITED STATES OF AMERICA

Series Introduction

The Nutraceuticals Science and Technology series provides a comprehensive and authoritative source of the most recent information for those interested in the field of nutraceuticals and functional foods. There is now a growing body of knowledge, sometimes arising from epidemiological studies and often substantiated by preclinical and clinical studies, demonstrating the relationship between diet and health status. Many of the bioactives present in foods, from both plant and animal sources, have been shown to be effective in disease prevention and health promotion. The emerging findings in the nutrigenomics and proteomics areas further reflect the importance of diet in a deeper sense, and this, together with the increasing burden of prescription drugs in treatment of chronic diseases such as cardiovascular ailments, certain types of cancer, diabetes, and a variety of inflammatory diseases, has raised interest in functional foods and nutraceuticals to a new high. This interest is quite widespread, from producers to consumers, regulatory agencies, and health professionals.

In this series, particular attention is paid to the most recent and emerging information on a range of topics covering the chemistry, biochemistry, epidemiology, nutrigenomics and proteomics, engineering, formulation, and processing technologies related to nutraceuticals, functional foods, and dietary supplements. Quality management, safety, and toxicology, as well as disease prevention and health promotion aspects of products of interest, are addressed. The series also covers relevant aspects of preclinical and clinical trials, as well as regulatory and labeling issues.

This series provides much needed information on a variety of topics. It addresses the needs of professionals, students, and practitioners in the fields of food science, nutrition, pharmacy, and health, as well as leads conscious consumers to the scientific origin of health-promoting substances in foods, nutraceuticals, and dietary supplements. Each volume covers a specific topic of related foods or prevention of certain types of diseases, including the process of aging.

Fereidoon Shahidi

Preface

Science and its applications to biotechnology today are facing the greatest opportunities in the history of mankind. Biological systems of virtually all sorts can be controlled in ways not thought possible as recently as a decade ago. The genomics revolution in the study of biological organisms is empowering all the life sciences. The use of genomics and functional genomics in disease target identification and drug discovery is propelling the pharmaceutical industry into a new era of successful intervention in human disease, promising individual health through therapeutics. In the view of many scientists and economists, innovation in agriculture will enrich virtually every human activity—from food and energy production to communication to polymer design to human habitation. With such unprecedented knowledge of living organisms, application of this knowledge to biological productivity can begin to address the great challenges of modern societies: starvation and food shortages, global energy, pollution, and safety. The inherent efficiencies of biology will continue to revolutionize and empower the lives of individuals by enhancing quality of life, preventing disease, and extending human performance capabilities. In no field is the promise of innovation in agriculture from biotechnology so vivid as that of food.

Ironically, at the precise moment that biotechnology is poised to revolutionize every aspect of food, the consuming public, including scientists, has lost faith in modern science to improve our food supply. The world is turning its back on science and the application of biotechnology to food at a time when scientific knowledge has become most predictive and useful in food applications. With the challenges facing the world's immense population, do we dare slow the progress of science addressing our most essential human need?

The contributors to this book have taken on the challenge of addressing this problem directly. An important underlying cause in this loss of confidence by the public is a real or perceived disconnection within the scientific community and a sense that biotechnology as big business is leaving mainstream scientists behind. Such a perception is incorrect but emerges from a lack of knowledge and communication. This book is a clear statement of clarification.

An international group of scientists from academic, governmental, and industrial research settings have addressed the problem directly. These individuals have shown unusual vision in their writing and the potential of modern biological science to revolutionize the biotechnology of foods. Each chapter articulates the contributor's view of the possible future of food biotechnology and how science will realize that promise within his or her respective specialization. We are pleased to present a broad spectrum of research perspectives that not only illustrate the power and safety of biotechnological research but should serve as a blueprint for the progress of the science of foods.

Part I addresses biological organisms in which scientific research illustrates how powerfully biotechnology can improve all aspects of traditional food commodity production. As the scope of the many agricultural commodities is extremely wide, this book specifically includes areas that have not been well addressed in most other texts on biotechnology applications to food and agriculture, which have focused solely or mostly on plant foods. Consequently, the reader should refer to other reviews if specifically interested by plant foods.

In this first parts, animal products are examined. Chapter 1 describes in detail poultry and egg production. As a magnificent example of a modern bioreactor, the laying hen represents an astonishingly productive organism delivering one of agriculture's most nutritious products. The poultry industry has become one of the most successful and valuable contributors to the global food supply. Chapter 2 addresses the dairy industry and its myriad product offerings. Few commodities are more linked to food traditions around the world, and dairy products are fast becoming the chosen carriers of innovative nutritional values.

In Part II, microbial products are examined. Microbial systems provide almost limitless potential for introducing biotechnology. From their origins thousands of years ago, as perhaps the most primitive biological means to process agricultural commodities, microbial systems are being re-examined for producing specific food ingredients.

Two separate chapters provide an overview of the scientific strength and application potential of microbiology to the future value of the food supply. Finally, the health potential of probiotic organisms, bacteria that are ingested for the purpose of directly affecting the consumer's own intestinal health, is the subject of a fascinating chapter. That biotechnology can impact the lives and health of consumers through the consumption of bacteria designed specifically for this purpose is extremely exciting.

Part II also addresses (a) which health effects can be expected from specific food ingredients and (b) how such ingredients can be produced for further addition in food products. Both aspects are presented in some of the chapters, whereas other authors have developed only one of these two questions, depending on the class of the functional ingredient.

Chapter 6 focuses on prebiotic carbohydrates from lactose and plant polysaccharides. Here, health effects and production processes are equally reported. The well-documented fructo-oligosaccharides (FOS) are not presented here due to the numerous reports already existing on these prebiotic ingredients. Chapter 7 deals with dextrans and gluco-oligosaccharides, which should be regarded more as colonic foods than as prebiotics. Both questions of health effects and enzymatic technologies for production of these carbohydrates are discussed. These important reviews on non-digestible carbohydrates are followed by Chapter 8, which is entirely focused on human and mechanistic studies aimed at measuring the effects of a prophylactic usage of prebiotics to prevent gut disorders.

Chapter 9 deals with questions related to the addition of recombinant milk proteins and peptides to infant formula. Such polypeptides may be produced in transgenic animals or, alternatively, in microorganisms or plants. These aspects are discussed in great detail, as are the questions related to biochemical assessment, digestibility, and in vivo evaluation of these ingredients. Chapter 10 is restricted to the use of enzymes as food ingredients, especially for functional foods. Production guidelines are also presented.

In Chapters 11 and 12, the health effects of plant metabolites of two classes are reported: isoflavones and anandamides. Analytical aspects, biological effects, and intervention trials are thoroughly presented and discussed.

In Part IV, chapters address the vital issues that will promote or retard the applications of biotechnology in our lifetime. How does the consumer perceive biotechnology, its benefits, and its risks? How can the consumer be educated on the appropriate assessment of risk and benefit? Legal implications of globalization of biotechnology remain important issues and have been addressed from the perspective of the principles of both biotechnology and international law.

It has been said that the nineteenth century saw the industrialization of chemistry to produce chemicals, leading to the chemical industry and all the improvements in the human condition that followed. Similarly, the twentieth century saw the industrialization of chemistry to enhance biology; the commercialization of everything from fertilizers and pesticides guided the human condition through its greatest century ever, successfully addressing issues from infectious diseases to agricultural productivity. The twenty-first century will see the industrialization of biology, which will drive another quantum leap forward in the human condition. The scientific community has produced the biological tools, and these tools are accelerating knowledge of biotechnology and its myriad applications. It is now up to the imaginations of scientists and industrialists to create opportunities for utilization biotechnological innovation throughout the human experience. This book provides a glimpse into that future and how science will enable it.

Jean-Richard Neeser
Bruce J. German

Contents

Contributors

Kevin J. Acheson Nestlé Research Center, Lausanne, Switzerland

Yuriko Adkins Department of Nutrition, University of California, Davis, Davis, California, U.S.A.

Daniel Auriol Centre de Bioingénierie Gilbert Durand, INSA, Toulouse, France

A. Berger Nestlé Research Center, Lausanne, Switzerland

Linda F. Bisson Department of Viticulture and Enology, University of California, Davis, Davis, California, U.S.A.

Stephanie Blum Nestlé Research Center, Lausanne, Switzerland

Christine M. Bruhn Center for Consumer Studies, Department of Food Science and Technology, University of California, Davis, Davis, California, U.S.A.

Ross G. Crittenden Food Science Australia, Werribee, Victoria, Australia

Willem M. de Vos Wageningen University and Wageningen Center for Food Sciences, Wageningen, The Netherlands

V. Di Marzo Istituto per la Chimica di Molecole di Interesse Biologico, Naples, Italy

M. Eric Gershwin Division of Rheumatology, Allergy and Clinical Immunology, University of California, Davis, Davis, California, U.S.A.

Glenn R. Gibson The University of Reading, Reading, England

Ryota Haba Tokyo University of Agriculture, Tokyo, Japan

Barry Halliwell National University of Singapore, Singapore

Todd R. Klaenhammer Department of Food Science, North Carolina State University, Raleigh, North Carolina, U.S.A.

Chang Y. Lee Cornell University, Geneva, New York, U.S.A.

Cherl-Ho Lee Center for Advanced Food Science and Technology, The Graduate School of Biotechnology, Korea University, Seoul, Korea

Bo Lönnerdal Department of Nutrition, University of California, Davis, Davis, California, U.S.A.

Katherine Mace Nestlé Research Center, Lausanne, Switzerland

Annick Mercenier Nestlé Research Center, Lausanne, Switzerland

Pierre F. Monsan Centre de Bioingénierie Gilbert Durand, INSA, Toulouse, France

Mark A. Morse Division of Environmental Health Sciences, School of Public Health and Comprehensive Cancer Center, The Ohio State University, Columbus, Ohio, U.S.A.

Jun Ogawa Division of Applied Life Sciences, Kyoto University, Kyoto, Japan

Martin J. Playne Melbourne Biotechnology, Hampton, and RMIT University, Melbourne, Victoria, Australia

Pedro A. Prieto Abbott Laboratories, Columbus, Ohio, U.S.A.

Robert A. Rastall School of Food Biosciences, The University of Reading, Reading, England

Eduardo J. Schiffrin Nestlé Research Center, Lausanne, Switzerland

Sakayu Shimizu Division of Applied Life Sciences, Kyoto University, Kyoto, Japan

Judith S. Stern University of California, Davis, Davis, California, U.S.A.

Gary D. Stoner Division of Environmental Health Sciences, School of Public Health and Comprehensive Cancer Center, The Ohio State University, Columbus, Ohio, U.S.A.

Luc Tappy Physiology Institute, Lausanne, Switzerland

Sayo Uesugi Tokyo University of Agriculture, Tokyo, Japan

Rosemary L. Walzem Department of Poultry Science, Texas A&M University, College Station, Texas, U.S.A.

Shaw Watanabe Tokyo University of Agriculture, Tokyo, Japan

Miriam H. Watson University of California, Davis, Davis, California, U.S.A.

Detlef Wilke Dr. Wilke & Partner Biotech Consulting GmbH, Wennigsen, Germany

G. Crozier Willi Nestlé Research Center, Lausanne, Switzerland

Gerald N. Wogan Biological Engineering Division, Massachusetts Institute of Technology, Cambridge, Massachusetts, U.S.A.

Steven H. Yoshida Consultant, Davis, California, U.S.A.

Introduction
The Role of Biotechnology in Functional Foods

The term *physiologically functional food* first appeared in *Nature* news (1993) with the headline "Japan Explores the Boundary between Food and Medicine" (1). The news introduced hypoallergenic rice as an example of this new food produced by biotechnology, with special emphasis on its medical significance.

Neither the terminology nor the concept of "functional food" had existed until nine years earlier. In 1984, an ad hoc research group started a systematic, large-scale national project under the sponsorship of the Ministry of Education, Science, and Culture (MESC) to explore the interface between food and medical sciences (2). It seems that the initiation of this research event in Japan reflects an underlying thought in the minds of Japanese people throughout history, which originated in the ancient Chinese saying "Medicine and food are isogenic."

However, after the advent of modern food science in Japan, which took place about 100 years ago, subsequent research proceeded toward a purely nutritional rather than medical science. Especially in the early twentieth century, the nutritive value of foods was of great academic concern. A number of scientists in the field of chemistry as well as food science made particular efforts to discover new nutrients. One of the best examples is offered by oryzanin (vitamin B_1), the first vitamin discovered, which was isolated from rice bran by the outstanding agricultural chemist Umetaro Suzuki, professor at the University of Tokyo. Initially, he found that it had a role in combatting the serious disease beriberi. However, further experimentation made him recognize this factor as an essential nutrient rather than a specific antiberiberi medicine (3). The recognition triggered the establishment of vitaminology and also contributed in a large measure to the development of general food nutrition, which was critically important for the health of people during the time throughout the pre–, mid–, and post–World War II periods. Probably, a similar situation existed in most countries of the world.

In the meantime, nutritional problems due to food shortage were nearly solved in many countries and a new period characterized by high economic growth began. That condition prevailed in the time encompassing the 1960s, when the social climate focused on food preference and aversion. Such a trend of hedonism led to academic studies on sensory properties of foods. Epoch-making advances in instrumental analysis assisted the studies greatly. Japan, among many developed countries, thus contributed to two mainstays of food science: nutrition and hedonics. In accordance with this, food industries were activated to supply a great deal of nutritionally and sensorily acceptable products to the market. We enjoyed our rich dietary life until almost 20 years ago, when a new, antithetical situation arose.

In the 1980s, as the aging of society began to manifest itself in many countries of the world, prompt increases in incidence of so-called lifestyle-related diseases became a matter of public concern. Growing awareness of the need for nutrition to beat the odds close. The purpose was to prevent lifestyle-related diseases such as diabetes, arteriosclerosis, osteoporosis, allergies, cancer, and even some kinds of infectious diseases through improved dietary practices in our daily life. This gave a strong impetus to food science in Japan. In 1984, the MESC project, "Systematic Analysis and Development of Food Function," commenced. Food value criteria were defined by three categories:

1. *Primary function*, identified as the function of ordinary nutrients that exert purely nutritional effects on the body
2. *Secondary function*, referring to the function of tastants, odorants, and other elements in relation sensory organs, cells, receptors, etc.
3. *Tertiary function*, newly defined as the body-modulating function of food factors involved directly or indirectly in the prevention of lifestyle-related diseases

Factors with a tertiary function are often called physiologically functional food factors or simply functional factors. They are sometimes called functional components. It is noted, however, that, as mentioned, although vitamin B_1 was used to prevent beriberi, it is not recognized as a functional factor because the disease results from a type of malnutrition due to a deficiency in vitamin B_1 itself. On the other hand, vitamin C, if used for the purpose of preventing cancer, is not a nutrient but a functional factor. Incidentally, Clydesdale (4) has commented that the concept of tertiary function is a most interesting notion and certainly has merit, since it is broad in scope and obviously based on science. The MESC project (1984–87) proposed at the same time the term *functional food* as a new entity that can be designed and produced according to the concept of the tertiary function. It was followed by a second project (1988–91), "Analysis of Body-Modulating (i.e., Tertiary) Functions of Foods." Shortly afterwards, the last in a series of MESC "Functional Foods" projects occurred in 1992, it focused on "Analysis and Molecular Design of Functional Foods" with the following programs:

1. Analysis and design of body-regulating food factors

Factors whose active forms function
- To induce endogenous substances, e.g., hormones, in the body
- To act as biomimetics for endogenous substances, e.g., neurotransmitters, in the body

Factors whose precursor forms function
- At the preabsorptive stage (e.g., to modulate the intestine) after activation by intraluminal digestion
- At the postabsorptive stage (e.g., to control hypertension) after activation by intraluminal digestion

2. Analysis and design of body-defending food factors

Factors involved in immunological defenses
- Immunostimulants, e.g., those that activate immunocompetent cells in the body
- Immunosuppressants, e.g., those that induce the state of energy in the body

Factors involved in nonimmunological defenses
- Anti-infection factors, e.g., antiviral substances
- Antitumor factors, e.g., phytochemicals with anti-initiation and anti-promotion effects

3. Development of a technological basis for the design of functional foods

Design of microscopic structures
- By molecular breeding to produce, e.g., deoxyribonucleic acid (DNA) recombinants
- By molecular tailoring to produce, e.g., hypoallergenic proteins with removed epitopes

Design of macroscopic structures
- By newly developed technologies for structural conversion, e.g., microcapsulation
- By newly developed technologies for structural analysis, e.g., nondestructive testing for functional factors

A total of 60 academic professionals participated, most from medical as well as food science fields. Representative research reports, all of which resulted from the use of modern biotechnology, are summarized as follows:

1. A highly hydrophobic molecular fraction from an enzymatic hydrolysate of soybean glycinin functions at the preabsorptive stage to bind tightly bile acids, with a resultant decrease in the serum cholesterol level. Also, construction of a random peptide library by enzymatic hydrolysis of wheat gluten and screening for functional components yielded Gly-Tyr-Tyr-Pro-Thr, which is effective in accelerating the secretion of insulin (5). Thus, a food protein is not only an amino acid supplier to the body (primary function) but also a functional factor (tertiary function). The development of basic studies along this line was directly or indirectly applied to producing a variety of hypotensive oligopeptides as functional components contained in "foods for specified health uses" (FOSHUs) described later.

2. Reinvestigation of conventional food components for anticarcinogenicity disclosed that α-carotene is more potent than β-carotene. On the other hand, research into novel phytochemicals elucidated the occurrence of new antipromoters, e.g., fucosterols in seaweed and auraptenes in citrus fruits (6). The discovery and characterization of potent antioxidants of plant origin, e.g., curcuminoids in ginger, were also reported. Another unique example was offered by oryzacystatin, a cysteine proteinase inhibitor found in the rice *Oryza sativa* L., which was molecularly cloned (7) and demonstrated to inhibit the proliferation of human herpesvirus in infected animal cells (8).

3. Rice-associated allergy in the form of atopic dermatitis is continually increasing in Japan, a major rice-consuming country, and countermeasures have been urgently needed in recent years. In response to such a social need, an experiment was undertaken to design and produce hypoallergenic rice grains by enzymatic hydrolysis of the main allergenic proteins that occur in the globulin fraction. The development of well-controlled conditions for the hydrolysis yielded an immunologically and clinically satisfactory product (9) and the production process was industrialized for its manufacture. "Fine Rice" was approved as the first FOSHU item after careful intervention tests were made in many medical centers. These basic and applied studies were followed by the discovery and characterization of soybean and wheat allergens, which led to development of hypoallergenic soybean and flour products (10).

The orthodox way of producing functional foods is to maximize beneficially functional factors that are responsible for the modulation of immune, endocrine, nerve, circulatory, or digestive systems in the body. Sometimes, however, a unique approach can be adopted. Uniqueness is found in minimizing nonbeneficial functional factors by enzyme technology or even genetic engineering. The importance of using this concept for the design and production of functional foods has been emphasized since 1984, when the first MESC project began. Noting the significance of this, Roberfroid (11) has summarized approaches to developing functional foods as follows:

1. Eliminating a component known or identified to cause deleterious effects to the consumer, e.g., allergenic protein
2. Increasing the concentration of beneficial components naturally present in food
3. Adding a beneficial component that is not normally present in most foods
4. Replacing a component, usually a macronutrient, whose intake is usually excessive, with a component for which beneficial effects have been demonstrated

How should we then define the position of functional foods? Early in 1984, a proposal was made to define the position of functional foods in relation to ordinary foods and medicines or pharmaceuticals. As illustrated in Fig. 1, we take nutrients from foods every day to promote our health; even completely healthy boys and girls must do so, since otherwise they are unable to lead healthy lives. If for some reason we get sick, it is necessary to take medicines for the purpose of treating the disease. However, a recent medical concept suggests that there is a semi-healthy state between health and a particular disease. For example, when a man has a systolic blood pressure of

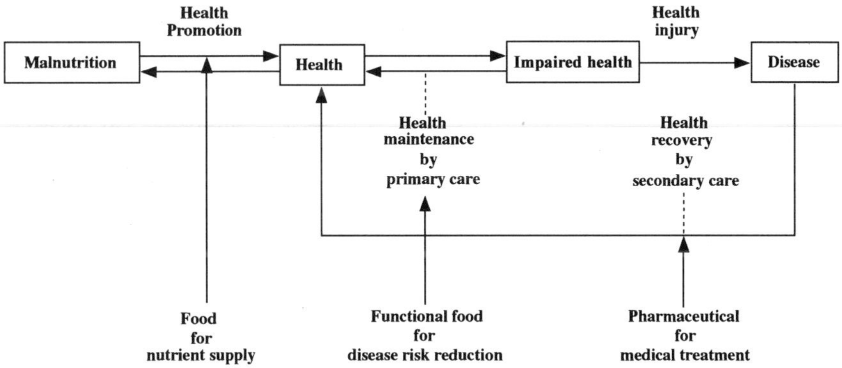

Figure 1 Proposal for positioning a functional food in relation to an ordinary food and a medicine.

130 mmHg, he may not be diagnosed as suffering from hypertension at that moment. However, if his pressure tends to go up month by month or year after year, the present status is thought to be a predrome to this disease. The same may be true for most lifestyle-related diseases. It is at such a stage that functional foods should be taken for primary care to reduce the risk of development of a particular disease, whereas pharmaceuticals are used for secondary care in medical treatment.

With this notion as a background, the Japan Ministry of Health and Welfare (MHW) in 1991 initiated the World's first policy of legally permitting the commercialization of selected functional foods in terms of FOSHU, as mentioned. The new policy is defined by new legislation and also characterized by approval of the presentation of a health claim for each FOSHU product. Such a legal framework is also expected to hinder the presentation of ill-defined and misleading advertisements in commercial products. Thus, since 1993, some selected food products have been approved to claim a certain degree of medical representation never before permitted for any food. The first was hypoallergenic rice grains (commercial name "Fine Rice"), developed after extensive immunological studies in the MESC project, produced by enzyme technology, and industrialized after clinical intervention tests (Fig. 2) (1,9).

A great many FOSHU candidates have been evaluated so far, and more than 300 approved by September 30, 2003. Table 1 summarizes the state of the art. It deserves note that all have been designed and produced on the basis of science. Examples are offered by, among others, the following two items. The first is a sterilized lactic acid bacterium drink (commercial name "Ameal-S"), with a blood pressure–modulating or hypotensive function. This is produced from milk by fermentation under controlled conditions, in which β-casein is microbially converted to peptides, including two lactotripeptides, Ile-Pro-Pro and Val-Pro-Pro, which can inhibit angiotensin-converting enzyme. Apparently, these are derived from bovine β-casein 74IPP76 and 84VPP86, respectively (Fig. 3). Continuous intake of the drink, one bottle a day, is expected to decrease systolic and diastolic pressures to a reasonable degree. The second example is a fermented soybean product, *natto*, which has enhanced vitamin K_2 (menaquinone-7) content. Though ordinary natto products per se contain a certain amount of menaquinone-7 synthesized from soybean constituents by fermentation with *Bacillus natto*, the FOSHU product (commercial name "Honegenki") is characterized by having 1.5-fold more of this substance as a result of using a genetically improved *Bacillus* species. Historically, vitamin K was identified as a blood-coagulating factor but now close attention is paid to its contribution to the synthesis of the 4-carboxyglutamic acid residues in the protein osteocalcin, which transports calcium to the bone. Actually, data showing that daily intake of the natto

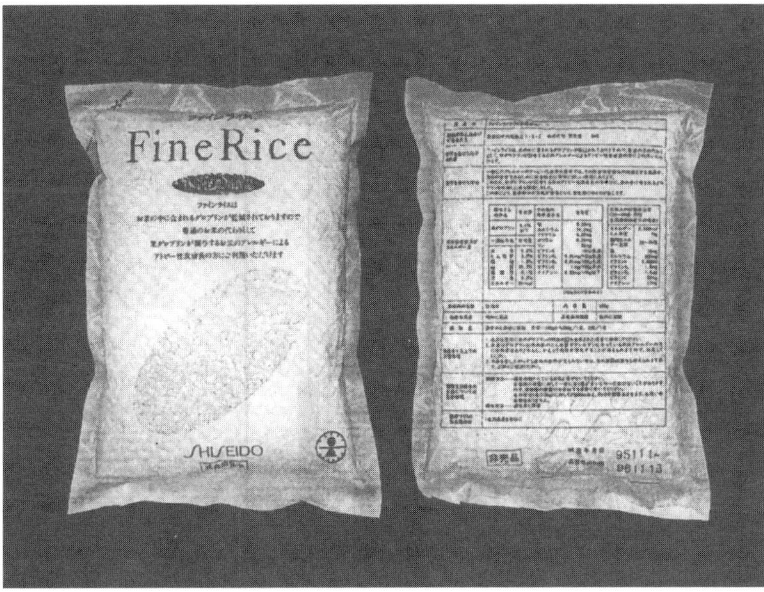

Figure 2 Hypoallergenic rice as the first approved food for specified health use and its health claim. This rice, because its globulin has been reduced, can be used, in place of normal rice, for those who are allergic (atopic dermatitis) to proteins of the rice globulin class.

product with enhanced menaquinone-7 can regulate the progression of osteoporosis, especially in aged women whose normal estrogen function has been jeopardized, are available.

Apparently, the implementation of the MHW FOSHU policy as well as the initiation of functional science in Japan had a strong impact on many countries of the world, particularly on Europe. Early in 1995, the U.K. Ministry of Agriculture, Fisheries, and Food temporarily defined *functional foods* as foods that have components incorporated to yield specific medical or physiological benefits, other than purely nutritional effects (12).

Deeper attention was paid by the European body of the International Life Science Institute (ILSI Europe), which addressed the present status by claiming that we stand today at the threshold of a new frontier in nutritional science and also that the concept of food is changing from a past emphasis on eating to end of hunger into a current emphasis on promising uses of foods to reduce the risk of chronic illness. This new concept is of extreme importance in view of the demands of the elderly population for improved quality of their

Table 1 Data on 195 Foods for Specified Health Uses (31 March 2001)[a]

Health claim[b]	Functional factor	Number of products approved	Forms of products on the market
I	Prebiotics	58	
	Lactosucrose	21	Soft drink, yogurt, biscuit, cookie, table sugar
	Fructo-oligo saccharides	15	Table sugar, tablet candy, pudding, soybean curd
	Soybean oligosaccharides	8	Soft drink, table sugar
	Xylo-oligo saccharides	5	Soft drink, vinegar, chocolate, tabet candy
	Galacto-oligo saccharides	4	Table sugar, vinegar
	Isomalto-oligo saccharides	3	Table sugar
	Raffinose	1	Powdered soup
	Lactulose	1	Soft drink
	Probiotics	36	
	Lactobacillus casei Shiota	22	Lactic acid bacterium drink
	L. delbrueckii and *Steptococcus salivarius*	6	Fermented milk
	Bifidobacterium breve Yakult	3	Fermented milk
	L. GG	2	Fermented milk
	B. longum	2	Fermented milk
	L. acidophilus and *B. longum*	1	Fermented milk
	Luminacoids	41	
	Indigestible (resistant) dextrin	19	Sausage, cookie, powdered drink
	Psyllium husks	19	Powdered drink, noodle, cereal
	Wheat bran	2	Cereal
	Galactomannan	1	Jelly
	Partially hydrolyzed guar gum	1	Powdered drink
II	Soy protein	9	Soft drink, humberg, meat ball, sausage
	Low-molecular-weight sodium alginate	5	Soft drink
	Chitosan	3	Biscuit
III	Peptides	12	
	Sardine peptides (containing VY)	4	Soft drink, soup
	Lacto-tripeptides (VPP and IPP)	3	Lactic acid bacterium drink

Table 1 Continued

Health claim[b]	Functional factor	Number of products approved	Forms of products on the market
	Dried-bonito peptides (containing IKP)	2	Fermented soybean soup
	Casein dodecapeptide (FFVAPFPEVFGK)	2	Soft drink
	Eucommia leaf glycoside (geniposidis acid)	2	Soft drink
IV	Diacylglycerol	5	Cooking oil
	Diacylglycerol and β-sitosterol	4	Cooking oil
V	Casein phosphopeptide	5	Soft drink, chewing gum, soybean curd
	Calcium citrate malate	3	Soft drink
	Heme	3	Soft drink
	Menaquinone-7 producing *Bacillus subtilis*	2	Fermented soybean (natto)
VI	Manitol, palatinose, and tea polyphenols	3	Chocolate, chewing gum
	Manitol	2	Candy
VII	Wheat albumin	2	Powdered soup
	Globin digest	1	Soft drink
	Guava polyphenols	1	Soft drink

[a] For more information contact jhnfa@mxl.alpha-web.ne.jp.
[b] Claim I, foods that improve gastro-intestinal conditions; II, foods for those who have high serum cholesterol; III, foods for those who have high blood pressure; IV, foods for those who have high serum triacylglycerol; V, foods related to mineral absorption and transport; VI, non-cariogenic foods; and VII, foods for those who begin to feel concerned about their blood sugar level.

late life, a continuous increase in life expectancy, increasing costs for health care, and technical advances in the food industry.

As recognized in the Japan MESC research project, up-to-date knowledge in biosciences including biochemistry and molecular biology, together with sophisticated biotechnologies based on these biosciences, support the hypothesis that foods can modulate various functions in the body and thus participate in the maintenance of a state of health that reduces the risk of lifestyle-related diseases. That hypothesis is the origin of the concept of functional food and the development of a new discipline, functional food science, in Europe as well as in Japan.

```
RELEELNVPGEIVESLSSSE   20
ESITRINKKIEKFQSEEQQQ   40
TEDELQDKIHPFAQTQSLVY   60
PFPGPIPNSLPQNIPPLTQT   80
PVVVPPFLQPEVMGVSKVKE  100
AMAPKHKEMPFPKYPVQPFT  120
ESQSLTLTDVENLHLPPLLL  140
QSWMHQPHQPLPPTVMFPPQ  160
SVLSLSQSKVLPVPEKAVPY  180
PQRDMPIQAFLLYQQPVLGP  200
VRGPFPIIV             209
```

Figure 3 Primary structure of bovine β-casein, with Ile-Pro-Pro and Val-Pro-Pro sequences underlined to indicate their positions.

Functional food science in Europe (13) aims in particular

1. To identify beneficial interactions between the presence or absence of a food component (whether a nutrient or a nonnutrient) and its specific function(s) in the body
2. To understand their mechanisms so as to support the hypothesis to be tested in protocols relevant to human studies

The demonstration, in human subjects, of a specific interaction with one or a limited number of functions in the body supports a claim of functional effects (i.e., function claim) or a disease risk reduction (i.e., health claim). Clydesdale (7) also emphasizes the importance of these two kinds of claims to obtain public acceptance for functional foods.

ILSI Europe stresses the significance of relaying the claims to the public by addressing characteristic points of functional food science. It insists that the science is part of nutrition in a broad sense, as the objective is to improve well-being by creating the conditions for disease risk reduction. Therefore, this science is quite distinct from medical and pharmaceutical sciences that are involved in curing or treating diseases. As a working definition, ILSI Europe proposes that a food can be regarded as *functional* if it is satisfactorily demonstrated to affect beneficially one or more target functions in the body, beyond adequate nutritional effects, in a way that is relevant to either an improved state of health and well-being and/or a reduction of the risk of diseases. This institute also requires functional foods to

1. Remain as foods
2. Demonstrate their effects in amounts that can normally be expected to be consumed in the diet
3. Be consumed as part of a normal food pattern (i.e., not as pills or capsules)

What may be most important at present is to evaluate the function. Since any functional food must be based on science, its evaluation should essentially depend on data from biochemistry, physiology, molecular and cell biology, and most other modern biosciences. A key approach to the development of a functional food entails the identification and validation of relevant markers, including biomarkers, that can predict potential benefits related to a target function in the body. If the markers represent an event directly involved in the process, these should be considered as functional factors. On the other hand, if the markers represent correlated events, they should be considered as indicators. In detail, possible markers have been tentatively listed in relation to target functions (Table 2) (14).

There are many options for selecting appropriate markers. A promising option may be the use of accumulated knowledge about genomics. As soon as the human genome project was completed, the so-called postgenomic era for the use of genomic data followed; it may not be unrelated to functional food science. A number of sophisticated biotechnologies will become available, a most useful one of which is DNA micro- and macroarray technology (15). It is helpful because of its high throughput potency to assess possible effects and safety of functional food factors by total gene expression profiles. A brief but typical example can be provided by the DNA tip technology applied to carotenoids as possible anticarcinogens. Conventionally, experiments have been carried out to look one by one at the expression profiles of genes, e.g., *c-myc*. For this, however, a great deal of time and effort must be invested. In contrast, the use of DNA arrays, whether microscopic or macroscopic, will solve the problem to a great extent. Actually, it is found that β-cryptoxanthin, for example, positively or negatively affects the expression of various genes (16). The same technique can also be applied to detecting single nucleotide polymorphisms (SNPs) as topical DNA sequence variations due to point mutations. Genetically, each SNP can take place at a site on the DNA in which a single base pair varies from person to person. Thus, gene responses triggered by ingested functional food factors must vary, depending on the SNPs at the individual basis. Detailed investigations of different responses in different individuals will eventually lead to the design of tailor-made functional foods.

Current genome programs on food organisms are proceeding as well. These include rice, wheat, corn, soybean, and other major plant seeds for food use. A variety of postgenomics, sciences to be born shortly before and after the completion of the genomics, will lead to the development of new biosciences such as nutrigenomics and functional genomics. These developments will support advances in proteomics, providing key information about the molecular design of functional food proteins. It will thus be possible to define easily their specific functional regions involved in the pre-

Table 2 Six Categories of Target Functions and Possible Markers

Function	Possible markers
I. Functions related to growth, development and differentiation	
Maternal adaptation during pregnancy and lactation	Maternal weight Body fat Infant birth weight Milk volume and quality
Skeletal development	Ultrasound measures Anthropometric measures Bone mineral density (e.g. dual-energy X-ray absorptiometry)
Neural tube development	Ultrasound measurements
Growth and body composition	Anthropometry Body fat mass Total body water Procollagen propeptide excretion
Immune function	Cellular and non-cellular immune markers
Psychomotor and cognitive development	Tests of development, behaviour, cognitive function and visual acuity (electro-) physiological measurements
II. Functions related to substrate metabolism	
Maintenance of desirable body weight	Body mass index, body fat content anthropometric indices and imaging techniques as indirect and direct measures of "central obesity" Respiratory quotient Resting metabolic rate
Control of blood glucose levels and insulin sensitivity	Fasting glucose Postprandial glucose Glucose tolerance test Glucosylated hemoglobin Measures of insulin dynamics Fasting plasma insulin
Control of triacylglycerol metabolism	Fasting plasma triacylglycerol Postprandial plasma triacylglycerol
Optimal performance during physical activity	Body temperature Performance testing Muscle mass Muscle protein synthesis
Fluid homeostasis	Water balance Electrolyte balance
III. Functions against reactive oxidative species	
Preservation of structural and functional activity of DNA	Measurement of damaged DNA components
Preservation of structural and functional activity of polyunsaturated fatty acids	Measurement of lipid hydroperoxides or derivatives
Preservation of structural and functional activity of lipoproteins	Measurement of lipid hydroperoxides and oxidized apoproteins

Table 2 Continued

Preservation of structural and functional activity of proteins	Measurement of damaged proteins or components
IV. Functional related to the cardiovascular system	
Lipoprotein homeostasis	Lipoprotein profile, including: plasma LDL-cholesterol plasma HDL-cholesterol plasma triacylglycerol (fasting and postprandial response) lipoprotein particle size
Arterial integrity	Some prostaglandins and transforming growth factors Adhension molecules Cytokines Platelet function Activated clotting factors and activation peptides
Control of hypertension	Systolic and diastolic blood pressure
Control of homocysteine levels	Plasma homocysteine levels
V. Functions related to intestinal physiology	
Optimal intestinal function and stool formation	Stool consistency Stool weight Stool frequency Transit time
Colonic flora composition	Composition Enzyme/metabolic activities
Control of gut-associated lyphoid tissue function	IgA secretion Cytokines
Control of fermentation products	Short-chain fatty acid
VI. Functions related to behaviour and psychological functions	
Appetite, satiety and satiation	Reliable direct measure of food intake Indirect measures of food intake Subjective ratings of appetite and hunger
Cognitive performance	Objective performance tests Standardized mental function tests Interactive games Multifunction tests
Mood and vitality	Quality and length of sleep Attitudes (questionnaires) Standardized mental function tests Ratings of subjective state
Stress (distress) management	Blood pressure Blood catecholamines Blood opioids Skin electrical impendance Heart rate continuous monitoring

Source: Adapted from Ref. 14.

vention of such lifestyle-related diseases as diabetes, hypertension, hypercholesterolemia, allergies, cancer, and even infectious diseases. Among a number of trials, proteomics for the design of antiviral proteins may serve as an example. One such study involves a three-dimensional nuclear magnetic resonance (NMR) analysis of solution structures of oryzacystatin, an antiherpes protein of rice origin described earlier (7,8). The analysis provides information about the minimal effective region (17), the use of which will contribute to molecular breeding of a new rice cultivated with an antiviral function.

As reviewed in this introduction, the importance of functional foods is internationally accepted and the science and technology are enthusiastically pursued for further development. However, we personally think that it is necessary to propose some modification of a current practice in functional food science, because its policy is apparently approaching medicine and pharmacology. A main reason for this trend is lack of consideration of the importance of the secondary function, which per se is the most representative attribute of any food. All foods must be sensorily acceptable, even functional foods. The science should include studies on the senses of taste and smell. The significance of this foods is emphasized since such sensations have been found to be more or less related to the functions of endocrine and exocrine systems as well as to those of digestive systems. An interesting example can be found in red pepper, whose unique irritative component, capsaicin, has a "hot" flavor in seasoning (secondary function). A 1997 paper reports that this substance, shortly after entering the body per os, interacts with vanilloid receptors such as calcium channels in sensory neurons (18), inducing a specific dynamic action, possibly with an antiobesity result (tertiary function). Capsaicin can thus be regarded as a functional factor with such a bilateral effect. Sensory abnormalities and disorders are other targets of functional food science, because insights into syndromes of these kinds will supply patients with organoleptically acceptable tailor-made foods that also have tertiary functions.

The design and development of functional foods should not be carried out independently of primary (purely nutritional) and secondary (sensory or cognitive) functions of foods. Future advances in relevant sciences for inducing to foods with integral attributes will distinguish functional foods from medicines. Research along this line will add a new dimension, and biotechnologies related to nutritional, sensory, and functional sciences are academically and socially understood as having great potential for the development of new food science, technology, and industry. From a marketing point of view, an even larger increase in the world turnover of functional foods can thus be expected in the early twenty-first century.

Readers are also advised to consult recently published volumes to learn more about the state of the art concerning functional foods (19–23).

Soichi Arai
Professor
Department of Nutritional Science
Tokyo University of Agriculture
Tokyo, Japan

Masao Fujimaki
Emeritus Professor
Faculty of Agriculture
University of Tokyo
Tokyo, Japan

REFERENCES

1. Swinbanks, D. and O'Brien, J.: Japan explores the boundary between food and medicine. *Nature* 364, 180 (1993).
2. Arai, S.: Studies on functional foods in Japan: states of the art. *Biosci. Biotechnol. Biochem.* 60, 9–15 (1996).
3. Suzuki, U., Shimamura, U., and Okada, S.: †ber Oryzanin, ein Bestandteil der Reiskleie und seine physiologische Bedeutung. *Biochem. Z* . 43, 89–153 (1912).
4. Clydesdale, F. M.: A proposal for the establishment of scientific criteria for health claims for functional foods. *Nutr. Rev.* 55, 413–422 (1997).
5. Fukudome, S., and Yoshikawa, M.: Opioid peptides derived from wheat gluten: their isolation and characterization. *FEBS Lett.* 296, 107–111 (1992).
6. Murakami, A., Wada, K., Ueda, N., Sasaki, K., Haga, M., Kuki, W., Takahashi, Y., Yonei, H., Koshimizu, K., and Ohigashi, H.: In vitro absorption and metabolism of a citrus chemopreventive agent, auraptene, and its modifying effects on xenobiotic enzyme activities in mouse livers. *Nutr. Cancer* 36, 191–199 (2000).
7. Abe, K., Emori, Y., Kondo, H., Suzuki, K., and Arai, S.: Molecular cloning of a cysteine proteinase inhibitor of (rice oryzacystatin). *J. Biol. Chem.* 262, 16793–16797 (1987).
8. Aoki, H., Akaike, T., Abe, K., Kuroda, M., Arai, S., Okamura, R., Negi, A., and Maėda, M.: Antiviral effect of oryzacystatin, a proteinase inhibitor in rice, against herpes symplex virus type 1 in vitro and in vivo. *Antimicrob. Agent Chemother.* 39, 846–849 (1995).
9. Watanabe, M., Yoshizawa, T., Miyakawa, J., Ikezawa, Z., Abe, K., Yanagisawa, T., and Arai, S.: Production of hypoallergenic rice by enzymatic decomposition of constituent proteins. *J. Food Sci.* 55, 781–783 (1990).
10. Tanabe, S., Tesaki, S., Yanagihara, Y., Mita, H., Takahashi, K., Arai, S. and

Watanabe, M.: Inhibition of basophil histamine release by a haptenic peptide mixture prepared by chymotryptic hydrolysis of wheat flour. *Biochem. Biophys. Res. Commun.* 223, 492–495 (1996).

11. Roberfroid, M.B.: The European concept of functional food. *VTT Symposium* 177, 161–170 (1996).
12. Richardson, D. P.: Functional foods—shades of gray: An industry perspective. *Nutr. Rev.* 54, S174–S185 (1996).
13. Bellisle, F., Diplock, A. T., Hornstra, G., Koletzko, B., Roberfroid, M., Salminen, S., and Saris, W. H. M.: Functional food science in Europe. *Br. J. Nutr.* 80, S1–S193 (1998).
14. Diplock, A. T., Aggett, P. J., Ashwell, M., Bornet, F., Fern, E. B., and Roberfroid, M. B.: Scientific concepts of functional foods in Europe. Consensus Document. *Br. J. Nutr.* 81, S1–S27 (1999).
15. Marton, M. J., DeRisi, J. L., Bennett, H. A., Iyer, V. R., Meyer, M. R., Roberts, C. J., Stoughton, R., Burchard, J., Slade, D., Dai, H., Bassett, D. E., Hartwell, L. H., Brown, P. O., and Friend, S. H.: Drug target validation and identification of secondary drug target effects using DNA microarrays. *Nature Med.* 4, 1293–1301 (1998).
16. Narisawa, T., Fukaura, T., Oshima, S., Inakuma, T., Yano, M., and Nishino, H.: Chemoprevention by the oxgenated carotenoid β-cryptoxanthin of *N*-methylnitrosourea-induced colon carcinogenesis in F344 rats. *Jpn. J. Cancer Res.* 90, 1061–1065 (1999).
17. Nagata, K., Kudo, N., Abe, K., Arai, S., and Tanokura, M.: Three dimensional solution structure of oryzacystatin-I, a cysteine proteinase inhibitor of the rice, *Oryza sativa* L. japonica. *Biochemistry* 39, 14753–14760 (2000).
18. Caterina, M. J., Schumacher, M. A., Tominaga, M., Rosen, T. A., Levine, J. D., and Julius, D.: The capsaicin receptor, a heat-activated ion channel in the pain pathway. *Nature* 389, 816–824 (1997).
19. Childs, N. M.: Science, nutrition and clinical practice: Bringing expanded focus and expertise to the *Journal of Nutraceuticals, Functional Medical Foods*. *J. Nutraceuti. Functional Med. Foods* 3, 1–3 (2000).
20. Gibson, G. R., and Williams, C. M.: Functional Foods: Concept to Product. Woodhead Publishing, Cambridge, England (2000).
21. Wildman, R. E.: Handbook of Nutraceuticals and Functional Foods. CRC Press, Boca Raton, London, New York, Washington, D.C. (2001).
22. Kwak, S., and Jukes, D. J.: Functional foods. Part 1: the development of a Regulatory Concept. *Food Control* 12, 99–107 (2001).
23. Arai, S., Osawa, T., Ohigashi, H., Yoshikawa, M., Kaminogawa, S., Watanabe, M., Ogawa, T., Okubo, K., Watanabe, S., Nishino, H., Shinohara, K., Esashi, T., and Hirahara, T.: A mainstay of functional food science in Japan: History, present status, and future outlook. *Biosci. Biotechnol. Biochem.* 65, 1–13 (2001).

1
Poultry, Eggs, and Biotechnology

Rosemary L. Walzem
Texas A&M University, College Station, Texas, U.S.A.

I. INTRODUCTION

The term *poultry* refers to domesticated species of birds valued for their meat and eggs. The most frequently encountered examples, chickens and turkeys, belong to the order Galliformes, as do pheasants, quail, and grouse. Notably, two other orders of birds are included in the term *poultry*, Columbiformes, doves and pigeons (e.g., squab), and Anseriformes, ducks and geese. Each of the many species of birds within each order has been highly valued for many thousands of years for both their beauty and their contribution to the diet (1). Within each species there is a wide array of strains and types, varying in many aspects of plumage, including feather color and shape. Body type and size, growth rate, egg production, and disease resistance vary among individual types of poultry. Thus the term encompasses a highly diverse group of birds of broadly differing habits, genetic diversity, environmental requirements, and nutritional needs.

It is a widely proposed that during prehistory, poultry consumption was an adventitious event, the outcome of a successful hunt or fortunate discovery of a nest of eggs. Poultry and eggs are noted sources of essential nutrients, including energy, protein, fatty acids, vitamins, and minerals (2). Presumably, an even more diverse collection of bird species was consumed in prehistory—essentially whatever could be caught. Humans evolved within this pattern of food intake and inadvertently benefited from what is now termed *biostreaming*: namely, that certain desirable or essential nutrients consumed by birds, retained and concentrated within their bodies, were made available, or available in greater concentrations, to the humans who consumed those birds. As such, poultry have made important nutritional contributions to humans through biostreaming and through converting plant and insect foods that are indigestible or unpalatable for humans to highly digestible and nutritious food. An interesting possibility is that certain bioactive phytochemicals or xenobiotics have only been consumed by humans as components of poultry during our evolutionary history (3–6). Recognition that poultry possessed the capability to acquire flavor and texture attributes through their diets and environment is likely the basis for traditional feeding strategies. These strategies include the addition of herbs, particular juiciness (due to enhanced subcutaneous and intramuscular fat deposits), promotion of feed components such as corn, and other manipulations to alter the final nutrient and chemical composition to suit the consumer (7,8).

As hunter–gatherer societies transformed into agrarian-based societies, plant and animal species that proved tractable to human cultivation were encouraged. Food supplies stabilized and increased in abundance. Under these conditions, sheer need or hunting and gathering skill influenced human dietary choices less while hedonistic, intellectual, and philosophical considerations influenced them more. These bits of sociological assumption are noted to emphasize that our species has physiological and historical inclinations to be omnivorous. Moreover, humans actively cultivate and fabricate the foods they desire. Biotechnology provides another set of options to improve food quality. Within this context, *biotechnology* is defined as the use of microorganisms, plant cells, animal cells, or parts of cells such as enzymes, immunoglobulins, or genes to make products or carry out processes (9).

One objective of this chapter is to provide factual information on biotechnological approaches that can enhance the nutritional value of poultry and eggs for use in human dietary supplement or contribute to other nonfood products that enhance health or well-being (10). An example of a nonfood application is the use of egg membranes to "bandage" ocular burn patients (11). Another objective is to describe the types of enhancements that might prove desirable within modern dietaries delivered by modern food supply systems. This directed focus somewhat limits description of the benefits that biotechnology will confer to continued interactions between poultry and humans. Table 1 provides a listing of many healthful components of eggs or poultry that may be isolated, stabilized, or augmented by biotechnology. Table 2 provides a generalized classification of modifications or applications according to groups most likely to benefit. From this listing it is clear that many of the biotechnological improvements that have the greatest priority at present are those that improve the bird's ability to digest and assimilate feed and thus produce less waste. The environmental gains in water or soil quality and sanitation realized through these efforts will improve human health in indirect but meaningful ways (12,13). Similarly, biotechnological approaches

Table 1 Healthful Egg or Poultry Components That May Be Isolated, Stabilized, or Augmented by Biotechnology

Components found in or suitable for delivery by poultry or eggs that have health or biotechnological applications
Proteins
Native: immunoglobulins, lysoyzme, angiotensin-converting enzyme–(ACE)-inhibitory oligopeptides, CCK-gastrin immunoreactive protein, phosvitin, transferrin, ovomucoid, ovomucin, Cystatin, riboflavin-binding protein, avidin
Engineered: insulin, growth hormone, human serum albumin, humanized immunoglobulins, monoclonal antibodies, α-interferon, spider silk
Lipids
Choline, lecithin, cephalin, betaine, cholesterol, sphingomyelin, α-tocopherol, carotenoids, xanthophylls, lutein, lycopene, n-3 fatty acids, vitamin K, vitamin D
Miscellaneous
Sialic acid, $CaCO_3$, shell membranes

to enhance disease resistance and reduce mortality rate within the production unit will improve human health through removal of antibiotics from feed and absence of pathogenic organisms in poultry meat or eggs offered for sale.

Suggesting a fundamental change to the nutrient composition of poultry meat or eggs is a more speculative endeavor. At present, there is a lack of sufficient physiological understanding to make unequivocal statements regarding what foods constitute an optimal human dietary. However, biotechnology provides additional tools to enhance nutritional value of foods as such information becomes available. Moreover, nutritional optimization is likely to be highly individual (14,15). In this regard, the inherent genetic diversity, short generation time, and emerging cloning strategies for poultry provide the flexibility needed to provide consumers with eggs and meat specifically tailored to their physiological characteristics, organoleptic preferences, and eating patterns.

Applicability is a concern in any scientific endeavor, and biotechnology is no exception. Data from the Food for Thought II study conducted by the International Food Information Council in 1997 showed that 70% of consumers in Canada, Portugal, Japan, and the United States were likely to purchase foods enhanced by biotechnology (16). In the Netherlands, United Kingdom, Italy, and Sweden, at least half of consumers were so inclined. A quarter to one-third of consumers in Austria and Germany indicated that they would be likely to purchase such foods. In the fourth biannual tracking survey of food and health news, "Food for Thought IV, 2001," found that biotechnology was the most reported single food and health issue, although

Table 2 Groups Benefiting from Modifications or Applications of Biotechnology

Types of modifications and applications
Benefit to the producer
Increased resistance to disease organisms
Enhanced feed digestion and assimilation capabilities
Improved control of food intake
Improved livability
Benefit to the processor
Increased resistance to processing-related contamination
Reduced incidence of processing-sensitive phenotypes
Improved compatibility of starter materials for complementary approaches to enhance safety, flavor, texture, and stability
Reduced trimming waste through improved methods of further processing
Benefit to the consumer
Improved sanitation
Enhanced vitamin or trace mineral content in whole products, or foods made from those whole products
Enhanced bioavailability of nutrients, or redistribution of existing components such as decreased total fat, and increased light to dark meat
Improved flavor or texture of products including, or formulated with, poultry or eggs
Eggs containing protein or lipid functionality—flavor, texture, therapeutics (see Table 1)
Benefit to society
Reduced fecal waste production through enhanced digestibility, leading to reduced environmental impacts
Reduced medical costs due to malnutrition and conditions for which poultry or eggs, or parts thereof, act as therapeutic agents or serve in the manufacture of those agents
Reduced costs due to reduced food-borne illness as a result of improved live and processing contamination resistance; improved bird welfare

much of the commentary was cautionary and not placed within a consumer context (17). Despite these aspects of reporting, a poll of 1000 representative American consumers above 18 years of age found that only 2% wanted to know or were concerned about foods that were modified by biotechnology (17). This same survey found that 33% of consumers believed modified foods were currently in supermarkets, that more (65% compared with 52%) were likely to purchase foods identified as modified by biotechnology for purposes of reduced pesticides/antibiotics than for flavor enhancement per se, and that if foods of improved nutritional value were available through biotechnological modification, that factor would encourage (36%) or have no effect (41%)

on the purchase of that product. Overall, these consumers believed that biotechnology would improve their health and nutrition (39%) the quality, taste, and variety of foods available to them (33%), but fewer expected reduced food costs (9%). These data suggest that foods enhanced by biotechnology are generally acceptable to consumers. The data also suggest that consumers are becoming sophisticated about how and when biotechnology can be used for particular purposes to their advantage.

II. ROLE OF POULTRY IN THE HUMAN DIET

Consumption of poultry and eggs is no longer an adventitious event. It has also moved beyond being an infrequent luxury meal, such as the Sunday dinner chicken, Thanksgiving turkey, or Christmas goose. Total per capita poultry consumption in the year 2000 varied from a high of 57.4 kg per year in Hong Kong to a low of 3.7 kg per year in Romania (18). Total poultry production for all major countries was expected to be 59.6 million tons, and total consumption of poultry meat to be 58.5 million tons in 2001. People in the United States and China consume the most poultry. In the United States, total poultry consumption increased from 15.6 kg per person per year in 1960 to 45.3 kg estimated for 2002 (19). Of this amount, 37.0 kg of chicken and 8.3 kg of turkey were consumed. This nearly threefold increase in consumption was largely due to the vertical integration of the poultry industry, which has made it capable of being exquisitely responsive to consumer demands in terms of price, quality, and final product form.

Consumers do demand poultry. Market data from 1997 showed that 85% of restaurants offered one or more poultry entrées, and that 12% of all main-course entrées in 1997 contained chicken (20). Similarly, 12.3% of all meals or snacks consumed in the United States contained chicken or poultry. The top two appetizers in 2001 were chicken strips and chicken wings. In addition to extremely popular nuggets, strips, fingers, or fried chicken, chicken is increasingly used as a topping for salads. In 1997, chicken constituted 19% of all main-dish salad, and 55% of all menus offered chicken on a salad. The total value of U.S. poultry production in 2000 was $16.9 billion and clearly had significant impact on the spectrum of nutrients available within the diet (3,5,21–24).

Egg consumption in the United States followed a different path, decreasing from 403 per capita per year in 1945 to a low of 234 in 1991. That decline has been widely attributed to concerns over cholesterol and changes in breakfast meal habits (25). From 1970 to 1994, processed egg consumption rose from 33 to 61 per person per year. In 2000, 198.4 million cases (360 eggs per case) of shell eggs were produced in the United States, and approximately a third were further processed to egg products (26). The term *egg products* refers to processed or convenience forms of eggs obtained by breaking and processing shell eggs. Egg products include any of various whole eggs, egg whites, and egg yolks in frozen, refrigerated liquid, and dried forms, as well as specialty egg products. Specialty egg products include prepeeled hard-cooked eggs, egg rolls or "long eggs," omelets, egg patties, quiches, quiche mixes, scrambled eggs, and fried eggs. The newest category within the American Egg Board's Egg Product Reference Guide is "Eggs as Nutraceuticals" (27). Several publications now available (28–30) describe novel uses and functionalities for whole eggs and for egg components in particular. These emerging areas of investigation suggest that potential is to develop novel health promoting products with egg components is primarily limited by imagination of health scientists, processors, and engineers working in the area (Table 2). Applications include food, cosmetics, and medical products. Thus, eggs and poultry also provide starting materials for further processing that employs biotechnological methods, as well as being targets for direct modification by biotechnological techniques.

Several epidemiological studies or meta-analyses of dietary intervention studies relating dietary fat and cholesterol to plasma cholesterol concentration (31–33) have led medical authorities such as the American Heart Association to state that periodic egg consumption is unlikely to influence plasma cholesterol concentrations (34). This shift in dietary advice was among the factors that caused egg consumption to increase to 258 per person per year by 2000. Annual per capita egg consumption in a survey of 36 countries in 2000 varied from a low of 34 in India to a high of 320 in Japan. Despite the decreasing confidence in egg cholesterol risk, cholesterol reduction in eggs remains as a much sought after target (35–39), as does reduction of oxide formation during processing (40).

Poultry meat and eggs are nutritious foods. Eggs in particular must possess concentrated amounts of nutrients in order to support incubation and hatch. Indeed, despite providing less than 1.3% of daily U.S. calories, eggs provide 3.9% of daily protein, and similar or greater amounts of vitamin B_{12}, vitamin A, folate, vitamin E and riboflavin (21). Egg protein is considered the highest-quality protein and contains an ideal spectrum of essential amino acids in a highly digestible form (41). Data from NHANES III showed that egg consumers were better nourished than nonconsumers and had no higher plasma cholesterol concentrations (21). Thus, even if cholesterol content cannot be decreased (42), eggs remain a desirable vehicle for added nutritional functionality.

Poultry meat is also nutritious (43), although the exact contribution it makes to the diet is more variable (2,44–49). Consumer selections, including choice of white or dark meat and method of preparation, such as roasting or deep fat frying, greatly influence the net diet contribution. Biotechnology offers options in food preparation that will support increasingly sophisticated consumer demand for healthy functional foods even when consumed in forms that are traditionally viewed as less healthy. For example, it was noted that the industry must appreciate that "it's not going to be as easy as simply reducing the fat percentage in a fried chicken patty, it will have to be made with omega-3 enriched meat, using an estrogenic soy binder, and fried in a non-absorbable fat that is fortified with antioxidants" (50). Each component mentioned in these chicken patties may be the product of biotechnological processes.

III. HISTORY OF GENETIC AND ENVIRONMENTAL MANIPULATION OF POULTRY

Because poultry and eggs are so nutritious and are such versatile components of the diet, the goal of first-generation biotechnological approaches was primarily to produce more. The approaches used in these endeavors included traditional methods of genetic selection for desired traits, controlled environments, nutritionally optimized diets, and feeding programs (51). At present, poultry breeders offer strains of birds produced through traditional selection methods that are highly suited to particular climates and rearing systems (52). Poultry strains are also highly selected to achieve specific production goals; thus birds that are grown for meat production are physically and physiologically distinct from those used for egg production (53). As a point of comparison, dogs are perhaps the only other commonly encountered species that demonstrate equivalent diversity of body types, sizes, and colors inherent within a given genome due to sustained selective breeding.

Whereas most poultry strains are quite handsome, extreme differences in external appearance (phenotype) do result in extreme preferences among fanciers and raise practical issues for poultry breeders (54–58). It is important to note that all this diversity was obtained by traditional shuffling of genes by means of cross-mating. Table 2 provides a nonexhaustive list of targets within selective breeding programs. Most traits are the result of coordinate expression of multiple genes, and selection programs are directed toward several simultaneous targets. Biotechnology will allow these complex expression profiles to be described and more readily manipulated by traditional (59–61) as well as nontraditional (62–64) approaches. The ability to manipulate phenotype-defining patterns of gene expression provides the means to optimize bird biological characteristics. In contrast, traditional breeding approaches are limited by coexpression of desirable and undesirable traits.

IV. HISTORY OF NUTRITIONAL MANIPULATION OF POULTRY

Application of scientific methods to the study of nutrition started in the mid-19th century. An interesting side note is that poultry, particularly chickens, provided much of the data that shape our basic understanding of purpose and metabolism of vitamins, (particularly thiamin, folate, and vitamin D), minerals (particularly calcium), lipotropes (choline and betaine), and macronutrients (65). Moreover, the widespread use of poultry throughout the world has generated and continues to generate applied nutrition knowledge regarding diet optimization in different environments and at different stages of the life cycle. Many of the principles that underlie optimization of nutrition for poultry health and production are implicit in efforts to develop similar strategies for humans to combat chronic disease consumption of functional foods.

Beginning in the early 1960s, health messages about eggs shifted from wholly positive to cautionary on the basis of associations among plasma cholesterol concentrations, the incidence of atherosclerotic cardiovascular disease, and the cholesterol content of eggs (66). Messages regarding cholesterol risk have grown equivocal as more has been learned about human cholesterol metabolism (33,34,66,67), but the criticism spurred efforts to describe and improve other health-building egg components. As a result, a variety of specialty eggs have become available. These eggs typically are enriched in specific fatty acids, are reduced in cholesterol content, and possess increased vitamin E, vitamin A, or iodine content (Table 3). Other nutritional improvements include "nonessential" but perceived health-promoting nutrients such as phytoestrogens, catechins (also known as "tea" eggs), and carotenoids such as lutein and lycopene.

Further improvements in egg and poultry nutritional contents are limited by the existing physiological traits of the birds. As such, they represent attractive targets for biotechnological improvement. For example, a single egg from a hen reared to produce vitamin E–enriched eggs can currently provide 20–25% of the daily value recommended, or about 3–5 mg (68). In contrast, vitamin E prophylaxis doses are usually in the range of 200–400 mg per day (69,70). Further improvements in egg vitamin E content cannot be realized by dietary approaches as a result of the essential, but limiting, role of tocopherol transport protein in directing vitamin E to egg yolk (71). Overexpression of this protein by genetically modified birds could increase yolk vitamin E concentration markedly. Moreover, the endogenous stereospecificity of tocopherol transport protein for the alpha form of vitamin E will ensure that the most bioactive form of the vitamin will be deposited into egg yolk. Increased tissue tocopherol content could stabilize

Table 3 Genetic Manipulation to Produce Specialty Birds and Eggs

Meat-type birds
Growth rate
Feed efficiency
Disease resistance
Frame functional properties, skeletal development, body shape
Carcass muscling distribution, proportion of light and dark meat
Carcass fat content and distribution
Market size
Feather color
Egg-type birds
Egg size
Shell quality and color
Feed efficiency
Body size
Time to first egg and duration of egg laying
Frame functional properties, skeletal development, body shape
Disease resistance

poultry meat during processing to prevent development of warmed over and off-flavors (72).

Among animal production systems, the poultry industry has been the most proactive in coupling selective breeding to nutritional programs to optimize bird performance in a variety of environments at each stage of the life cycle. Through its use of complementary approaches, the poultry industry has improved human nutrition and health through the provision of high-quality animal protein rich in vitamins and minerals (zoonutrients). Biotechnology will further enable poultry and eggs to serve this role in human dietaries.

V. BIOTECHNOLOGY

A. Complementary Approaches

Biotechnology is defined as the use of microorganisms, plant cells, animal cells, or parts of cells, such as enzymes, immunoglobulins, or genes, to make products or carry out processes. Given the diverse opportunities afforded by biotechnology to improve the food supply, multiple approaches to solve a particular problem will be employed, with market forces driving most efficient solutions. Because the poultry industry is highly vertically integrated, it has a history of using complementary approaches to optimize production outcomes

(73). Thus, the full scope of potential for benefit from biotechnology will likely be first realized in highly integrated production systems, such as are found within the poultry industry.

Significant concerns of consumers are food safety and sanitation. Although the primary concern is microbial safety, the presence of pesticides, hormones, and antibiotics also causes concern (74). These concerns are well addressed by vertical integration of complementary approaches afforded by biotechnology. For example, chicks and poults can be provided with neonatal feed products containing probiotic bacterial cultures in combination with prebiotic compounds in order to establish a healthy intestinal flora that excludes pathogenic organisms from the intestine (75,76). Similar dietary approaches are known to improve human health and well-being (77,78) and are no less appropriate for poultry. Moreover, biotechnology can be used to optimize the mixture of organisms, their ability to attach to the intestinal mucosa and competitively exclude pathogens, and their ability to utilize specific prebiotic nutrients in the food and so synergize the protective effects. These newly hatched birds may well be vaccinated by using recombinant deoxyribonucleic acid (DNA) vaccines (79). The DNA vaccines are more flexible and allow a rapid response to changes in pathogens, and safety issues with regard to use in food animals are being addressed (79).

Biotechnology allows the very eggs young birds start in to be selectively enriched in protective immunoglobulins. These protective proteins are deposited into the yolk by the hen in response to maternal vaccination against neonatal pathogens (80). Such responses occur spontaneously in nature in response to pathogen exposure but are haphazard, and many hatchlings die before the hen begins depositing effective amounts of antibodies. Biotechnology allows this endogenous protective mechanism to be used proactively (81,82). This same natural process is also used to develop high-quality antibodies for use in various biotechnological applications (83–88). When placed in the grower house, these young birds can continue to be protected by vaccines present in the corn they eat (89). These steps will decrease or eliminate the need for antibiotics. Notably, the flora initiated in the birds at hatching may be further engineered to contain strains that competitively exclude pathogens from the skin or that produce proteins lethal to disease-causing microbes such as *Salmonella* or *Listeria* species but that are harmless to humans and other animals (90–95). The bird itself may be modified to produce such proteins in the skin or muscles. During processing, such meat would be resistant to accidental contamination. Egg whites have long been used to "fine" wines in order to remove undesirable compounds (96). Eggs containing specific antibodies or binding proteins could provide biotechnological prophylaxis to protect portions of the food supply from overt (terrorist) efforts to contaminate it.

Poultry diets could contain other proteins or compounds produced with biotechnological methods to enhance feed digestibility or ability to transport and retain nutrients. For example, at present poultry diets must be formulated with inorganic phosphorus to ensure an adequate intake of this essential mineral element. The plant materials that constitute the bulk of poultry diets also have phosphorus in the form of phytate. This organic form of phosphorus is indigestible by poultry, and as a result, it passes through the bird and is excreted as waste, is broken down by microbes, and can ultimately enter the water supply. At present, feed companies have products that contain microbial enzymes called *phytases* that release phosphorus from phytate, thus making it available to the bird (97–99). This feed amendment decreases the amount of inorganic phosphorus added to the diet and ultimately the amount of waste phosphorus (100,101). The molecular biological attributes of phytase enzymes from various sources have been studied extensively, and effective isoforms have been cloned and expressed in cells and plants suitable for use as feed ingredients (102). A second approach is to endow the birds themselves with the ability to digest phytate; this has been done in pigs (103)and mice (104). A third biotechnological approach to this topic is the selection of low-phytate grains (105), or engineering of the plants to enhance the availability of the minerals they contain (99,106,107). Biotechnological improvements in poultry diets may also lead to the formulation of grains that contain compounds used to enhance growth or promote lean muscle gains (108–110). Such nutritional strategies decrease the value of steroidal anabolic agents and could increase the nutrient content of poultry and eggs. Similarly, strategies such as these would support enhanced biostreaming of nutrients into human dietaries. Modification of the birds themselves may also be used to achieve production goals more efficiently while diminishing the use of exogenous anabolic agents. Market forces will influence what becomes the most persistent strategy.

1. Modifying the Bird Itself

Biotechnological approaches hold the potential to alter specific avian physiological features that improve product quality or functionality. There are two types of objectives sought after by methods that directly manipulate the genome of birds. The first type seeks targets identical with those of traditional genetics. These include fundamental alterations in body composition to reduce total fat, increase lean muscle mass, and increase the proportion of white to dark meat (Table 3). As noted previously, poultry is a highly diverse genetic population, and individual strains often possess certain desirable traits to a great extent. The advantage that biotechnology offers is selectivity in that desirable traits are usually present in combination with undesirable traits and require many generations of selection to capture, if indeed the undesirable and desirable traits will segregate while remaining highly heritable. The second type of objective is the introduction of novel functions within the bird. Overexpression of tocopherol transport protein to enhance vitamin E content of meat and eggs is an example of this second type. The advantage here is that objectives insurmountable by alternate approaches may be solved by introduction of novel genes or novel patterns of gene expression within existing genomes.

A number of examples that only begin to frame the enhancement in nutritional content or functional properties afforded by altered gene expression can be suggested (Table 4). For example, as has been done in plants (100,106), coordinate overexpression of metal transport and storage genes could be used to enhance concentrations of iron, zinc, or copper in poultry meat. Introduction of novel genes or reprogramming of existing genomes could overcome functional defects such as the myopathy that underlies pale soft exudates in turkeys (111). Novel ligand binding domains could be expressed to add combinatorial capabilities in processing applications (112). Such capability could allow, for example, stabilization of specific nutrients within foods, addition or enhancement of flavor or color components in poultry products, or engineering capabilities to enhance texture in final products. Altered glycoprotein expression within muscle tissue could alter water-holding capacity, pathogen resistance, browning ability, or fat penetration with frying (113–116). Each contemplated modification requires an exquisite understanding of biology as well as functional aspects of poultry and eggs as biomaterials. Some

Table 4 Companies Dedicated to Transgenic Poultry Production

Company	Focus
AviGenics, Athens, GA *http://www.avigenics.com*	Biopharmaceutical protein production Agronomic traits, including disease resistance Improved feed efficiency and muscle growth
Origen Therapeutics, Burlingame, CA *http://www.origentherapeutics.com*	Biopharmaceutical protein production Somatic chimeras from elite stock
Sima Biotechnology, Minneapolis, MN	Biopharmaceutical protein production
TranXenoGen, Shrewsbury, MA *http://tranxenogen.com*	Biopharmaceutical protein production
Vivalis, Nantes, France *http://vivalis.com*	Vaccine development with Aventis Pastuer

of this knowledge will only be acquired through iterative attempts toward the desired outcome.

2. Altering Gene Expression

As attractive as many of these changes are, there remains concern that altering the genome or altering patterns of gene expression is somehow unsafe—that birds or eggs arising from transgenic strategies are unnatural. Careful studies of the avian genome have found natural examples of ancient retroviral gene insertions. These discoveries suggest that exogenous modification of endogenous genomes has been partially responsible for mutations that drive evolutionary change (62). Hen feathering is a trait in some male birds that arises from an approximately 40-fold increase in the enzyme aromatase within the feather follicle of homozygous dominant mutants (117). Aromatase is one of the enzymes responsible for the conversion of androgens to estrogens. Elevated estrogen concentrations in the feather follicles of mutant males change local gene expression such that feathers growing from these follicles take on a rounded, characteristically "hen" shape. Matsumine and associates (118) reported that retroviral insertion into a regulator sequence of the aromatase gene removes the normal restriction of gene expression in extragonadal tissues. The mutation is highly prized in Bantam birds because of its effects on plumage phenotype. Another spontaneous retrovirally mediated alteration in gene expression causes slow feathering (62). This gene has practical utility within the poultry industry as a simple visual method to sex chickens. Rapid feathering males (k+/k+) are bred to slow-feathering females (K/−) to produce slow feathering males (K/k+) and rapid feathering females (K+/−). As this method of sexing is widely used within the industry, nearly everyone has *safely* consumed poultry containing retroviral elements. Studies of these ancient gene insertions provide a pattern for the molecular detail needed to create safe and stable constructs for intentional gene insertions.

Traits such as egg production, growth, or body fatness result from complex pattern expression from many genes (54,56,58). In such situations, adding new genes may not be as desirable as controlling the pattern of endogenous gene expression to positive effect. Although alteration of gene expression is still in its infancy, small molecules capable of altering gene expression profiles are being created (119–121). This approach holds much promise but will require substantial expansion of our understanding of gene interactions and development of appropriate delivery technologies to be employed effectively. Early embryonic development—interactions between developing tissue layers—has inductive and suppressive effects that control tissue phenotype (122–124). Antibodies can be employed to alter such interactions and thus the development of specific tissue types. Antibodies, monoclonal antibodies in particular, provide the means for quite specific and selective knock-out of proteins involved in tissue–tissue interactions or cell surface receptors for signaling molecules. A patent has already been issued to use antibodies to suppress adipose tissue development in poultry, and practical application of the technique is being pursued (125,126).

Several approaches exist to insert novel genes into poultry. It is beyond the scope of this chapter to describe these methods in detail, and the interested reader is directed to relevant reviews (63,64,127–130). Progress in method development is ongoing as poultry provide unique challenges for transgenic methodology (131). For the most part, methods are directed toward the germ line in order that inserted genes are stably passed on to progeny. However, given the extensive breeding programs that produce highly developed strains, and the limited production interval (40 days for broilers and 60–156 weeks for layers), it seems that somatic transfer, such as is used in gene therapy for human disease (132,133), could be developed as a viable alternative or adjunct approach. Bird embryos develop atop a large fragile yolk that defies routine microinjection techniques used in most mammalian systems (134). Thus, manipulation and culture of modified embryos remain ongoing and active areas of investigation (135).

Leading objectives in the field of avian transgenesis are method development for cultivation of embryonic stem cells and development and testing of vector constructs that allow homologous recombination (64,136). A third requirement, the ability to produce germ line chimeras, is established, but not perfected. Somatic chimeras can be routinely produced by hand (130), and mechanization of the process is being vigorously pursued (137). At present, intense commercial interest in poultry transgenesis makes searches of patent literature key to understanding advances in the field (136,138,139), although academic publications provide good general outlines (140). Because methodologies in poultry lag behind those of plants and some other animal species, the poultry industry is in a position to benefit from the experience of commodity groups in both strategy selection and consumer acceptance issues.

There are several companies dedicated to transgenic poultry production (Table 4). Most are engaged in producing therapeutic proteins or immunoglobulin-rich eggs that can be further processed into diagnostics and therapeutic agents because of the premium price these products—as opposed to food products—command. Thus, the animal systems and methods to produce nutritional and processing functionality are in active development but not yet employed in food production per se. The proteins produced in eggs are high quality and suitable for food grade and pharmaceutical uses. Egg-derived proteins are likely to improve nutritional and organoleptic properties of processed foods in a variety of products.

Other products that are likely to reach the market in the foreseeable future are somatic chimeras produced from elite broiler stock, such as those currently being developed by Origen Therapeutics (137). These birds are the product of a biotechnological strategy similar to that used to propagate strawberries and grapes and, as a result, do not carry novel genes that could slow approval of other nearly ready birds (Table 4). Briefly, fertilized eggs are collected from prolific egg laying strains and their development halted until blastodermal cells from donor embryos can be injected into the recipient embryo. The donor embryos are derived from strains with superior growth but whose egg production rates are typically low (58). The resulting hatchling has heritage from both parental lines, but the bird itself typically exhibits superior growth characteristics. This technology will ultimately be combined with embryonic stem cell propagation and culture techniques to produce somatic clones of modified animals.

VI. CONCLUSIONS

Increased knowledge about avian biological characterstics, coupled with various biotechnological methodologies, provides the means to supply consumers with safe, wholesome poultry products. Current goals of improved disease resistance and improved yield or growth rates will ultimately mature into genuine improvements in utility and nutritional values of eggs and poultry products. Such improvement will be further enhanced by complementary approaches that incorporate biotechnology at appropriate points in the production system. These points will be found throughout the continuum of animal rearing to food delivery or fabrication. Moreover, biotechnology is already supporting the enhancement and harvest of functional components from eggs to improve human health through novel medicines, diagnostics, and cosmetics (28,29). Great new and healthy poultry and egg foods and products that last longer and taste better can be expected.

The most likely immediate improvement in human health through biotechnological improvements in poultry and eggs will be in decreased waste and enhanced water quality and decreased reliance on antibiotics. These efforts are already in use within the industry, and the efficacy of these methods will continue to improve. However, as methods and understanding of avian biological traits are advancing rapidly, biotechnological improvement of both nutritional and functional properties of poultry and eggs should be realized within the next several years. As the ability to influence nutrient composition in the whole bird through transgenic methods becomes practical, manipulations will always be tempered by basic biological characteristics. Particular changes that are incompatible with healthy, disease-resistant birds are unlikely to be acceptable to consumers or producers.

REFERENCES

1. RJ Etches. Genetic approach to physiology. (Colloquium Lake Tahoe, California, March 1991). In: RJ Etches, AM Verrinder Gibbins, eds. Manipulation of the Avian Genome. Boca Raton: CRC Press, 1993, pp 1–13.
2. AP Simopoulos. New products from the agri-food industry: The return of n-3 fatty acids into the food supply. Lipids 34:S297–S301, 1999.
3. Y Ito, R Sasaki, S Suzuki, K Aoki. Relationship between serum xanthophyll levels and the consumption of cigarettes, alcohol or foods in healthy inhabitants of Japan. Int J Epidemiol 20:615–620, 1991.
4. V Ollilainen, M Heinonen, E Linkola, P Varo, P Koivistoinen. Carotenoids and retinoids in Finnish foods: Dairy products and eggs. J Dairy Sci 72:2257–2265, 1989.
5. GJ Handelman, ZD Nightingale, AH Lichtenstein, EJ Schaefer, JB Blumberg. Lutein and zeaxanthin concentrations in plasma after dietary supplementation with egg yolk. Am J Clin Nutr 70:247–251, 1999.
6. MB Abou-Donia, CM Lyman. Metabolic fate of gossypol: The metabolism of gossypol-14C in laying hens. Toxicol Appl Pharmacol 17:160–173, 1970.
7. R Bou, F Guardiola, A Grau, S Grimpa, A Manich, A Barroeta, R Codony. Influence of dietary fat source, alpha-tocopherol, and ascorbic acid supplementation on sensory quality of dark chicken meat. Poult Sci 80:800–807, 2001.
8. LM Poste. A sensory perspective of effect of feeds on flavor in meats: Poultry meats. J Anim Sci 68:4414–4420, 1990.
9. National Science and Technology Council. Biotechnology for the 21st century: New horizons: A report from the Biotechnology Research Subcommittee, committee on fundamental Science. Washington, DC: US Government Printing Office, 1995.
10. SG Uzogara. The impact of genetic modification of human foods in the 21st century: A review. Biotechnol Adv 18:179–206, 2000.
11. JL Diedler, A Benoit. Prevention of symblepharon in severe ocular burns. Apropos of 3 cases. J Fr Ophtalmol 7:31–33, 1984.
12. ER Haapapuro, ND Barnard, M Simon. Review—animal waste used as livestock feed: Dangers to human health. Prev Med 26:599–602, 1997.
13. CM Williams, JC Barker, JT Sims. Management and utilization of poultry wastes. Rev Environ Contam Toxicol 162:105–157, 1999.
14. SM Watkins, BD Hammock, JW Newman, JB German. Individual metabolism should guide agriculture toward foods for improved health and nutrition. Am J Clin Nutr 74:283–286, 2001.

15. M Roberts, W Geiger, JB German. The revolution in microanalytic chemistry: A macro-opportunity for clinical nutrition. Am J Clin Nutr 71:434–437, 2000.
16. Meager portions of food news. Prepared Foods, May 1998. (*preparedfoods.com*, accessed May 2002).
17. International Food Information Council Foundation. Food for thought. IV. A report from the Nation's Media. Food Insight, January/February, 2002. (*http://www.ific.org*, accessed May 2002).
18. Foreign Agricultural Service. Per capita poultry meat consumption. Washington, DC: United States Department of Agriculture, Commodity and Marketing Programs, 2001. (*http://www.fas.usda.gov/*, accessed May 2002).
19. Economic Research Service, US Department of Agriculture. Per capita consumption of poultry and livestock 1960 to estimated 2002. (*http://eatchicken.com/statistics/consumption_pounds_60-02.cfm*, accessed May, 2002).
20. G Han. Poultry takes flight. Washington, DC: National Restaurant Association, November 1998. (*http://www.restaurant.org/rusa/magArticle.cfm?ArticleID=319*, accessed May, 2002).
21. WO Song, JM Kerver. Nutritional contribution of eggs to American diets. J Am Coll Nutr 19:556S–562S, 2000.
22. SM Moeller, PF Jacques, JB Blumberg. The potential role of dietary xanthophylls in cataract and age-related macular degeneration. J Am Coll Nutr 19:522S–527S, 2000.
23. SH Zeisel. Choline: Needed for normal development of memory. J Am Coll Nutr 19:528S–531S, 2000.
24. C Hotz, RS Gibson. Complementary feeding practices and dietary intakes from complementary foods amongst weanlings in rural Malawi. Eur J Clin Nutr 55:841–849, 2001.
25. Committee on Technological Options to Improve the Nutritional Attributes of Animal Products, National Research Council. Designing Foods: Animal Product Options in the Marketplace. Washington, DC: National Academy Press, 1988.
26. US Department of Agriculture. US Agriculture—linking consumers and producers. Agriculture Fact Book. Office of Communications. Washington, DC: United States Government Printing Office, 2001, pp 1–13.
27. American Egg Board. Eggs as nutraceuticals. Egg Products Reference Guide. Park Ridge, IL: American Egg Board. (*http://www.aeb.org/*, accessed May 2002).
28. JS, Sim, S Nakai, W Guenter, eds. Egg Nutrition and Biotechnology. Tucson, AR: CABI Publishing, 2000.
29. T, Yamamoto, LR Juneja, H Hatta, M Kim, eds. Hens Eggs, Their Basic and Applied Science. Boca Raton, FL: CRC Press, 1996.
30. RR Watson. Eggs and Health Promotion. Ames: Iowa State Press, 2002.
31. FB Hu, MJ Stampfer, EB Rimm, JE Manson, A Ascherio, GA Colditz, BA Rosner, D Spiegelman, FE Speizer, FM Sacks, CH Hennekens, WC Willett. A prospective study of egg consumption and risk of cardiovascular disease in men and women. JAMA 281:1387–1394, 1999.

32. WH Howell, DJ McNamara, MA Tosca, BT Smith, JA Gaines. Plasma lipid and lipoprotein responses to dietary fat and cholesterol: A meta-analysis. Am J Clin Nutr 65:1747–1764, 1997.
33. SB Kritchevsky, D Kritchevsky. Egg consumption and coronary heart disease: An epidemiologic overview. J Am Coll Nutr 19:549S–555S, 2000.
34. RM Krauss, RH Eckel, B Howard, LJ Appel, SR Daniels, RJ Deckelbaum, JW Erdman Jr, P Kris-Etherton, IJ Goldberg, TA Kotchen, AH Lichtenstein, WE Mitch, R Mullis, K Robinson, J Wylie-Rosett, S St Jeor, J Suttie, DL Tribble, TL Bazzarre. AHA Dietary guidelines: Revision 2000: A statement for healthcare professionals from the Nutrition Committee of the American Heart Association. Circulation 102:2284–2299, 2000.
35. B Mohan, R Kadirvel, M Bhaskaran, A Natarajan. Effect of probiotic supplementation on serum/yolk cholesterol and on egg shell thickness in layers. Br Poult Sci 36:799–803, 1995.
36. SM Abdulrahim, SY Haddadin, EA Hashlamoun, RK Robinson. The influence of Lactobacillus acidophilus and bacitracin on layer performance of chickens and cholesterol content of plasma and egg yolk. Br Poult Sci 37:341–346, 1996.
37. AC Awad, MR Bennink, DM Smith. Composition and functional properties of cholesterol reduced egg yolk. Poult Sci 76:649–653, 1997.
38. RG Elkin, Z Yan, Y Zhong, SS Donkin, KK Buhman, JA Story, JJ Turek, RE Porter Jr, M Anderson, R Homan, RS Newton. Select 3-hydroxy-3-methylglutaryl-coenzyme A reductase inhibitors vary in their ability to reduce egg yolk cholesterol levels in laying hens through alteration of hepatic cholesterol biosynthesis and plasma VLDL composition. J Nutr 129:1010–1019, 1999.
39. S Kaya, T Kececi, S Haliloglu. Effects of zinc and vitamin A supplements on plasma levels of thyroid hormones, cholesterol, glucose and egg yolk cholesterol of laying hens. Res Vet Sci 71:135–139, 2001.
40. GP Savage, PC Dutta, MT Rodriguez-Estrada. Cholesterol oxides: Their occurrence and methods to prevent their generation in foods. Asia Pac J Clin Nutr 11:72–78, 2002.
41. Committee on Dietary Allowances, Food and Nutrition Board, Division of Biological Sciences, Assembly of Life Sciences, National Research Council. Recommended Dietary Allowances. 9th ed. Washington, DC: National Academy of Sciences, 1980.
42. HD Griffin. Manipulation of egg yolk cholesterol: A physiologist's view. World Poult Sci J 48:101–112, 1992.
43. Agricultural Research Service. USDA Nutrient Database for Standard Reference Release 14: Nutrient Data Laboratory. Agricultural Research Service, Beltsville Human Nutrition Research Center: Riverdale, MD, 2001.
44. I Cowin, A Emond, P Emmett. Association between composition of the diet and haemoglobin and ferritin levels in 18-month-old children. Eur J Clin Nutr 55:278–286, 2001.
45. SL Booth, JA Pennington, JA Sadowski. Dihydro-vitamin K1: Primary food

sources and estimated dietary intakes in the American diet. Lipids 31:715–720, 1996.
46. RT Paterson, N Joaquin, K Chamon, E Palomino. The productivity of small animal species in small-scale mixed farming systems in subtropical Bolivia. Trop Anim Health Prod 33:1–14, 2001.
47. D Li, A Ng, NJ Mann, AJ Sinclair. Contribution of meat fat to dietary arachidonic acid. Lipids 33:437–440, 1998.
48. KL Tucker, GE Dallal, D Rush. Dietary patterns of elderly Boston-area residents defined by cluster analysis. J Am Diet Assoc 92:1487–1491, 1992.
49. KM Koehler, WC Hunt, PJ Garry. Meat, poultry, and fish consumption and nutrient intake in the healthy elderly. J Am Diet Assoc 92:325–330, 1992.
50. A Sams. Irradiation and other hang ups. WATT Poul USA 3:20–22, 2002.
51. R EMoreng, JS Aven. Commercial Poultry Industry Related to Agricultural Food Production, RE Moreng, JS Avens, eds. Reston, VA: Reston Publishing, 1985, pp 1–13.
52. WA Rishell. Breeding and genetics—historical perspective. Poult Sci 76:1057–1061, 1997.
53. National Research Council (US) Subcommittee on Poultry Nutrition, Committee on Animal Nutrition, Board on Agriculture. Nutrient Requirements of Poultry, 9th ed. Washington, DC: National Academy Press, 1994.
54. RH Hammerstedt. Symposium summary and challenges for the future. Poult Sci 78:459–466, 1999.
55. DL Pollock. A geneticist's perspective from within a broiler primary breeder company. Poult Sci 78:414–418, 1999.
56. A Emsley. Integration of classical and molecular approaches of genetic selection: Egg production. Poult Sci 76:1127–1130, 1997.
57. DA Emmerson. Commercial approaches to genetic selection for growth and feed conversion in domestic poultry. Poult Sci 76:1121–1125, 1997.
58. GF Barbato. Genetic relationships between selection for growth and reproductive effectiveness. Poult Sci 78:444–452, 1999.
59. JB Dodgson, HH Cheng, R Okimoto. DNA marker technology: A revolution in animal genetics. Poult Sci 76:1108–1114, 1997.
60. HH Cheng. Mapping the chicken genome. Poult Sci 76:1101–1107, 1997.
61. J Hillel. Map-based quantitative trait locus identification. Poult Sci 76:1115–1120, 1997.
62. RJ Etches. From chicken coops to genome maps: Generating phenotype from the molecular blueprint. Poult Sci 80:1657–1661, 2001.
63. MM Perry, HM Sang. Transgenesis in chickens. Transgenic Res 2:125–133, 1993.
64. JN Petitte, L Karagenc, M Ginsburg. The origin of the avian germ line and transgenesis in birds. Poult Sci 76:1084–1092, 1997.
65. ML Scott. Nutrition of Humans and Selected Animal Species. New York: Wiley-Interscience, 1986.
66. WA McIntosh. The symbolization of eggs in American culture: A sociologic analysis. J Am Coll Nutr 19:532S–539S, 2000.

67. DJ McNamara. The impact of egg limitations on coronary heart disease risk: Do the numbers add up? J Am Coll Nutr 19:540S–548S, 2000.
68. SM Michella, BT Slaugh. Producing and marketing a specialty egg. Poult Sci 79:975–976, 2000.
69. I Jialal, CJ Fuller, BA Huet. The effect of alpha-tocopherol supplementation on LDL oxidation: A dose-response study. Arterioscler Thromb Vasc Biol 15: 190–198, 1995.
70. MJ Stampfer, CH Hennekens, JE Manson, GA Colditz, B Rosner, WC Willett. Vitamin E consumption and the risk of coronary disease in women. N Engl J Med 328:1444–1449, 1993.
71. A Hosomi, M Arita, Y Sato, C Kiyose, T Ueda, O Igarashi, H Arai, K Inoue. Affinity for alpha-tocopherol transfer protein as a determinant of the biological activities of vitamin E analogs. FEBS Lett 409:105–108, 1997.
72. BW Sheldon, PA Curtis, PL Dawson, PR Ferket. Effect of dietary vitamin E on the oxidative stability, flavor, color, and volatile profiles of refrigerated and frozen turkey breast meat. Poult Sci 76:634–641, 1997.
73. BL Sheldon. Research and development in 2000: Directions and priorities for the world's poultry science community. Poult Sci 79:147–158, 2000.
74. CM Bruhn. Consumer perceptions and concerns about food contaminants. Adv Exp Med Biol 459:1–7, 1999.
75. C Gusils, SN Gonzalez, G Oliver. Some probiotic properties of chicken lactobacilli. Can J Microbiol 45:981–987, 1999.
76. M Vanbelle, E Teller, M Focant. Probiotics in animal nutrition: A review. Arch Tierernahr 40:543–567, 1990.
77. C Duggan, J Gannon, WA Walker. Protective nutrients and functional foods for the gastrointestinal tract. Am J Clin Nutr 75:789–808, 2002.
78. GR Gibson, KC Mountzouris, AL McCartney. Intestinal microflora of human infants and current trends for its nutritional modulation. Br J Nutr 87: 405–420, 2002.
79. S van Drunen Littel-van den Hurk, V Gerdts, BI Loehr, R Pontarollo, R Rankin, R Uwiera, LA Babiuk. Recent advances in the use of DNA vaccines for the treatment of diseases of farmed animals. Adv Drug Deliv Rev 43:13–28, 2000.
80. M Sakaguchi, H Nakamura, K Sonoda, H Okamura, K Yokogawa, K Hira, K Hira. Protection of chickens with or without maternal antibodies against both Marek's and Newcastle diseases by one-time vaccination with recombinant vaccine of Marek's disease virus type 1. Vaccine 16:472–479, 1998.
81. N Eterradossi, D Toquin, H Abbassi, G Rivallan, JP Cotte, M Guittet. Passive protection of specific pathogen free chicks against infectious bursal disease by in-ovo injection of semi-purified egg-yolk antiviral immunoglobulins. Zentralbl Veterinarmed [B] 44:371–383, 1997.
82. C Rollier, C Charollois, C Jamard, C Trepo, L Cova. Maternally transferred antibodies from DNA-immunized avians protect offspring against hepadnavirus infection. J Virol 74:4908–4911, 2000.
83. D Carlander, J Stalberg, A Larsson. Chicken antibodies: A clinical chemistry perspective. Ups J Med Sci 104:179–189, 1999.

84. D Carlander, H Kollberg, PE Wejaker, A Larsson. Peroral immunotherapy with yolk antibodies for the prevention and treatment of enteric infections. Immunol Res 21:1–6, 2000.
85. CH Kweon, BJ Kwon, SR Woo, JM Kim, GH Woo, DH Son, W Hur, YS Lee. Immunoprophylactic effect of chicken egg yolk immunoglobulin (Ig Y) against porcine epidemic diarrhea virus (PEDV) in piglets. J Vet Med Sci 62: 961–964, 2000.
86. DJ Smith, WF King, R Godiska. Passive transfer of immunoglobulin Y antibody to Streptococcus mutans glucan binding protein B can confer protection against experimental dental caries. Infect Immun 69:3135–3142, 2001.
87. M Tini, UR Jewell, G Camenisch, D Chilov, M Gassmann. Generation and application of chicken egg-yolk antibodies. Comp Biochem Physiol A Mol Integr Physiol 131:569–574, 2002.
88. EM Akita, EC Li-Chan. Isolation of bovine immunoglobulin G subclasses from milk, colostrum, and whey using immobilized egg yolk antibodies. J Dairy Sci 81:54–63, 1998.
89. SJ Streatfield, JM Jilka, EE Hood, DD Turner, MR Bailey, JM Mayor, SL Woodard, KK Beifuss, ME Horn, DE Delaney, IR Tizard, JA Howard. Plant-based vaccines: Unique advantages. Vaccine 19:2742–2748, 2001.
90. V Portrait, S Gendron-Gaillard, G Cottenceau, AM Pons. Inhibition of pathogenic Salmonella enteritidis growth mediated by *Escherichia coli microcin* J25 producing strains. Can J Microbiol 45:988–994, 1999.
91. N Natrajan, BW Sheldon. Efficacy of nisin-coated polymer films to inactivate Salmonella Typhimurium on fresh broiler skin. J Food Prot 63:1189–1196, 2000.
92. C Zhao, T Nguyen, L Liu, RE Sacco, KA Brogden, RI Lehrer. Gallinacin-3, an inducible epithelial beta-defensin in the chicken. Infect Immun 69:2684–2691, 2001.
93. JM Rodriguez, MI Martinez, J Kok. Pediocin PA-1, a wide-spectrum bacteriocin from lactic acid bacteria. Crit Rev Food Sci Nutr 42:91–121, 2002.
94. GA Dykes, SM Moorhead. Combined antimicrobial effect of nisin and a listeriophage against *Listeria monocytogenes* in broth but not in buffer or on raw beef. Int J Food Microbiol 73:71–81, 2002.
95. R Brewer, MR Adams, SF Park. Enhanced inactivation of Listeria monocytogenes by nisin in the presence of ethanol. Lett Appl Microbiol 34:18–21, 2002.
96. M Castellari, A Versari, A Fabiani, GP Parpinello, S Galassi. Removal of ochratoxin A in red wines by means of adsorption treatments with commercial fining agents. J Agric Food Chem 49:3917–3921, 2001.
97. V Ravindran, S Cabahug, G Ravindra, PH Selle, WL Bryden. Response of broiler chickens to microbial phytase supplementation as influenced by dietary phytic acid and non-phytate phosphorous levels. II. Effects on apparent metabolisable energy, nutrient digestibility and nutrient retention. Br Poult Sci 41:193–200, 2000.
98. K Zyla, DR Ledoux, M Kujawski, TL Veum. The efficacy of an enzymic cocktail and a fungal mycelium in dephosphorylating corn-soybean meal-based feeds fed to growing turkeys. Poult Sci 75:381–387, 1996.

99. DM Denbow, EA Grabau, GH Lacy, ET Kornegay, DR Russell, PF Umbeck. Soybeans transformed with a fungal phytase gene improve phosphorus availability for broilers. Poult Sci 77:878–881, 1998.
100. N Grotz, ML Guerinot. Limiting nutrients: An old problem with new solutions? Curr Opin Plant Biol 5:158–163, 2002.
101. XG Lei, CH Stahl. Biotechnological development of effective phytases for mineral nutrition and environmental protection. Appl Microbiol Biotechnol 57:474–481, 2001.
102. AB Zhang, ET Kornegay, JS Radcliffe, DM Denbow, HP Veit, CT Larsen. Comparison of genetically engineered microbial and plant phytase for young broilers. Poult Sci 79:709–717, 2000.
103. SP Golovan, RG Meidinger, A Ajakaiye, M Cottrill, MZ Wiederkehr, DJ Barney, C Plante, JW Pollard, MZ Fan, MA Hayes, J Laursen, JP Hjorth, RR Hacker, JP Phillips, CW Forsberg. Pigs expressing salivary phytase produce low-phosphorus manure. Nat Biotechnol 19:741–745, 2001.
104. SP Golovan, MA Hayes, JP Phillips, CW Forsberg. Transgenic mice expressing bacterial phytase as a model for phosphorus pollution control. Nat Biotechnol 19:429–933, 2001.
105. V Raboy. Seeds for a better future: 'Low phytate' grains help to overcome malnutrition and reduce pollution. Trends Plant Sci 6:458–462, 2001.
106. T Gura. Biotechnology. New genes boost rice nutrients. Science 285:994–995, 1999.
107. TC Verwoerd, PA van Paridon, AJ van Ooyen, JW van Lent, A Hoekema, J Pen. Stable accumulation of Aspergillus niger phytase in transgenic tobacco leaves. Plant Physiol 109:1199–1205, 1995.
108. JD Palombo, SJ DeMichele, JW Liu, BR Bistrian, YS Huang. Comparison of growth and fatty acid metabolism in rats fed diets containing equal levels of gamma-linolenic acid from high gamma-linolenic acid canola oil or borage oil. Lipids 35:975–981, 2000.
109. VC Knauf, D Facciotti. Genetic engineering of foods to reduce the risk of heart disease and cancer. Adv Exp Med Biol 369:221–228, 1995.
110. MW Pariza, Y Park, ME Cook. The biologically active isomers of conjugated linoleic acid. Prog Lipid Res 40:283–298, 2001.
111. CM Owens, EM Hirschler, SR McKee, R Martinez-Dawson, AR Sams. The characterization and incidence of pale, soft, exudative turkey meat in a commercial plant. Poult Sci 79:553–558, 2000.
112. A Skerra. Lipocalins as a scaffold. Biochim Biophys Acta 1482:337–350, 2000.
113. E El Maradny, N Kanayama, H Kobayashi, B Hossain, S Khatun, S Liping, T Kobayashi, T Terao. The role of hyaluronic acid as a mediator and regulator of cervical ripening. Hum Reprod 12:1080–1088, 1997.
114. H Qin, T Ishiwata, G Asano. Effects of the extracellular matrix on lumican expression in rat aortic smooth muscle cells in vitro. J Pathol 195:604–608, 2001.
115. JM Ervasti, AL Burwell, AL Geissler. Tissue-specific heterogeneity in alpha-dystroglycan sialoglycosylation: Skeletal muscle alpha-dystroglycan is a latent

receptor for Vicia villosa agglutinin b4 masked by sialic acid modification. J Biol Chem 272:22315–22321, 1997.
116. SG Velleman. The role of the extracellular matrix in skeletal muscle development. Poult Sci 78:778–784, 1999.
117. FW George, H Matsumine, MJ McPhaul, RG Somes Jr, JD Wilson. Inheritance of the henny feathering trait in the golden Campine chicken: Evidence for allelism with the gene that causes henny feathering in the Sebright bantam. J Hered 81:107–110, 1990.
118. H Matsumine, MA Herbst, SH Ou, JD Wilson, MJ McPhaul. Aromatase mRNA in the extragonadal tissues of chickens with the henny-feathering trait is derived from a distinctive promoter structure that contains a segment of a retroviral long terminal repeat: Functional organization of the Sebright, Leghorn, and Campine aromatase genes. J Biol Chem 266:19900–19907, 1991.
119. AK Mapp, AZ Ansari, M Ptashne, PB Dervan. Activation of gene expression by small molecule transcription factors. Proc Natl Acad Sci USA 97:3930–3935, 2000.
120. Y Zhou, YP Ching, KH Kok, HF Kung, DY Jin. Post-transcriptional suppression of gene expression in Xenopus embryos by small interfering RNA. Nucleic Acids Res 30:1664–1669, 2002.
121. P Svoboda, P Stein, RM Schultz. RNAi in mouse oocytes and preimplantation embryos: effectiveness of hairpin dsRNA. Biochem Biophys Res Commun 287:1099–1104, 2001.
122. H Eyal-Giladi. The gradual establishment of cell commitments during the early stages of chick development. Cellular Differentiation 14:245–255, 1984.
123. JA Porter, KE Young, PA Beachy. Cholesterol modification of hedgehog signaling proteins in animal development. Science 274:255–259, 1996.
124. PW Ingham. Hedgehog signaling: A tale of two lipids. Science 294:1879–1881, 2001.
125. YJ Wu, M Valdez-Corcoran, JT Wright, AL Cartwright. Abdominal fat pad mass reduction by in ovo administration of anti-adipocyte monoclonal antibodies in chickens. Poult Sci 79:1640–1644, 2000.
126. YJ Wu, JT Wright, CR Young, AL Cartwright. Inhibition of chicken adipocyte differentiation by in vitro exposure to monoclonal antibodies against embryonic chicken adipocyte plasma membranes. Poult Sci 79:892–900, 2000.
127. RM Shuman. Production of transgenic birds. Experientia 47:897–905, 1991.
128. H Sang. Transgenic chickens—methods and potential applications. Trends Biotechnol 12:415–420, 1994.
129. MJ Federspiel, SH Hughes. Retroviral gene delivery. Methods Cell Biol 52: 179–214, 1997.
130. RJ Etches, RS Carsience, ME Clark, RA Fraser, A Toner, AM Verrinder Gibbins. Chimeric chickens and their use in manipulation of the chicken genome. Poult Sci 72:882–889, 1993.
131. AM Gibbins. The chicken, the egg, and the ancient mariner. Nat Biotechnol 16:1013–1014, 1998.
132. JP Katkin, RC Husser, C Langston, SE Welty. Exogenous surfactant enhances

the delivery of recombinant adenoviral vectors to the lung. Hum Gene Ther 8:171–176, 1997.

133. MA Hickman, RW Malone, Kk Lehmann-Bruinsma, TR Sih, D Knoell, FC Szoka, R Walzem, DM Carlson, JS Powell. Gene expression following direct injection of DNA into liver. Hum Gene Ther 5:1477–1483, 1994.
134. RE Hammer, VG Pursel, CE Rexroad Jr, RJ Wall, DJ Bolt, KM Ebert, RD Palmiter, RL Brinster. Production of transgenic rabbits, sheep and pigs by microinjection. Nature 315:680–683, 1985.
135. G Speksnijder, R Ivarie. A modified method of shell windowing for producing somatic or germline chimeras in fertilized chicken eggs. Poult Sci 79:1430–1433, 2000.
136. JN Petitte, I Chang. Method of producing an undifferentiated avian cell culture using avian primordial germ cells. US Patent No. 6,333,192, December 25, 2001.
137. J Kureczka. Mass producing superior poultry. NIST Advanced Technology Program, 2001. (*http://www.atp.nist.gov/awards/00004402.htm*, accessed May 2002).
138. CP Hodgson. Method of transferring a DNA sequence to a cell in vitro. Omaha, NE: Nature Technology Corporation, 2001.
139. FA Ponce de Leon, C Blackwell, XY Gao, JM Robl, SL Stice, DJ Jerry. Prolonged culturing of avian primordial germ cells (PGCs) using specific growth factors, use thereof to produce chimeric avians. University of Massachusetts Office of Vice Chancellor for Research at Amherst, 2000.
140. X Yang, XC Tian, Y Dai, B Wang. Transgenic farm animals: Applications in agriculture and biomedicine. Biotechnol Annu Rev 5:269–292, 2000.

2

Modern Biotechnology for the Production of Dairy Products

Pedro A. Prieto
Abbott Laboratories, Columbus, Ohio, U.S.A.

I. INTRODUCTION

A. Early and Modern Biotechnology

For practical reasons it is important to distinguish between the traditional or early methods of biotechnology and the modern approaches and techniques of this field. This is particularly relevant in the case of dairy science and technology because regulations, risk–benefit analyses, and perceptions of processes involving the use of genetic engineering differ from those that do not involve recombinant technology. Animal husbandry and food technology have provided solutions to the challenges and problems encountered during the production of milk and milk-derived products; the tools and methods of these fields constitute "early biotechnology." This is in contrast to modern biotechnology, which is constituted by methods based on recombinant deoxyribonucleic acid (DNA) techniques (1) and novel approaches for the purification of materials, selection of microbial strains, fermentation and manufacturing processes, and analysis of foods. The specialist who implants embryos while aiming to expand a desirable characteristic in a herd of dairy cows and the cheese maker who inoculates curd with a naturally occurring starter culture are indeed using the tools and methods of traditional or early biotechnology. On the other hand, the molecular biologist who attempts to insert a gene fragment at a specific site of the bovine genome with the purpose of producing dairy products containing human milk proteins is clearly using modern methodologies. Likewise, the task of genetically modifying a microbial strain to speed up the maturation process in a cheese requires modern techniques. This review focuses on processes and technologies that utilize genetic engineering and that, in most cases, produce genetically modified organisms (GMOs) or their derivatives. The present account constitutes a brief overview of modern biotechnology as it applies to the production and modification of dairy products. Its purpose is to provide a catalogue of techniques involving the use of recombinant nucleic acids in the context of animal productivity and milk and milk-derived product remodeling or improvement. Ancillary aspects are also discussed. The reader is referred to previously published reviews that address general aspects of modern biotechnology (2–4).

Table 1 Biotechnology Applications That Affect the Production of Milk and Milk-Derived Products

Aspect of dairy production	Technology	Impact on milk production or dairy product	References
Feeding	Use of GMO[a] to produce animal food	Mostly quantitative	Krishna 1998 (18), Crooker 1996 (5)
Improving feed utilization, bioconversion	Use of GMO as probiotics or GMO-derived enzymes	Mostly quantitative, increase in feed efficiency	Bera-Maillet et al., 2000 (6) Blum et al. 1999 (7) Gregg et al. 1998 (8) Ziegelhogger 1999 (9)
Improvement of growth rate, milk yields, altering of reproductive cycle	Use of recombinant hormones	Quantitative	Bauman, 1999 (10)
Improvement of milk yields, synthesis of novel compounds in milk, alteration of milk composition	Production of transgenic animals and targeted mutants	Qualitative and quantitative	Prieto et al. 1999 (11)
Identification and production of improved cultures and production of novel strains with custom designed characteristics	Production and identification of microbial strains and GMO-derived enzymes to process milk and milk-derived products	Qualitative	Henriksen et al., 1999 (12)

[a] GMO, genetically modified organism.

B. Scope of the Present Review

Table 1 lists some of the steps involved in the production of dairy products and technologies that have been used or are being investigated for each step. The table indicates whether the main target of a particular biotechnology is focused on a quantitative effect, such as milk yield, or a qualitative effect, such as the presence or increase of a particular compound in milk. Other technical approaches that are closely related to those listed in Table 1, but are not covered in this account are (a) the transgenic expression of hormones targeted to the modification of carcass composition and (b) the use of animals as bioreactors to produce biomolecules for pharmaceutical applications. These subjects have been reviewed by Pursel and Solomon (13), Velander and associates (14), and Ziomek (15). The technologies listed in Table 1 can also be grouped in two categories: (a) those aimed at the dairy animal in regard to its productivity and (b) those aimed at modifying milk or its derivatives. This classification is useful to explain the aims and potential consequences of particular technologies and the problems they intend to solve. The use of genetically modified microorganisms in food products is analyzed in depth in another chapter of the present volume; however, specific applications to dairy products are briefly discussed in Sec. V to illustrate novel technologies as they pertain to fermented milk and cheese manufacturing.

II. USE OF GENETICALLY MODIFIED ORGANISMS TO ENHANCE FEED EFFICIENCY FOR DAIRY COWS

Crooker (5) states, "In animal agriculture, feed generally represents the major input component while tissue gain (growth) or milk yield are the primary useful outputs." In addition, Crooker indicates that approximately 30% of the calories and proteins in the diet of a dairy cow find their way into productive functions such as milk synthesis. Biotechnology can impact the manufacturing of animal feed in several ways, such as (a) improvement of microbial strains for the synthesis of diet components, (b) production of recombinant enzymes to improve digestion of diet components and improve feed utilization, (c) production of recombinant microorganisms that act as probiotics in the rumen of dairy cows, and (d) production of transgenic plants as feed constituents.

A. Improved Microbial Strains

Microbial fermentation processes produce key amino acids used to supplement feed. For example, the genera *Corynebacterium* and *Brevibacterium* are used to synthesize lysine and threonine. Studies with these microbes elucidated their metabolic pathways and key control points in amino acid synthesis. This information was then applied to design and implement strategies to circumvent metabolic bottlenecks such as mechanisms of feedback inhibition. An example of these studies and their applications is described in a report by Malumbres and Martin (16) in which the genetic control of the key enzymatic steps involved in amino acid synthesis is modified by redirecting the flow of carbon skeletons. This was accomplished by amplifying genes encoding feedback-resistant aspartokinase and homoserine dehydrogenase.

B. Genetically Modified Organism–Produced Enzymes and Genetically Modified Organisms as Probiotics for Dairy Cows

Hydrolytic enzymes such as lactase (β-galactosidase) are used by lactose-intolerant humans to aid in the digestion of lactose. Similarly, enzymes can be used to increase feed efficiency in dairy animals and ruminants in general. Some enzymes used as feed additives or pretreatments are made naturally by microbes. For instance, the white rot fungus *Coprinus fimetarius* enzymatically degrades lignin, thus releasing cellulose and hemicellulose from plant-based feed (17,18); the released polysaccharides become susceptible to the action of other hydrolytic enzymes such as cellulases and hemicellulases, thus providing glucose, which was unavailable before the fungal pretreatment. Many naturally occurring enzymes have been cloned and are now produced efficiently through fermentation; examples of these are glycosidases such as cellulases and xylanases (6,7). A more novel strategy is colonization of the farm animal's rumen with recombinant microorganisms that have been modified to acquire new metabolic capabilities that in turn benefit the host animal (5). These probiotic organisms for ruminants are designed to support specific functions that go beyond the overexpression of hydrolytic enzymes. For example, Gregg and colleagues (8) developed strains of *Butyrivibrio fibrisolvens* expressing a gene encoding fluoroacetate dehalogenase. These GMOs colonize the rumen of sheep and reduce symptoms of fluoroacetate poisoning.

C. Transgenic Manipulation of Plants for Enhancement of Animal Feed

Another biotechnological option to affect the efficiency of feed is the use of transgenic plants. For instance, alfalfa- and tobacco-expressing cellulase genes have been produced (9). These cultivars promote efficient use of feed because their recombinant cellulases are aids in the hydrolysis of plant components. Genetically modified plants can also be used to provide specific nutrients; transgenic canola with increased methionine content exemplifies the potential of genetically modified plants as enhanced sources of amino acids in the feed. This advantage is easily exploited because some of these plants are already constituents of animal feed (19). Therefore, the expression of nutrients in transgenic plants may affect both the quantity and the quality of milk without costly purification of recombinant products (20). Another example of transgenic synthesis of targeted nutrients in plants is the production of specific fatty acids. Enzymes that participate in the biosynthesis of polyunsaturated fatty acids (PUFAs) have been cloned by Knutzon and coworkers (21,22) and Huang and associates (23), and transgenic seeds that contain oils enriched in PUFA have been produced. On the other hand, it has also been demonstrated that dairy cows fed fish oil increased, albeit inefficiently, the PUFA content of their milk (24). Feeding dairy cows with these modified canola seeds or with PUFA-supplemented feed could result in large-scale production of PUFA-enriched milk.

III. USES OF RECOMBINANT AND SYNTHETIC HORMONES TO IMPROVE MILK YIELDS AND AFFECT REPRODUCTIVE CYCLES

Hormone treatment of dairy cows is now used as an alternative method to improve bioconversion of feed into milk. Since the 1930s scientists have known that growth hormone or somatotropin from pituitary glands increases milk yield in dairy cows (25). The commercial application of recombinant bovine somatotropin (rBST) started in 1994 and today rBST is considered to be the first major biotech product applied to animal agriculture (26). Referring to productivity improvement attained by the use of rBST, Bauman (10) states: “The magnitude of the gain efficiency of milk production was equal to that normally achieved over a 10- to 20-year period with artificial insemination and genetic selection technologies.” This statement illustrates the potential for rapid impact that characterizes modern biotechnology. Today the use of rBST to improve milk yields is common in the United States and has been studied from several perspectives. Other hormones or hormone analogues have been used in dairy cows to modify reproductive cycles with particular aims. For example, a synthetic leuteinizing hormone–releasing hormone analogue is used to increase conception rates (27) and estradiol and progesterone are used to induce lactation in prepubertal animals (28). This application of hormones may or may not have commercial impact; however, it unquestionably aids in the development of transgenic animal technology targeted to synthesis of novel milk compounds in the dairy cow. Hormones can be added exogenously as part of the management schedule of a dairy farm, or they can also be produced in situ and in excess of their natural concentrations by remodeling the genetic makeup of an animal (29–31).

IV. TRANSGENIC ANIMALS

Transgenic animals (TAs) constitute perhaps one of the most tantalizing and powerful modern technologies for the manipulation of quantitative and qualitative aspects of dairy production. Most TAs and targeted mutants (TMs) produced so far are experimental models in laboratory animals. These are used to study the effects of gene expression in whole organisms or are prototypes for the production of modified tissues or biological fluids for industrial purposes. The potential of genetically modified dairy cows for the manufacture of food and specialized nutritional products resides mainly in three aspects of the production of TAs and TMs: (a) technologies available for the production of TAs, (b) current technical hurdles and limitations of these technologies, and (c) applications of TAs and TMs to the fields of human health and nutrition. Understanding these aspects of the technology is also useful to evaluate the safety and potential environmental impact of TAs and TMs.

A. Production of Transgenic Animals

Briefly, a transgenic animal (TA) is a GMO that has acquired or lost a function or functions with respect to the wild type (11). A function is gained in a transgenic animal by inserting exogenous genetic information into its genome, usually during its embryonic or proembryonic stage. Loss of function is achieved by inserting a modified nonexpressible gene construct in place of its homologous wild-type functioning gene. In these animals the original gene is "knocked out" and the animals themselves are referred to as *targeted mutants* (TMs), *gene disrupted animals*, or *knockout animals*. Particular aspects of the technologies and strategies for the production of TAs and TMs for dairy production have been reviewed by Wall and colleagues (32), Karatzas and Turner (33), Murray (34), and Pintado and Gutiérrez Adán (35). The present section focuses on modified or remodeled milk produced by TAs or TMs. The production of TAs requires several steps: (a) design and construction of the fusion gene (also called *transgene*) that is commonly inserted into an embryo, (b) physical introduction of the fusion gene, and (c) its incorporation into the genome.

The most common method to introduce fusion genes into cells for generation of TAs or TMs is microinjection into the pronuclei of embryos

(36). Alternatively, the fusion gene can be electroporated directly into sperm (37) or into the testis of an animal, thus propagating the transgene in sperm (38). There are other technologies that are suitable for production of TAs and TMs in which the fusion gene is not introduced into pronuclear stage embryos; Chan and coworkers (39) describe the production of transgenic cattle by injecting oocytes with replication-defective retroviral vectors containing transgenes. Increasingly, multicellular embryos and somatic cells such as fibroblasts are used as recipients of fusion genes to produce transgenic cows (39–42). These technologies are important because they result in the production of cloned transgenic animals. Briefly, if a transgene is incorporated into the genome of a fibroblast and a cell line is established from this fibroblast, cells can be tested for the presence and in some cases for the expression or function of the transgene. As a last step, nuclei from this cell line can be transferred to enucleated oocytes. In this fashion, several pseudopregnant females can be implanted with cloned transgenic oocytes. Once a transgenic embryo is produced, it is usually implanted into pseudopregnant cows. The resulting calves are analyzed for the presence of the acquired transgene, and if the transgene is found in the tissues of a calf, then it is allowed to mature for further assessment. The resulting transgenic cow is then mated and its milk is analyzed for the desired modifications. Alternatively, as discussed previously, induction of prepubertal lactation can abbreviate the time from implantation of the embryo to the determination of successful production of "remodeled" milk (28). This example illustrates how modern biotechnology methods and tools combine to accelerate the development of prototype animals during the feasibility stages of a biotechnology-based project.

All the mentioned steps are common in the production of targeted mutants, with the exception of the first step. The genetic element used to produce a TM contains homologous elements that hybridize with endogenous DNA, thus allowing its insertion into the precise site occupied by the gene that is intended to be disrupted. It also contains fragments engineered to prevent its transcription into a functional messenger ribonucleic acid (mRNA). An additional method to control gene expression without obliterating endogenous gene expression is transgenic expression of ribozymes, which are RNA molecules that have the ability to hydrolyze RNA sequences in their midst (43,44). When a ribozyme is expressed in a particular tissue it interferes with the translation of mRNA transcripts, thus reducing the overall synthesis of a given protein. This technology is useful in controlling the synthesis of α-lactalbumin in lactating mammary glands (45) and enzymes involved in the synthesis of lipids (46). Both instances are discussed later in the context of targeted modifications to the lactating mammary glands of dairy cows.

A key problem in the construction of a fusion gene is targeting its expression to selected tissues or organs. If the desired outcome is the modification of milk, the fusion gene requires a transcriptional regulatory element

(TRE) that promotes its expression in the lactating mammary gland. Lothar Hennighausen and coworkers pioneered the development of such a TRE (47,48). This group expressed human tissue plasminogen activator in the lactating mammary glands of mice by using the TRE that directs the expression of the gene encoding the most abundant whey protein in murine milk, the whey acidic protein. Since then, several other lactation-specific TREs have been isolated and characterized. Examples of milk protein TREs are those of bovine κ-casein (49), goat β-casein (50), and bovine α-lactalbumin (51).

B. Technical Problems and Limitations in the Production of Dairy Transgenic Animals and Targeted Mutants and Approaches to Their Solution

Transgenic dairy cows expressing exogenous components in their milk have been produced by microinjection as described (41). However, the introduction of fusion genes into embryos, gametes, or gonads has several constraints that limit the applications of this method for large-scale production of transgenic cows and goats. Examples of these hurdles are summarized in Table 2. Perhaps the most frequent problem experienced in generating transgenic animals is the lack of specific insertion of the fusion gene into the genome. For instance, a fusion gene can be integrated into a transcriptionally inactive region of a chromosome, thus severely limiting the expression of a transgene. Piedrahita has reviewed the consequences of nonspecific insertion of fusion genes and points out that the technologies for targeted genome modification are still in the early developmental stage (52). The strategies to circumvent this "random insertion" problem have been either to shield the transgene from the surrounding chromatin or to target insertions to specific sites of the genome. A transgene can be shielded by dominant regulatory sequences known as *locus control regions* (LCRs) or *matrix attachment regions* (MARs) (53,54). Alternatively, a transgene can be directed to a specific genomic site by homologous recombination, the most frequently used approach for TM generation (55). Homologous recombination has not yet been perfected for the production of TAs, and the technique yields a low frequency of successful transgenic events (52).

Experimental animals such as rabbits and mice have short gestational times, relatively large litters, and relatively low maintenance cost for large quantities of animals. For these reasons it has been acceptable to use fusion gene transfer techniques that are inefficient and produce relative low numbers of useful transgenic events while developing prototypes in laboratory animals. In large domestic animals such as cattle, these inefficiencies cause a significant increase in the cost and time necessary to produce useful transgenic founders. Several techniques are being developed to increase transfer efficiency. For

Table 2 Hurdles for the Production of Transgenic Dairy Animals and Examples of Techniques Explored to Circumvent Them[a]

Problem	Consequence	Desired technical outcome	Approaches
Nonspecific integration of fusion gene into genome	Transgene expression affected by surroundings—unpredictable control of transgene expression	Transgene expression controlled by TRE in fusion gene or regulatory elements present in targeted site	*Transgene shielding*: Fujiwara et al. (53), McKnight et al. (54) *Homologous recombination*: Riele et al. (55) Piedrahita (52)
Low frequency of transgenic events due to low number of live births Low transfer efficiency inherent in microinjection technique	Reliance of efforts to produce transgenic founders and herds on low-probability events; transgenic animals not always obtained	Higher percentage of live births Higher percentage of transgenic animal production	*Microinjection of embryos at two- or four-cell stage*: Echelard et al. (41) *Transgenic production through retroviral transfection*: Chan (39)
Long reproductive cycle in cattle; significant lag time between fusion gene transfer and verification of success	Necessity for long-term maintenance of animals before their usefulness is determined	Shorter periods to determine useful transgenic founders	*Lactaction induction*: Ball et al. (28); *retroviral transfer through teat canal*: Archer et al. (56)
Low frequency of transgenic events due to lack of ability to express gene products	Many nontransgenic embryos implanted and later nontransgenic animals maintained	Determination of embryo potential before implantation	*Embryo genotyping*: Hyttinen et al. (57), Jura et al. (58), and Saberivand et al. (59)
Slow transgenic propagation into herd	Wait time required for commercialization of acquired or lost trait to establish productive herd	Fast establishment of transgenic herd	*Cloning of TA through nuclear transfer*: Cibelli et al. (42)

[a] TRE, transcriptional regulatory elements; TA, transgenic animal.

example, Chan and associates (39) have used viral vectors to transfer fusion genes via reverse transcription to produce transgenic cattle. Also, Echelard and colleagues (41) have microinjected cow embryos at the two- and four-cell stages and compared, favorably, the number of live births obtained by this technique with those resulting from microinjection into pronuclear stage embryos. Another limitation inherent to the relatively long reproductive cycle of cattle is the lag time between the fusion gene transfer and the production of milk from a transgenic animal. An animal can have detectable elements of a fusion gene in its genome, but their presence does not guarantee that the gene product will be expressed in its milk; therefore, the true usefulness of a transgenic animal cannot be determined until the animal produces milk. As mentioned, one technique that shortens this wait period is to induce lactation in prepuber heifers (28); another is to produce adult transgenic animals that have been exposed to retroviral particles containing fusion genes that are injected through the teat canal (56). This technique has the added advantage that the fusion gene is not incorporated into the germ line, thus providing short-term verification of the efficacy of a particular fusion gene. Several groups have also developed methods to determine whether transfected embryos can express transgene-encoded elements before implanting the embryos into pseudopregnant animals (57–59).

Finally, perhaps the most obvious limitation for commercialization of milk and milk derived products for massive consumption is the establishment of productive transgenic herds. Cibelli and coworkers (42) have produced cloned transgenic calves by transferring the nuclei of stable transgenic fetal fibroblasts into enucleated oocytes. The fusion gene contained elements encoding β-galactosidase and neomycin resistance; this allowed for selection of neomycin-resistant fibroblasts that were also determined to express β-galactosidase activity. Nuclei from this cell line were used to generate transgenic oocyte clones. Although the techniques to obtain transgenic animals are rapidly evolving, commercial applications for the production of dairy products have mostly been demonstrated in laboratory animals, and there are only a few reports of the actual production of transgenic dairy cows. However, the prototypes produced in the laboratory and the fast changing environment of functional foods and dietary supplements are providing a basis to postulate potentially useful transgenic animals that produce remodeled milk.

C. Applications of Transgenic Animals and Targeted Mutants to the Production of Dairy Products

Bovine milk and its constituents are mostly used for general nutrition for the population at large. For this reason, remodeled milk with added value is a plausible commercial target. On the other hand, transgenically remodeled milk is an attractive target for applications in special nutrition such as products for nutritional disease management or for preterm infants and the elderly. Production of transgenic animals that yield modified milk may also benefit the manufacturer of dairy products such as cheese, cream, and yogurt or the dairy farm entrepreneur without an apparent direct advantage for the consumer other than potential reductions in cost and increased availability. Several reviews have listed potential modifications to milk through transgenic technology (32,34,60). Table 3 is a summary of proposed modifications for commercial applications of TAs and TMs to the production of dairy products. It is important to reiterate that the use of the lactating mammary gland as a bioreactor for the production of pharmaceutical compounds is not being considered here. Only potential applications that are relevant for the production of dairy products or other nutritional products such as dietary supplements or medical foods are listed.

Perhaps the easiest way to obtain commercially viable dairy products from genetically modified dairy cows is by eliminating native milk constituents, such as bovine β-lactoglobulin and lactose, that cause adverse reactions in consumers. A TM that does not produce β-lactoglobulin is desirable for at least two reasons: (a) its milk is more suitable for cheese manufacturing, and (b) it is devoid of a major allergen of bovine milk (61).

Lactose-free and low-lactose products are now occupying a niche in functional foods markets to satisfy the needs of consumers with actual or perceived lactose intolerance. For a 1998 account on lactose intolerance the reader is referred to de Vrese and associates (62). The extent to which lactose intolerance plays an actual role in the commercial success of low-lactose or lactose-free products in different markets is debatable. However, evidence indicates that non-lactose-intolerant individuals are being driven away from milk and dairy products by the perception that gastric symptoms of discomfort are attributable to lactose. This occurs even when lactose concentrations in these products would not be enough to cause illness or discomfort due to lactose intolerance or maldigestion (63,64). Nevertheless true lactose-intolerants and lactose maldigesters require low-lactose milk and dairy products as sources of good quality protein and calcium. The reduction or elimination of lactose is currently being achieved by treating milk or whey with β-galactosidase. On the other hand, several strategies have been explored to produce low-lactose milk in TAs or TMs. L'Huillier and colleagues (45) reported the reduction of α-lactalbumin, which is a component of the lactose synthetase complex and is necessary for lactose synthesis. Other methods to eliminate lactose or its effects are transgenic synthesis of β-galactosidase to hydrolyze the disaccharide after its synthesis (65) and utilization of lactose as a raw material to build larger sugars that have lower osmotic values and, therefore, are less prone to cause some of the symptoms of lactose intolerance. The latter

Table 3 Examples of Desirable Acquired or Lost Traits in Milk Produced by Transgenic Animals and Targeted Mutants[a]

Trait	Example of desired outcome	Potential approaches	Potential commercial application
Lactose synthesis	Reduced lactose	Deletion or down-regulation of α-lactalbumin or glactosyltransferase gene TA expression of glycosidases; Building of larger, less osmotic lactose-based carbohydrates	Low-lactose milk, formulas for lactose intolerance
Synthesis of immunogenic proteins	β-Lactoglobulin-free milk	Knockout lactoglobulin gene, introduction of rybozymes to reduce β-lactoglobulin synthesis	Hypoallergenic milk and formulas, improved cheese manufacturing
Lipid synthesis	Reduced fat milk, modified fatty acid profile	Transgenic expression of rybozymes against acetyl-CoA-carboxylase	Milk and dairy products for weight management
Synthesis of human milk glycoconjugates	Milk with human oligosaccharides	TA expression of exogenous glycosyltransferases	Infant formulas, supplements with prebiotics
Synthesis of oligo- and polysaccharides	Milk containing fiber or slow-digest starches	Expression of transgenes encoding glycosidases and glycosyltransferases from plants and microbes	Milk and dairy products for weight management
Immunoglobulin synthesis	Milk containing antibodies against childhood diseases	TA expression of antibodies or fragments of antibodies	Milk products to prevent infectious diseases
Synthesis of bacteriostatic compounds	Milk containing lysozyme, defensins	TA expression of antibacterial peptides or proteins	Increased product shelf life
Synthesis of human milk proteins	Milk containing human caseins, lactoferrin	TA expression of human milk proteins	Formulas that emulate human milk

[a] TA, transgenic animal; acetyl-CoA, acetyl-coenzyme A.

is achieved by expressing glycosyltransferases that utilize lactose as substrate for the synthesis of elongating oligosaccharides (66). These two strategies may have the advantage of allowing for normal expression of milk because lactose is present at some point in the lactating mammary gland, thus osmotically driving the movement of water from the plasma compartment (65). Lipid content in milk can also be reduced. Ha and Kim (46) demonstrated that the synthesis of lipids can be down-regulated by coexpression with ribozymes directed against mRNA sequences encoding acetyl-coenzyme A (acetyl-CoA) carboxylase, which catalyzes the key step in fatty acid synthesis. In a transgenic cow this inhibition has at least two potential effects: (a) production of low-fat milk and (b) increase in milk production efficiency. Because the reaction catalyzed by the enzyme consumes energy, its inhibition prevents the consumption of adenosine triphosphate (ATP) molecules that are then available to drive other synthetic reactions.

Another potential application of transgenic technology is the modification of the fatty acid profile of milk. Medical foods that contain specific profiles of long-chain polyunsaturated fatty acids (PUFAs) are used as adjunct therapy to treat inflammatory processes (67), and the literature is replete with reports on the health benefits derived from the consumption of PUFAs [examples found in Kenler and coworkers (68) and Whelan and associates (69)]. As stated, Knutzon and colleagues, and Huang and coworkers (21–23) have cloned and expressed enzymes involved in the reduction and elongation of fatty acids and have produced transgenic plants that synthesize high levels of PUFA. It is conceivable that through transgenic expression of the enzymes identified by this group cow milk containing a beneficial PUFA profile could become available. This would be an example of the transgenic synthesis of secondary gene products in which the aim is the expression of a transgene-encoded enzyme, which in turn acts on substrates available in the lactating mammary gland.

Another example is the synthesis of human milk oligosaccharides. We have expressed several glycosyltransferases in the lactating mammary glands of mice (11,66) and pigs and rabbits (70). The expression of these enzymes results in the synthesis of oligosaccharides by elongating lactose, thus simulating the process that in human mammary glands leads to the synthesis of the varied repertoire of oligosaccharides that can be found in breast milk (71). Several biological activities have been associated with these molecules; Kunz and Rudloff and Kunz and associates (72,73) reviewed their potential function, and most recently Zopf and Roth (74) have discussed their anti-infective properties. The presence of these structures could add value to milk-derived dietary supplements and medical foods. In addition, nonmammalian glycosyltransferases, glycosidases, or enzymes such as epimerases can be expressed in the lactating mammary gland. Raw materials such as phosphorylated monosaccharides are pluripotent building blocks, which are used by plants to synthesize fibers and starches. Milk with such molecules would provide novel benefits for consumers who already have dairy products in their diets.

An additional strategy to supplement milk for massive consumption or for specific applications is to express transgenically antibodies against either known human pathogens or microbes that cause spoilage in dairy products. Cordle and his group (75,76) have produced antibodies against the human rotavirus in milk of hyperimmunized cows. However, these antibodies had to be concentrated because their relatively low concentration in the milk was not sufficient for human applications. It is conceivable to overexpress transgene-encoded antibodies directly into cow's milk, thus providing a nutritional product with activity against pathogens that are common in day care centers or nursing homes. In this fashion the milk itself has anti-infective properties but does not generate typical antibiotic resistance. Although this application may be limited to pathogens that act on gastrointestinal (GI) mucosal surfaces, it may be of value especially for persons with compromised immune systems. In addition, cytokines have been transgenically produced in rabbit milk (77). Such milk could be used in medical foods to attenuate the immune response in patients with chronic inflammatory processes or to boost immunity in persons with immature or depressed immune systems such as preterm infants and the elderly. Antibacterial proteins and peptides such as lysozyme and defensins (78–80) not only may help the consumer of dairy products to fight mucosal pathogens but may have a significant effect in increasing shelf life of milk and its derivatives.

Finally, an area that has been reviewed by several workers in the field is the transgenic expression of human milk proteins in the milk of transgenic animals (31,81,82). This research has been focused on simulating human milk, which is considered the gold standard of infant nutrition. Particular proteins of breast milk have been associated with different biological activities (83). For example, a variant of human α-lactalbumin has antibacterial activity (84) and β-casein generates, upon proteolytic hydrolysis, peptides that have neurotropic effects and potential immunostimulatory properties (85,86). Although the transgenic expression of breast milk proteins in animal milk seem to be a natural target for companies that manufacture infant formula, the applications are not limited to pediatric nutrition. These functional human-milk proteins may also be less allergenic than those naturally present in bovine milk. The use of these proteins in the manufacturing of dairy products could make accessible cheeses, fermented milks, and other products for those who are allergic to bovine milk proteins. Conceivably, any protein including enzymes can be expressed in the lactating mammary glands of dairy cows. The examples described are only a few for which some level of technical feasibility has been demonstrated in transgenic experimental animals or plants. Modern biotechnology can contribute to the health of farm animals and the qualitative and quantitative improvement of milk and dairy products through the application of additional technologies briefly mentioned in the next section.

V. ADDITIONAL TECHNOLOGIES

There are additional recombinant-DNA technologies being developed that could have a significant impact in the manufacturing of dairy products. For example, DNA vaccines are being explored for the treatment and prevention of diseases in farm animals, including cattle (87). Additionally, significant developments have taken place in the design and use of DNA-based diagnostics for cattle, specifically for dairy cows, as exemplified by the work of Zimmer and Tegtmeier (88,89). Other technologies that have developed significantly since 1995 are the design and implementation of polymerase chain reaction (PCR)-based determinations of microbial contamination in dairy products. The assay developed by Nogva and coworkers for the detection of *Listeria monocytogenes* in milk (90) is a prime example. Modern techniques that do not involve recombinant DNA and have relied on the development of advanced instrumentation are currently used to monitor biochemical parameters of dairy herds. These techniques are also becoming tools for DNA-based technologies. For example, biosensors have been developed for the measurement of progesterone in milk and other fluids (91). Real-time measures obtained with these instruments can be used to optimize the delivery of recombinant hormones or measure the impact of transgenic expression of hormones and other factors that affect physiological parameters of lactation. The convergence of powerful analytical biochemistry methods and production-oriented biotechnology is also illustrated by the selection and use of microbial strains for the production of fermented milks and cheese. Although the topic of "food microbes" is discussed in depth elsewhere in this volume, it is worthwhile to exemplify this marriage of technologies with the work of Desmasures and colleagues (92) and Fortina and Carminati (93). The first group characterized *Lactococcus* spp. strains from raw milk on the basis of biochemical parameters and genotyped and classified them by using gene restriction analysis, thus providing an objective basis for identifying a strain and predicting its effect on milk processing. Fortina and his group (93) embarked on a systematic and comprehensive study of the biochemical parameters of *Lactobacillus helveticus* strains in natural cheese starters. They determined the extent of DNA homology by DNA hybridization and several biochemical parameters such as proteolytic and peptidase activity, lactic acid production, antibiotic and lysozyme resistance, acidification properties, and cell surface protein synthesis. These analyses, which involve characterization of bio- and genotypes, provide the basis for quality control of strains and for appropriate selection and mixing of strains to obtain specific organoleptic and biochemical outcomes. Reviews of the uses of biotechnology in the selection, biochemical characterization, production and development of milk-processing microbes can be found in Henriksen and associates (12) and Palva (94).

The production and use of enzymes to process milk products are also benefiting from the tools provided by biotechnology. Techniques for the immobilization of enzymes and entire microbes are being employed to hydrolyze lactose from milk or whey streams (95). Likewise, several naturally occurring and recombinant sources of milk clotting enzymes are being developed (96–98). Another case of converging modern technologies is presented by Kim and colleagues (99); because the hydrolysis of lactose results in the release of monosaccharides, milk treated with lactase acquires a sweet taste that is objectionable to some consumers. This group cleverly enclosed lactase in liposomes that can be dried and reconstituted in water or milk. The liposomes were able to withstand a simulated gastric hydrolytic environment but released their contents upon contact with bile salts. This technology represents a novel approach because health and organoleptic effects of a milk containing liposomes would not be realized until consumption; the bile salts of the consumer carry out the last phase of the lactose reduction process by releasing lactase from the liposomes. Enzymes can also be used as process aids to modify endogenous milk components. An example of the use of recombinant enzymes to alter physical–chemical properties of milk proteins is the overphosphorylation of caseins using recombinant protein phosphorylases (100). This process may be desirable because additional phosphate in proteins enhances calcium binding and because the properties of caseins (such as curd formation) for dairy product manufacturing can be modulated by their degree of phosphorylation. Though all the examples cited offer an overview of what modern biotechnology has to offer for the improvement of dairy products and their manufacturing processes, this analysis of applications is an incomplete description of the technical field. Better understanding of a technology requires that it be studied from the standpoint not only of its potential benefits but also of its historical and social contexts.

VI. CONCLUSIONS

Regulatory, safety, and perceptual aspects of modern biotechnology are being discussed in a polarized debate. A published account of this debate, although slanted toward transgenic crops, can be found in the April 2001 issue of *Scientific American* (101–103) Salient features of this issue are two interviews that present metaphorical mirror images of the same topic; one interview is with Robert Horsch (vice president of Product and Technology Cooperation of Monsanto) and the other with Margaret Mellon (director of the Agricultural and Biotechnology Program of the Union of Concerned

Scientists). A theme that permeates Margaret Mellon's statements is the need for study and analysis of genetically modified foods in order to assess their risks, benefits, and safety adequately. The rational fundament of her views stems from a knowledgeable assessment of the current situation of biotechnology. She points out the urgent need of workers in the field of biotechnology to inform peers of the progress and risk assessments of nascent technologies and their applications. Furthermore, accounts such as the present review have the intrinsic purpose of discussing the future and evolving strengths and weaknesses of biotechnology in addition to the current realities. Similarly, some detractors of biotechnology point to its potential harm and not to scientifically established risks. Both are views of the future, and to some extent both are missing a third important stream of ideas, which is precisely a discussion of methods needed to determine the safety and risk-versus-benefit analysis of biotechnological applications. The extent to which a particular application of biotechnology is close to or distant from the consumer is a consideration that already has practical consequences. For example, transgenic corn is approved in the United States for animal feed but not for human consumption (102). Fig. 1 is a diagrammatic representation of the way technologies affect the discrete units of the manufacturing process for dairy products. Certainly, transgenically expressed PUFA in a plant component of

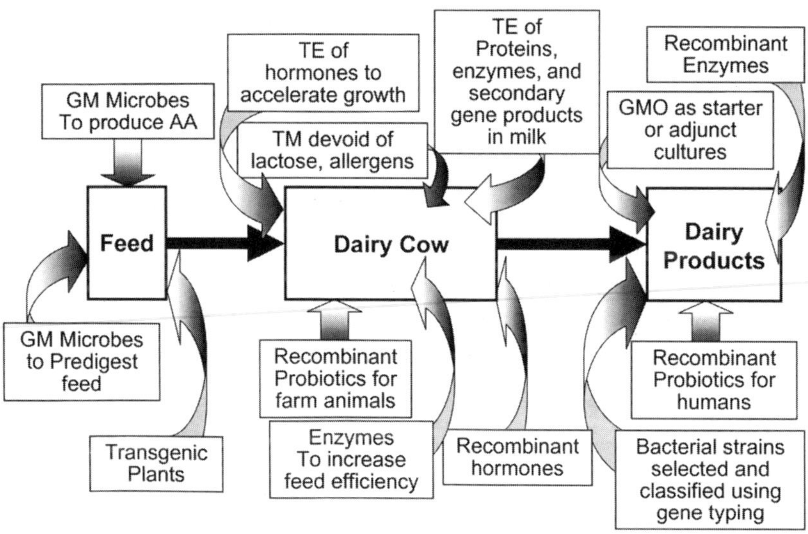

Figure 1 Effects of GMO and GMO-derived products on different elements and different stages of dairy product manufacturing. TE, transgenic expression; TM, targeted mutant; GMO, genetically modified organism.

animal feed that eventually finds its way into milk is far removed from the consumer as compared with a transgenic tomato. This distance does not address the potential ecological impact of any of the two transgenic cultivars, but it illustrates a practical approach, already in use, to design operational regulations. Mollet (104) provides another set of considerations to ascertain the extent to which organisms are genetically modified. A first group constitutes the genetic modifications that can occur spontaneously in nature such as mutations; a second group represents the introduction of recombinant DNA of a different strain of the same species; and the third group involves the transfer of genetic information from one species to the other. Yet another consideration is the permanence of a genetic change in the food chain or in nature at large. Although not devoid of conceivable risks, a transgenic cow producing remodeled milk as a result of the expression of a fusion gene inserted through the teat canal will not transmit the genetic modification through the germ line, and, theoretically, the resulting GMO cannot expand the acquired or lost traits encoded by the transgene. It appears that an important challenge for biotechnologists is to confront in an open and self-scrutinizing manner the challenges posed by legitimate safety concerns and perceptual distortions of the current nature and potential of biotechnology. However, current biological problems and forces of the marketplace present other challenges. In certain countries the future of the entire dairy industry is in jeopardy as a result of epidemics such as bovine spongiform encephalopathy (BSE) and hoof and mouth disease. Piedrahita (52) suggests that a targeted mutant, in which the expression of a gene encoding a "normal prion protein" is impeded, could be less susceptible to contract BSE. Similarly, the receptor sites for viruses and bacteria can be removed by targeted mutation, thus generating animals that are resistant to viral infection through the loss of a function. Last, the reader is referred to two volumes that discuss in depth several aspects of biotechnology and agriculture: *Transgenic Animals in Agriculture*, edited by Murray and coworkers (105), and *Agricultural Biotechnology*, edited by Altman (106). Both publications contain chapters that deal with ethical, regulatory, and strategic issues of transgenic technologies.

REFERENCES

1. National Science and Technology Council, USA in *www.wabio.com/definition_biotech.htm*, 1995.
2. NG Gregory. Intensive farming of animals. Outlook Agric 29:15–23, 2000.
3. G Bulfield. Farm animal biotechnology. TIBTECH 18:10–13, 2000.
4. DT Galligan. The economics of optimal health and productivity in the commercial dairy. Rev Sci Tech 18:512–519, 1999.

5. B Crooker. Feeding the high producing dairy cow: Biotechnology, body condition and reproduction. Bovine Pract 31:34–36, 1996.
6. C Bera-Maillet, V Broussolle, P Pristas, JP Girardeau, G Gaudet, E Forano. Characterisation of endogluconases EGB and EGC from Fibrobacter succinogenes. Biochim Biophys Acta 1476:191–202, 2000.
7. DL Blum, XL Li, H Chen, LG Ljungdajl. Characterization of an acetyl Xylan esterase from the anaerobic fungus Orpinomyces Sp. Strain PC-2. Appl Environ Microb 65:3990–3995, 1999.
8. K Gregg, B Hamford, K Henderson, J Kopecmy, C Wong. Genetically modified ruminal bacteria protect sheep from fluoroacetate poisoning. Appl Environ Microbiol 64(9):3496–3498, 1998.
9. T Ziegelhogger, J Will, S Austin-Phillips. Expression of bacterial cellulase genes in transgenic alfalfa (Medicago sativa) potato (Solanum tuberosum L.) and tobacco (Nicotiana tabacum L.). Mol Breed 5(4):309–318, 1999.
10. DE Bauman. Bovine somatotropin and lactation: From the basic science to commercial application. Domest Anim Endocrinol 17:101–116, 1999.
11. PA Prieto, JJ Kopchick, B Kelder. Transgenic animals and nutrition research. J Nutr Biochem 10(12):682–695, 1999.
12. C Henriksen, D Nilsson, S Hansen, E Johansen. Industrial applications of genetically modified microorganisms: Gene technology at Chr. Hansen A/S. Int Dairy J 9:17–23, 1999.
13. V Pursel, M Solomon. Alteration of carcass composition in transgenic swine. Food Rev Int 9(3):423–439, 1999.
14. W Velander, H Lubon, W Drohan. Transgenic livestock as drug factories. Sci Am 276:70–74, 1997.
15. C Ziomek. Commercialization of proteins produced in the mammary gland. Theriogenology 49:139–144, 1998.
16. M Malumbres, JF Martín. Molecular control mechanisms of lysine and threonine biosynthesis in amino acid-producing corynebacteria redirecting carbon flow. FEMS Microbiol Lett 143(2/3):102–114, 1996.
17. F Zadrazil, AK Puniya, S Kishan. Biological upgrading of feed and feed components. In: R Wallae, A Chesson, ed. Biotechnology in Animal Feeds and Animal Feeding. Weinheim: Wiley, VCH, 1995, pp 55–69.
18. N Krishna, D Mohan, E Rao. Biotechnology in livestock feeding. Indian J Anim Sci 68(8):837–842, 1998.
19. SB Altenbach, CC Kuo, LC Starace, KW Pearson, C Wainwright, A Georgescu, J Tonsend. Accumulation of a Brazil nut albumin in seeds of transgenic canola results in enhanced levels of seed protein methionine. Plant Mol Biol 18(2):235–245, 1992.
20. OJM Goddijin, J Pen. Plants as bioreactors. Trends Biotechnol 13(9):379–387, 1995.
21. DS Knutzon, JM Thurmond, YS Huang, S Chaudhary, EG Bobik, GM Chan, SJ Kirchner, P Mukerji. Identification of delta-5-desaturase from Mortierella alpina by heterlogous expression in baker's yeast and canola. J Biol Chem 273(45):29360–29366, 1998.
22. DS Knutzon, GM Chan, P Mukerji, JM Thurmond, S Chaudhary, J-S Huang.

Genetic engineering of seed oil fatty acid composition. In: A Altman, M Ziv, S Izhar, eds. Plant Biotechnology and In Vitro Biology In the 21st Century. Shrub Oak, NY: Agritech Publications, 1999, pp 575–578.

23. Y-S Huang, P Mukerji, T Das, D Knutzon. Transgenic production of long-chain polyunsaturated fatty acids. In: T Hamazaki, H Okuyama, eds. World Review of Nutrition and Dietetics. Basel, Switzerland: Karger AG, 2001, pp 343–248.
24. NW Offer, M Mardsen, J Dixon, BK Speake, FE Thacker. Effect of dietary fat supplements on levels of n-3 poly-unsaturated fatty acids, trans acids and conjugated linoleic acid in bovine milk. Anim Sci 69(3):613–665, 1999.
25. GJ Asimov, NK Krouze. The lactogenic preparations from the anterior pituitary and the increase in milk yield from cows. J Dairy Sci 20:289–306, 1937.
26. W Lesser, J Bernard, K Billah. Methodologies for ex ante projections fo adoption rates for agbiotech products: Lessons learned from rBST. Agribusiness 15(2):149–162, 1999.
27. JL Nappier, GA Hoffman, TS Arnold, TD Cox, DR Reeves, VL Hubbard. Determination of the tissue distribution and excretion of 14C-fertirelin acetate in lactating goats and cows. J Agric Food Chem 46(11):4563–4567, 1998.
28. S Ball, K Poison, W Eyestone, RM Akers. Induced lactation in prepubertal Holstein heifers. J Dairy Sci 83:2459–2463, 2000.
29. J Kopchick, J Jura, P Mukerji, B Kelder. Transgenic technology as it applies to animal agriculture. Biotechnologia 2(33):31–51, 1996.
30. JJ Kopchick, LL Bellush, KT Coschigano. Transgenic models of growth hormone action. Annu Rev Nutr 19:437–461, 1999.
31. F Siewerdt, EJ Eisen, JD Murray, IJ Parker. Response to 13 generations of selection for increased 8-week body weight in lines of mice carrying a sheep growth hormone-based transgene. J Anim Sci 78(4):832–845, 2000.
32. R Wall, D Kerr, K Bondioli. Transgenic dairy cattle: Genetic engineering on a large scale. J Dairy Sci 80:2213–2224, 1997.
33. C Karatzas, J Turner. Toward altering milk composition by genetic manipulation: Current status and challenges. J Dairy Sci 80:2225–2232, 1997.
34. JD Murray. Genetic modification of animals in the next century. Theriogenology (51), 149–159, 1999.
35. B Pintado, A Gutierrez-Adan. Transgenesis in large domestic species: Future development for milk modification. Reprod Nutr Dev 39:535–544, 1999.
36. JW Gordon, GA Scangos, DJ Plotkin, JA Barbosa, FH Ruddle. Genetic transformation of mouse embryos by microinjection or purified DNA. Proc Natl Acad Sci USA 77:7380–7384, 1980.
37. HJ Tsai, CH Lai, HS Yand. Sperm as a carrier to introduce an exogenous DNA fragment into the oocyte of Japanese abalone (Haliotis divorsicolor suportexta). Transgenic Res 6:85–95, 1997.
38. K Chang, A Ideda, K Hayashi, Y Furuhata, M Nishihara, A Ohata, S Ogawa, MM Takahashi. Production of transgenic rats and mice by the testis-mediated gene transfer. J Reprod Dev 45:30–36, 1999.
39. A Chan, E Homan, L Ballou, JC Burn, RD Bremel. Transgenic cattle produced

by reverse-transcribed gene transfer in oocytes. Proc Natl Acad Sci USA 95: 14028–14033, 1998.

40. W Eyestone. Production and breeding of transgenic cattle using in vitro embryo production technology. Theriogenology 51:509–517, 1999.
41. Y Echelard, W Groen, M Destrempes. Transgenic cow production from microinjection into in vivo-derived embryos. Theriogenology 53:513, 2000.
42. J Cibelli, S Stice, P Golueke, JJ Kane, J Jerry, K Blackwell, A. Ponce de Leon, J.M. Robl. Cloned transgenic calves produced from nonquiescent fetal fibroblasts. Science 280:1256–1611, 1998.
43. J Haseloff, WL Gerlach. Simple RNA enzymes with new and highly specific endoribonuclease activities. Nature 334(6183):585–591, 1988.
44. G Cantor, T McElwain, T Birkebak, G Palmer. Ribozyme cleaves rex/tax mRNA and inhibits bovine leukemia virus expression. Proc Natl Acad Sci USA 90(23):10932–10936, 1993.
45. PJ L'Huillier, S Soulier, MG Stinnakre, L Lepourry, SR Davis, JC Mercier, JL Vilotte. Efficient and specific ribozyme-mediated reduction of bovine alpha-lactalbumin expression in double transgenic mice. Proc Natl Acad Sci USA 93(13):6698–6703, 1996.
46. J Ha, KH Kim. Inhibition of fatty acid synthesis by expression of an acetyl-CoA carboxylase-specific ribozyme gene. Proc Natl Acad Sci USA 91:9951–9955, 1994.
47. CW Pittius, L Hennighausen, E Lee, H Westphal, E Nicols, J Vitale, KA Gordon. Milk protein gene promoter directs the expression of human tissue plasminogen activator cDNA to the mammary gland in transgenic mice. Proc Natl Acad Sci USA 85(16):5874–5878, 1988.
48. CW Pittius, L Sankaran, YJ Topper, L Hennighausen. Comparison of the regulation of the whey acidic protein gene with that of a hybrid gene containing the whey acidic protein gene promoter in transgenic mice. Mol Endocrinol 2(11):1027–1032, 1988.
49. M Rijnkels, P Kooiman, P Krimpenfort, H De Boer, F Pieper. Expression analysis of the individual bovine β-, αS2-, and k-casein genes in transgenic mice. Biochem J 311:929–937, 1995.
50. M Persuy, M Stinnakre, C Printz, M Mahe, J Mercier. High expression of the caprine β-casein gene in transgenic mice. Eur J Biochem 205:887–893, 1992.
51. S Jeng, G Bleck, M Wheeler, R Jimenez-Flores. Characterization and partial purification of bovine αlactalbumin and β-casein produced in milk of transgenic mice. J Dairy Sci 80:3167–3175, 1997.
52. JA Piedrahita. Targeted modification of the domestic animal genome. Theriogenology 53:105–116, 2000.
53. Y Fujiwara, M Miwa, R Takahashi, M Hirabayashi, T Suzuki, M Ueda. Position-independent and high-level expression of human alpha-lactalbumin in the milk of transgenic rats carrying a 210-kb YAC DNA. Mol Reprod Dev 47(2):157–163, 1997.
54. R McKnight, M Spencer, R Wall, L Hennighausen. Severe position effects imposed on a 1 kb mouse whey acidic protein gene promoter are overcome

by heterologous matrix attachment regions. Mol Reprod Dev 44(2):179–184, 1996.

55. HT Riele, ER Maandag, A Berns. Highly efficient gene targeting in embryonic stem cells through homologous recombination with isogenic DNA constructs. Proc Natl Acad Sci USA 89:5128–5132, 1992.
56. J Archer, W Kennan, M Gould, R Bremel. Human growth hormone (hGH) secretion in milk of goats after direct transfer of the hGH gene into the mammary gland by using replication-defective retrovirus vectors. Proc Natl Acad Sci USA 91(15):6840–6844, 1994.
57. J Hyttinen, T Peura, M Tolvanen, J Aalto, L Alhonen, R Sinervirta, M Halmekyto, S Myohanen, J Janne. Generation of transgenic dairy cattle from transgene-analyzed and sexed embryos produced *in vitro*. Biotechnology 12: 606–608, 1994.
58. J Jura, J Kopchick, W Chen, T Wagner, J Modlinski, M Reed, J Knapp, Z Smorag. In vitro and in vivo development of bovine embryos from zygotes and 2-cell embryos microinjected with exogenous DNA. Theriogenology 41(6): 1259–1266, 1994.
59. A Saberivand, PM Outteridge. The use of embryo genotyping in the propagation of genes involved in the immune response. Immunol Cell Biol 74(2): 109–120, 1996.
60. A Clark. Gene expression in the mammary glands of transgenic animals. Biochem Soc Symp 63:133–140, 1998.
61. J Ducheteau, A Michilis, J Lambert, B Gossart, G Casimir. Anti-betalactoglobulin IgG Antibodies bind to a specific profile of epitopes when patients are allergic to cow's milk proteins. Clin Exp Allergy 28(7):824–833, 1998.
62. M de Vrese, R Sieber, M Stransky. Laktose in der menschlichen Ernahrung. Schweiz Med Wochenschr 128(38):1393–1400, 1998.
63. A Carroccio, G Montalto, G Cavera, A Notarbatolo. Lactose intolerance and self-reported milk intolerance: Relationship with lactose maldigestion and nutrient intake: Lactase Deficiency Study Group. J Am Coll Nutr 17(6):631–636, 1998.
64. LD McBean, GD Miller. Allaying fears and fallacies about lactose intolerance. J Am Diet Assoc 98(6):671–676, 1998.
65. B Jost, JL Vilotte, I Duluc, JL Rodeau, JN Freund. Production of low lactose milk by ectopic expression of intestinal lactase in the mouse mammary gland, (comment). Nat Biotechnol 17(2):135–136, 1999.
66. PA Prieto, P Mukerji, B Kelder, R Erney, D Gonzalez, JS Yun, DF Smith, KW Moremen, C Nardelli, M Pierce, et al. Remodeling of mouse milk glycoconjugates by transgenic expression of a human glycosyltransferase. J Biol Chem 270(49):29515–29519, 1995.
67. JE Gadek, S DeMichele, MD Karlstad, ER Pacht, M Donahoe, TE Albertston, CV Hoozen, AK Wennberg, JL Nelson, M Noursalehi, Enteral Nutrition in ARDS Study Group. Effect of enteral feeding with eicosapetaenoic acid, γ-linolenic acid, and antioxidants in patients with acute respirator distress syndrome. Crit Care Med 27(8):1409–1420, 1999.

68. AS Kenler, WS Swails, DF Driscoll, et al. Early enteral feeding in postsurgical cancer patients: Fish oil structured lipid-based polymeric formula versus a standard polymeric formula. Ann Surg 223:316–333, 1995.
69. J Whelan, KS Broughton, JE Kinsella. The comparative effects of dietary alpha-linlenic acid and fish oil on 4- and 5- series leukotriene formation in vivo. Lipids 26:119–126, 1991.
70. Prieto et al. Unpublished obvservations, 2001.
71. RM Erney, WT Malone, MB Skelding, AA Marcon, KM Kelman-Leyer, ML O'Ryan, G Ruiz Palacios, MD Hilty, LK Pickering, PA Prieto. Variability of human milk neutral oligosaccharides in a diverse population. J Pediatr Gastroenterol Nutr 30(2):181–192, 2000.
72. C Kunz, S Rudloff. Biological functions of oligosaccharides in human milk. Acta Pediatr 82:903–912, 1993.
73. C Kunz, M Rodriguez-Palmero, B Koletzko, R Jensen. Nutritional and biochemical properties of human milk. Part I. General aspects, proteins, and carbohydrates. Clin Perinatol 26(2):307–333, 1999.
74. D Zopf, S Roth. Oligosaccharide anti-infective agents. Lancet 347:1017–1021, 1996.
75. CT Cordle, JP Schaller, TR Winship, EL Candler, MD Hilty, KL Smith, LJ Saif, EM Kohler, S Krakowka. Passive immune protection from diarrhea caused by rotavirus or *E. coli*: An animal model to demonstrate and quantitate efficacy. Adv Exp Med Biol 310:317–327, 1991.
76. JP Schaller, LJ Saif, CT Cordle, E Candler Jr, TR Winship, KL Smith. Prevention of human rotavirus-induced diarrhea in gnotobiotic piglets using bovine antibody. J Infect Dis 165(4):623–630, 1992.
77. TA Buhler, T Bruyere, DF Went, G Stranzinger, K Burki. Rabbit β-casein promoter directs secretion of human interleukin-2 into the milk of transgenic rabbits. Biotechnology 8:143, 1995.
78. EA Maga, GB Anderson, MC Huang, JD Murray. Expression of human lysozyme mRNA in the mammary gland of transgenic mice. Transgenic Res 3(1):36–42, 1994.
79. EA Maga, GB Anderson, JD Murray. The effect of mammary gland expression of human lysozyme on the properties of milk from transgenic mice. J Dairy Sci 78(12):2645–2652, 1995.
80. A Weinberg, S Krisanaprakornkit, BA Dale. Epithelial antimicrobial peptides: Review and significance for oral applications. Crit Rev Oral Biol Med 9(4):399–414, 1998.
81. B Lonnerdal. Recombinant human milk proteins—an opportunity and a challenge. Am J Clin Nutr 63:622S–626S, 1996.
82. B Lonnerdal, J Carver, J Buts, O Koldovsky. Annales Nestle: Bioactive Factors in Milk. Nestec Ltd., Vevey, Switzerland: Nestle Nutrition Services, Vol. 54 Number 3, 1996.
83. M Hamosh. Protective function of proteins and lipids of human milk. Biol Neonate 74(2):163–176, 1998.
84. A Hakansson, M Svensson, AK Mossberg, H Sabharwal, S Linse, I Lazou, B

Lonnerdal, C Svanborg. A folding variant of alpha-lactalbumin with bactericidal activity against Streptococcus pneumoniae. Mol Microbiol 35(3):589–600, 2000.

85. D Migliore-Samour, F Floc'h, P Jolles. Biologically active casein peptides implicated in immunomodulation. J Dairy Res 56(3):357–362, 1989.
86. EM Blass. Mothers and their infants: Peptide-mediated physiological, behavioral and affective changes during suckling. Regul Pept 66(1–2):109–112, 1996.
87. S Van Drunen-van den Hurk, V Gerdts, BI Oler, R Pontarollo, R Rankin, R Uwiera, LA Babiuk. Recent advances in the use of DNA vaccines for the treatment of diseases of farmed animals. Adv Drug Delivery Rev 43:13–28, 2000.
88. K Zimmer, KG Drager, W Klawonn, RG Hess. Contribution to the diagnosis of Johne's disease in cattle: Comparative studies on the validity of Ziehl-Neelsen staining, faecal culture and a commercially available DNA-probe test in detecting Mycobacterium paratuberculosis in faeces from cattle. J Vet Med Ser B 46(2):137–140, 1999.
89. C Tegtmeier, O Angen, P Ahrens. Comparison of bacterial cultivation, PCR, in situ hybridization and immunohistochemistry as tools for diagnosis of Haemophilus somnus pneumonia in cattle. Vet Microbiol 76(3):385–394, 2000.
90. HK Nogva, K Rudi, K Naterstad, A Holck, D Lillehaug. Application of 5'-nuclease PCR for quantitative detection of Listeria monocytogenes in pure cultures, water, skim milk, and unpasteurized whole milk. Appl Environ Microbiol 66(10):4266–4271, 2000.
91. R Claycomb, M Delwiche, C Munro, R BonDurant. Rapid enzyme immunosassay for measurement of bovine progesterone. Biosens Bioelecton 13:1165–1171, 1998.
92. N Desmasures, I Mangin, D Corroler, M Guengen. Characterization of lactococci isolated from milk produced in the camembert region of Normandy. J Appl Microbial 85(6):999–1005, 1998.
93. MG Fortina, N Carminati. *Lactobacillus helveticus* heterogeneity in natural cheese starters: the diversity in phenotypic characteristics. J Appl Microbiol 84:72–80, 1998.
94. A Palva. Contribution of modern biotechnology of lactic acid bacteria to development of health-promoting foods. Agric Food Sci Finland 7:267–282, 1998.
95. J Szczodrak. Hydrolysis of lactose in whey permeate by immobilized B-galactosidase from penicillium notatum. Acta Biotechnol 3:235–250, 1999.
96. S Seker, H Beyenal. Production of microbial rennin from *Mucor miehei* in a continuously fed fermenter. Enzyme Microb Technol 23:469–474, 1998.
97. A Hashem. Purification and properties of a milk-clotting enzyme produced by *Penicilliem oxalicum*. Biosource Technol 75:219–222, 2000.
98. S Chitpinityol, D Goode, JC Crabbe. Sites-specific mutations of calf chymosin B which influence milk-clotting activity. Food Chem 62:133–139, 1998.
99. CK Kim, HS Chung, MK Lee, LN Choi, MH Kim. Development of dried liposomes containing beta-galactosidase for the digestion of lactose in milk. Int J Pharm 183(2):185–193, 1999.

100. E Pitois, T Chardot, JC Meunier. Overphosphorylation of milk caseins by a recombinant protein kinase CK2 catalytic subunit. J Agric Food Chem 47: 3996–4002, 1999.
101. K Brown. Seeds of concern. Sci Am 284(4):52–57, 2001.
102. K Hopkin. The risks on the table. Sci Am 284(4):60–61, 2001.
103. S Nemecek. Does the world need GM foods? Sci Am 284(4):62–65, 2001.
104. B Mollet. Genetically improved starter strains: Opportunities for the dairy industry. Int Dairy J 9:11–15, 1999.
105. JD Murray, GB Anderson, AM Oberbauer, MM McGloughlin, eds. Transgenic Animals in Agriculture. New York: CABI Publishing, 1999.
106. A Altman, ed. Agricultural Biotechnology. New York: Marcel Dekker, 1998.

3

Bacterial Food Additives and Dietary Supplements

Detlef Wilke
Dr. Wilke & Partner Biotech Consulting GmbH, Wennigsen, Germany

I. INTRODUCTION

Microbes and microbial biomass are typically not considered as food although they significantly contribute to food properties and food quality. Traditional spontaneous or controlled fermentations by lactic and acetic acid bacteria of vegetables as well as milk strongly modify the taste, texture, and consistency of such food raw materials, but foremost are preservation measures: removal of fermentable sugars and concomitant acidification are the preferred technological methods for natural conservation of noncooked foodstuff with high water content. Similarily alcoholic fermentation of beverages improves both organoleptic properties as well as shelf life of sugar containing fruit juices and starch containing crop mashes. In any case food processing is the underlying reason for such microbial fermentation processes, and not the nutritional or dietary value of the food bacteria and metabolites that inevitably remain in the final product.

However, it has long been recognized that fermented milk products such as yogurt and kefir exhibit some health benefits that are supposed to be due to a positive direct or indirect influence on the composition of the human and animal bacterial gut flora. This recognition also led to the concept of probiotics, predominantly lactic acid bacteria, which are consumed as living organisms in fermented milk products and are considered to pass the stomach in order to colonize the gut transiently (see Chapter 4).

Beyond food bacteria and probiotics all other contributions to food properties, taste, and nutritional qualities are provided by isolated bacterial primary metabolites, enzymes, and other secretion products that are manufactured by industrial fermentation processes.

II. REGULATORY AND SAFETY ISSUES

Conceptually, ignoring any specific national legislation, food manufacturing and distribution are subject to the responsibility and liability of the manufacturer and distributor, who must ensure that the consumption of such food products does not risk or harm the health of the consumer. In other words, there is no manufacturing or marketing approval policy for food products comparable to a worldwide standard for pharmaceuticals. The administrative measure used to ensure food safety is rigid governmental inspection of manufacturing plants and products in the marketplace for general conformity of raw materials, process technologies, and finished products, and particularly hygiene. Among food control health issues is the search for any harmful contamination, such as that by heavy metals, pesticides, or toxins. The latter refers, for example, to aflatoxins and other mycotoxins carried forward from contaminated nut and cereal raw materials to processed foods, the negative aspects of microbiological contributions to food production: health damaging contamination and spoilage. The lack of regulatory approval policies in food production reflects the historically proven safety of food products but also reflects practical considerations, since food production is still mainly performed by local small and midsized manufacturers with a virtually infinite number of different consumer end products.

Food bacteria used as starter cultures are exempted from approval and labeling requirements, as are enzymes that are intra- or extracellular components of such food bacteria. The underlying logic is clear: Traditionally applied food bacteria and their enzymatic components used as processing aids, including single isolates or improved mixed cultures, which are typically used in starter cultures, have been proved safe. The use of any recombinant food microorganism, however, depends on specific national product approvals since they fall under not only the respective food safety directives but also the more general issue of deliberate environmental release of genetically modified organisms (GMOs).

No specific approval under the respective genetic engineering directives is typically needed for microbial enzymes that are manufactured with recombinant production strains as their manufacture is state of the art now in the enzyme industry, provided that the commercial enzyme product is free of GMOs and replicative nucleic acids. Enzyme production has a proven safety record for both industrial manufacturing and enzyme application for food, household, and technical purposes. In the past enzyme yields of microbial production strains have been optimized by multiple mutation and selection programs, and since 1990 the majority of all production strains are further improved by specific overexpression of the respective enzyme gene through genetic engineering. Most enzymes commercially available are secreted from the microbial cell and can be easily purified to a cell-free homogeneous protein concentrate by large-scale industrial downstream processing technology.

The approval exemption for food microorganisms and enzymes applies only to traditionally used known organisms, such as lactic acid bacteria and yeast. Any introduction of unknown organisms or their enzyme products into the food chain depends on extended safety tests that have to be presented to the national authorities as part of the approval procedure. This would also refer, for example, to a bacterial strain isolated from a local fermented food not currently used in Western countries, as has been generally stipulated in the "novel food" legislation that was part of the approval policies for "transgenic food."

Indeed many food processing enzymes are derived from and manufactured with bacterial and fungal strains not particularly known as food microorganisms. The introduction and use of such enzymes for food processing purposes have required detailed safety and toxicity study of the enzyme itself as well as of the microbial host. In case of the U.S. approval procedure the Food and Drug Administration (FDA) may subsequently grant so-called GRAS (generally recognized as safe) status. The certificate issued refers to the particular product, e.g., an enzyme, manufactured with the particular strain and does not automatically allow the use in food processing of the same product derived from or manufactured in another microbial strain, species, or genus. On the other hand, GRAS status of a compound manufactured with a particular strain significantly eases the subsequent approval of other fermentation products manufactured with the same strain or species.

Besides enzymes, other microbial fermentation products are used as processing aids for food production, typically referred to as *food additives*, or for nutritional purposes, often referred to as *food* or *dietary supplements*. The regulatory status of such ingredients as organic acids, amino acids, vitamins, and polymers depends on the specific national food and dietary legislation. The use of these ingredients, though absolutely natural and in most cases part of regular food and even a human biochemical constituent, depends on the particular compound approval, which may be restricted in dosage or even to certain kinds of food, such as bakery goods, milk products, or wine only, and must be labeled in convenience products.

Although certain fermentation products exhibit both functionalities—technological as well as nutritional—their use is not at the discretion of the manufacturer: Approved dosages of additives for processing purposes, such as vitamin C as an antioxidant, are typically too low to exhibit nutritional benefits, and higher dosages and health claims are only allowed if such "nutraceuticals" are approved by the respective food supplement or dietary food directives.

The introduction of any new chemical, biochemical, or semisynthetic compound, whether "natural" or "synthetic," into the food chain depends on previous safety analysis of the compound and, in the case of biotechnological manufacturing processes, on the safety analysis of the production strain. Some manufacturerś intentions to get approval for pharmacologically active compounds, even natural compounds, under food additive or food supplement regulations are highly questionable since the authorities are forced to impede any circumvention of the rigid drug approval and prescription legislation by launching health food products with pharmacologically active ingredients.

The principle of current drug approval legislation is to provide marketing licenses to each individual applicant per single finished dosage form of a respective drug active ingredient. This very tough safety concept would—for the reasons outlined—not fit food products or food additives and dietary supplements. Here the general use of a compound by any food processor is approved but is specified for dosage range, food categories, and labeling requirements. Current drug legislation, however, goes far beyond the marketing approval of finished products with defined compositions: The drug approval process also concerns the detailed process description of primary and secondary manufacturing that leads to the drug ingredient and the finished dosage form. The objective that the pharmaceutical manufacturer has to attain is defined as good manufacturing practice (GMP). The GMP directives basically describe how a manufacturing process must be developed, validated, operated, controlled, and documented in order to exclude any voluntary or unintended deviation from the manufacturing protocol, the drug master file, that could lead to any unknown toxic contamination or any change in product specification. Consequently the drug master file is part of the marketing license, and any modification of the process by the manufacturer requires a notification or even a formal approval by the authorities in advance of its commercial implementation.

In the early 1980s a new *Bacillus subtilis* fermentation process for the amino acid L-tryptophan as a dietary supplement was introduced. Subsequently the production strain has been improved by genetic engineering of the tryptophan operon, leading to a further yield improvement of L-tryptophan synthesized from an externally fed metabolic precursor. This material regrettably caused a number of severe and even lethal side effects and had to be withdrawn from the market. Upon inquiry it was discovered that because of the improved fermentor yield with the genetically modified production strain the recovery and purification process of the final product was shortened. However, the manufacturer did not check the safety of the product that resulted from this revised process and overlooked a minor, but highly toxic by-product that under the original purification regimen was safely removed from the product. This regrettable experience, which was due to a lack of process validation and general "GMP compliance," but by no means a result of genetic engineering, drew to the attention of the industry the urgent need for strict process control, for change validation and change documentation of any compound targeted to human or animal use regardless of whether it falls under the formal GMP requirements of drug legislation: "The process is the product" (McCormick, 1992).

III. AMINO ACIDS AND DERIVATIVES

L-Amino acids are the largest bacterial biotech products serving the nutrition markets—food, feed, and clinical nutrition. Fermentative amino acids as single compounds have broadly substituted amino acids purified from hydrochloric acid hydrolysates of animal and vegetable proteins or from other agroindustrial by-products, such as molasses. L-cystein, which is applied as a dough conditioner in the baking industry, as an antioxidant in cosmetic products, and as the starting material for *S*-carboxymethyl- and *N*-acetyl-L-cystein, leading mucolytic ingredients in cough medicines, is manufactured by protein hydrolysis and extraction from hair, bristles, and feathers. Meanwhile recombinant *Escherichia coli* strains have been described with deregulated L-cystein biosynthesis and overproduction of the amino acid (Leinfelder and Heinrich, 1997), and commercial production has been commenced. The industrial fermentation product will strongly compete against extracted L-cystein because of the common complaint that the proteinaceous feedstocks are not in line with the application of L-cystein in food products and cosmetics. Indeed, a recent directive by the European Commission (2000/63/EC) has banned use of L-cystein of human origin in the latter two market segments.

L-Glutamic acid, or more precisely its salt, monosodium glutamate (MSG), is the leading amino acid by volume. More than 800,000 tons/year of MSG are manufactured worldwide by fermentation of *Corynebacterium glutamicum* and related organisms. The bulk of this material is used as a flavor enhancer. Monosodium glutamate has partially replaced soya and other vegetable hydrolysates as a seasoning and has strongly benefited from the development of convenience food in Western countries. The compound is manufactured by a few industrial biochemical companies in dedicated plants. Scale of operation and advanced fermentation and recovery technology result in production costs typical for semicommodities in the chemical marketplace (Wilke, 1999).

The second largest L-amino acid is L-lysine, which is used predominantly as a feed additive for fortifying the amino acid composition of feed grains and other low- or suboptimal protein sources. L-Lysine strongly competes with soya protein, which is a high-protein feed crop well balanced with respect to L-lysine and other essential amino acids such as L-threonine and L-tryptophan. L-Lysine is manufactured by fermentation of *Corynebacterium glutamicum* and *Escherichia coli*. Its market volume is in the range of 400,000 tons/year worldwide, and the recent ban of animal protein feeds in the European Union (2001/999/EC) which must be compensated by domestic and imported feed crops, will certainly support the use of fermentative feed amino acids.

All other L-amino acids are considerably smaller by volume (50–200 tons/year each) and are used as food supplements, in enteral and parenteral clinical nutrition formulations, and as building blocks for drug synthesis. Two amino acids, L-phenylalanine and L-aspartic acid, are the building blocks of *N*-L-aspartyl-L-phenylalanine methyl ester, aspartame, a widely used low-caloric sweetener. Aspartame is an excellent example of a biotech food additive since not only are the two amino acids derived from industrial biotechnology processes but the coupling reaction can be performed enzymatically (Cheetham, 1995).

IV. NUCLEOTIDES

The two nucleotides disodium inosine monophosphate (IMP) and disodium guanosine monophosphate (GMP) are used in processed meat products as flavor enhancers in combination with monosodium glutamate. The nucleotides can be obtained by enzymatic hydrolysis and chemical modification of yeast nucleic acids, or preferentially by direct fermentation with *Bacillus* or *Brevibacterium* species.

V. VITAMINS

Vitamin C, ascorbic acid, is the most important compound of its class manufactured in a combination of synthetic and biotechnological steps. The annual production volume of vitamin C exceeds 60,000 tons worldwide and clearly reflects its significance as a food additive (antioxidant) as well as a dietary ingredient (mono- and multivitamin formulations). The Reichstein process, which starts from glucose, includes a microbial transformation step, the oxidation of sorbitol to sorbose with *Acetobacter suboxidans*. In this well-established semisynthetic route a further chemical oxidation reaction has now been replaced by a bacterial fermentation step, the oxidation of sorbose to L-ketogulonic acid with *Gluconobacter* species. Some research programs are focused on entirely biotechnological synthesis of ketogulonic acid from glucose. Such concepts are getting more and more realizable as a result of the genetic rearrangement of enzymes from different organisms in one optimized production strain and the various kinds of targeted genetic improvements of enzyme functionality, which together result in metabolic pathway engineering of microbial (and plant) production systems.

Vitamin B_2, riboflavin, is manufactured by fungal as well as bacterial fermentation (*Ashbya gossypii* and *Bacillus subtilis,* respectively). The world-wide annual production volume of ca. 3,000 tons is predominantly used in the feed sector, followed by dietary formulations. Vitamin B_{12}, cyanocobalamin, is derived from *Propionibacterium* species or *Pseudomonas denitrificans* fermentation. Its annual production volume of only 10 tons indicates that vitamin B_{12} is exclusively used as a dietary supplement and in pharmaceutical vitamin formulations.

VI. ORGANIC ACIDS

Industrial lactic acid fermentation with lactobacilli is significantly growing since new applications of this organic acid beyond its important use as food acidulant are emerging. Highly purified thermostable lactic acid is the building block for a novel generation of polymers, polylactides, allowing the entry of this renewable feedstock into food packaging and various technical markets. Its present production volume of approximately 150,000 tons/year therefore has disproportionate growth potential.

VII. ENZYMES

The largest class of bacterial enzymes by production volume, alkaline *Bacillus* spp. proteases, is used outside the food industry, as a performance additive in household detergents. *Bacillus* spp. and fungal proteases, however, also have a small application in the manufacturing of vegetable, whey, meat, and fish protein hydrolysates as seasonings or nonallergenic and readily resorbable peptide sources in infant formulas, sports foods, etc.

Bacillus licheniformis heat-stable α-amylase is a bulk enzyme used in industrial starch liquefaction, the first step of glucose production from corn and other starch feedstocks. Glucose is subsequently isomerized into a glucose–fructose mixture, high-fructose corn sirup, which is a major alternative to sucrose as a caloric sweetener in soft drinks since the 1970s. The isomerization reaction is performed with immobilized enzymes, glucose isomerases, derived from *Bacillus coagulans*, *Streptomyces rubiginosus*, *Actinoplanes missouriensis*, and other bacterial strains.

The majority of microbial enzymes used in food processing (dairy products, bakery, fruit juices, etc.) are of fungal origin.

VIII. POLYMERS

Xanthan, a bacterial exopolysaccharide manufactured with *Xanthomonas campestris*, has established a firm position as a thickener in dressings and convenience food products in general (Becker et al., 1998). Xanthan's wide application in the food industry is based on its outstanding rheological properties and broad temperature, pH, and salt tolerance, which result in an annual production volume for food and technical markets of ca. 40,000 tons.

Outside nutritional applications another bacterial polysaccharide may be mentioned, namely, hyaluronic acid derived from *Streptococcus equi* fermentation. This hydrocolloid is broadly used as a moisturizing agent in cosmetic products and as a pharmaceutical ingredient and medical device, respectively, for reconstitution of human synovial fluids in ophthalmic and joint disease indications. Originally hyaluronic acid was extracted from animal tissue, cockscombs, but the fermentative material has higher molecular weight (performance), is free of allergenic proteinaceous impurities (safety), and can easily and reproducibly be manufactured by an industrial process (economics). Hyaluronic acid is therefore a good example of the benefits of industrial biotech processes compared to natural extraction, particularly if animal-derived compounds are considered for food, feed, pharmaceutical, and cosmetic applications.

IX. MISCELLANEOUS BACTERIAL COMPOUNDS

Isomaltit, the sugar alcohol of isomaltulose, has been established as a noncariogenic, low-calorie sweetener in food, dietetic, and confectionary products. Isomaltulose is derived from sucrose by enzymatic conversion with sucrose mutase found in various gram-negative bacteria. The commercial conversion of sucrose into isomaltulose is performed with immobilized *Protaminobacter rubrum* cells (Vandamme and Soetaert, 1995).

L-Carnitine is a vitaminlike cofactor of fatty acid metabolism that can be isolated from meat and liver. The compound facilitates the transport of long-chain fatty acids through the mitochondrial membrane and is therefore considered to speed oxidation and energy generation from fat. Since its industrial production has been achieved, it has gained a significant position as a food and feed supplement. The annual production volume is in the range of 300 tons. The chemical synthesis includes as a final step a biotransformation reaction that can be performed with a variety of soil bacteria and special isolates (Yokozeki and Kubota, 1984; Kulla and Lehky, 1985) and that introduces the β-hydroxy group in the stereochemical L-configuration, the critical step in the entire synthesis. Alternatively L-carnitine can be obtained from racemic D,L-carnitine by optical resolution, generally a less attractive processing route. L-Carnitine may be considered a nutraceutical par excellence: It is a natural compound, but its concentration in regular diets is too low to effectuate any nutritional, degestive, or metabolic contributions. Its dietary and health supporting potential can only be utilized if infant formulas, sports drinks, and health foods (as well as animal feed formulations) are supplemented with the respective nutraceutical in the appropriate and regulatorily defined dosage. As has been mentioned, any pharmacological efficacy is strictly excluded from food ingredients, supplements, and additives. The natural carnitine derivative acetyl-L-carnitine is considered a neuro-supportive compound, for which the industry seeks approval as a food supplement with memory enhancing functionality, but also as a clinically documented anti–Alzheimer's disease drug. How the regulatory authorities will judge such strategies is open.

Trehalose, an α,α-disaccharide that is either extracted from yeast or derived by fermentative production as a secreted metabolite of *Brevibacterium* spp. can be used as an additive for cryopreservation of proteins and cellular systems and for improvement of freeze stability of processed food (Vandamme and Soetaert, 1995).

Cyclodextrins are cyclic oligosaccharides derived from starch by enzymatic degradation with *Bacillus* or *Klebsiella* spp. cyclodextrin glucosyl transferases. Because of the toruslike structure of cyclodextrins they can form inclusion complexes with a variety of compounds (Duchêne and Wouessidjewe, 1993). Their application ranges from stabilization of volatile, light-, or oxygen-sensitive molecules to "solubilization" of hydrophobic compounds such as vitamin E. In the United States, cyclodextrins are already approved for food applications.

X. CONCLUSIONS

Bacterial and fungal production of biochemicals has estabished a firm position in speciality chemical synthesis for food, feed, pharmaceutical, and technical applications. Its competitive position compared to those of chemical synthesis and natural product extraction has a clear technological basis: The molecular structure of such biochemicals and secondary metabolites such as amino acids and antibiotics is rather complicated and often characterized by one or more asymmetrical carbon centers, as a result, chemical synthesis is difficult or even impossible with respect to reasonable manufacturing costs. For well known natural compounds, such as ascorbic or citric acid, there is no agricultural feedstock available for direct extraction of bulk quantities that would meet present worldwide demands. Production feasibility and production economy are therefore the underlying principle for using "microbial" compounds for nutritional purposes, and only a few products, such as xanthan gum, are unique microbial constituents without an alternative production source and are used because of their unique functional characteristics.

During the second half of the 20th century some 40 bacterial fermentation and biotransformation products were launched for nutritional applications (Table 1), an average less than one "new" compound per year. Meanwhile many biotechnological manufacturing routes are based on recombinant production organisms; however, despite its high potential for generating new products by enzyme or metabolic pathway design, genetic engineering has not really increased product flow. Certainly manufacturing technology is no longer the bottleneck to launching new nutritional compounds of biotechnological origin; rather, the market itself and the significant investments in basic and applied research as well as regulatory requirements that are specified for introducing a new food additive or supplement are. At

Table 1 Bacterial Biotech Products for the Nutrition Markets

L-Glutamic acid	*Corynebacterium glutamicum*
L-Lysin	*C. glutamicum, Escherichia coli*
L-Cystein	*E. coli*
Other L-Amino acids	Various bacterial strains
Aspartame	Various bacterial strains
Guanosine monophosphate	*Bacillus* spp., *Brevibacterium* spp.
Inosine monophosphate	*Bacillus* spp., *Brevibacterium* spp.
Ascorbic acid	*Acetobacter suboxidans, Gluconobacter* spp.
Riboflavin	*Bacillus subtilis*
Lactic acid	*Lactobacillus* spp.
Numerous food enzymes	Various bacterial strains
Xanthan	*Xanthomonas campestris*
Isomaltulose	*Protaminobacter rubrum*
L-Carnitine	Various bacterial strains
Trehalose	*Brevibacterium* spp.
Cyclodextrins	*Bacillus* spp., *Klebsiella* spp.

least the difference in attitude compared to that of the pharmaceutical or even the agrochemical industry is striking: Those industries have performed intensive and very successful microbial compound screenings. As an example we may consider microbial enzymes as digestive aids and as neutraceuticals with further unexpected health benefits. Such screening work is nowadays strongly facilitated by microbial genomics data and expression libraries.

There are established neutraceuticals, such as vitamin E and polyunsaturated fatty acids, for which improved manufacturing technologies would be desirable. Vitamin E is a good example: A strong preference exists for the natural compound in food and dietary products versus the less expensive and unrestrictedly available synthetic material. Microbial production routes as potential technological alternatives are now facing competition from improved agricultural processes: Molecular plant genetics will soon give access to conventional and transgenic crops with significantly increased vitamin E content or specialty oil content that cut the manufacturing costs of such nutraceuticals (Wilke 1999). Presently advanced breeding technologies will also provide "new" plant ingredients with particular nutritional and health supporting properties, such as steroids or polysaccharides that may be further modified by enzymatic and other biotechnological techniques.

REFERENCES

A Becker, F Katzen, A Pühler, L Ielpi. Xanthan gum biosythesis and application: A biochemical/genetic perspective. Appl Microbiol Biotechnol 50:145–152, 1998.

PSJ Cheetham. Biotransformations: New routes to food ingredients. Chem & Ind 3 April 1995: 265–268.

D Duchêne, D Wouessidjewe. New possibilities for the pharmaceutical use of cyclodextrins and their derivatives. Chimicaoggi 1,2: 17–24, 1993.

H Kulla, P Lehky. Verfahren zur Herstellung von L-Carnitin auf mikrobiellem Wege. EP 0158194, 1985.

W Leinfelder, P Heinrich. Process for preparing *O*-acetylserine, L-cystein and L-cystein-related products. WO 97/15673, 1997.

D McCormick. L-Trp's lesson: The process is the product. Biotechnology 10:5, 1992.

EJ Vandamme, W Soetaert. Biotechnical modification of carbohydrates. FEMS Microbiol Rev 16: 163–186, 1995.

D Wilke. Chemicals from biotechnology: Molecular plant genetics will challenge the chemical and the fermentation industry. Appl Microbiol Biotechnol 52:135–145, 1999.

K Yokozeki, K Kubota. Method for producing L-carnitine. EP 0122794, 1984.

4

Genomics of Probiotic Lactic Acid Bacteria: Impacts on Functional Foods

Todd R. Klaenhammer
North Carolina State University, Raleigh, North Carolina, U.S.A.

Willem M. de Vos
Wageningen University and Wageningen Center for Food Sciences, Wageningen, The Netherlands

Annick Mercenier
Nestlé Research Center, Lausanne, Switzerland

I. INTRODUCTION

The period since the early 1990s has seen numerous developments aimed at improving the functionality of foods. This can be realized by appropriate selection of raw materials, specific physicochemical processing, or the addition of ingredients (including health-beneficial microorganisms). However, in many cases food functionality can also be enhanced via biological conversions such as by exploiting the activity of lactic acid bacteria or other microbes with a long history of safe use in the food industry. Lactic acid bacteria are used for the industrial production of fermented dairy, vegetable, and meat products and form a group of evolutionarily related low-GC-content gram-positive bacteria, comprising species of *Lactobacillus*, *Lactococcus*, or *Leuconostoc*. Related gram-positive bacteria such as strains of *Bacillus subtilis*, *Enterococcus faecalis*, or the somewhat more distantly related (high GC content)

Bifidobacterium or *Propionibacterium* spp. are used for specific food fermentations and can be applied similarly (Fig. 1). Whereas traditionally lactic acid and related bacteria have been applied as starter bacteria, most food innovations of lactic acid and related bacteria are correlated to their use in fermentations with a specific function or the production of ingredients that have potential as nutraceuticals. Yet another important application of these microorganisms is their inclusion as live cells in food and feed matrixes or nutritional complements for the design of health-promoting—probiotic or functional—food products.

The completion of many genomic sequences of lactic acid and related bacteria and the application of functional genomics approaches are greatly advancing the knowledge base for innovations (1–3). Moreover, food-grade approaches for high-throughput screening or specific genetic improvements such as self-cloning are continually being sophisticated while acceptable selection systems are being perfected (4). Hence, this chapter summarizes the recent advances in genomics, focusing on the innovations relating to probiotic lactic acid bacteria.

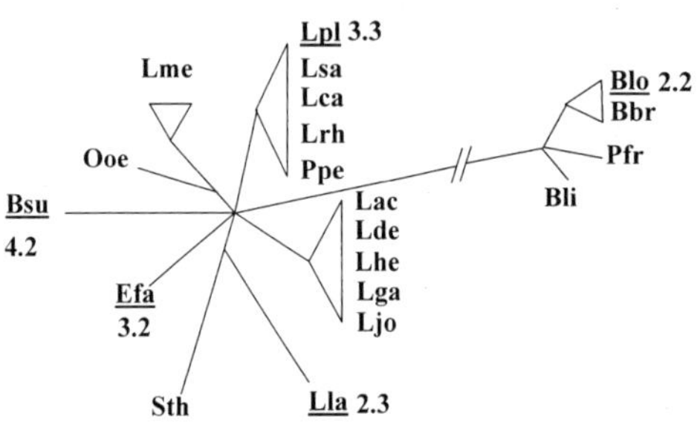

Figure 1 Phylogenetic tree based on 16S recombinant deoxyribonucleic acid (rDNA) sequences of lactic acid and related bacteria that have been completely or partially analyzed at the genomic level. Left, low GC bacteria; right, high GC bacteria. The size of the completely known genomes (underlined) is indicated in Mb. Bsu, *Bacillus subtilis* (21); Lla, *Lactococcus lactis* (20); Blo, *Bifidobacterium longum* (18); Lpl, *Lactobacillus plantarum* (19); Efa, *Enterococcus faecalis* (22); Sth, *Streptococcus thermophilus*; Lac, *Lactobacillus acidophilus*; Ljo, *Lactobacillus johnsonii*; Lga, *Lactobacillus gasseri;* Lbr, *Lactobacillus breve*; Lhe, *Lactobacillus helveticus;* Lde, *Lactobacillus delbrueckii;* Lsa, *Lactobacillus sake;* Lca, *Lactobacillus casei;* Lrh, *Lactobacillus rhamnosus;* Ppe, *Pediococcus pentosaeus*; Lme, *Leuconostoc mesenteroides*; Ooe, *Oenococcus oenus*; Pfe, *Propionibacterium freudenreichii*; Bbr, *Bifidobacterium breve*; Bli, *Brevibacteriujm linens* (2).

II. EVOLUTION OF THE PROBIOTIC CONCEPT

Since Eli Metchnikoff published *The Prolongation of Life: Optimistic Studies* the field of probiotics has become nearly 100 years old (5). However, today it is a predominantly empirical science that remains to be deciphered scientifically and mechanistically. In the first generation of probiotic cultures (1904 to 1970), there was little accurate information about the identity of the cultures or the specific health benefits that could be attributed to them. Strains were often chosen on the basis of poorly controlled screening experiments. Also, products reputed to contain probiotic cultures often lacked quality control over the identity of the culture(s) and their viability and levels (6,7).

Since the 1980s, major efforts have been invested in the proper identification and selection of probiotic strains. The evolution of molecular taxonomy, deoxyribonucleic acid (DNA) fingerprinting, and ribosomal DNA sequence analysis has allowed unequivocal identification of strains being used in both clinical trials and commercial products (8). Armed with defined strains, efforts began to correlate specific health traits (e.g., lactose tolerance, immunomodulation, antimicrobial activity, anticarcinogenic activity) with specific strains or strain combinations (7,9–11). However, most of these health-beneficial properties were studied primarily in vitro and in a variety of animal models. Even though the number of well-controlled human intervention trials is steadily increasing, there is still a lack of correlation between the different methods used to select probiotic strains (7). Nevertheless, the end result of these efforts is a rapidly accumulating database of well-conducted studies that are beginning to unravel the clinical manifestations of probiotic cultures (12). With the rapid progress over the past decade on the genetics of lactic acid bacteria and pending release of complete genome sequence information for major probiotic species, the field is now positioned to progress on a number of fronts. The genomic information available on both host and bacterial partners is expected to reveal the contributions of probiotics and commensals to general health and well-being and explicitly identify the mechanisms and corresponding host–microbe cross-talk that provide the basis for their positive roles.

This platform of information is currently leading to the selection of second-generation probiotics, those identified by their genomic content and expression levels for specific traits with known probiotic functionality. Rapidly growing insight into the structure and function of the target ecosystem, and the intestinal microbiota that include endogenous lactic acid bacteria (13), will also contribute to understanding and prediction of the functionality of probiotics (1,14).

Last, third-generation probiotics—genetically modified cultures that serve as delivery vehicles or production for biologicals such as antigens, enzymes, or complex carbohydrates—are on the immediate horizon (15,16).

III. GENOMIC DEVELOPMENTS OF PROBIOTIC LACTIC ACID BACTERIA

Exploration into the genomes of probiotic cultures is now well under way (2) and promises to contribute significantly to our understanding of the ecological and phylogenetic characteristics, metabolic capacity, and safety of probiotics. Hence genomics approaches are expected to lead the way to the development of even newer generations of probiotics or functional cultures. The International Scientific Association for Probiotics and Prebiotics (ISAPP) (http://www.isapp.net/) has recommended that the genomes for all commercial probiotic cultures be sequenced to ensure their identity, potential functionalities, and, most importantly, safety (17).

The genomes of two probiotic species, *Bifidobacterium longum* (18) and *Lactobacillus plantarum* (19), have been completed, and those of at least eight more are rapidly nearing completion: *L. johnsonii* (Pridmore et al., in preparation), *L. acidophilus*, *L. gasseri*, *L. casei* (two strains), *L. rhamnosus*, *B. longum* (a second strain), and *B. breve*. Draft genome information also became available in the public domain in 2002 with the publication and appearance of genome sequences for lactic acid bacteria (LAB) provided by the Joint Genome Institute (http://www.jgi.doe.gov/JGI_microbial/html/index.html) in collaboration with the Lactic Acid Bacteria Genomics Consortium (see also Fig. 1).

The complete size of the present paradigm genomes of *Lactococcus lactis* (20), *B. longum* (18), and *L. plantarum* (19) varies somewhat between 2 and 3 Mb, in line with their fastidious lifestyle and incapacity to live without specific medium additions. In contrast, the genome of *B. subtilis* is somewhat double that size and contains many genes for auxotrophic growth (21). However, in several cases the genome size of these food-grade bacteria is considerably enlarged by the presence of plasmids and other extrachromosomal elements that contribute to the genomic flexibility of the host (22). Notably, *Lactococcus* spp. are known to carry up to 20% of their DNA as plasmids, and this characteristic may provide these starters with additional flexibility. Moreover, even though bacteriophages are a threat to industrial fermentations, prophages correspond to a natural and diverse component of many genomes that may contain up to half a dozen different copies, including both complete and incomplete versions (23).

A. Functional Genomics and High-Throughput Approaches

Approximately one-third of the genes in a genome encode functions that have not been described yet and cannot be annotated. Moreover, several genes have multiple functions and many annotations are incorrect because they refer to circumstantial evidence and not to an experimentally verified function. Hence, substantial efforts are focusing on elucidating genome functions by a variety of approaches that include targeted or random gene inactivation, overexpression analysis, and high-throughput expression profiling by transcriptome or proteome analysis. In the case of probiotic bacteria, this approach remains rather difficult as a consequence of the lack of validated in vitro bioassays. Nevertheless, ongoing research aims at assigning a probiotic function to specific genes by conducting head-to-head comparison of selected sets of strains by a series of in vitro or in vivo models. Appropriate selection of strains for these studies can be based either on comparative genomic analyses or on isogenic pairs of strains corresponding to targeted mutants and their wild-type counterparts.

Noteworthily, the first proteomic analysis of a lactic acid bacterium, *L. lactis*, was completed in 2003 (24). Even though this species is generally not considered as a probiotic microorganism per se, similar studies are expected to be reported soon for *Lactobacillus* and/or *Bifidobacterium* spp.

B. Comparative Genomics and Strain Diversity

Comparative analysis of genomes may provide insight into evolutionary relations, reveal specific functions of genes within their context, and allow for the analysis of genome diversity. A detailed study of *L. plantarum* using DNA microarrays revealed high genomic diversity and identified specific lifestyle islands that are characterized by aberrant codon usage or GC content and at the same time seem to differ between strains (Molenaar et al., personal communication). Whereas some of these islands in fact are lysogenic bacteriophages or remnants thereof (Brüssow et al., personal communication), many appear to code for relevant functions such as the production of antimicrobial peptides or sugar transport and utilization systems (Molenaar et al., personal communication). Equivalent studies conducted with *L. johnsonii* strains revealed that specific chromosomal regions are restricted to the probiotic strain La1 and a few closely related isolates. These regions include prophages and genes that could reflect the adaptation of La1 to the upper gastrointestinal tract. Typically, the lack of amino acid biosynthetic pathways appears to be compensated for by the existence of a wealth of peptidases and monomeric nutrient uptake systems (Pridmore et al., in preparation).

The high diversity within species indicates that specific approaches are needed to exploit fully the genomic potential of food-grade and probiotic bacteria. One such approach is subtractive hybridization that aims to reveal new sequences that are unique to a specific strain in comparison with another (sequenced or well-studied) strain. A specific form of this approach is *representational display analysis*, which has been used to differentiate natural isolates of *Bacillus* spp. and is expected to be useful to capitalize on the diversity of other food-grade bacteria (25).

Although the completely sequenced genomes serve as the basis for understanding genome organization, evolution, and expression, it is evident that these and most of the unfinished genomes (discussed previously) will also be instrumental in the screening for target functions including the health-promoting properties attributed to probiotic organisms. An exciting set of discoveries are already apparent, revealing, for example, fimbriae-like structures on *B. longum* as well as the presence of a serpin-like gene, which had not been identified before in prokaryotes and may play a role in immunomodulation (18). In *L. plantarum* gene regions that appear to promote a lifestyle that is flexible in habitats ranging from green plants to the human gastrointestinal tract have been identified (19). This may seem a trivial way to exploit a genomic sequence, but it is undoubtedly very straightforward, effective, and immediate. A specific example is that before the publication of the complete genome sequence of *B. longum*, predicted to encode approximately 2200 genes, only a handful of bifidobacterial genes had been discovered (18). Timely public availability of genome information for various LAB species will catapult the field's collective efforts to carry on comparative and functional genomic analyses of probiotic species within the lactic acid bacteria. Features predicted by ISAPP (17) to be important for transient persistence, survival, functionality, and safety of probiotic bacteria are summarized in Table 1.

Genomics will help to decipher the mechanistic elements of these features and their probable roles in probiotic functionality. Soon, second-generation probiotics will be completely characterized for genetic content and potential functionality, on the basis of expression (transcriptomics, proteomics, metabolomics) of these traits and many more likely to be discovered through genomic analysis. The concomitant progress in the design of more relevant bioassays will contribute to defining rational probiotic selection criteria. Rather than remaining largely empirical and generic as they are at present, specific health-beneficial traits will be selectable, keeping the final product and specific consumer populations in mind. In addition, unique genetic signatures for probiotic strains are likely to be found in the genomes and could provide unequivocal and rapid strain identification through RAPDs, patterned gene content in microarrays, and transcriptional profiling. Last but not least, holistic approaches that can best be taken when the complete genome of the microorganism of interest is known could be applied to verify how processing and formulation affect the function of a selected probiotic bacterium, a field that remains virtually unexplored at present.

Table 1 Genetic Features Predicted to Be Important for Probiotic Bacteria

Acid, bile, and stress tolerance
Surface and extracellular proteins
Lipoteichoic acid biosynthesis
Exopolysaccharides
Adherence and aggregation factors
Bacteriocin production
Carbohydrate (prebiotic) utilization and metabolism
Quorum sensors and response regulators
Biofilm formation
Immune modulation
Putative virulence factor homologs
Siderophores, scavengers of Fe^{++}
Antibiotic resistance/sensitivity
Gene transfer potential
Mobile genetic elements
Prophages, prophage remnants, lysogenic conversion characters

Source: Adapted from Ref. 17.

IV. BIOTECHNOLOGY APPLICATIONS: THIRD-GENERATION PROBIOTIC STRAINS

Global views of probiotic genomes and knowledge of the elements and conditions that regulate gene expression within these microbes will establish the platform to isolate new probiotics. As mentioned, since the early 1990s, biotechnology applications for lactic acid bacteria have supported the development of genetic tools (e.g., transformation systems, cloning and expression vectors, food-grade expression systems, integration vectors, and systems for gene inactivation) in a selected number of probiotic cultures (26). Genetic accessibility is vital to establish the genotype–phenotype relationships that will ultimately define the mechanisms underlying probiotic effects.

Beyond the use of genetic approaches to investigate probiotic cultures, there is an exciting new field developing that will exploit probiotic cultures in new or improved applications. One application is the use of genetic modification to improve their inherent performance or activity. The first example of this approach was targeted inactivation of the D-lactate dehydrogenase (*D-ldh*) gene in *L. johnsonii* La1, a probiotic culture bacterium that ferments lactose to D- and L-lactate in a 60%:40% ratio. It has been reported that in rare cases, D-lactate may accumulate in the blood of patients suffering from some intestinal disorders. The Codex Alimentarius also prevents the addition of D-lactate-producing lactic acid bacteria to infant formulas. In order to minimize D-lactate production, the gene (*D-ldh*) of La1 was isolated and an 8-base-pair (8-bp) deletion was generated to inactivate its function. The truncated copy of that gene was then used to replace the native *D-ldh*, via homologous recombination within the genome. The resulting genetically modified (GM) probiotic culture produced only L-lactate, at increased levels (27). Even though the present legislation in some countries would not allow the use of GM strains in food and feed products, such a strain certainly corresponds to a probiotic with improved health properties.

Second, metabolic engineering allows for the rerouting of cellular pathways and the production of specific metabolites. Although not yet applied to probiotic cultures, metabolic engineering has allowed generation of *Lactococcus lactis* with increased potential health benefits that overproduce the antioxidants mannitol and sorbitol (28), immunostimulatory exopolysaccharides (29), or vitamins such as folic acid (30). Moreover, lactic acid bacteria that produce alanine instead of lactic acid that sequester ammonia, a potential toxic intestinal metabolite, have been constructed (31). Similarly, the construction of strains producing other compounds than lactic acid, such as the presumed beneficial butyrate, may be feasible.

Third, the use of lactic acid bacteria as delivery vehicles for biological compounds has been considered and actively investigated for a number of years (15,16,32–35). The generally recognized as safe (GRAS) status of the lactic acid bacteria and their suitability for oral consumption at levels as high as 10^9 cfu/g make them attractive candidates for this application in both human and animals. Probiotic cultures may offer additional advantages for enhanced delivery of biologicals to specific locations in the gastrointestinal tract, mouth, vagina, or other selected tissues.

The potential of lactic acid bacteria to act as a live mucosal delivery system was first investigated with protective antigens. The use of lactic acid bacteria represents an attractive alternative to the use of attenuated pathogenic microorganisms as vaccine carriers. Mucosal vaccines based on lactic acid bacteria would be particularly suited to alleviate the drawbacks linked to injectable vaccines: ease of administration, reduced side effects, no risk of cross-contamination by needles, and limited side effects. By nature, lactic acid bacteria are best adapted to passage through the stomach and would be ideal carriers for vaccination of less immunocompetent individuals such as infants and elderly. The most complete immunization studies have been carried out with the C subunit of the tetanus toxin (TTFC). Both persisting—i.e., *Streptococcus gordonii*, *L. plantarum*, and *L. casei*—and nonpersisting—i.e., *L. lactis*—species have been used as carriers. The strains producing sufficient levels of the antigen were shown to induce high levels of serum immunoglobulin G (IgG) after nasal or intragastric administration. In the best cases, these immune responses were shown to be protective against a tetanus toxin lethal challenge. In addition, it was demonstrated that local TTFC-specific IgA was elicited as well (for recent reviews see Refs. 15,16,32–35). A series of other antigens, including capsular polysaccharides (36), were successfully produced in lactic acid bacteria. Immunizations in mouse models mimicking the targeted mucosal infection are in progress in different laboratories (for reviews see Refs. 33,35). Interestingly, the mechanisms by which these recombinant lactic acid bacteria interact with the host immune cells have begun to be analyzed (37,38). It is foreseen that these immunization studies will shed a new light on the probiotic field, as the readouts in these studies are well defined and relatively easy to follow.

The concept of targeted mucosal delivery by lactic acid bacteria was extended from antigens to interleukins by Steidler and associates, who first demonstrated that immune responses could be enhanced by the codelivery of interleukin 2 Il-2 or IL-6 and TTFC (39). This approach was next applied to the design of new anti-inflammatory treatments targeting inflammatory bowel disease. These authors constructed a recombinant *L. lactis* strain secreting murine IL-10 and successfully demonstrated that, upon intragastric administration of repeated doses, this strain was able to prevent or treat inflammation in murine colitis models. Notably, this effect was obtained with much lower (10,000-fold) doses of IL-10 than those required when the interleukin was used as a single therapeutic agent (40). Steidler and associates (41) further constructed a safe (no antibioresistance marker and chromosomally integrated transgene) biologically contained strain secreting the human IL-10 that should allow completion of a phase 1 clinical trial in humans.

In the recent years, production and mucosal delivery of different types of health-promoting molecules such as ScFv antibodies, allergens, or digestive enzymes have been achieved with lactic acid bacteria. Targeted diseases included microbial infections such as vaginal candidiosis (42) and dental caries (43), allergies (44,45), autoimmune diseases (46), and metabolic defects such as pancreatic insufficiency (47). In most cases, the targeted biological effect was indeed demonstrated in animal models mimicking the corresponding human disease. Notably, this process often relied on the use of live bacterial cells. Efforts are presently being devoted to improve the efficiency of lactococci or lactobacilli as a delivery system. Therefore, mutants that release intracellular compounds more efficiently (48) or that present improved delivery properties have been generated (Grangette, Hols, Delcour, and Mercenier, unpublished).

Even though recombinant bacterial strains would not be accepted today in functional foods, their future use in therapeutic approaches can be foreseen, provided that the benefit/risk balance is positive. This novel approach is rather promising as grafting the gene encoding for a therapeutic molecule could enhance the natural health-beneficial properties of a strain (49). Genomic information and genetic tools will be critically important to furthering the development of these applications. Systems for tailored gene expression (regulated promoters; intracellular, anchored, secreted proteins) have already been used for targeted and regulated delivery of specific biological compounds. It is anticipated that these systems will become more sophisticated in the near term and promote an explosion of biotechnological applications with probiotic cultures.

V. CONCLUSIONS

Genomics is revolutionizing our view of microorganisms and the beneficial processes that they carry out. Global and detailed analyses of host–microbial interactions are becoming feasible (50). Concurrently, the probiotic field is gaining increased scientific attention and research efforts are beginning to unravel the mechanistic basis of the ways probiotic cultures contribute to health and well-being. Genomic analysis of probiotic cultures will now play a pivotal role in scientifically understanding the 100-year-old concepts of Metchnikoff. The impact of the genomic revolution on probiotic cultures will be staggering, allowing a full view of their genetic composition, and an understanding of the molecular mechanisms by which they interact with the host and providing the knowledge base and tools to allow them to be exploited in novel biotechnological applications.

REFERENCES

1. WM de Vos. Advances in genomics for microbial food fermentations and safety. Curr Opin Biotechnol 12(5): 493–498, 2001.
2. TR Klaenhammer, E Altermann, F Arigoni, A Bolotin, F Breidt, J Broadbent, R Cano, S Chaillou, J Deutscher, M Gasson, M van de Guchte, J Guzzo, A Hartke, T Hawkins, P Hols, R Hutkins, M Kleerebezem, J Kok, O Kuipers, M Lubbers, E Maguin, L McKay, D Mills, A Nauta, R Overbeek, H Pel, D Pridmore, M Saier, D van Sinderen, A Sorokin, J Steele, D O'Sullivan, W de Vos, B Weimer, M Zagorec, R Siezen. Discovering lactic acid bacteria by genomics. Antonie Van Leeuwenhoek 82: 59–71, 2002.
3. OP Kuipers, A de Jong, RJS Baerends, SAFT van Hijum, AL Zomer, HA Karsens, CD den Hengst, NE Kramer, G Buist, J Kok. Transcriptome analysis and related databases of *Lactococcus lactis*. Antonie Van Leeuwenhoek 82: 113–122, 2002.

4. WM De Vos. Safe and sustainable systems for food-grade fermentations by genetically modified lactic acid bacteria. Int Dairy J 9: 3–10, 1999.
5. E Metchnikoff. The Prolongation of Life: Optimistic Studies. PC Mitchell, ed., 1907, William Heinemann, London.
6. ME Sanders, J Huis in 't Veld. Bringing a probiotic-containing functional food to the market: Microbiological, product, regulatory and labeling issues. Antonie Van Leeuwenhoek 76: 293–315, 1999.
7. A Mercenier, S Pavan, B Pot. Probiotics as biotherapeutic agents: Present knowledge and future prospects. Curr Pharm Des 9: 175–191, 2003.
8. TR Klaenhammer, WM Russell. Species of the *Lactobacillus acidophilus* complex. In: RK Robinson, C Batt, PD Patel, eds. Encyclopedia of Food Microbiology, Vol. 2. Academic Press, 2000, pp 1151–1157.
9. GW Tannock. Probiotics: A Critical Review. Horizon Scientific Press, 1999.
10. GW Tannock. Probiotics and Prebiotics: Where Are We Going? Caister Academic Press, 2002.
11. AC Ouwehand, S Salminen, E Isolauri. Probiotics: An overview of beneficial effects. Antonie van Leeuwenhoek 82: 279–289, 2002.
12. C Dunne, L O'Mahony, L Murphy, G Thornton, D Morrissey, S O'Halloran, M Feeney, S Flynn, G Fitzgerald, C Daly, B Kiely, GC O'Sullivan, F Shanahan, JK Collins. In vitro selection criteria for probiotic bacteria of human origin: Correlation with in vivo findings. Am J Clin Nutr 73(Suppl 2): 386S–392S, 2001.
13. EE Vaughan, MC de Vries, EG Zoetendal, K Ben-Amor, ADL Akkermans, WM de Vos. The intestinal LAB's. Antonie Van Leewenhoek 82: 341–352, 2002.
14. GW Tannock. Analysis of the intestinal microflora using molecular methods. Eur J Clin Nutr 56(Suppl 4): S44–S49, 2002.
15. A Mercenier, H Muller-Alouf, C Grangette. Lactic acid bacteria as live vaccines. Curr Issues Mol Biol 2: 71–25, 2000.
16. PH Pouwels, A Vriesema, B Martinez, FJ Tielen, JF Seegers, RJ Leer, J Jore, E Smit. Lactobacilli as vehicles for targeting antigens to mucosal tissues by surface exposition of foreign antigens. Methods Enzymol 336: 369–389, 2001.
17. G Reid, ME Sanders, HR Gaskins, GR Gibson, A Mercenier, R Rastall, M Roberfroid, I Rowland, C Cherbut and TR Klaenhammer. New scientific paradigms for probiotics and prebiotics. J Clin Gastroenterol 37:105–118, 2003.
18. MA Schell, M Karmirantzou, B Snel, D Vilanova, B Berger, G Pessi, MC Zwahlen, F Desiere, P Bork, M Delley, RD Pridmore, F Arigoni. The genome sequence of *Bifidobacterium longum* reflects its adaptation to the human gastrointestinal tract. Proc Natl Acad Sci USA 99: 14422–14427, 2002.
19. M Kleerebezem, J Boekhorst, R Van Kranenburg, D Molenaar, OP Kuipers, R Leer, R Tarchini, SA Peters, HM Sandbrink, MW Fiers, W Stiekema, RM Lankhorst, PA Bron, SM Hoffer, MN Groot, R Kerkhoven, M De Vries, B Ursing, WM De Vos, RJ Siezen. Complete genome sequence of *Lactobacillus plantarum* WCFS1. Proc Natl Acad Sci U S A 100: 1990–1995, 2003.
20. A Bolotin, P Wincker, S Mauger, O Jaillon, K Malarme, J Weissenbach, SD Erlich, A Sorokin. The complete genome sequence of the lactic acid bacterium *Lactococcus lactis* ssp. *Lactis* IL1403. Genome Res 11: 731–753, 2001.

21. F Kunst, N Ogasawara, I Moszer, AM Albertini, G Alloni, V Azevedo, MG Bertero, P Bessieres, A Bolotin, S Borchert, R Borriss, L Boursier, A Brans, M Braun, SC Brignell, S Bron, S Brouillet, CV Bruschi, B Caldwell, V Capuano, NM Carter, SK Choi, JJ Codani, IF Connerton, A Danchin, et al. The complete genome sequence of the gram-positive bacterium *Bacillus subtilis*. Nature 390(6657): 249–256, 1997.
22. IT Paulsen, L Banerjei, GS Myers, KE Nelson, R Seshadri, TD Read, DE Fouts, JA Eisen, SR Gill, JF Heidelberg, H Tettelin, RJ Dodson, L Umayam, L Brinkac, M Beanan, S Daugherty, RT DeBoy, S Durkin, J Kolonay, R Madupu, W Nelson, J Vamathevan, B Tran, J Upton, T Hansen, J Shetty, H Khouri, T Utterback, D Radune, KA Ketchum, BA Dougherty, CM Fraser. Role of mobile DNA in the evolution of vancomycin-resistant *Enterococcus faecalis*. Science 299(5615): 2071–2074, 2003.
23. A Chopin, A Bolotin, A Sorokin, SD Ehrlich, M Chopin. Analysis of six prophages in *Lactococcus lactis* IL1403: Different genetic structure of temperate and virulent phage populations. Nucleic Acids Res 29(3): 644–651, 2001.
24. A Guillot, C Gitton, P Anglade, M-Y Mistou. Proteomic analysis of *Lactococcus lactis*, a lactic acid bacterium. Proteomics 3: 337–354, 2003.
25. A Felske. Streamlined representational difference analysis for comprehensive studies of numerous genomes. J Microbiol Methods 50(3): 305–311, 2002. Erratum in J Microbiol Methods 51(3):425, 2002.
26. MJ Kullen, TR Klaenhammer. Genetic modification of intestinal lactobacilli and bifidobacteria. Curr Issues Mol Biol 2: 41–45, 2000.
27. L Lapierre, J-E Germond, A Ott, M Delley, B Mollet. D-lactate dehydrogenase gene (*ldhD*) inactivation and resulting metabolic effects in the *Lactobacillus johnsonii* strains La1 and N312. Appl Environ Microbiol 65: 4002–4007, 1999.
28. J Hugenholtz, EJ Smid. Nutraceutical production with food-grade microorganisms. Curr Opin Biotechnol 13(5): 497–507, 2002.
29. IC Boels, R van Kranenburg, MW Kanning, BF Chong, WM de Vos, M Kleerebezem. Increased exopolysaccharide production in *Lactococcus lactis* due to increased levels of expression of the NIZO B40 *eps* gene cluster. Appl Environ Microbiol. 69: 5029–5031, 2003.
30. W Sybesma, M Starrenburg, M Kleerebezem, I Mireau, WM de Vos, J Hugenholtz. Increased production of folate by metabolic engineering of *Lactococcus lactis*. Appl Environ Microbiol. 69: 3069–3076, 2003.
31. P Hols, M Kleerebezem, AN Schanck, T Ferain, J Hugenholtz, J Delcour, WM de Vos. Conversion of *Lactococcus lactis* from homolactic to homoalanine fermentation through metabolic engineering. Nat Biotechnol 17(6): 588–592, 1999.
32. JE Thole, PJ van Dalen, CE Havenith, PH Pouwels, JF Seegers, FD Tielen, MD van der Zee, ND Zegers, M Shaw. Live bacterial delivery systems for development of mucosal vaccines. Curr Opin Mol Ther 2: 94–99, 2000.
33. JM Wells, K Robinson, LM Chamberlain, KM Schofield, RW Le Page. Lactic acid bacteria as vaccine delivery vehicles. Antonie Van Leeuwenhoek 70: 317–330, 1996.

34. JF Seegers. Lactobacilli as live vaccine delivery vectors: Progress and prospects. Trends Biotechnol 20: 508–515, 2002.
35. JM Wells, A Mercenier. Lactic acid bacteria as mucosal delivery vehicles. In: BJB Wood and PJ Warner, eds. Genetics of Lactic Acid Bacteria. Kluwer Academic/Plenum Publishers, 2003, pp 261–290.
36. C Gilbert, K Robinson, RW Le Page, JM Wells. Heterologous expression of an immunogenic pneumococcal type 3 capsular polysaccharide in *Lactococcus lactis*. Infect Immun 68: 3251–3260, 2000.
37. M Rescigno, C Citterio, M Théry, M Rittig, D Medaglini, G Pozzi, S Amigorena, P Ricciardi-Castagnoli. Bacteria-induced neo-biosynthesis, stabilization, and surface expression of functional class I molecules in mouse dendritic cells. Proc Natl Acad Sci U S A 95: 5229–5234, 1998.
38. S Corinti, D Medaglini, C Prezzi, A Cavani, G Pozzi, G Girolomoni. Human dendritic cells are superior to B cells at presenting a major histocompatibility complex class II–restricted heterologous antigen expressed on recombinant *Streptococcus gordonii*. Infect Immun 68: 1879–1883, 2000.
39. L Steidler, K Robinson, L Chamberlain, KM Schofield, E Remaut, RWF Le Page, JM Wells. Mucosal delivery of murine interkeukin-2 (IL-2) and IL-6 by recombinant strains of *Lactococcus lactis* coexpressing antigen and cytokine. Infect Immun 66: 3183–3189, 1998.
40. L Steidler, W Hans, L Schotte, S Neirynck, F Obermeier, W Falk, W Fiers, E Remaut. Treatment of murine colitis by *Lactococcus lactis* secreting interleukin-10. Science 289: 1352–1355, 2000.
41. L Steidler, S Neirynck, N Huyghebaert, V Snoeck, A Vermeire, B Goddeeris, E Cox, JP Remon, E Remaut. Biological containment of genetically modified *Lactococcus lactis* for intestinal delivery of human interleukin-10. Nat Biotechnol 21:785–789, 2003.
42. C Beninati, MR Oggioni, M Boccanera, MR Spinoza, T Maggi, S Conti, W Magliani, F De Bernardis, G Teti, A Cassone, G Pozzi, L Polonelli. Therapy of mucosal candidiasis by expression of an anti-idiotype in human commensal bacteria. Nat Biotechnol 18: 1060–1064, 2000.
43. C Krüger, Y Hu, Q Pan, H Marcotte, A Hultberg, D Delwar, PJ van Dalen, PH Pouwels, RJ Leer, CG Kelly, C van Dollenweerd, JK Ma, L Hammarström. In situ delivery of passive immunity by lactobacilli producing single-chain antibodies. Nat Biotechnol 20: 702–706, 2002.
44. A Kruisselbrink, M-J Heijne den Bak-Glashouwer, CEG Havenith, JER Thole, R Janssen. Recombinant *Lactobacillus plantarum* inhibits house dust mite–specific T-cell responses. Clin Exp Immunol 126: 2–8, 2001.
45. JM Chatel, P Langella, K Adel-Patient, J Commissaire, JM Wal, G Corthier. Induction of mucosal immune response after intranasal or oral inoculation of mice with *Lactococcus lactis* producing bovine beta-lactoglobulin. Clin Diagn Lab Immunol 8: 545–551, 2001.
46. CBM Maassen, JD Laman, MJ Heijne den Bak-Glashouwer, FJ Tielen, JCPA van Holten-Neelen, L Hoogteijling, C Antonissen, RJ Leer, PH Pouwels, WJA Boersma, DM Shaw. Instruments for oral disease–intervention strategies: Re-

combinant *Lactobacillus casei* expressing tetanus toxin fragment C for vaccination or myelin proteins for oral tolerance induction in multiple sclerosis. Vaccine 17: 2117–2128, 1999.
47. S Drouault, C Juste, P Marteau, P Renault, G Corthier. Oral treatment with *Lactococcus lactis* expressing *Staphylococcus hyicus* lipase enhances lipid digestion in pigs with induced pancreatic insufficiency. Appl Environ Microbiol 68: 3166–3168, 2002.
48. CA Walker, TR Klaenhammer. Leaky *Lactococcus* cultures that externalize enzymes and antigens independently of culture lysis and secretion and export pathways. Appl Environ Microbiol 67: 251–259, 2001.
49. L Steidler. Genetically engineered probiotics. In: Y Bega, ed. Best Practice and Research Clinical Gastroenterology 17:861–876, 2003.
50. J Xu, MK Bjursell, J Himrod, S Deng, LK Carmichael, HC Chiang, LV Hooper, JI Gordon. A genomic view of the human–*Bacteroides thetaiotamicron* symbiosis. Science 299: 2074–2076, 2003.

5

Biotechnological Modification of *Saccharomyces cerevisiae*: Strategies for the Enhancement of Wine Quality

Linda F. Bisson
University of California, Davis, Davis, California, U.S.A.

I. INTRODUCTION

Saccharomyces cerevisiae is a premier research organism, serving as a genetically tractable model system in which to dissect the biological properties of all eukaryotes. Indeed, a battery of biotechnological tools has been developed for the genetic manipulation and analysis of this yeast. *Saccharomyces cerevisiae* is also a major participant in the bioconversion of foods and beverages intended for human consumption. It is viewed as the first true domestic organism, and its uses in the production of bread and fermented beverages predate recorded history.

Throughout the millennia *S. cerevisiae* and its fungal cousins have been the salvation of the human race during periods of disease and famine. This yeast was an important source of micronutrients in many early human diets. Fermented beverages produced by this organism were essential in numerous cultures to assure a safe water supply and to preserve other foods. *S. cerevisiae* is still used today to modify flavor and to boost the nutritional content of many modern foods in addition to filling its traditional roles in bread, beer and wine production. Thus, this organism has had and continues to have a crucial role in the provision of safe foods and beverages for humankind.

It is only natural to ask whether contemporary molecular strategies can be used to engineer *S. cerevisiae* strains better suited for the bioconversion processes in which this organism is involved or that can extend the range of uses of this yeast, not only in the food system, but in the production of enzymes, proteins, and renewable energy sources. What are the risks of release of genetically engineered strains of an organism so central to our food supply? What are the benefits? Do the benefits outweigh any risks? These questions are addressed in this chapter. In particular, the goals, strategies, and risks for the genetic engineering of yeast strains involved in wine production are discussed. The bioconversion of grape juice to wine represents the yeast natural environment and poses many unique challenges to the genetic engineer.

II. PERCEIVED AND REAL RISKS OF GENETIC ENGINEERING

In the physical world, for every action there is a reaction. The same premise holds true for the biological world: For every genetic action there are thousands of possible genetic reactions. It is our inability to predict all possible outcomes of a directed genetic manipulation that is the underlying cause of anxiety over genetic engineering of organisms and their subsequent use in our food system. Will it be possible to predict the genetic responses to genomic modification of *S. cerevisiae*? The answer to this question is yes. The completion of the sequence of the yeast genome has led to the development of tools to monitor global changes in gene expression patterns and protein profiles in response to environmental or genetic manipulation (1–4). The application of these genomic and proteomic tools is leading to an unprecedented understanding of the physiological characteristics and biological responses of this organism (5–8). Well-designed studies will indeed allow the determination of the possible biological consequences of a directed genetic change.

Sophisticated molecular tools that have been developed permit analysis of the extent of lateral transfer of any modified genetic trait across a population of different strains of *S. cerevisiae* and to other yeasts and bacteria with which this organism shares an environmental niche (9). Rapid screening techniques exist and continue to be optimized that negate the need to use and maintain selectable markers to detect desired recombinant strains (10). This eliminates the need to use drug resistance or any heterologous dioxyribonucleic acid (DNA) as a marker in the development of industrial strains of *S. cerevisiae* and thereby prevents the transfer of such genes to other less benign organisms in the environment. These kinds of strategies can and should be developed and used for other organisms as well. Scientists are well aware of the risks imposed by release of drug, pesticide, fungicide, and herbicide resistance genes to a wild population and the potential for lateral gene transfer. Convenience and expediency do not and need not outweigh valid concerns of risk of lateral transfer, especially when those concerns can be addressed experimentally.

Consumers raise another issue, that the process of genetic exchange and mutation is unnatural and therefore undesired. This view is of course not correct. Lateral gene transfer occurs in yeast as in most microbial populations. Indeed, the yeast propensity for incorporating native as well as alien DNA into its genome and the resulting ease of genetic manipulation are the reasons this organism has gained such prominence as an experimental system. The process of sexual reproduction and haploid spore formation in fungi serves the sole purpose of reassortment of the genome. Reassortment also occurs during vegetative growth, though at a lower frequency. Spontaneous mutation frequencies are high in native *Saccharomyces* populations (11–16), suggesting that this organism, like many others, is a natural tinkerer of its own genome. Lateral gene transfer has been documented among different *Saccharomyces* species (9). The role of transposable elements such as the yeast Ty element in host genomic evolution is only now becoming appreciated (12,17, 18). It is imperative that this native tendency to exchange genetic information be anticipated in the design strategy for any engineered organism destined to be released into the food chain. Although *Saccharomyces* is not pathogenic to the general human population, strains causing potentially lethal infections of severely immunocompromised individuals have been identified (19,20). These strains appear to have several different genes that lead to enhanced pathogenesis not found in the native *Saccharomyces* population (21,22). This demonstrates that this organism can indeed obtain and stabilize genes via natural processes from other microbes in an environment, becoming more fit for that environment. The potential to enhance pathogenicity of a normally nonpathogenic organism must also be assessed in light of the ever-growing immunocompromised subpopulation. Thus, *Saccharomyces* engages in frequent genetic exchange and modification of existing traits. Rates of mutation and genomic stability need to be evaluated across the wild or native populations of *Saccharomyces* in order to understand the true potential risks of release of a recombinant organism.

III. STRATEGIES FOR THE IMPROVEMENT OF *SACCHAROMYCES CEREVISIAE* FOR WINE PRODUCTION

Saccharomyces cerevisiae and *S. bayanus* are responsible for the conversion of grape juice into wine. It has been argued that given their long association with humans and the fact that wine has been produced for more than 7000 years, *Saccharomyces* evolved along with the practice of the production of wine to the modern species in existence today (23). Thus anaerobic fermentation of grape juice is the environmental niche to which *Saccharomyces* has adapted over the millennia. Credence for this view arises from the lack of dominance of this organism on the surface of fruit (23–25) and its relatively high level of sensitivity to ultraviolet irradiation and extremes of temperature when compared to those of other yeast residents of fruit surfaces. Heavy inoculation of vineyards with specific *Saccharomyces* strains has demonstrated the poor survivability of this organism in the wild. This trait greatly reduces the risk of environmental contamination by a recombinant *Saccharomyces* strain. However, because of the close association of yeast strains with humans, populations may serve as vectors of transfer of yeast from one production facility to another. In numerous examples in the wine industry the source of a yeast contaminant has been an employee rather than any biologically active materials entering the winery.

There are several properties of yeast that, if modified, would enhance the quality of the wine produced or eliminate processing problems leading to economic losses for producers (Fig. 1). Yeast metabolism is responsible for the appearance of several key organoletpic components of wine. The importance of the yeast contribution depends upon the style of wine to be produced. Yeast generate numerous detectable compounds in wine: higher alcohols and their esters, volatile fatty acids, aldehydes, sulfur-containing volatiles, esters derived from amino acid metabolism, and volatile phenols generated from the enzymatic modification of plant phenolic compounds (10,26,27). Depending upon the wine produced, enhancement or reduction of these activities may be desired.

Researchers are also generating recombinant *Saccharomyces* strains that would facilitate wine processing or the achievement of wine stability. Protein present in wine post fermentation can denature, leading to the formation of a visible haze. Yeast strains optimized for acid protease production can reduce wine protein content, thereby eliminating haze (28). This will only be a viable process if the peptides produced do not confer any unwanted characters, such as bitterness, on the wine. Similarly, wine filterability is improved by degradation of grape polysaccharides. *Saccharomyces* strains with pectinase or cellulase activity that have been constructed offer an alternative to the use of commercial enzyme preparations (28–36). The advantage of a recombinant strain would be the elimination of any potentially undesired side activity that is present in the commercial preparations. However, an enzyme addition may be easier to control and monitor than a recombinant strain.

Finally, yeast strains that possess enzymatic activities enhancing the release of grape flavor and aroma characters are also being developed and tested. These strains express higher levels of native enzymes that impact flavor release, contain mutations of native genes leading to the release of metabolic

Improved Fermentation Performance

- High Ethanol/Acetaldehyde/Acetate Tolerance
- Low/High Temperature Tolerance
- Efficient Utilization of Nitrogen Resources
- Stress Resistance

Enhanced Organoleptic Qualities

- Elimination of Off-Character Production
- Increased Glycerol Production
- Increased Release of Yeast Esters
- Enhanced Release of Flavorants from Glycoside-Glucose Precursors
- Heterologous Expression of Flavorant Biosynthesis Enzymes

Facilitation of Wine Processing

- Elimination of Wine Hazes
- Degradation of Grape Pectins/Polysaccharides
- Regulated Flocculation Ability
- Sulfite Tolerance
- Production of Bacteriocins

Medically Beneficial Modifications

- Elimination of Ethyl Carbamate
- Enhancement of Vitamin Content
- Removal of Pesticide Residues/Mycotoxins/Heavy Metals
- Removal of Inhibitory Phenolic Compounds/Naturally Occurring Plant Carcinogens
- Enhancement of Levels of Bioactive/Health Promoting Antioxidants

Figure 1 Objectives of genetic engineering of *Saccharomyces cerevisiae* for wine production.

intermediates with aromatic character, or produce heterologous enzymes or pathways for the synthesis of grape components (37–41). Whether flavor-modifying strains gain consumer acceptance is another matter. Surveys that have been conducted to date suggest that consumers regard such innovations as correcting problems associated with poor starting quality of fruit or incompetent processing decisions. In this case it is not the fact that the strain is recombinant that is objectionable, but rather why such a strain is needed in the first place.

More important genetic changes are those that will make wine safer for human consumption or improve its nutritional content. An example are our efforts to engineer genetically strains that reduce the appearance of ethyl carbamate in the fermentation environment. Ethyl carbamate is a naturally occurring carcinogen formed spontaneously from the reaction of carbamyl group–containing compounds such as urea and ethanol and found in virtually all fermented foods and beverages (42–44). Although ethyl carbamate has been a part of the human diet for centuries, the current view is that every step possible should be taken to reduce the level of this and other naturally occurring carcinogens in our food supply.

It is also possible to engineer strains that would elevate the nutritional value of wine by increasing the content of vitamins and cofactors in limited supply in a typical unhealthy diet of developed countries. Nutritional enrichment of a dietary staple would be quite beneficial in the reduction of the incidence of disease. Finally, efforts to generate recombinant yeast strains that will increase the levels of important phytochemicals such as resveratrol are under way (45). The aim of this work is to increase the level of bioactive phenolic compounds that have been shown to have positive dietary outcomes in the prevention of disease. Such engineered strains may be of more value to the population as a whole than those leading to subtle differences in wine aroma and flavor or the elimination of economic loss for producers.

A. The *S. cerevisiae* Native Environment

S. cerevisiae can be found in nature on grape surfaces (23–25). Although only a minor member of the flora of the grape surface, this organism dominates the fermentation of crushed grapes. Initial levels of *Saccharomyces* in freshly prepared grape juice are in the range of 1 to 100 cells/mL, vastly outnumbered by other yeasts and bacteria present on the order of 10^6 to 10^8 cfu/mL. By the end of fermentation, these numbers are reversed, and a nearly pure culture of *Saccharomyces* exists (24). *Saccharomyces* is so well adapted to this environmental niche that even when outnumbered millions to one in the initial juice, it rapidly becomes the predominant organism present. Several biological properties are thought to allow the dominance of *Saccharomyces*. Principal among these are the production of and resistance to ethanol. Concentrations of 6% to 7% (v/v) of ethanol are inhibitory to most organisms, wild and commercial *Saccharomyces* strains tolerate concentrations as high as 12% to 17%. Sensitivity to ethanol is enhanced at low pH, and the pH of grape juice is typically between 3.0 and 4.0. Sensitivity is also enhanced at higher temperatures, and rapid fermentation leads to elevation of juice temperature, contributing to the toxicity of ethanol. *Saccharomyces* is also adept at sequestration of micronutrients from the environment. This appears to be a more pronounced trait in other yeasts with which *Saccharomyces* must compete (46). This yeast also releases sulfite at levels inhibitory to bacteria. *Saccharomyces* evolved to resist the inhibitory factors, such as acetic acid, that are the end products of metabolism of other organisms. As a consequence, any genetically engineered organism has to be fully competitive with other microbes.

The grape juice environment is high in osmolarity, ranging from 20% to 28% or higher in sugar concentration, an equimolar mixture of glucose and fructose. *Saccharomyces* tolerates broad changes in medium specific gravity as sugar is consumed and ethanol produced. The environment is most frequently limiting in nitrogen, and functional genomic analyses that we have conducted have shown that commercial strains are efficient nitrogen recyclers (47). Wine strains of *Saccharomyces* rapidly strip grape juice of amino acids; actively growing cells display large vacuoles typically associated only with stationary phase in laboratory strains. In most cases, however, nutrients are in sufficient supply and net yeast cell division ceases as a result of the attainment of terminal cell density. Nongrowing cells of *Saccharomyces* conduct most of the conversion of sugar to ethanol during a typical fermentation (Fig. 2). A sustained rate of fermentation requires the presence of specific nutrients called *survival factors* (48). The survival factors are required for maintenance of fermentative rates at high population densities and in the presence of high

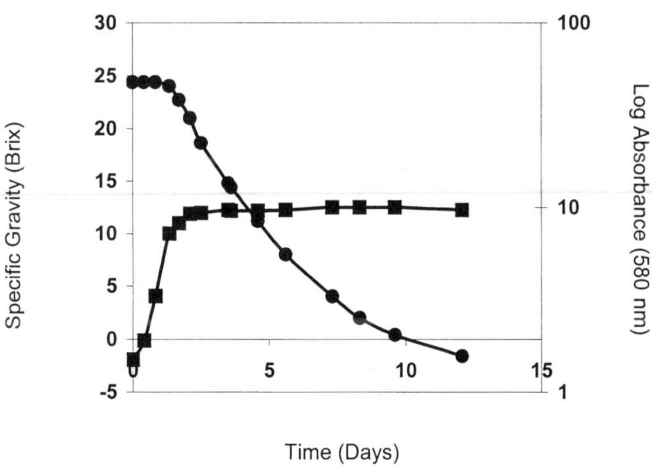

Figure 2 Fermentation of a Chardonnay grape juice by *S. cerevisiae*. (●), specific gravity; (■), log absorbance, 580 nm.

concentrations of ethanol. Most of the survival factors have been associated with the generation of an ethanol-tolerant cell surface and the maintenance of protein function in a high-ethanol environment. Tolerance of ethanol poses great constraints on a biological system. Yeast persists under conditions not conducive to the growth or survival of most organisms. Any genetic alteration that impacts fitness for the environment will be of little practical value.

B. Challenges Impacting Genetic Engineering Strategies of Wine Strains of *Saccharomyces*

The production of wine is a natural fermentative process dependent upon the activity of bacteria and non-*Saccharomyces* yeasts in addition to *Saccharomyces*. Delicate flavors and aromas must be retained and protected at all stages of the process. The limitations of the environment noted and the inability to impact strain fitness pose substantial challenges for the genetic modification of wine strains. Further, since the production of wine does not occur under sterile conditions, any introduced strain may become an established winery resident and undergo genetic exchange with other members of the same species. Thus, any strategy for the genetic construction of a wine strain must consider the risk of lateral gene transfer and cross-contamination of fermentation lots. Changes made must not alter the ability of the strain to dominate the native environment, maintain ethanol tolerance, or complete the fermentation. Several issues in addition to lateral gene transfer must therefore be taken into account in the development of genetically altered wine strains of *Saccharomyces* (Fig. 3).

The persistence of the modified organism in the winery environment is a very crucial issue. The ideal strain would be one that is able to dominate a fermentation when used as an inoculum but not dominate the winery flora. This may be quite difficult to achieve in practice. Persistence in this case means not only presence during the entire course of the fermentation but also the ability to form biofilms and displace other winery resident species. If the modified organism is only desired in a subset of the wines being produced, biofilm establishment can be a serious problem for the winery. As discussed

- Lateral Gene Transfer
- Persistence of Strain
- Dominance of Strain
- Unwanted Side Activities
- Adaptive Secondary Mutations
- Stability of Genetic Change
- Impact on Other Microbes in Environment

Figure 3 Issues in the genetic engineering of wine yeast.

later, several genetic modifications that have already been constructed in wine strains are suitable only for a subset of wines being produced. The risk of problems associated with cross-contamination of other fermentations where these novel biological activities are not desired will prevent the widespread use of these organisms. Success of persistence can be as much of a problem as failure of persistence. The engineered *Saccharomyces* strain will have to be able to compete with native organisms of the same species, as these will be unpreventably present in the fermentation, but not compete with them for the dominance of winery biofilms. More information is clearly needed on the ability of yeast strains to form and displace biofilms to elucidate fully the risks of persistence.

The engineered strain will also need to maintain all of the biological features that permit dominance of the native environment over non- *Saccharomyces* yeasts and bacteria. We have found that a seemingly inconsequential change in expression of a single glucose transporter gene (one of eight expressed during fermentation) that displayed no phenotype in pure culture studies led to the loss of dominance of the altered strain when cocultivated with the lactic acid bacteria commonly associated with grapes (49). Subtle changes in the ability to maintain energy production therefore can have a drastic impact on competitiveness of the strain.

Another critical challenge to the genetic modification of a yeast stain for wine production is the need to be certain that the alteration does not lead to the generation of any unwanted characters in the wine, particularly those that would impart an off-odor or off-flavor or result in the absence of important organoleptic compounds. *Saccharomyces* produces hydrogen sulfide and several other types of objectionable sulfur containing compounds with low thresholds of detection. These odiferous compounds are considered defects or faults in the wine and are undesirable. The production of most of these characters increases with increased nutritional or environmental stress. Any genetic change that directly or indirectly increased the level of production of these compounds would be useless commercially. Further, the biological activities of *Saccharomyces* impact those of the other yeasts and bacteria present in the fermenting medium. It is important that any change in the genetic properties of *Saccharomyces* not lead to off-character production by or otherwise inhibit the other microbes present.

It is also important to understand all of the possible side activities of any heterologous gene used for genetic enhancement of wine strains. Grape juice is a complex mixture of numerous classes of chemical compounds, the majority of which have not been identified. Any novel biochemical activity might lead to undesirable modifications of other wine components that would impact the final characters or their stability in the wine. Finally, if the novel trait poses any selective disadvantage to the yeast there will be strong selective pressure against its maintenance. This may mean the appearance of secondary mutations or simple loss of gene function. Both processes may negate the benefits of the initial change.

The limitations imposed by the process of wine production on the construction of recombinant *Saccharomyces* strains may seem insurmountable. Success will be attained and risks be inconsequential only if the biological characteristics and responses of the organism and other residents of the same environmental niche are well understood and predictable. In the case of wine, while much work remains to be done; the capacity to anticipate the impact of a directed change will be possible once the genomes of the major species have been sequenced. Lessons learned from the biotechnological manipulation of *Saccharomyces* to alter the expression of traits in the natural environment will serve as a paradigm for alteration of the genomes of other important food organisms.

C. Goals and Strategies for the Genetic Engineering of Wine Strains

Reviews in 1996 and 2000 catalogued the many ways in which wine strains of *Saccharomyces* might be modified to eliminate production of off-odors, reduce incidence of arrest of fermentation, improve flavorant properties of wine, or facilitate wine processing (10,50). Therefore, this chapter takes a different approach: analyzing the existing and proposed biotechnological modifications from the perspective of the type of recombinant DNA technology involved. Molecular tools can be used to modify the existing native genomic DNA and thereby affect properties of the yeast or to create truly recombinant organisms expressing nonnative sequences.

There are two broad types of goals for the construction of wine strains of *Saccharomyces*: alteration of existing traits and generation of novel characteristics. The genes used may be naturally occurring in other strains of *Saccharomyces*, may be alterations of a native gene, or may be artificial genes constructed from native genes. Alteration of regulation via promoter exchange and construction of novel enzymes by fusion of functional and regulatory domains of existing genes are examples of constructs from native genetic material. Most consumers would consider alteration of alleles (allele replacement) or generation of a novel allele from native sequences to be fairly benign and of low risk for the food supply. Indeed, given appropriate selective pressure, such changes can be expected to occur in the population naturally. However, this characteristic highlights the hazards of banning a technology altogether rather than its specific uses.

Of greater concern are the cases of heterologous gene transfer, or the incorporation of DNA that encodes a unique metabolic activity from a different or foreign organism into the yeast genome. But here also the source of the DNA needs to be considered. If it is derived from another *Saccharomyces* species or a closely related genus, such a change can be expected to pose a low risk for the use of the recombinant organism, again because such exchanges occur readily in nature. The DNA from more distantly related organisms may present greater risks, but this is dependent upon the specific gene that is being incorporated. Allele replacement, development of engineered alleles, and heterologous gene expression strategies have all been employed or are under active development for the generation of improved wine strains. Examples of the uses of each of these strategies, the risks and benefits, are discussed in the following sections.

1. Allele Replacement

Allele replacement has several uses for the construction of improved wine strains. We are using allele replacement strategies to eliminate the production of two undesired compounds by wine yeasts: ethyl carbamate and hydrogen sulfide. Naturally occurring species of *S. cerevisiae* have that produce little to no ethyl carbamate been identified. Our work has shown that the amount of ethyl carbamate formed by yeast during fermentation is directly correlated with the amount of urea released (51,52). Urea that is not immediately metabolized can be released from the cells into the surrounding medium. The amount of urea released is a function of the relative activities of arginase and urea amidolyase (52). Urea amidolyase is a bifunctional enzyme containing urea carboxylase and allophanate hydrolase activities on a single polypeptide. Strains with a relatively higher ratio of urea amidolyase to arginase degrade rather than release urea to the environment, whereas those with a lower ratio appear to have a limited capacity for urea catabolism and do release this compound (Fig. 4). Allele replacement strategies can be used to adjust the ratio of these two enzymes in any commercial strain, thus reducing the amount of ethyl carbamate formed.

A similar strategy is being employed to define the enzymatic activities leading to a release of hydrogen sulfide. Hydrogen sulfide is made from the reduction of sulfate in the biosynthesis of the sulfur containing amino acids. Low-hydrogen-sulfide–releasing strains appear to have higher activity of the enzymes downstream of sulfite reductase, particularly of *O*-acetylserine *O*-acetylhomoserine sulfhydrolase, that fix reduced sulfur in organic molecules (Fig. 5). The technique of allele replacement can therefore be used to reduce the production level of hydrogen sulfide, by adjusting the ratio of activities of sulfide production and consumption. Allele replacement strategies can be used to address other comparable goals either to increase production of a desirable compound or to reduce production of the undesirable.

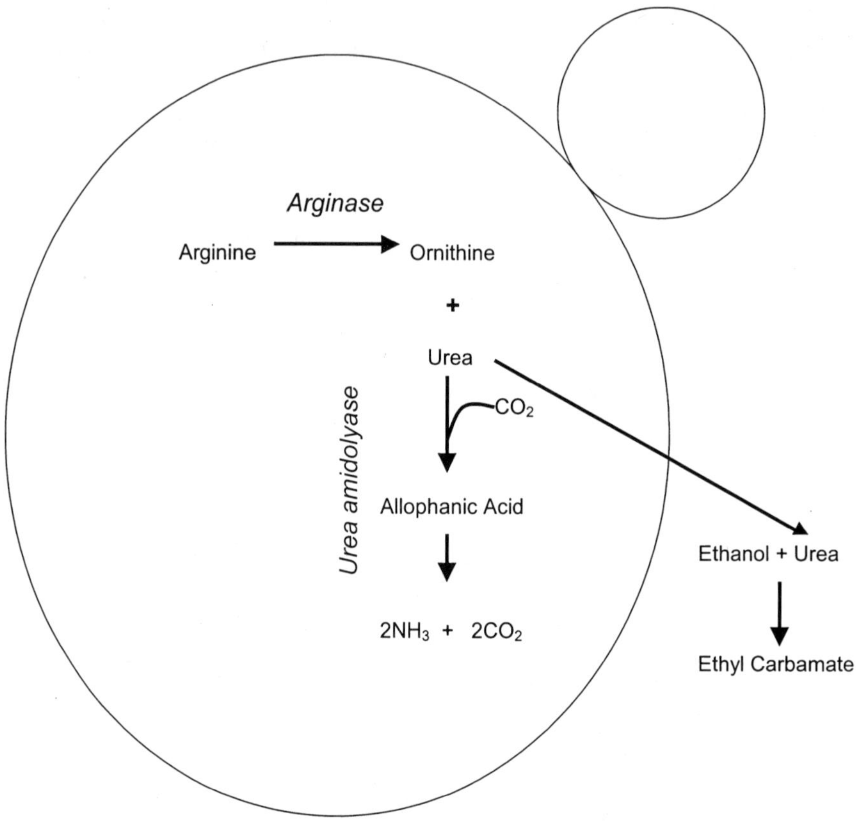

Figure 4 Ethyl carbamate production by yeast strains. Urea produced from the degradation of arginine is excreted from the cells to the medium, whereupon a spontaneous chemical reaction with ethanol occurs, forming ethyl carbamate.

In these cases the biotechnological tools being used result in benign genomic modifications. The specific examples given rely on naturally occurring alleles in the population to eliminate the production of a carcinogen or an objectionable off-odor. Each requires a thorough biochemical understanding of the source of the compound and the regulation of the enzymatic steps involved. The ability to change a single allele or a small set of alleles will have minimal impact on the biological activities of the organism, especially if those changes already occur in nature. Biotechnological modification with the goal of allele replacement can lead to commercial strains with improved properties. Such strains pose little to no risk to the consumer.

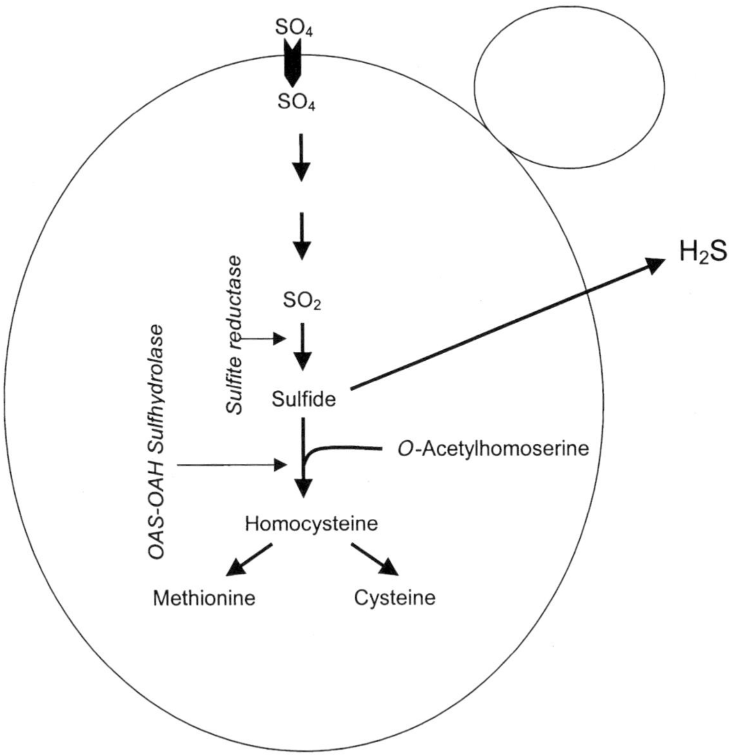

Figure 5 Hydrogen sulfide production by yeast. Sulfide is produced from sulfite via the sulfate reduction pathway. Sulfide not incorporated into homocysteine can be released into the medium.

2. Engineered Alleles

There are also potential benefits of the modification of alleles to forms that currently do not exist in nature or that have not yet been identified. Such changes may alter the regulation of a gene or pathway or modify the coding sequence directly to change the properties of an enzyme or pathway. Since natural processes are being modified, these changes can also be considered relatively harmless to the environment as a whole. Strategies for strain improvement that involve generation of null or knockout mutations are also being employed, but it is important to know whether the loss of function will lead to selection of suppressor mutations in the strain. The appearance of suppressors may not pose an environmental threat but may eliminate the effectiveness of the initial genomic modification. Specific examples of null mutations are elimination of the decarboxylase responsible for a phenolic off-flavor (10,53), mutation of farnesyl-diphosphate synthetase (*ERG20*) to introduce terpene production to wine yeast (54), deletion of *ALD6* (acetaldehyde dehydrogenase) and concomitant reduction of acetic acid production (55). Null mutation of sulfite reductase has been posed as a strategy to eliminate the appearance of hydrogen sulfide (56), but this then requires that the cells be provided with methionine in the medium. Methionine can itself be degraded to objectionable sulfur containing compounds, so this is not a viable strategy for the elimination of off-aromas. Alteration of regulation of nitrogen catabolism by mutation of transcriptional repressors has been shown to improve nitrogen consumption of wine strains (57). This can lead to sustained or elevated fermentation rates under limiting nitrogen conditions.

Swapping a native promoter for one leading to a higher level of expression of the genes involved in glycerol production has also been undertaken (58,59). However, in this case large amounts of undesired components, succinate, acetate, 2,3-butanediol, were produced. The high acetaldehyde level noted during the growth phase led to a reduction of maximal biomass (58,59). It is not clear whether such a strain would be sufficiently competitive with wild strains of *Saccharomyces* and be able to dominate a wine fermentation.

An interesting example of the problems of tailoring a strain for a specific style of wine concerns yeast glycosidase activity. Strains with enhanced activity of this enzyme are being generated for white wine production and the enhancement of wine flavor as a result of release of volatile components from bound sugar moieties (60). At the same time strains with reduced activity of this enzyme are being produced to stabilize wine color losses as the enzymes also catalyze the loss of sugar moieties from anthocyanins (61). However, in the case of Sparkling wine production, color loss is desired (62). If cross-contamination and persistence are problems, wineries attempting to use both strains could suffer from reduced flavor evolution in white wines or excessive color loss in red wines.

As with allele replacement, allele engineering creates no risk other than the problem of cross-contamination of wine production lots. Such cross-contamination, if it negatively impacts the organoleptic character (aroma, flavor, color, or mouth feel) of the wine, may lead to economic losses for the winery but not pose a health or environmental hazard.

3. Engineering Novel Traits

The ease of stable incorporation of heterologous genetic material into the genome of *Saccharomyces* has resulted in tremendous research interest in modifying the metabolic properties of this organism. Indeed, much is being learned about the ability to engineer nonnative biochemical pathways in this yeast (63). What has emerged from these attempts is an appreciation of the difficulty of making a stable change in an organism highly adapted to a specific environment. Yeast strains that possess novel or heterologous traits important in wine production are also being developed.

The earliest attempts to engineer a wine strain genetically focused on generation of a strain stably producing the bacterial enzyme that catalyzes the conversion of malic acid to lactate. Malate and tartrate represent the two major acid species of grape. In cooler growing regions, malate levels may be unacceptably high at the time of harvest, leading to excessively tart wines. The conversion of malate to lactate and the accompanying reduction in the tartness of wines are caused by members of the lactic acid bacteria. If such a reduction in malic acid is desired, proliferation of the lactic acid bacteria must be encouraged. Yeast and the lactic acid bacteria compete for nutrients during fermentation and produce compounds inhibitory toward the other species. It is frequently quite challenging to obtain both alcoholic fermentation and malolactic conversion in wines. The lactic acid bacteria produce a spectrum of organoleptic characters in addition to the reduction in tartness. Some of these traits, such as the cream and buttery notes, are desired in some wines, whereas other traits (the mousy and other animal characters) are not desired. Therefore, the incentive to obtain a strain of *Saccharomyces* capable of reducing malic acid levels exists.

The first approach taken by Williams and associates (64), was the construction of a strain of *Saccharomyces* expressing the bacterial malolactic enzyme (Fig. 6). However, the level of activity obtained was insufficient to lead to significant metabolism of malate. The efficacy of use of the malolactic enzyme from different species of lactic acid bacteria was also evaluated with limited success (65,66). Many of these attempts have led to undesirable side reactions such as increases in the level of acetic acid produced by the yeast. A novel strategy was employed by van Vuuren and colleagues (67,68). The yeast *Schizosaccharomcyes pombe* degrades malate to ethanol in a multistep process. Malate is converted to pyruvate, which is subsequently degraded to ethanol and CO_2 (67). Although *S. cerevisiae* possesses the same pathway, the substrate affinity of the *Saccharomyces* malic enzyme for malate is so low as to preclude efficient conversion of malate to pyruvate under enological conditions. Further, *Saccharomyces* does not possess an efficient malate permease. Instead malate is transported into the cell by passive diffusion. The cytoplasmic level of malate is well below the K_m of the enzyme. The genes encoding the *S. pombe* malate permease (*mae1*) and the malic enzyme (*mae2*) were both used to transform wine strains of *S. cerevisiae,* resulting in more efficient catabolism of malate (67). Strains expressing the bacterial *mleS* gene were also transformed with the *mae1* transporter gene; in much higher efficiency of conversion of malate to lactate by the recombinant yeast resulted

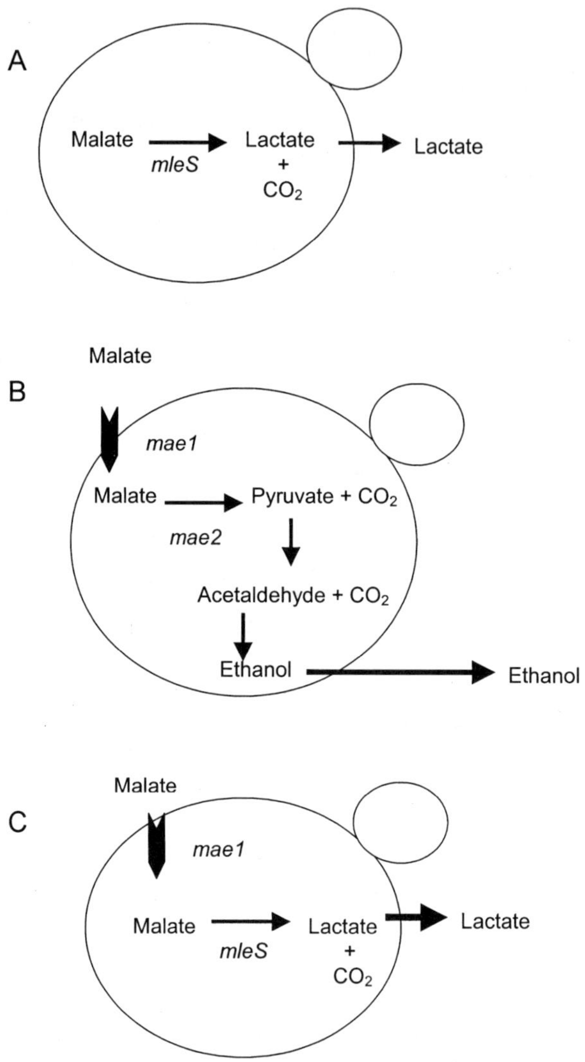

Figure 6 Strategies for the construction of malate utilizing strains of *Saccharomyces cerevisiae*. Panel A, Introduction of the bacterial malolactic (*mleS*) gene, resulting in the production of lactate and CO_2 from malate; panel B, use of the *S. pombe mae1* and *mae2* genes, resulting in the production of ethanol and carbon dioxide from malate; panel C, addition of the *mae1* malate transporter to a strain expressing the *mleS* gene.

(68). Efficient metabolism of malate was only obtained in strains able to transport this compound into the cell efficiently. Although these efforts have been successful in the generation of *Saccharomyces* strains capable of degrading malate, other concerns have been raised. In addition to reducing tartness and sourness, the lactic acid bacteria consume many nutrients left by the yeast that support bacterial growth. The activities and metabolites of the lactic acid bacteria are relatively benign when compared to those of other spoilage organisms. The degradation of malate by yeast may lead to an increased risk of microbial spoilage of the wine.

Wine strains of *Saccharomyces* that produce bacteriocins have also been constructed (69). Bacteriocin production by yeast strains offers many advantages over the current practice of use of sulfur dioxide. Individuals deficient in sulfite oxidase display a strong hypersensitivity to sulfite. In severe cases, exposure can lead to death. The role of sulfite in elimination of unwanted bacteria can be effectively replaced by bacteriocins. In addition, bacteriocins show more selectivity than sulfite and can be used to inhibit specific classes of bacteria, not harming those that are desired in the wine.

Finally, much attention has been paid to the generation of yeast strains possessing glycosidase activity. Many grape aroma compounds are present in the fruit in the form of bound glycoside precursors. The terpenes are a prime example. Unbound or free terpenes impart floral and intense fruity notes to grapes and grape juice. The attachment of a glycoside group prevents volatilization and therefore detection of these compounds. Bound terpenes are hydrolyzed over time, thereby maintaining the aroma characteristics of the fruit. The grape glycosidase is inhibited by glucose and is therefore not active in juice or during the early stages of fermentation. Once fermentation has been completed and the sugar content decreased, these enzymes can slowly hydrolyze the precursor compounds over time. Yeast strains expressing a fungal β(1-4) endoglucanase have been generated (60,70–72). These strains result in the release of more intense fruity character in the wine. However, the use of these strains may impact the long-term aging potential of the wine and lead to the loss of too many characters too quickly.

The traits selected for heterologous expression in wine strains of *Saccharomyces* are likewise expected to be environmentally benign. None of these yeasts is being used commercially largely because of concerns about cross-contamination of fermentations where such activity is not desired. There are also concerns among winemakers of potentially unanticipated side effects of such strains, such as a decrease in aging potential of the wine or impacts on wine microbial stability. Finally, if heterologous enzymes are used, it is important to understand any potential allergenic reactions to the protein that might exist in the population.

IV. CONCLUSIONS

Biotechnological modification of *Saccharomyces* can be used to produce strains with a better combination of native characteristics, with altered traits leading to the enhancement of the nutritional quality of the product or to the removal of naturally occurring carcinogens. The introduction of specific heterologous genes can lead to improved organoleptic properties of the wine or facilitate downstream processing. Lateral gene transfer and mutational change of genomic information are processes that occur naturally in yeast populations that must be taken into account in the risk assessment of any genetic alteration of a population to be released. The risks to human health and well-being of the generation of such modified strains of *Saccharomyces* for food and beverage production are minimal, however. Risks to processing and desired product composition are more real but can be adequately evaluated. Thus, astute biotechnological modification of *Saccharomyces* can indeed result in improved strains for the beverage industries that pose little risk to the consumer and that may provide a more wholesome product.

REFERENCES

1. M Schena, D Shalon, RW Davis, PO Brown. Quantitative monitoring of gene expression patterns with a complementary DNA microarray. Science 270:467–470, 1995.
2. JL DeRisi, VR Iyer, PO Brown. Exploring the metabolic and genetic control of gene expression on a genomic scale. Science 278:680–686, 1997.
3. A Shevchenko, ON Jensen, AV Podtelejnikov, F Sagliocco, M Wilm, O Vorm, P Mortensen, H Boucherie, M. Mann. Linking genome and proteome by mass spectrometry: Large-scale identification of yeast proteins from two-dimensional gels. Proc Natl Acad Sci USA 93:14440–14445, 1996.
4. B Futcher, GI Latter, P Monardo, CS McLaughlin, JI Garrels. A sampling of the yeast proteome. Mol Cell Biol 19:7357–7368, 1999.
5. MB Eisen, PT Spellman, PO Brown, D Botstein. Cluster analysis and display of genome-wide expression patterns. Proc Natl Acad Sci USA 95:14863–14868, 1998.
6. J DeRisi, B van den Hazel, P Marc, E Balzi, P Brown, C Jacq, A Goffeau. Genome microarray analysis of transcriptional activation in multidrug resistance yeast mutants. FEBS Lett 470:156–160, 2000.
7. AP Gasch, PT Spellman, CM Kao, O Carmel-Huvel, MB Eisen, G Storz, D Botstein, PO Brown. Genomic expression programs in the response of yeast cells to environmental changes. Mol Biol Cell 11:4241–4257, 2000.
8. JJ Kang, RM Watson, ME Fisher, R Higuchi, DH Gelfand, MJ Holland. Transcript quantitation in total yeast cellular RNA using kinetic PCR. Nucleic Acids Res 28:1–8, 2000.

9. G Marinoni, M Manuel, RF Peterson, J Hvidtfeldt, P Sulo, J Piskur. Horizontal transfer of genetic material among *Saccharomyces* yeasts. J Bacteriol 181:6488–6496, 1999.
10. IS Pretorius. Tailoring wine yeast for the new millennium: Novel approaches to the ancient art of winemaking. Yeast 16:675–729, 2000.
11. I Miklos, T Varga, A Nagy, M Sipiczki. Genome instability and chromosome rearrangements in a heterothallic wine yeast. J Basic Microbiol 37:345–354, 1997.
12. AC Codon, T Benitez, M Korhola. Chromosomal polymorphism and adaptation to specific industrial environments of *Saccharomyces* strains. Appl Microbiol Biotechnol 49:154–163, 1998.
13. EA Winzeler, DR Richards, AR Conway, AL Goldstein, S Kalman, MJ McCullough, JH McCusker, DA Stevens, L Wodicka, DJ Lockhart, RW Davis. Direct alleleic variation scanning of the yeast genome. Science 281:1194–1197, 1998.
14. S Puig, A Querol, E Barrio, JE Perez-Ortin. Mitotic recombination and genetic changes in *Saccharomyces cerevisiae* during wine fermentation. Appl Environ Microbiol 66:2057–2061, 2000.
15. D Cavalieri, JP Townsend, DL Hartl. Manifold anomalies in gene expression in a vineyard isolate of *Saccharomyces cerevisiae* revealed by DNA microarray analysis. Proc Natl Acad Sci USA 97:12369–12374, 2000.
16. JR Johnston, C Baccari, RK Mortimer. Genotypic characterization of strains of commercial wine yeasts by tetrad analysis. Res Microbiol 151:583–590, 2000.
17. N Rachidi, P Barre, B Blondin. Multiple Ty-mediated chromosomal translocations lead to karyotype changes in a wine strain of *Saccharomyces cerevisiae*. Mol Gen Genet 261:841–850, 1999.
18. MG Kidwell, DR Lisch. Transposable elements and host genome evolution. Trends Ecol Evol 15:95–99, 2000.
19. KV Clemons, JH McCusker, RW Davis, DA Stevens. Comparative pathogenesis of clinical and nonclinical isolates of *Saccharomyces cerevisiae*. J Infect Dis 169:859–867, 1994.
20. JH McCusker, KV Clemons, DA Stevens, RW Davis. *Saccharomyces cerevisiae* virulence phenotype as determined with CD-1 mice is associated with the ability to grow at 42°C and form pseudohyphae. Infect Immun 62:5447–5455, 1994.
21. KV Clemons, P Park, JH McCusker, MJ McCullough, RW Davis, DA Stevens. Application of DNA typing methods and genetic analysis to epidemiology and taxonomy of *Saccharomyces* isolates. J Clin Microbiol 35:1822–1828, 1997.
22. EA Winzeler, B Lee, JH McCusker, RW Davis. Whole genome genetic-typing in yeast using high-density oligonucleotide arrays. Parasitology 18:S73–S80, 1999.
23. A Vaughan, A Martini. Facts, myths and legends on the prime industrial microorganism. J Ind Microbiol 14:514–522, 1995.
24. RB Boulton, VL Singleton, LF Bisson, RE Kunkee. Principles and Practices of Winemaking. 1st ed. New York: Chapman Hall, 1996.
25. R Mortimer, M Polsinelli. On the origins of wine yeast. Res Microbiol. 150:199–204, 1999.
26. MG Lambrechts, IS Pretorius. Yeast and its importance to wine aroma—a review. S Afr J Enol Vitic 21:97–129, 2000.

27. T Tominaga, C Peyrot des Gachons, D Dubordieu. A new type of flavor precursors in *Vitis vinifera* L. cv. Sauvignon blanc: S-cysteine conjugates. J Agric Food Chem 46:5215–5219, 1998.
28. LS Lagace, LF Bisson. Survey of yeast acid proteases for effectiveness of wine haze reduction. Am J Enol Vitic 41:147–155, 1990.
29. B Cantwell, G Brazil, J Hurley, D McConnell. Expression of the gene for the endo-β-1,3-1,4- glucanase from *Bacillus subtilis* in *Saccharomyces cerevisiae* from *CYC1* and *ADH1* promoters. Curr Genet 11:65–70, 1986.
30. E Laing, IS Pretorius. Co-expression of an *Erwinia chrysanthemi* pectate lyase-encoding gene (pelE) and an *Erwinia carotovora* poly-galacturonase-encoding gene (peh1) in *Saccharomyces cerevisiae*. Appl Microbiol Biotechnol 39:181–188, 1993.
31. T Ooi, K Micamiguchi, T Kawaguchi, H Okada, S Murao, M Arai. Expression of the cellulase (FI-CMCase) gene of *Aspergilllus aculeatus* in *Saccharomyces cerevisiae*. Biotechnol Biochem 58:954–956, 1994.
32. MD Templeton, KR Sharrock, JK Bowen, RN Crowhurst, EHA Rikkerink. The pectin lyase encoding gene (pn1) family from *Glomerella cingulata*: Characterization of pn1A and its expression in yeast. Gene 142:141–146, 1994.
33. JM Crous, IS Pretorius, WH Van Zyl. Cloning and expression of an *Aspergillus kawachii* endo 1,4-β-xylanase gene in *Saccharomyces cerevisiae*. Curr Genet 28:467–473, 1995.
34. C Lang, AC Looman, Efficient expression and secretion of *Aspergillus niger* RH5344 polygalacturonase in *Saccharomyces cerevisiae*. Appl Microbiol Biotechnol 44:147–156, 1995.
35. DC LaGrange, IS Pretorius, WH Van Zyl. Expression of the *Trichoderma reesei* β-xylanase gene (*XYN2*) in *Saccharomyces cerevisiae*. Appl Environ Microbiol 62:1036–1044, 1996.
36. M Luttig, IS Pretorius, WH Van Zyl. Cloning of two β-xylanase-encoding genes from *Aspergillus niger* and their expression in *Saccharomyces cerevisiae*. Biotechnol Lett 19:411–415, 1997.
37. L Bardi, C Crivelli, M Marzona. Esterase activity and release of ethyl esters of medium chain fatty acids by *Saccharomyces cerevisiae* during anaerobic growth. Can J Microbiol 44:1171–1176, 1998.
38. K Fukuda, N Yamamoto, Y Kiyokawa, T Yanagiuchi, Y Wakai, K Kitamoto, Y Inoue, A Kimura. Balance of activities of alcohol acetyltransferase and esterase in *Saccharomyces cerevisiae* is important for production of isoamyl acetate. Appl Environ Microbiol 64:4076–4078, 1998.
39. L Bardi, C Cocito, M Marzona. *Saccharomyces cerevisiae* cell fatty acid composition and release during fermentation without aeration and in absence of exogenous lipids. Int J Food Microbiol 47:133–140, 1999.
40. MA Ganga, F Pinaga, S Valles, D Ramon, A Querol. Aroma improving microvinification processes by use of a recombinant yeast strain expressing the *Aspergillus nidulans xlnA* gene. Int J Food Microbiol 47:171–178, 1999.
41. M Lilly, MG Lambrechts, IS Pretorius. Effect of increased yeast alcohol acetyltransferase activity on flavor profiles of wine and distillates. Appl Environ Microbiol 66:744–753, 2000.

42. CS Ough. Ethyl carbamate in fermented beverages and foods. I. Naturally occurring ethyl carbamate. J Agric Food Chem 24:323–327, 1976.
43. CS Ough, EA Crowell, BR Gutlove. Carbamyl compound reactions with ethanol. Am J Enol Vitic 39:139–142, 1988.
44. FF Monteiro, EK Trousdale, LF Bisson. Ethyl carbamate formation in wine: Use of radioactively labeled precursors to demonstrate the involvement of urea. Am J Enol Vitic 40:1–8, 1989.
45. L Gonzalez-Candelas, JV Gil, PR Lamuela-Raventos, D Ramon. The use of transgenic yeast expressing a gene encoding a glycosyl-hydrolase as a tool to increase resveratrol content in wine. Int J Food Microbiol 59:179–183, 2000.
46. M Bataillon, A Rico, JM Sablayrolles, JM Salmon, P Barre. Early thiamine assimilation by yeasts under enological conditions: Impact on alcoholic fermentation kinetics. J Ferm Bioeng 82:145–150, 1996.
47. LE Backhus, J DeRisi, PO Brown, LF Bisson. Functional genomic analysis of a commercial wine strain of *Saccharomyces cerevisiae* under differing nitrogen conditions. FEMS Yeast Res 1:111–125, 2001.
48. S Lafon-Lafourcade, F Larue, P Ribereau-Gayon. Evidence for the existence of "survival factors" as an explanation for some peculiarities of yeast growth especially in grape must of high sugar concentration. Appl Environ Microbiol 38:1069–1073, 1979.
49. AA Heisey. Glucose uptake in *Saccharomyces cerevisiae* grown under vinification conditions: Effect of null mutations in the glucose transporter *HXT1* and *HXT2* structural genes. MS Thesis, University of California, Davis, 1991.
50. A Querol, D Ramon. The application of molecular techniques in wine microbiology. Trends Food Sci Technol 7:73–78, 1996.
51. D An, CS Ough. Urea excretion by wine yeasts as affected by various factors. Am J Enol Vitic 44:35–40, 1993.
52. D An. Urea transport and metabolism by three wine yeast strains of *Saccharomyces cerevisiae*. PhD Thesis, University of California, Davis, 1993.
53. CJ Van Wyk, IM Rogers. A "phenolic" off-odour in white table wines: Causes and methods to diminish its occurrence. S Afr J Enol Vitic 31:52–57, 2000.
54. C Javelot, P Griard, B Colonna-Ceccaldi, B Vladescu. Introduction of terpene-producing ability in a wine strain of *Saccharomyces cerevisiae*. J Biotechnol 21: 239–252, 1991.
55. F Remize, E Andrieu, S Dequin. Engineering of the pyruvate dehydrogenase bypass in *Saccharomyces cerevisiae*: Role of the cytosolic Mg^{2+} and mitochondrial K^{+} acetaldehyde dehydrogenases Ald6p and Ald4p in acetate formation during alcoholic fermentation. Appl Environ Microbiol 66:3151–3159, 2000.
56. V Jiranek, P Langridge, PA Henschke. Determination of sulphite reductase activity and its response to assimilable nitrogen status in a commercial *Saccharomyces cerevisiae* wine yeast. J Appl Bacteriol 81:329–336, 1996.
57. JM Salmon and P Barre. Improvement of nitrogen assimilation and fermentation kinetics under enological conditions by derepression of alternative nitrogen-assimilatory pathways in an industrial *Saccharomyces cerevisiae* strain. Appl Environ Microbiol 64:3831–3837, 1998.
58. S Michnick, JL Roustan, F Remize, P Barre, S Dequin. Modulation of glycerol

and ethanol yields during alcoholic fermentation in *Saccharomyces cerevisiae* strains overexpressed or disrupted for *GPD1* encoding glycerol 3-phosphate deyhdrogenase. Yeast 13:783–793, 1997.

59. F Remize, JL Roustan, JM Sablayrolles, P Barre, S Dequin. Glycerol overproduction by engineered *Saccharomyces cerevisiae* wine yeast strains leads to substantial changes in by-product formation and to stimulation of fermentation rate in stationary phase. Appl Environ Microbiol 65:143–149, 1999.
60. JA Perez-Gonzalez, R Gonzalez, A Querol, J Sendra, D Ramon. Construction of a recombinant wine yeast strain expressing β-(1,4)-endoglucanase and its use in microvinification processes. Appl Environ Microbiol 59:2801–2906, 1993.
61. T Benitez, JM Gasent-Ramirez, F Castrejon, AC Codon. Development of new strains for the food industry. Biotechnol Prog 12:149–163, 1996.
62. JM Barcenilla. Microorganisms in the manufacture of sparkling wines. Aliment Equipos Technol March:73–75, 1993.
63. S Ostergaard, L Olsson, J Nielsen. Metabolic engineering of *Saccharomyces cerevisiae*. Microbiol Mol Biol Rev 64:34–50, 2000.
64. SA Williams, RA Hodges, TL Strike, R Snow, RE Kunkee. Cloning the gene for the malolactic fermentation of wine from *Lactobacillus delbruekii* in *Escherichia coli* and yeasts. Appl Environ Microbiol 47:288–293, 1988.
65. V Ansanay, S Dequin, B Blondin, P Barre. Cloning, sequence and expression of the gene encoding the malolactic enzyme from *Lactococcus lactis*. FEBS Lett 332:74–80, 1993.
66. M Denayrolles, M Aigle, A Lonvaud-Funel. Functional expression in *Saccharomyces cerevisiae* of the *Lactococcus lactis mleS* gene encoding the malolactic enzyme. FEMS Microbiol Lett 125:37–44, 1995.
67. H Volschenk, M Vilijoen, J Grobler, B Petzold, F Bauer, RE Subden, RA Young, A Lonvaud, M Denayrolles, HJJ Van Vuuren. Engineering pathways for malate degradation in *Saccharomyces cerevisiae*. Nat Biotechnol 15:253–257, 1997.
68. H Volschenk, M Vilijoen, J Grobler, F Bauer, A Lonvaud-Funel, M Denayrolles, RE Subden, HJJ Van Vuuren. Malolactic fermentation in grape must by a genetically engineered strain of *Saccharomyces cerevisiae*. Am J Enol Vitic 48: 193–197, 1997.
69. H Schoeman, MA Vivier, M Du Toit, LMT Dicks, IS Pretorius. The development of bactericidal yeast strains by expressing the *Pediococcus acidilactici* pediocin gene (pedA) in *Saccharomyces cerevisiae*. Yeast 15:647–656, 1999.
70. P Van Rensburg, WH Van Zyl, IS Pretorius. Co-expression of a *Phanerochaete chrysosporium* cellobiohydrolase gene and a *Butyrivibrio fibrisolvens* endo-β-1,4-glucanase gene in *Saccharomyces cerevisiae*. Curr Genet 30:246–250, 1996.
71. P Van Rensburg, WH Van Zyl, IS Pretorius. Over-expression of the Saccharomyces cerevisiae exo-β-1,3-glucanase gene together with the Bacillus subtilis endo-β-1,3-1,4 glucanase gene and the *Butyrivibrio fibrisolvens* endo-β-1,4-glucanase gene in yeast. J Biotechnol 55:43–53, 1997.
72. P Van Rensburg, WH Van Zyl, IS Pretorius. Engineering yeast for efficient cellulose degradation. Yeast 14:67–76, 1998.

6

Prebiotics from Lactose, Sucrose, Starch, and Plant Polysaccharides

Martin J. Playne
Melbourne Biotechnology, Hampton, and RMIT University, Melbourne, Victoria, Australia

Ross G. Crittenden
Food Science Australia, Werribee, Victoria, Australia

I. INTRODUCTION

The human gastrointestinal tract contains a complex ecosystem of microorganisms with more than 400 different bacterial species, and potentially thousands of different strains. These bacteria are highly important to our health. They provide us with a barrier to infection by exogenous bacteria (1) and much of the metabolic fuel for our colonic epithelial cells (2) and contribute to normal immune function (3). Disturbances to this ecosystem leave us more vulnerable to exogenous and endogenous intestinal infections. Intestinal bacteria have also been implicated in causation of some chronic diseases of the gut such as Crohn's disease and ulcerative colitis (4,5). As we age, changes occur in the composition of the intestinal microbiota that may contribute to an increased level of undesirable microbial metabolic activity and subsequent degenerative diseases of the intestinal tract (6,7). Manipulating the intestinal microbiota to restore or maintain a beneficial population of microorganisms is a reasonable approach to maintaining intestinal health in cases when a deleterious or less than optimal population of microorganisms has colonized the gut.

In this chapter, we consider the role that oligosaccharides and polysaccharides play as prebiotics and as ingredients in functional foods. We also assess their ability to modulate gut microbial populations and activity, and their ability to improve human health. Fructooligosaccharides and inulin are reviewed in other chapters, as are human milk oligosaccharides.

A. Probiotics

> Probiotics are microbial cell preparations that have a beneficial effect on the health and well-being of the host, by modulating mucosal and systemic immunity, as well as improving the nutritional and microbial balance in the intestinal tract (104).

Since being proposed as a health promoting technique by Metchnikoff in 1907 (8), the oldest and still most widely used method to increase the numbers of advantageous bacteria in the intestinal tract is direct consumption of these live bacteria in foods. These bacteria, called *probiotics* (9), have to date been predominantly selected from the genera *Lactobacillus* and *Bifidobacterium*, both of which form part of the normal human intestinal microbiota. In the probiotic approach, the ingested bacteria are selected to survive gastrointestinal transit and to contribute positively to the activity of the intestinal microbiota. Proposed mechanisms of health action include increasing colonization resistance to exogenous and endogenous intestinal pathogens, reducing putrefactive and genotoxic intestinal reactions, modulating the innate immune system, contributing to lactose hydrolysis, and producing short-chain fatty acids (SCFAs) and micronutrients such as vitamins.

B. Prebiotics

> Prebiotics are nondigestible food ingredients that beneficially affect the host by selectively stimulating the growth and/or activity of one or a number of bacteria in the colon, thus improving health (12).

These ingredients are normally restricted to certain carbohydrates (particularly oligosaccharides) but could include certain proteins, peptides, and lipids. Oligosaccharides are defined as glycosides composed of between 3 and 10 monomer sugar units. Nondigestible oligosaccharides (NDOs) can be distinguished from other carbohydrates on the basis of being resistant to digestion in the stomach and small intestine. Not all oligosaccharides are NDOs.

Prebiotics represent a second strategy to improve the balance of intestinal bacteria. Rather than introducing exogenous strains into an individual's intestinal tract, prebiotics aim to selectively stimulate the proliferation and/or activity of advantageous groups of bacteria already present in the intestinal microbiota. To date, prebiotics have primarily been oligosaccharides and other indigestible carbohydrates that increase the population of strains of *Bifidobacterium* spp. in humans and animals. Hence, they are often referred to as *bifidogenic* or *bifidus factors*. The mechanism(s) by which prebiotics promote the relatively specific proliferation of bifidobacteria among the other major genera of intestinal bacteria remains speculative. It is possibly due to the relatively efficient utilization of these carbohydrates as carbon and energy sources by bifidobacteria and tolerance of these bacteria to the SCFAs and acidification resulting from fermentation. Inulin and fructooligosaccharides remain the most intensively studied prebiotics. However, there are a range of other indigestible di-, oligo-, and polysaccharides that have been shown to have potential to act as prebiotics, and on which this chapter focuses.

The prebiotic approach offers some advantages over the probiotic strategy. Technologically, prebiotics are stable and can be incorporated into a wider range of foods and beverages with longer shelf lives than probiotics. They also promote the proliferation of indigenous microbiota, therefore eliminating concerns about host specificity and colonization by exogenous strains. Studies by Tannock and coworkers (10,11) have indicated that indi viduals may carry their own unique complement of dominant strains of bifidobacteria. Prebiotics may therefore overcome the difficulty facing exogenous probiotics of trying to persist in an environment in which there is already an established microbiota and possibly host factors working against them. Prebiotics also modify the activity of the microbiota by providing a source of readily fermentable carbohydrate. Indeed, it may be this dietary fiber–like characteristic of modifying the fermentative activity of the existing microbiota that is the important factor in providing a number of health benefits to consumers.

C. Synbiotics

> A synbiotic is a mixture of a probiotic bacterium and a prebiotic carbohydrate that beneficially affects the host by improving the survival and persistence of live microbial dietary supplements in the gut, by selectively stimulating the growth and or metabolic activity of one or a limited number of health-promoting bacteria, but especially the added probiotic bacterial strain or strains.

There is an obvious potential for synergy between prebiotics and probiotics. Hence, foods containing both prebiotic and probiotic ingredients have been termed *synbiotics* (12). The prebiotics used in synbiotic combinations are often currently selected on the basis of food technology issues and may not in all cases be the optimal complement for the probiotic strains. However, if synergy between the probiotic and prebiotic ingredients is to be exploited, the probiotic strain should be able to utilize the prebiotic as a carbon substrate in the gut efficiently. Not all probiotic species and strains can utilize fructans as a carbon and energy source. Other nondigestible carbohydrates may be the most appropriate choice to include with many probiotics, to provide the best synergistic combination to benefit the probiotic's survival and/or functionality in the gastrointestine. Since it is unlikely a particular prebiotic carbohydrate will be fermented solely by the added probiotic strain in the intestinal tract, it is desirable that the prebiotic selectively promote the growth and activity only of beneficial populations within the intestinal microbiota.

Synbiotic technology is relatively new and continues to evolve (Fig. 1). Ultimately, manufacturers want prebiotics that can provide a health benefit to the consumer and complement their probiotic strains in the product as well as in the gastrointestinal tract by promoting stability, viability, and functionality. The prebiotic should also add to the taste and physicochemical properties of food products (14) (see Sec. VIII).

D. Why Induce Higher Numbers of Bifidobacteria and Lactobacilli in the Intestinal Microbiota?

The answer to the question of why to induce higher numbers of bifidobacteria and lactobacilli in the intestinal microbiota is still largely theoretical since studies conclusively linking naturally high levels of intestinal bifidobacteria or lactobacilli with improved health are yet to be published. The idea that increasing the number of these particular genera in the intestinal tract might be beneficial to health stems largely from studies of the changes in the composition of the intestinal microbiota with age. Bifidobacteria dominate

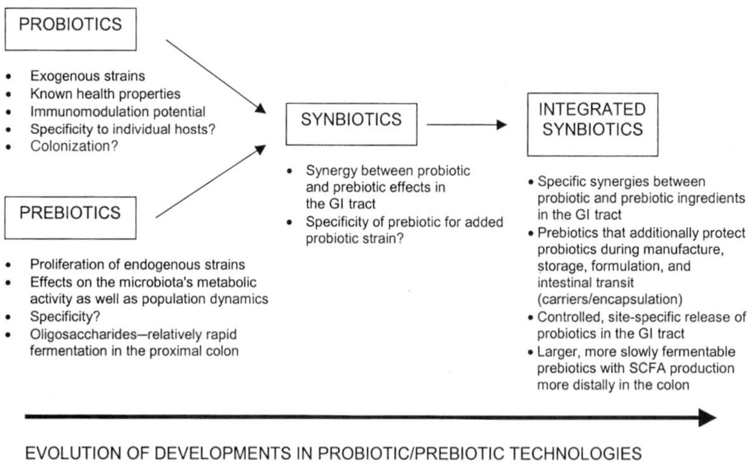

Figure 1 Evolution of the development of synbiotics. GI, gastrointestinal; SCFA, short-chain fatty acid.

the microbiota of breast-fed infants and, after remaining relatively stable as one of the dominant intestinal genera throughout adulthood, reduce in numbers in the elderly (6,13). In the elderly, the decrease in bifidobacterial numbers is concurrent with an increase in potentially putrefactive organisms such as clostridia (6). Intestinal lactobacilli and bifidobacteria are nonpathogenic, nonputrefactive, nontoxigenic, saccharolytic organisms that appear from available knowledge to provide little opportunity for deleterious activity in the intestinal tract. As such, they are reasonable candidates to target in terms of restoring a favourable balance of intestinal species in cases when the intestinal microbiota of the host contains a high proportion of potentially harmful bacteria. The safe use of these bacteria as probiotics and emerging evidence of benefits from consumption of probiotic lactobacilli and bifidobacteria add weight to the idea of enhancing these populations in the intestinal tract by using prebiotics. However, little is known of the role of many genera and species residing within the intestinal tract in health and disease, or how the microbiota composition and activity change with age, lifestyle, and culture, and how the changes influence human health. Other genera may also be important targets for prebiotics, in terms of both promoting beneficial bacteria and suppressing populations and activity of harmful bacteria. Further studies of colonic microecology and interactions between the microbiota and the host in health and disease are essential prerequisites to designing appropriate prebiotic strategies.

II. TYPES OF NEW PREBIOTICS AND THEIR BIFIDOGENIC EFFECTS

In terms of prebiotic effects on the composition of the colonic microbiota, studies of inulin and fructooligosaccharides (108) currently dominate the scientific literature. A number of studies have demonstrated that these fructans can increase the proportion of bifidobacteria in the fecal flora. On the strength of these studies, inulin and fructooligosaccharides are now the most commonly used prebiotics in functional foods in Europe, Japan, and Australia (14). However, fructans are not the only carbohydrates that have attracted the interest of researchers studying the effects of dietary components on the intestinal microbiota's activity and composition. A number of other oligo- and polysaccharides have been reported to act as bifidogenic factors and have been produced and marketed commercially for this purpose since 1985 (15). These include lactulose, galacto-, xylo-, isomalto-, and soybean oligosaccharides; lactosucrose; and a range of indigestible cereal polysaccharides and resistant starches. Carbohydrates (other than fructooligosaccharides) marketed for their bifidogenic potential are listed in Table 1. In most cases, the amount of published research is limited thus far to in vitro or animal studies, or small and limited human pilot trials. In many cases, too, feeding trials have been conducted without adequate monitoring of fluctuations in the baseline levels of microbial populations before intervention. However, all these carbohydrates have shown at least some potential to act as prebiotics that may be borne out in future, well-designed human feeding studies.

A. Galactooligosaccharides

Next to fructooligosaccharides and inulin, galactooligosaccharides are perhaps the most studied NDOs for prebiotic effects of the commercially available oligosaccharides. Galactooligosaccharides are fermented by a range of human *Bifidobacterium* species that utilize these sugars relatively well compared to many other intestinal bacteria when grown in vitro (16,17). However, in terms of eliciting significant bifidogenic effects in humans, the results of feeding trials have been mixed. Some trials have indicated that galactooligosaccharides can induce increases in fecal *Bifidobacterium* sp. levels (18–20), whereas others have observed no significant bifidogenic effect (21–23). This difference in the observed responses between the trials was not due to a dose effect since relatively low levels of galactooligosaccharides (2.5 g/day) were used in the trial by Ito and associates (1993) (19), whereas doses up to 15 g/day produced no significant response in the trial of Alles and colleagues (1999) (22). A range of factors may influence the size of any increase in *Bifidobacterium* sp. numbers, one of which is the initial size of the population. In comparing different trials conducted using fructooligosaccharides, Rao (1999) (24) observed that the size of the bifidogenic response was inversely proportional to the size of the initial *Bifidobacterium* sp. population rather than there being a strong dose response. Prebiotics produce bifidogenic responses only in individuals with relatively low initial populations of these bacteria, where there is room for an effect to occur. In the case of galactooligosaccharides, a consistently strong bifidogenic response in humans remains to be demonstrated.

B. Soybean Oligosaccharides

Composed mainly of raffinose and stachyose, soybean oligosaccharides are fermented well and relatively selectively by many human species of bifidobacteria (25). Soybean oligosaccharides have been trialed in a number of small human feeding studies that have demonstrated increases in fecal bifidobacterial populations (26–29). Hence, they are promising bifidogenic candidates that warrant further investigation to determine their impact on the wider intestinal microbiota population in well-designed human feeding trials.

Table 1 Typical Compositions of Commercial Oligosaccharides[a]

Soybean oligosaccharide	
SOYAOLIGO	Stachyose 24%; raffinose 8%; sucrose 39%, glucose/fructose 16%; other saccharides 13%
[Calpis Food Industry, Japan]	
Galactooligosaccharide	
CUP OLIGO	Minimum 70% of solids as galactooligosaccharides (available as syrup and powder) type: 4′-galactosyl lactose
[Nissin Sugar, Japan]	
TOS 85 POWDER	85% Galactooligosaccharides; 0.3% galactose; 3.3% glucose; 10.7% lactose
TOS Elix'or SYRUP	58.5% galactooligosaccharides of solids, 0.12% galactose; 19.7% glucose; 20.65 lactose; dry weight 73.9%
[Borculo-Domo, Netherlands]	
OLIGOMATE 55 SYRUP	>55% Galactooligosaccharides of solids; <45% monosaccharides and lactose; 75% solids
OLIGOMATE 55 POWDER	>55% Galactooligosaccharides; <45% monosaccharides and lactose
TOS pure POWDER [Yakult Honsha, Japan]	99% Galactooligosaccharides (<5% moisture)
Lactosucrose	
NYUKA ORIGO LS-40L	43% Lactosucrose of solids; 42% sucrose; 6% lactose; 3% monosaccharides; 6% other oligosaccharides
NYUKA ORIGO LS-55L	58% Lactosucrose of solids; 23% sucrose; 8% lactose; 3% monosaccharides; 8% other oligosaccharides
NYUKA ORIGO LS-55P	57% Lactosucrose; 7% sucrose; 22% lactose; 4% monosaccharides; other oligosaccharides 10%; <5% moisture
PET-OLIGO L55	45% Lactofructose;[b] 10% sucrose; 10% lactose; 3% monosaccharides; 6.5% other oligosaccharides; 24% moisture
PET-OLIGO P55	57% lactofructose;[b] 7% sucrose; 20% lactose; 3% monosaccharide; 7% other oligosaccharide
NEWKA OLIGO LS-35 [Ensuiko Sugar Refining Company, Japan and Hayashibara Shoji, Japan]	>35% Lactosucrose of solids; moisture <28%

Table 1 Continued

Lactulose	
CONCENTRATE (Duphalac, Bifiteral, Chronulac, Cephalac)	67% Lactulose; <6% lactose; <11% galactose; <5% epilactose; <1.4% tagatose; <0.7% fructose
CRYSTALLINE FORM	>95% Lactulose; < 2% lactose; <2.5% galactose; <3% tagatose; <1.5% epilactose
[Solvay, Germany]	
MLS-50 SYRUP	52% Lactulose; 19% mono- and disaccharides; 31% moisture
MLS-40 POWDER	41% Lactulose
MLC-A CRYSTALLINE ANHYDRATE	98% Lactulose
MLC-H CRYSTALLINE HYDRATE	86% Lactulose
[Morinaga, Japan]	
Xylooligosaccharide	
XYLO-OLIGO 95 POWDER	>95% Oligosaccharide
XYLO-OLIGO 70 SYRUP	>70% Oligosaccharide; <30% monosaccharide; 25% moisture
XYLO-OLIGO 35 POWDER	>35% Oligosaccharide
XYLO-OLIGO 20 POWDER	>20% Oligosaccharide
[Suntory Ltd, Japan]	
Isomaltooligosaccharide	
ISOMALTO-500 SYRUP	<25% Moisture; 40% monosaccharides; 29% disaccharide;18% trisaccharide; 13% tetrasaccharide plus (>50% isomaltooligosaccharides)
ISOMALTO-900 SYRUP	<25% Moisture; 4% monosaccharides; 47% disaccharide; 27% trisaccharide; 22% tetrasaccharide plus (>85% IMOs)
ISOMALTO-900 POWDER	<5% Moisture; saccharides similar to the ISOMALTO-900 syrup
[Showa Sangyo Co Ltd, Japan]	
PANORUP SYRUP	<25% Moisture; 4% monosaccharides; 47% disaccharide; 27% trisaccharide; 22% tetrasaccharide plus (>85% IMOs)
[Hayashibara Group, Japan]	
BIOTOSE 50 SYRUP	<25% Moisture; 41% glucose; 7% maltose; 20% isomaltose; 9% panose; 9% isomaltotriose; 14% others

Table 1 Continued

PANORICH SYRUP	<26% Moisture; 23% glucose; 17% maltose; 9% isomaltose; 30% panose; 3% maltotriose; 2% isomaltotriose; 16% branched others
[Nihon Shokuhin Kako Ltd, Japan]	
GENTOSE 50 SYRUP	<30% Moisture; 2% fructose; 51% glucose; 30% gentiobiose; 11% gentiotriose; 5% gentiotetraose
GENTOSE 80 SYRUP	<28% Moisture; 2% fructose; 6% glucose; 51% gentiobiose; 28% gentiotriose; 14% gentiotetraose
GENTOSE 80 POWDER	<5% Moisture; <5% saccharides
[Nihon Shokuhin Kako Ltd, Japan]	
PALATINOSE/ MALTULOSE	>90% Gentio-oligosaccharides
ICP	100% Palatinose powder
ICO	45% Palatinose powder; 45% palatin-oligosaccharides; 10% saccharides
[Mitsui Sugar Co, Japan]	
IOS	Mixed syrup

[a] Saccharides are expressed as percentage of total anhydrous saccharides present in all cases.
[b] Lactofructose is structurally identical to lactosucrose.

C. Xylooligosaccharides

Xylooligosaccharides (XOS) are also fermented well by many bifidobacteria (30–33) and have been proposed as potential prebiotics. Suntory Ltd, who are the main producers of xylooligosaccharide products, claim that xylooligosaccharides are very stable at high temperatures and in acidic conditions compared to other oligosaccharides. They also found that their effect on the intestinal flora is three to four times greater, and consequently lower doses can be used. In feeding trials in rodents (31) and in a small linear uncontrolled trial in humans (30), feeding of XOS appeared to increase the numbers of bifidobacteria in feces relatively selectively. Larger and more rigorously controlled crossover trials are required to confirm whether xylooligosaccharides act as selective prebiotics in humans.

D. Isomaltooligosaccharides

Unlike the other NDOs discussed, isomaltooligosaccharides are partially digested and absorbed in the small intestine. However, breath hydrogen tests indicate that a proportion of isomaltooligosaccharides fed to adult humans (25g dose) remains available for fermentation by intestinal bacteria (34). Isomaltooligosaccharides are well fermented by bifidobacteria, but also by many species from other major intestinal genera, including bacteroides and clostridia (35). In three sequential human feeding trials, isomaltooligosaccharide consumption resulted in small increases in fecal *Bifidobacterium* sp. numbers at doses of 10–20 g/day (35–37). In the two trials that examined effects on other groups of intestinal bacteria (35,36), the bifidogenic effect appeared to be relatively selective. However, further controlled human "crossover" studies are required to confirm the prebiotic nature of isomaltooligosaccharides.

E. Lactulose

Lactulose was first recognized to be bifidogenic in 1957 (38). It was not until 1971 that other oligosaccharides were also found to be able to promote the growth of bifidobacteria. Lactulose has been shown to be a relatively selective bifidogenic factor in two studies using healthy adult subjects (39,40) and promoted the development of a microbiota rich in bifidobacteria in a controlled study of an infant milk formula containing this disaccharide (41). Hence, lactulose has found commercial applications as a bifidogenic factor in infant milk formulas and foods in addition to its use as a therapeutic to relieve chronic constipation and hepatic encephalopathy (42).

F. Lactosucrose

Together with lactulose, and fructo-, soybean, isomalto-, and galactooligosaccharides, lactosucrose (β-D-galactopyranosyl-(1→4)-α-D-glucopyranosyl β-D-fructofuranoside) is recognized by the Japanese Foods for Specified Health Use (FOSHU) regulatory system as a bifidogenic factor. This oligosaccharide is sold in Japan in a range of drinks, confectionaries, and pet foods and as a table sugar. Bifidobacteria ferment lactosucrose well in vitro relative to most enterobacteriaceae (43) and have been observed to increase the size of the fecal *Bifidobacterium* sp. population in small trials on healthy adults (44–46) and patients with inflammatory bowel disease (47). As with the other oligosaccharides described in this chapter, lactosucrose shows good potential as a prebiotic that may be borne out in further human studies.

III. EMERGING PREBIOTICS

In addition to lactulose and the oligosaccharides discussed thus far, a number of other saccharides have been investigated for their effects on the human intestinal microbiota. These include lactitol, which produced similar effects to lactulose but was slightly less potent (40), and glucono-δ-lactone, which induced a selective increase in fecal *Bifidobacterium* sp. numbers in humans at a dosage of 3 or 9 g/day in a small sequential study (48). The prebiotic effects of lactitol were summarized in 2000 (49). Depolymerized pyrodextrin (50), partially hydrolysed guar gum (51), palatinose polycondensates (52), and α-glucooligosaccharides (53) have also been demonstrated to have potential as bifidogenic factors, although all require further research to confirm these effects.

Bacterial exopolysaccharides have to date remained largely unexplored as a source of prebiotic substrates. A polysaccharide produced by a strain of *Streptococcus macedonicus* was reported in 2001 to possess a structure with repeated oligosaccharide units that form the backbone to human milk oligosaccharides (54) and could potentially find applications in infant milk formulas. In the future, transglycosylation activity of glycosidases from specific probiotic bacterial strains might be exploited to produce oligosaccharides for specific synbiotic combinations.

A. Cereal Polysaccharides

The effect of cereal polysaccharides on microbial population dynamics in the intestinal tract has not been intensively studied, but a few reports have indicated that they may have potential to act as prebiotics. Arabinoxylans and β-glucans are the major indigestible cereal fiber constituents that are fermentable by bacteria in the human gastrointestinal tract. Cereal β-glucans are not fermented by most lactobacilli and bifidobacteria (55), but β-gluco-oligosaccharides produced by controlled hydrolysis of cereal β-glucans can be utilized by some bifidobacteria and lactobacilli (33,56), although studies of their selectivity and prebiotic effect in vivo have not yet been reported. Arabinoxylan and arabinoxylooligosaccharides appear to be good candidates for further study to assess their prebiotic potential. In vitro, these carbohydrates are relatively selectively fermented by *Bifidobacterium longum* and *Bifidobacterium adolescentis* and are poorly fermented by other intestinal species, including bacteroides, enterococci, *Clostridium difficile*, *Clostridium perfringens*, and *Escherichia coli* (31,33,55).

B. Resistant Starches as Prebiotics

In recent years, it has been recognized that resistant starches have the potential to act as prebiotics, enhancing counts of bifidobacteria in fecal samples in animal studies (57–61). Additionally, resistant starch can provide protection for probiotic bacteria during transit to the colon (60) and may act as an "adhesion site" in the colon for the bacteria (62). This physical association between probiotic bacterium and prebiotic substrate may be able to be exploited for strain-specific synbiotics. Furthermore, resistant starches can be consumed in greater daily quantities (e.g., 40 g/day/adult human) without bowel distension by gas production or laxative effects.

Resistant starches differ from NDOs in the rate of colonic fermentation. They show a more sustained period of fermentation, whereas most NDOs are rapidly fermented. Sustained fermentation results in colonic fermentation activity's extending into the transverse and distal segments of the colon, rather than being largely restricted to the proximal colon. There is, therefore, a theoretical advantage in supplying the host with a combination of short-chain oligosaccharides with a longer-chain complex carbohydrate, such as inulin or resistant starch, to gain maximal prebiotic effects including suppression of protein metabolism in the distal colon. Preliminary data in pigs from our research indicate that the total pool of colonic bifidobacteria increases with diets containing both resistant starch and fructooligosaccharide (58).

IV. SUSTAINED DOSAGE AND "WASHOUT" EFFECTS

The ability of sustained prebiotic intake to minimize the frequency of dosage of a probiotic bacteria needed to maintain a desired colonic population is not well documented. The commercial culture supplier's view is usually that daily dosage of a probiotic is necessary to maintain adequate viable numbers in the colon. A few experiments have examined "washout" of probiotics after cessation of daily dosage (63–65). Probiotic strains usually persist in high numbers in the intestinal tract for little longer than the normal intestinal transit period. We have examined such effects in pigs and mice, in the presence and absence of continued dosage of an oligosaccharide (Oligofructose) and a resistant starch (high-amylose HiMaize) during the "washout" period. The studies were conducted using *Bifidobacterium animalis* CSCC 1941 as the dosed probiotic strain. Continued ingestion of the prebiotics after the cessation of probiotic dosage increased the time in which the *Bifidobacterium* spp. persisted in high numbers in the intestinal tract by several days. This result demonstrates that synbiotic effects are possible and can promote enhanced functionality (in this case, persistence) of complementary probiotic strains.

V. HEALTH EFFECTS OF PREBIOTICS

Studies on the health effects of prebiotics are in their infancy, and most of the proposed benefits remain to be thoroughly studied, in terms of both understanding mechanisms of action and demonstrating clinical efficacy. Although

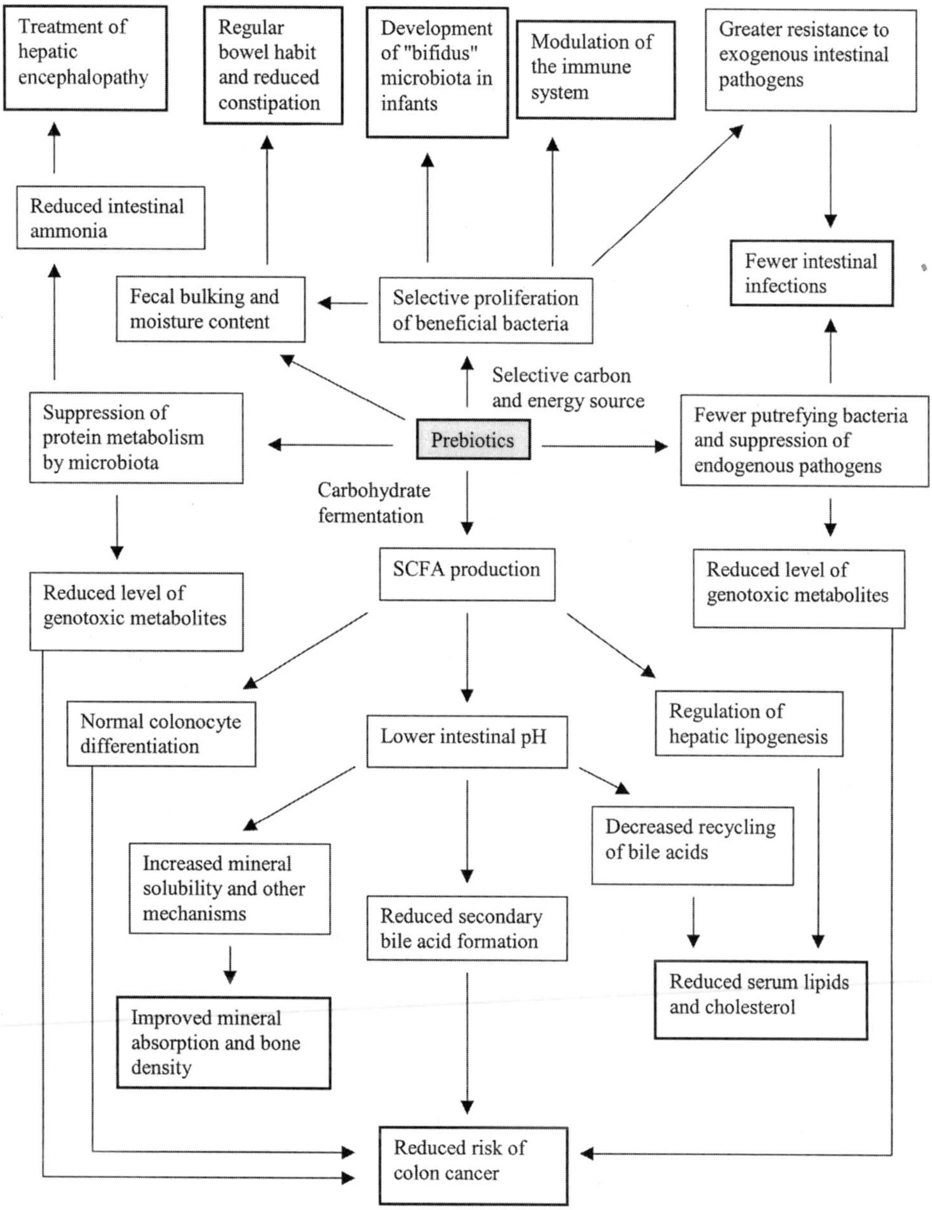

Figure 2 Potential health benefits proposed for prebiotics. SCFA, short-chain fatty acid.

largely conceptual at this early stage of development and our understanding of diet–microbiota–host interactions, there are a range of benefits to human health that could potentially be supplied by prebiotics (Fig. 2).

A. Management of the Intestinal Microbiota: Improvement of the Gut Barrier

Aiding the development of a "bifidus" intestinal microbiota in milk formula–fed infants to mimic microbiota development in breast milk–fed infants is one application for prebiotics that has already gained medical and commercial acceptance. Rapidly restoring a "healthy" microbiota in individuals with a perturbed intestinal flora to prevent exogenous and endogenous intestinal infections is another potential application that shows promise (66,67). Prebiotics may be especially applicable to the elderly, in whom *Bifidobacterium* sp. numbers decrease and are replaced by higher levels of putrefying bacteria. Use of prebiotics to treat inflammatory bowel diseases and irritable bowel syndrome through improvement of intestinal microbiota composition and activity has also been proposed (47,68,69) but not yet intensively studied. Through their action in fecal bulking, prebiotic oligosaccharides are effective in relieving constipation and maintaining normal stool frequency (70). Immunomodulation effects, as observed for probiotics (71–74), have thus far been scarcely investigated for prebiotics. Prebiotics may have applications in the prevention of food allergy in infants, as observed for probiotics (74), by contributing to the development of a "normal" microbiota.

Bifidogenic effects are the most discernable impact of prebiotics. Consequently, even in the absence of synbiotic treatments where an exogenous probiotic *Bifidobacterium* sp. strain is added together with a prebiotic, the results from any prebiotic addition will always be confounded in a physiological or health sense by the increases in numbers of resident bifidobacteria. Thus, the physiological and health effects shown by prebiotics will often overlap those found for bifidobacteria.

B. Prevention of Colon Cancer

Since they supply a source of fermentable carbohydrate to the colon, dietary fiber–like, anticarcinogenic effects have been proposed for prebiotics. Proposed mechanisms include the supply of the colonic epithelium with SCFA, particularly butyrate, and suppression of microbial protein metabolism, bile acid conversion, and other toxigenic bacterial reactions. A number of rodent studies [summarized by van Loo and associates 1999 (70)] have demonstrated reduced cancer risk in animals consuming prebiotics. However, in these studies, the animals often consumed oligosaccharide doses far in excess of those tolerable by humans (in some cases greater than 10% of the diet). Additionally, the rapid intestinal transit time in rodents compared to that in humans means that effects observed with rapidly fermented oligosaccharides in these models may not translate well to the distal colon and rectum in humans, the primary sites for colon carcinomas. Prebiotics that can supply a sustained source of fermentable carbohydrate to intestinal microbiota may prove the most effective in reducing the risk of colon cancer by extending SCFA production and suppression of toxigenic bacterial reactions through to the distal colon. Larger polysaccharides would have the added advantage of being tolerated in higher doses than smaller, rapidly fermented oligosaccharides.

The levels of toxigenic and mutagenic enzymes (azoreductase, β-glucuronidase, nitroreductase, etc.) and metabolites are often measured as an indication of the effectiveness of prebiotics in improving the intestinal environment. A number of studies have indicated that NDOs can reduce the levels of these enzymes and metabolites in feces, although conclusive links between these end points and reduced cancer risk remain to be established (105–107).

C. Improving Mineral Adsorption

Prebiotics have also been investigated for their effects on dietary mineral adsorption in a number of animal and human studies. Although some human studies have been neutral in their findings (75,76), a number have reported beneficial effects on calcium absorption (77–79). Further research is warranted to investigate links between long-term prebiotic consumption and improved bone density in humans at risk of development of osteoporosis.

D. Effects on Serum Lipid and Cholesterol Concentrations

A relatively large number of human studies have focused on the effects of oligosaccharide and inulin intake on lipid metabolism. These include eight human trials summarized by van Loo and colleagues (70) and more recent trials (80–84). Although some positive data have been reported, the results of these studies have not consistently demonstrated favorable effects of NDO and inulin consumption on serum triacylglycerol and cholesterol levels, although no deleterious effects have been reported.

VI. SAFETY AND RECOMMENDED DOSAGES

Galactooligosaccharides are considered safe food ingredients as they are minor constituents of human milk and can also be produced in the gut by intestinal bacteria from ingested lactose. Acute toxicity tests have shown in rats that ingestion of more than 15 g/kg body weight of β1-4 galactooligosaccharide was needed for LD_{50}. No adverse symptoms were recorded in a chronic toxicity test when 1.5 g/kg body weight was fed for 6 months. Nonmutagenicity was confirmed by AMES-Salmonella and Rec assay.

Excessive intake (>30 g/day in adults) of most oligosaccharides and lactulose causes osmotic diarrhea. Recommended intake of total oligosaccharides from all sources is therefore 10 to 15 g/day for adult humans (0.14–0.21 g/kg body weight). Naturally occurring fructooligosaccharides in foods such as onions are estimated to contribute between 1 and 9 g/day to most Western diets. Generally NDOs have laxative effects or cause flatulence and bowel discomfort when daily intakes exceed 25 g/day in a 70-kg human.

VII. QUANTITATIVE MODELING OF CARBON FLOWS IN THE HUMAN GUT

The contribution of NDOs to energy metabolism and carbon flow in humans has not received much attention. Calculations of this type in the human gastrointestinal tract have been made for ruminants and clearly would have value in improving our understanding of fermentation in the human gastro intestinal tract. For example, on active adult human of 70 kg body weight has an energy expenditure of about 12 MJ per day. One would normally expect a Western diet to contain 70 g fat (contributing 2772 kJ), 100 g protein (contributing 1860 kJ), and 420 g carbohydrate (contributing 7350 kJ) to make up the required 12 MJ. Up to 10% (or 42 g) of the carbohydrate sources supplied would be expected to be resistant to digestion and be available for fermentation in the colon. If this were all fermented, it would give rise to about 30 g or 428 mmole of SCFA. Clearly the more cellulosic and lignified components would not be fermented. An additional 15 g/day of NDOs would probably be completely fermented and theoretically give rise 143 mmole of SCFA. The amount of carbon needed for the synthesis of bacterial cells depends on the bacterial cell turnover rate and transit time in the colon. This diversion of carbon reduces the amount of SCFA produced from carbohydrates in the colon.

Questions to be answered include the following:

The volume of microbial biomass produced
The contribution of prebiotics to total available carbon pool in the colon
The contribution of epithelial cell material and mucins to the carbon pool

The site of prebiotic fermentation in the colon (proximal vs. transverse vs. distal)
The site and effects of carbon and other nutrient limitations on microbial fermentative activity
Production and absorption rates of SCFA (as opposed to concentrations of SCFA)

VIII. FUNCTIONAL PROPERTIES OF NONDIGESTIBLE OLIGOSACCHARIDES

There are a number of functional properties possessed by food grade oligosaccharides that although they do not confer direct physiological effects on the host are nevertheless important and make NDOs attractive for use in foods. Examples are sweetness, low cariogenicity, and low calorific value, all of which make them useful as a replacement for sugars and fats in foods. Their indigestibility and subsequent effect on glucose and insulin responses also make them suitable for diabetics. Obviously, there are factors affecting functionality of the oligosaccharides in a foodstuff. These include viscosity, water activity, moisture retention, color, osmolality, and storage stability.

Technological properties of galactooligosaccharides are the following:

Appearance: translucent/colorless
Sweetness: usually less than half that of sucrose
Water activity: similar to that of sucrose
Viscosity: similar to that of high-fructose glucose syrup
Heat stability (e.g., galactooligosaccharides are stable to 160°C for 10 min at pH 7 and to 100°C for 10 min at pH 2)
Acid stability: xylo- and galactooligosaccharides found to be more acid-stable than fructooligosaccharides (over the temperature span 5°C to 37°C)
Indigestibility: resistant to salivary and pancreatic alpha amylase and other carbohydrases and gastric juice
Energy value: about 50% of that of sucrose; suitable for diabetics
Cariogenicity: low
Other food technology advantages: increased viscosity, reduced Malliard reactions and crystal formation, and alteration of freezing points

Applications of NDOs in foods include the following:

Dairy products: milk, yogurt, infant milk powders, fermented milk drinks
Drinks: coffee drinks, soft drinks, soybean milk

Confectionery: candy, muesli bars, sports bars, energy bars, cakes, chocolate
Desserts: ice cream, pudding, jelly, sherbet
Bakery products: bread, buns
Jam: chilled jam, chocolate paste
Other products: health foods, diet sugar, processed meat products, processed marine products, honey products
Feed additive: animal feeds for cattle, pigs, poultry, companion animals

IX. PRODUCTION OF OLIGOSACCHARIDES

Food-grade oligosaccharides are made by three processes: (a) extraction and purification from plants, e.g., soybean oligosaccharides, inulin from chicory; (b) controlled enzymic degradation of polysaccharides, e.g., xylooligosaccharides, isomaltooligosaccharides, and some fructooligosaccharides; (c) enzymic synthesis from sugars, e.g., some fructooligosaccharides, galactooligosaccharides, and lactosucrose.

A. Enzymic Synthesis of Oligosaccharides

Much research has been conducted on the enzymic synthesis of oligosaccharides (85–87). Simple, cheap methods have been developed for production of food-grade oligosaccharide mixtures using enzymes. More complex and expensive methods requiring cofactors, such as uridine diphosphate (UDP), and expensive glycosyl transferase enzymes have been developed for production of structurally specific and longer-chain oligosaccharides. Lactose has been the substrate used for enzymic production of galactooligosaccharides, for isomerization to lactulose, and for chemical conversion to lactitol. Lactosucrose is produced by enzymic synthesis from lactose and sucrose. Consequently, much of this section concentrates on these processes using lactose.

1. Galactooligosaccharides

Galactooligosaccharides are formed by the enzyme β-galactosidase (EC 3.2.1.23) from lactose in a transgalactosylation reaction. This synthetic reaction occurs simultaneously with hydrolytic degradation of the lactose. Thus, a mixture of glucose, galactose, lactose, and tri-, tetra-, and pentaoligosaccharides are usually formed.

β-Galactosidases from different biological sources and even from different genera of bacteria have markedly different abilities to synthesize galactooligosaccharides (88) and show variation in their ability to produce particular chain lengths and structures. For example, several structures of trisaccharide can be formed (see Fig. 3). Both 4′-galactosyl lactose (Galβ1-4Galβ1-4Glc) and 6′-galactosyl lactose (Galβ1-6Galβ1-4Glc) predominate in the commercial products. However, 3′-galactosyl lactose is also commonly formed. A wide range of organisms have been studied for the suitability of their β-galactosidases, and organisms such as *Sterigmatomyces elviae* have given high oligosaccharide yields from lactose at 200 g/L at 85°C (89).

Manufacturing Process. Galactooligosaccharides are formed from lactose in a batch reactor, using one or more β-galactosidase enzymes (lactases) in sequence or together. Elevated temperatures are used (50°C–80°C). The principal reason for this method is to achieve a high solubility of lactose and thus reduce water activity and push the enzymic reaction toward synthesis of oligosaccharides rather than hydrolysis to monomer sugars. Lactose concentrations of up to 400 g/L are used. The choice of enzymes is important as some sources produce β1-4 linkages and others β1-6 linkages. Some enzymes also tend to have stronger hydrolytic activity than others; e.g., β-galactosidase from *Aspergillus niger* has a strong hydrolytic activity. The composition of the mixture of oligosaccharides produced by different enzymes varies; e.g., some produce more tetra-, penta- and hexasaccharides.

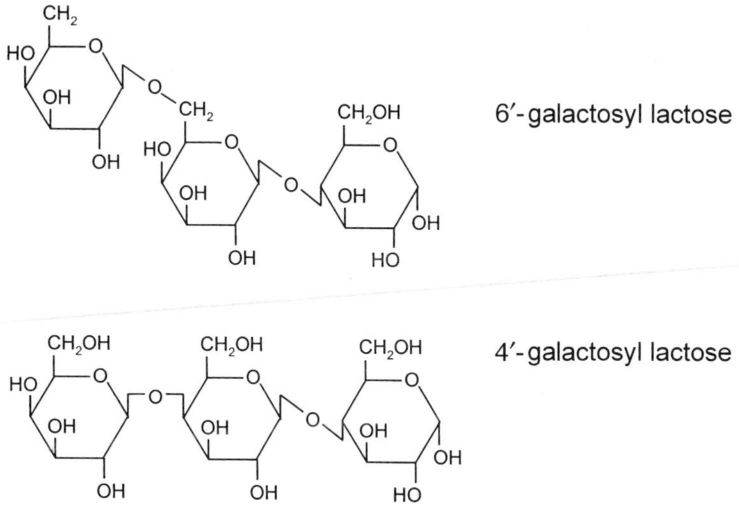

Figure 3 Chemical structures of the two predominant galatotrisaccharides present in commercial galactooligosaccharide products: α-glu–(1-4)-β-gal-(1-4)-β-gal and α-glu–(1-4)-β-gal-(1-6)-β-gal.

Generally, production of trisaccharides is dominant, forming more than 80% of the total oligosaccharides in the mixture. Enzyme derived from *Cryptococcus laurentii* and *Bacillus circulans* tend to produce β1-4 linked galactooligosaccharide, whereas *Aspergillus oryzae* and *Streptococcus thermophilus* produce predominantly β1-6 linked galactooligosaccharide (90). The former enzymes are used by Nissin to produce their Cup-Oligo product, which is a β1-4 galactosyl lactose, whereas Yakult and Snow Brand have used the latter enzymes for their products, which have predominantly β1-6 linkages. Yakult now also uses *Bacillus circulans* to produce β1-4 linked product. The Yakult products are named Oligomate 55 and TOS 100. Typical mixtures for the commercial products are shown in Table 1. Enzyme costs are an important consideration in these processes. Hence reuse and immobilization of enzymes have been researched. It is not thought that such techniques are yet used commercially.

After batch reaction, the product mix is decolorized and demineralized, filtered, and concentrated to produce either a syrup or a powder. When highly concentrated product is required, such as for Yakult's TOS100 product, then further processing is required and this usually involves chromatographic separation of the mono- and disaccharides from the longer-chain oligosaccharides (see Fig. 4).

Food-grade galactooligosaccharides, although mixtures, are well defined and consistent in composition. These characteristics are achieved by good quality control of the manufacturing process at each stage, namely, in choice of enzyme(s) source, temperature of process, lactose concentration, purification steps, and concentration procedures. The production, properties, effects, and uses of galactooligosaccharides were reviewed in 2000 (91).

2. Lactosucrose (β-D-Fructofuranosyl 4-*O*-β-D-Galactopyranosyl-α-D-Glucopyranoside)

Lactosucrose is a trisaccharide, the chemical structure of which is shown in Fig. 5. Ensuiko Sugar Refining Company is the main producer of lactosucrose. It is produced from lactose and sucrose feedstocks by an enzymic transfructosylation reaction to form a structure containing galactose-1-4-glucose-1-2-fructose. The compound is also called *lactosyl-fructoside*. The enzyme used is β-fructofuranosidase (invertase) (EC 3.2.1.26) from *Arthrobacter* species. Conditions required to produce high yields of lactosucrose product are similar to those required for galactooligosaccharide production, as the enzyme also hydrolyses the substrates.

Manufacturing Process. A mixture of sucrose and lactose (45:55) as a 38%–50% solution, at pH 5.8–6.2, is reacted in the presence of the enzyme β-

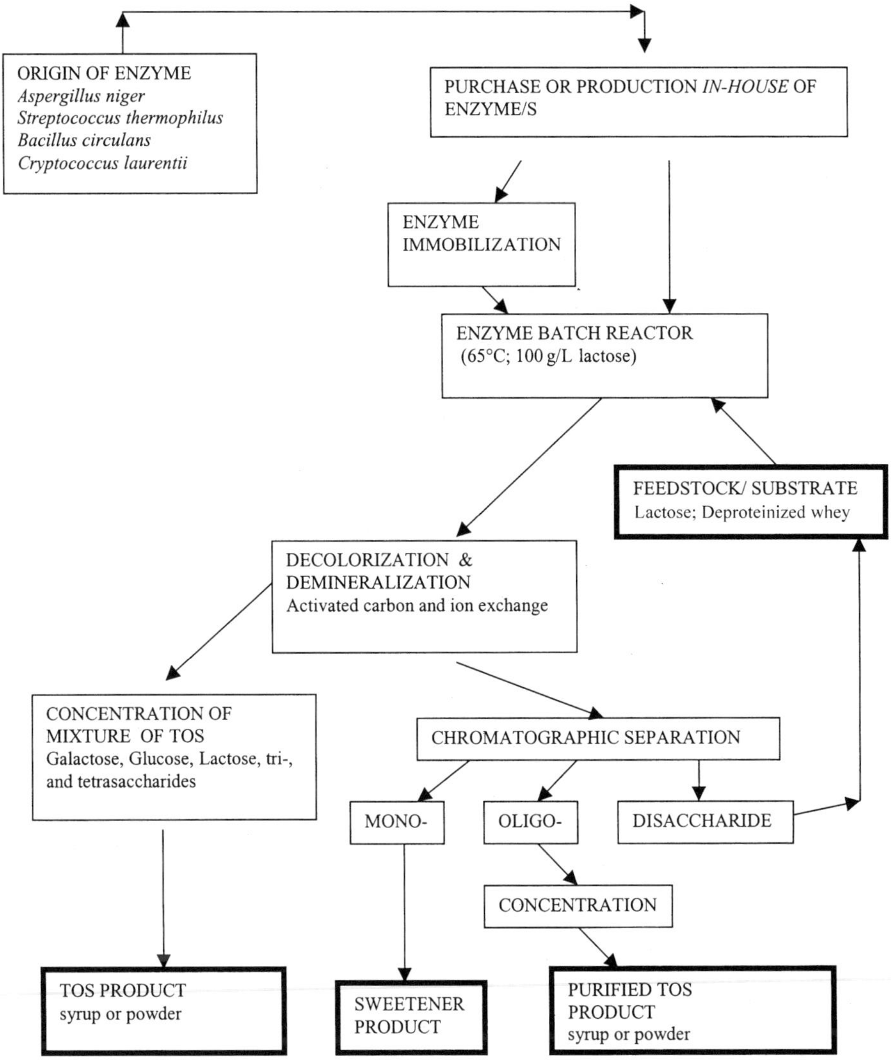

Figure 4 Commercial process for the production of galactooligosaccharides.

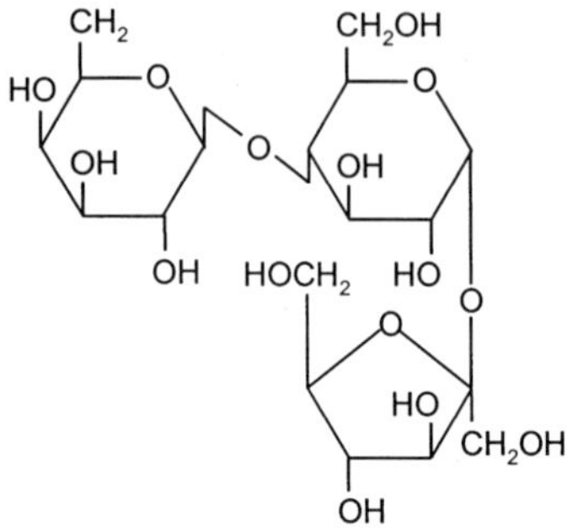

Figure 5 The chemical structure of lactosucrose.

fructofuranosidase at 55°C–60°C. After reaction, the enzyme is inactivated, and the solution decolorized with activated carbon, filtered, concentrated, purified, refiltered, deionized with ion-exchange resins, and concentrated (92).

3. Isomaltooligosaccharides

Isomaltooligosaccharides (IMOs), such as isomaltose, panose, isomaltotriose, and isomaltotetraose, have α-1-6 glucosidic linkages. Isomalto mixtures, such as Isomalto-900 produced by Showa Sangyo Co Ltd, are prepared from corn starch by the actions of alpha-amylase, pullulanase, and alpha-glucosidase. The IMOs occur naturally in foods such as miso, soy sauce, sake, and honey. These isomaltooligosaccharide mixtures have also been called *anomalously linked oligosaccharides mixtures* (ALOMs) (93).

Manufacturing Process. Yatake has described the process used by Showa Sangyo (93). A 30% slurry of starch at pH 6.0 is dextrinized (liquefied) using a thermostable bacterial α-amylase. The degree of hydrolysis expressed as a dextrose equivalent is kept between 6 and 10. Simultaneous saccharification and transglucosidation of the solution of dextrin take place using soybean β-amylase and fungal α-glucosidase (60°C, pH 5.0). The solution is then filtered, decolorized, demineralized, and concentrated. The concentrate is then separated into an oligosaccharide–rich fraction by moving bed liquid chromatography using Na-form cation-exchange resins, to produce IMO-900, and a sugar–rich fraction to produce IMO-500. A powdered version of IMO-900 is produced by spray drying.

Hayashibara Group produces Panorup, which is an isomaltooligosaccharide mixture with similar properties. The Mitsui Sugar Company produces palatinose (maltulose) products, including pure crystalline palatinose, and mixtures of palatinose oligosaccharides, which also contain palatinose polycondensates, other saccharides, trehalulose, and monosaccharides. It has been demonstrated to have a bifidogenic effect (52). Nihon Shokuhin Kako Co. Ltd produces Biotose 50, which is composed of isomaltose, panose, isomaltotriose, and glucose. This product is produced by a similar process. Nihon also produces Panorich, which is a panose syrup, containing 25% panose, maltose, glucose, and some IMOs. It is also claimed to have some bifidogenic effect, although like other maltooligosaccharides, it exhibits partial digestion in the stomach and small intestine. Along with maltooligosaccharides and glycosyl sucrose (coupling sugar), such oligosaccharide mixtures cannot be regarded as true prebiotics. The Nihon company produces Nisshoku Gentose, which because of the enzymes used results in a different product of transglucosidation , and the resulting gentiooligosaccharide has a β-1-6-glucosidic linkage, and not the usual α-1-4 and α-1-6 linkages. The product is claimed to be bifidogenic. Its composition is given in Table 1. Nakakuki and coauthors (94,103) have reviewed oligosaccharides derived from starch in comprehensive papers. The chemical structures of malto-, isomalto-, and gentiooligosaccharides are shown in Fig. 6.

4. Specific Long-Chain Oligosaccharides

Although food-grade oligosaccharides are invariably mixtures of oligosaccharides (both in chain length and in chemical structure), the pharmaceutical and chemical industries have extensively studied enzymic routes for the formation of specific oligosaccharides. Glycosidases (such as β-galactosidase) are less selective and consequently less regiospecific than glycosyltransferases. Glycosidases are widely available from plant cells, animal cells, and bacterial cells. Glycosyltransferases, however, require expensive cofactors (UDP), and their use can only be justified when longer-chain specific oligosaccharides are required. One application of the use of glycosyltransferases is the production of synthetic human milk oligosaccharides. A 1998 report (95) shows that it is possible to produce UDP-galactose and globotriose on a large scale by coupling metabolically engineered bacteria. This work could lead to the development of new methods to produce oligosaccharides by coupling sugar nucleotide production systems with glycosyltransferases.

B. Extraction from Plants and Controlled Enzymic Hydrolysis

1. Xylooligosaccharides

The structure of xylan is variable, ranging from a linear 1,4-β-linked polyxylose chain to highly branched heteropolysaccharides. The main chains consist of D-xylose, whereas the branches contain arabinofuranose and

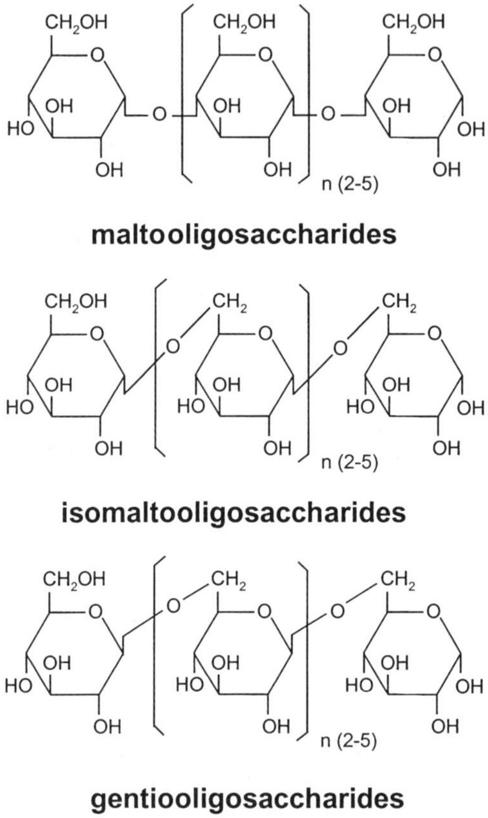

Figure 6 Chemical structures of malto-, isomalto-, and gentiooligosaccharides.

glucuronic acid residues. Xylan is a major component of hemicellulose and consequently is readily available from a range of lignocellulosic waste streams (e.g., corncob, sugarcane bagasse).

Manufacturing Process. Xylooligosaccharides (Fig. 7), produced by controlled hydrolysis of relatively unsubstituted xylan, are also well fermented by many bifidobacteria (30). The Suntory process uses pretreatment with alkali of corncobs to release hemicelluloses. After washing, saccharification takes place, using endo-1,4-β-xylanase to produce a solution of 75%–80% oligosaccharide. This is then ultrafiltered, followed by reverse osmosis (RO), which splits the product into two streams. The permeate stream, which contains 60%–70% oligosaccharide, receives a second RO treatment. The permeate from this is discarded. The concentrate, which contains 70% or

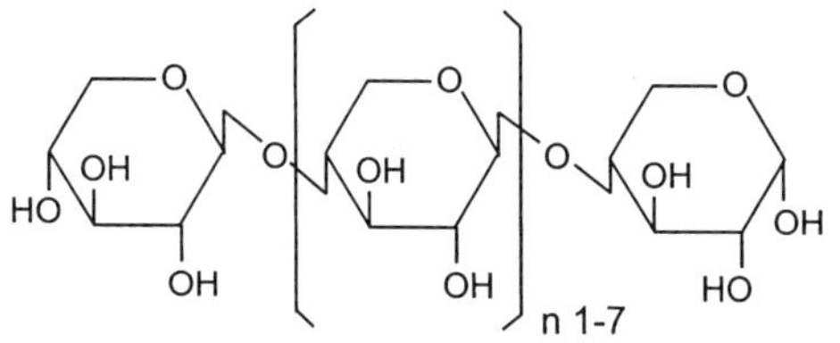

Figure 7 Chemical structure of xylooligosaccharides.

more oligosaccharide, is purified to yield the product Xylooligo 70. The concentrate from the first RO treatment, which contains more than 95% oligosaccharide, is purified and produces the Xylooligo 95 product.

2. Soybean Oligosaccharides

The Calpis Food Industry Company Ltd of Japan is the major producer of soybean oligosaccharides. The oligosaccharides produced are a mixture of stachyose, raffinose, and sucrose. Stachyose (four monomers) and raffinose (three monomers) are α-galactooligosaccharides. The chemical structures of raffinose and stachyose are shown in Fig. 8. The mixture produced by Calpis contains more than 26% galactooligosaccharides and less than 74% sucrose and other saccharides. A typical composition is listed in Table 1.

Manufacturing Process. Soybean oligosaccharides are produced by purification of soybean whey. The protein is removed by precipitation, and the filtrate is decolorized and demineralized and then concentrated.

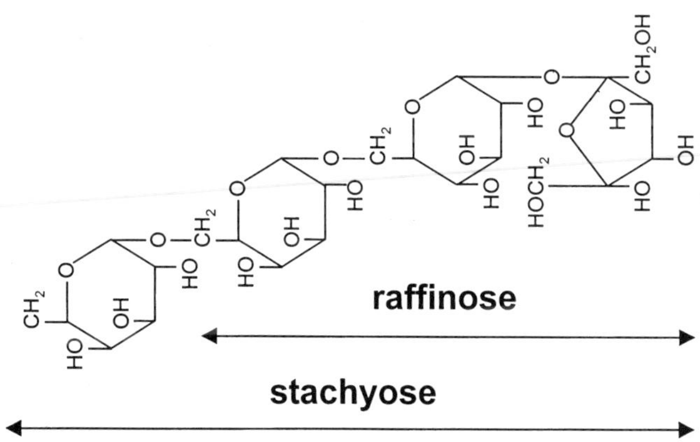

Figure 8 Structures of raffinose and stachyose, the predominant oligosaccharides in commercial soybean oligosaccharide products.

C. Chemical Conversion of Lactose

1. Lactulose

Lactulose is not a naturally occurring compound, although small amounts are found in some processed milk lines when heat has been applied. It was first synthesized in 1930 using calcium hydroxide as the alkali for the isomerization from lactose (see Fig. 9). Since then, many other alkalis have been tested, including alkaline borate reagents. The conversions have been reviewed by Mizota and associates (1987) (96). Lactulose is very soluble in water (76% at 30°C) and melts at 169°C.

Manufacturing Process. Purified lactose is dissolved, an alkali such as sodium hydroxide is added with or without a catalyst such as borate, and the mixture is heated to 100°C for the isomerization of lactose to lactulose. The solution is then deionized and decolorized, and unreacted lactose is removed as a precipitate after crystallization. After pasteurization, the product is concentrated and syrups, powders, or crystals are produced.

2. Lactitol (4-β-D-Galactopyranosyl-D-Sorbitol)

Lactitol is a sugar alcohol with very high water solubility (206% at 25°C) but is insoluble in ethanol (0.75% at 25°C). Lactitol is chemically derived from

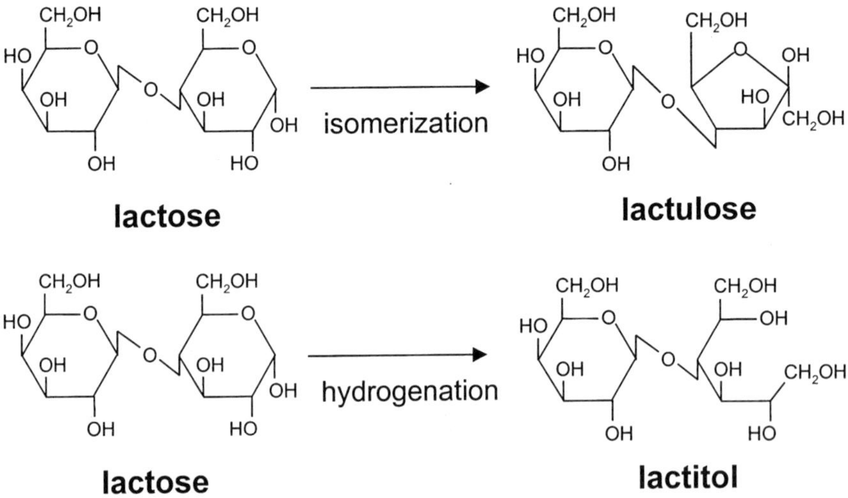

Figure 9 Conversion of lactose to lactulose and lactitol.

lactose by catalytic hydrogenation at high temperature and pressure. It does not occur naturally or in processed milk products.

Manufacturing Process. Lactitol is prepared from lactose (Fig. 9). The conversion of a sugar to a sugar alcohol involves the reduction of a carbonyl group. This can be achieved by catalytic hydrogenation. A typical manufacturing process converts a 30%–40% solution of lactose using Raney nickel as catalyst and hydrogen at 100°C to 200°C and 10,000- to 15,000-kPa pressure. The sediment produced is filtered, decolorized with activated carbon, demineralized by ion exchange, evaporated, and purified by crystallization. The reduction of the carbonyl group can also be achieved with sodium borohydride. The properties, production, and physiological effects of lactitol were summarized by Booy (1987) (97).

X. FUTURE DIRECTIONS AND CHALLENGES

A. Lectin Ligands, Soluble Adhesion Receptor Analogues, and Human Milk Oligosaccharides

The term *chemical probiosis* has been coined by workers at the Rowett Research Institute in Scotland to characterize the ability of specific oligosaccharides found in human milk and colostrum to block the adhesion of *Escherichia coli* K88. A Canadian company, Synsorb Biotech Inc., has developed oligosaccharides (Synsorb) with an inert carrier, which act as alternative receptors to absorb lectinlike toxins from toxigenic bacteria (verotoxigenic *Escherichia coli*, enterohaemorrhagic *Escherichia coli*, *Clostridium difficile*). Zopf and Roth (98) discussed the use of oligosaccharides as anti-infective agents, using a decoy oligosaccharide in the mucous layer to bind the microorganism's carbohydrate binding proteins. They claim that such oligosaccharides (as found in human milk) prevent pathogen attachment. An example is given by the Neose Technology's (USA) anti-*Helicobacter pylori* oligosaccharide, NE 0080. Idota and colleagues (99) have examined the utilization by bifidobacteria and by lactobacilli of certain human milk oligosaccharides. Sialyl lactose was only used by *Bifidobacterium infantis*, whereas *N*-acetyl-neuraminic acid was used by *Bifidobacterium longum*, *Lactobacillus casei*, and *Lactobacillus salivarius*, but not by *Lactobacillus acidophilus*. Human milk contains 3 to 6 g/L of oligosaccharide (100), whereas cow's milk contains very little (0.03–0.06 g/L) and most of that is sialyl lactose. The major oligosaccharides in human milk are lacto-*N*-tetraose and lacto-*N*-fucopentaose. Kunz and Radloff (100) regard the evidence as strong as these compounds are potent inhibitors of bacterial adhesion to epithelial cells: "Human milk oligosaccharides are considered to be soluble receptor analogues of epithelial cell surfaces, participating in the nonimmunological defence system of human milk-fed infants." In the development of any oligosaccharide to compete with pathogen adhesion and in the gut to impair colonization, care should to be taken to ensure that the oligosaccharide does not also interfere with colonization of the normal intestinal microbiota.

B. Synbiotics

The concept of increasing the survival, persistence, and/or activity of a probiotic strain with a complementary prebiotic is yet to be well demonstrated in humans. Certainly, the concept has merit and the molecular tools now exist to monitor individual strains and their activity in the intestinal microbiota (101). Exploiting physical associations between probiotics and particulate indigestible carbohydrates, such as between bifidobacteria and resistant starch (62), may provide a means to ensure probiotics can specifically benefit from an added prebiotic. Prebiotics to complement specific lactobacilli require development. The ultimate in synbiotic combinations will include prebiotics that can not only benefit the proliferation and activity of the probiotic strains in the colon, but also protect the strains during gastrointestinal transit and during manufacture, formulation, and storage.

C. Management of Microbiota Composition and Activity

Little is currently known of the subgenus changes in bifidobacterial populations that can be induced by prebiotics or of the importance of these changes in a health context. New molecular tools to investigate changes in population levels at the genus level, such as reported in 2001 by Satokari and coworkers (102), now allow an opportunity to investigate the effect of different prebiotics on individual *Bifidobacterium* species. These techniques can equally be applied at the genus, species, and strain levels and will allow a better understanding of how prebiotics affect microbial population dynamics in the intestinal tract.

Increasing the numbers of bifidobacteria or lactobacilli in the intestinal microbiota of individuals with an unfavorable intestinal microbial balance appears with our current understanding of the human intestinal microbiota to be a reasonable approach to promoting intestinal health. However, basic research into the composition and role of different microbial populations within the intestinal microbiota in health and disease is an essential prerequisite for the development of appropriate prebiotic strategies. A more profound understanding of what constitutes a "healthy" intestinal microbiota composition, and which microbial groups and activities are definitively involved in health and disease, will allow the development of prebiotics with specifically targeted health effects in the future.

D. Improving Human Health

The challenge remains to elucidate links between increased numbers of bifidobacteria within the intestinal microbiota induced by prebiotics and beneficial effects on the health of the host. Prophylaxis of intestinal infections through the rapid restoration of colonization resistance after perturbations to the intestinal microbiota is perhaps one of the simpler health benefits to examine and is being targeted in a number of research laboratories in Europe. Prebiotics may have a positive role to play in protection against colorectal cancer and in dietary mineral absorption and lipid metabolism. Further studies to elucidate mechanisms by which prebiotics act in these cases are certainly warranted.

XI. CONCLUSIONS

A range of nondigestible oligosaccharides, already commercially available, show strong potential as prebiotics and can increase the populations of bifidobacteria within the intestinal microbiota. New prebiotic carbohydrates continue to be developed with an emphasis on specific synbiotic interactions and on extension of fermentation to intestinal sites beyond the proximal colon. Synbiotic combinations will become more frequently adopted in probiotic functional foods, not only with added bifidobacteria, but also with new prebiotics designed to complement specific lactobacilli. The development of molecular methods to examine the composition and activity of the intestinal microbiota and interactions with the host is central to the future progression of prebiotic research.

REFERENCES

1. S Salminen, E Isolauri, T Onnela. Gut flora in normal and disordered states. Chemotherapy 41(Suppl 1):5–15, 1995.
2. G D'Argenio, G Mazzacca. Short-chain fatty acid in the human colon: Relation to inflammatory bowel diseases and colon cancer. Adv Exp Med Biol 472:149–158, 1999.
3. VJ McCracken, RG Lorenz. The gastrointestinal ecosystem: A precarious alliance among epithelium, immunity and microbiota. Cell Microbiol 3:1–11, 2001.

4. C Dunne, L Murphy, S Flynn, L O'Mahony, S O'Halloran, M Feeney, D Morrissey, G Thornton, G Fitzgerald, C Daly, B Kiely, EM Quigley, GC O'Sullivan, F Shanahan, JK Collins. Probiotics: From myth to reality: Demonstration of functionality in animal models of disease and in human clinical trials. Antonie van Leeuwenhoek 76:279–292, 1999.
5. M Schultz, RB Sartor. Probiotics and inflammatory bowel diseases. Am J Gastroenterol 95:S19–S21, 2000.
6. T Mitsuoka. Recent trends in research on intestinal flora. Bifidobacteria Microflora 1:3–24, 1982.
7. MJ Hopkins, R Sharp, GT Macfarlane GT. Age and disease related changes in intestinal bacterial populations assessed by cell culture, 16S rRNA abundance, and community cellular fatty acid profiles. Gut 48:198–205, 2001.
8. E. Metchinikoff. The Prolongation of Life. London: Heinemann, 1907.
9. R Fuller. Probiotics in man and animals. J Appl Bacteriol 66:365–378, 1989.
10. GW Tannock, AL McCartney, W Wenzhi. Molecular analysis of the composition of the bifidobacterial and lactobacillus microflora of humans. Appl Environ Microbiol 62:4608–4613, 1996.
11. GW Tannock, K Kimura, AL McCartney, MA McConnell. Analysis of fecal populations of bifidobacteria and lactobacilli and investigation of the immunological responses of their human hosts to the predominant strains. Appl Environ Microbiol 63:3394–3398, 1997.
12. GR Gibson, Roberfroid MB. Dietary modulation of the human colonic microbiota: Introducing the concept of prebiotics. J Nutr 125:1401–1412, 1995.
13. R Mackie, A Sghir, HR Gaskins. Developmental microbial ecology of the neonatal gastrointestinal tract. Am J Clin Nutr 69:1035S–1045S, 1999.
14. MJ Playne, RG Crittenden. Commercially-available oligosaccharides. Bull IDF 313:10–22, 1996.
15. RG Crittenden, MJ Playne. Production, properties and applications of food-grade oligosaccharides. Trends Food Sci Technol 7:353–361, 1996.
16. R Tanaka, H Takayama, M Morotomi, T Kuroshima, S Ueyama, K Matsumoto, A Kuroda, M Mutai. Effects of administration of TOS and *Bifidobacterium breve* 4006 on the human faecal flora. Bifidobacteria Microflora 2:17–24, 1983.
17. M Dombo, H Yamamoto, H Nakajima. Production, health benefits and applications of galacto-oligosaccharides. In: M Yalpani, ed. New Technologies for Healthy Foods and Neutraceuticals. Shrewsbury, MA: ATL Press, 1997, pp 143–156.
18. M Ito, Y Deguchi, A Miyamori, K Matsumoto, K Kikuchi, K Matsumoto, Y Kobayashi, T Yajima, T Kan. Effects of administration of galacto-oligosaccharides on the human faecal microflora, stool weight and abdominal sensation. Microbial Ecol Health Dis 3:285–292, 1990.
19. M Ito, Y Deguchi, K Matsumoto, M Kimura, N Onodera, T Yajima. Influence of galacto-oligosaccharides on human faecal microflora. J Nutr Sci Vitaminol 39:635–540, 1993.
20. Y Bouhnik, B Flourie, L D'Agay-Abensour, P Pochart, G Gramet, M Durand,

JC Rambaud. Administration of transgalacto-oligosaccharides increases fecal bifidobacteria and modifies colonic fermentation metabolism in healthy humans. J Nutr 127:444–448, 1997.

21. U Teuri, M Karkkainen, C Lamberg-Allardt, R Korpela. Addition of inulin to breakfast does not acutely affect serum ionized calcium and parathyroid hormone concentrations. Ann Nutr Metab 43:356–364, 1999.
22. MS Alles, R Hartemink, S Meyboom, JL Harryvan, KM van Laere, FM Nagengast, JG Hautvast. Effect of transgalactooligosaccharides on the composition of the human intestinal microflora and on putative risk markers for colon cancer. Am J Clin Nutr 69:980–991, 1999.
23. M Alander, J Mättö, W Kneifel, M Johansson, B Kögler, R Crittenden, T Mattila-Sandholm, M Saarela. Effect of galacto-oligosaccharide supplementation on human faecal microflora and on survival and persistence of *Bifidobacterium lactis* Bb-12 in the gastrointestinal tract. Int Dairy J 11:817–825, 2001.
24. AV Rao. Dose-response effects of inulin and oligofructose on intestinal bifidogenesis effects. J Nutr 129(Suppl 7):1442S–1445S, 1999.
25. R Hartemink, BJ Kok, GH Weenk, FM Rombouts. Raffinose-bifidobacterium (RB) agar, a new selective medium for bifidobacteria. J Microbiol Methods 27:33–43, 1996.
26. Y Benno, K Endo, N Shiragani, K Sayama, T Mitsuoka. Effects of raffinose intake on human fecal microflora. Bifidobacteria Microflora 6:59–63, 1987.
27. K Hayakawa, J Mizutain, K Wada, T Masai, I Yoshihara, T Mitsuoka T. Effects of soybean oligosaccharides on human faecal microflora. Microb Ecol Health Dis 3:293–303, 1990.
28. K Wada, J Watabe, J Mizutani, M Tomoda, H Suzuki, Y Saitoh. Effects of soybean oligosaccharides in a beverage on human fecal flora and metabolites. J Agric Chem Soc Jpn 66:127–135, 1992.
29. T Hara, N Ikeda, K Hatsumi, J Watabe, H Iino, T Mitsuoka. Effects of small amount ingestion of soybean oligosaccharides on bowel habits and fecal flora of volunteers. Jpn J Nutr 55:79–84, 1997.
30. M Okazaki, S Fujikawa, N Matsumoto. Effect of xylo-oligosaccharide on the growth of bifidobacteria. Bifidobacteria Microflora 9:77–86, 1990.
31. H Yamada, K Itoh, Y Morishita, H Taniguchi. Structure and properties of oligosaccharides from wheat bran. Cereal Foods World 38:490–492, 1993.
32. J Jaskari, P Kontula, A Siitonen, H Jousimies-Somer, T Mattila-Sandholm, K Poutanen. Oat beta-glucan and xylan hydrolysates as selective substrates for *Bifidobacterium* and *Lactobacillus* strains. Appl Microbiol Biotechnol 49:175–181, 1998.
33. KM van Laere, R Hartemink, M Bosveld, HA Schols, AG Voragen. Fermentation of plant cell wall derived polysaccharides and their corresponding oligosaccharides by intestinal bacteria. J Agric Food Chem 48:1644–1652, 2000.
34. T Kohmoto, K Tsuji, T Kaneko, M Shiota, F Fukui, H Takaku, Y Nakagawa, T Ichikawa, S Kobayashi. Metabolism of ^{13}C-isomaltooligosaccharides in healthy men. Biosci Biotechnol Biochem 56:937–940, 1992.
35. T Kohmoto, F Fukui, H Takaku, Y Machida, M Arai, T Mitsuoka. Effect of

isomalto-oligosaccharides on human facal flora. Bifidobacteria Microflora 7:61–69, 1998.
36. T Kaneko, T Kohmoto, H Kikuchi, M Shiota, T Yatake, H Iino, K Tsuji. Effects of isomaltoologosaccharides intake on defecation and intestinal environment in healthy volunteers. J Home Econ Jpn (in Japanese) 44:245–254, 1993.
37. T Kaneko, T Kohmoto, H Kikuchi, M Shiota, H Iino, T Mitsuoka. Effects of isomaltooligosaccharides with different degrees of polymerisation on human fecal bifidobacteria. Biosci Biotechnol Biochem 58:2288–2290, 1994.
38. A Terada, H Hara, M Kataoka, T Mitsuoka. Effect of lactulose on the composition and metabolic activity of the human faecal flora. Microbial Ecol Health Dis 5:43–50, 1992.
39. F. Petuely. Bifidusflora bei flaschenkindern durch bifidogene substanzen (Bifidusfaktor). Z Kinderheilkunde 79:174–179, 1957.
40. J Ballongue, C Schumann, P Quignon. Effects of lactulose and lactitol on colonic microflora and enzymatic activity. Scand J Gastroenterol 32 (Suppl 222): 41–44, 1997.
41. R Nagendra, S Viswanatha, SA Kumar, BK Murthy, SV Rao. Effect of feeding milk formula containing lactulose to infants on faecal bifidobacterial flora. Nutr Res 15:15–24, 1995.
42. T Mizota. Lactulose as a growth promoting factor for *Bifidobacterium* and its physiological aspects. Bull IDF 313:43–48, 1996.
43. Y Minami, K Yazawa, Z Tamura, T Tanaka, T Yamamoto. Selectivity of utilization of galactosyl-oligosaccharides by bifidobacteria. Chem Pharm Bull 31:1688–1691, 1983.
44. M Yoneyama, T Mandai, H Aga, K Fujii, S Sakai, Y Katayama. Effects of 4-β-D-galactosylsucrose (lactosucrose) intake on intestinal flora in healthy humans. J Jpn Soc Nutr Food Sci 45:101–107, 1992.
45. K Fujita, K Hara, S Sakai, T Miyake, M Yamashita, Y Tsunstomi, T Mitsuoka. Effects of 4-β-D-galactosylsucrose (lactosucrose) on intestinal flora and its digestibility in humans. J Jpn Soc Starch Sci 38:249–255, 1991.
46. T Ohkusa, Y Ozaki, C Sato, K Mikuni, H Ikeda. Long-term ingestion of lactosucrose increases *Bifidobacterium* sp. in human fecal flora. Digestion 56: 415–420, 1995.
47. F Teramoto, K Rokutan, Y Kawakami, Y Fujimura, J Uchida, K Oku, M Oka, M Yoneyama. Effect of 4G-beta-D-galactosylsucrose (lactosucrose) on fecal microflora in patients with chronic inflammatory bowel disease. J Gastroenterol 31:33–39, 1996.
48. T Asano, K Yuasa, K Kunugita, T Teraja, T Mitsuoka T. Effects of gluconic acid on human faecal bacteria. Microbial Ecol Health Dis 7:247–256, 1994.
49. P Mesters, S Brokx. Lactitol: A functional prebiotic. In: F Angus, C Miller, eds. Functional Foods 2000. Surry, England: Leatherhead Publishing, 2000, pp. 173–189.
50. M Satouchi, S Wakabayashi, K Ohkuma, K Tsuji. Effect of depolyerised pyrodextrin on human intestinal flora. Biosci Microflora 15:93–101, 1996.

51. T Okubo, N Ishihara, H Takahashi, T Fujisawa, M Kim, T Yamamoto, T Mitsuoka. Effects of partially hydrolyzed guar gum intake on human intestinal microflora and its metabolism. Biosci Biotechnol Biochem 58:1364–1369, 1994.
52. J Kashimura, Y Nakajima, Y Benno, K Endo, T Mitsuoka. Effects of palatinose and its condensate intake on human fecal microflora. Bifidobacteria Microflora. 8: 45–50, 1989.
53. Z Djouzi, C Andrieux, V Pelenc, S Somarriba, F Popot, F Paul, P Monsan, O Szylit. Degradation and fermentation of alpha-gluco-oligosaccharides by bacterial strains from human colon: In vitro and in vivo studies in gnotobiotic rats. J Appl Bacteriol 79:117–127, 1995.
54. SJ Vincent, EJ Faber, JR Neeser, F Stingele, JP Kamerling. Structure and properties of the exopolysaccharide produced by *Streptococcus macedonicus* Sc136. Glycobiology 11:131–139, 2001.
55. R Crittenden, S Karppinen, S Ojanen, M Tenkanen, R Fagerström, J Mättö, M Saarela, T Mattila-Sandholm, K Poutanen. In vitro fermentation of cereal dietary fibre carbohydrates by intestinal bacteria. J Sci Food Agric 82:781–789, 2002.
56. P Kontula, A von Wright, T Mattila-Sandholm. Oat bran beta-gluco- and xylo-oligosaccharides as fermentative substrates for lactic acid bacteria. Int J Food Microbiol 45:163–169, 1998.
57. B Kleessen, G Stoof, J Proll, D Schmiedl, J Noack, M Blaut. Feeding resistant starch affects fecal and cecal microflora and short-chain fatty acids in rats. J Anim Sci 75:2453–2462, 1997.
58. I Brown, M Warhurst, J Arcot, M Playne, RJ Illman, DL Topping. Fecal numbers of bifidobacteria are higher in pigs fed *Bifidobacterium longum* with a high amylose cornstarch than with a low amylose cornstarch. J Nutr 127:1822–1827, 1997.
59. IL Brown, X Wang, DL Topping, MJ Playne, PL Conway. High amylose maize starch as a versatile prebiotic for use with probiotic bacteria. Food Aust 50:603–610, 1998.
60. X Wang, IL Brown, AJ Evans, PL Conway. The protective effects of high amylose maize (amylomaize) starch granules on the survival of *Bifidobacterium* spp. in the mouse intestinal tract. J Appl Microbiol 87:631–639, 1999.
61. S Silvi, CJ Rumney, A Cresci, IR Rowland. Resistant starch modifies gut microflora and microbial metabolism in human flora-associated rats inoculated with faeces from Italian and UK donors. J Appl Microbiol 86:521–530, 1999.
62. R Crittenden, A Laitila, P Forssell, J Mättö, M Saarela, T Mattila-Sandholm, P Myllärinen. Adhesion of bifidobacteria to granular starch and implications in probiotic technologies. Appl Environ Microbiol 67:3469–3475, 2001.
63. Y Bouhnik, P Pochart, P Marteau, G Arlet, I Goderel, JC Rambaud. Fecal recovery in humans of viable *Bifidobacterium* sp ingested in fermented milk. Gastroenterology 102:875–878, 1992.
64. T Mattila-Sandholm, S Blum, JK Collins, R Crittenden, W de Vos, C Dunne, R

Fondén, G Grenov, E Isolauri, B Kiely, P Marteau, L Morelli, A Ouwehand, R Reniero, M Saarela, S Salminen, M Saxelin, E Schiffrin, F Shanahan, E Vaughan, A von Wright. Probiotics: Towards demonstrating efficacy. Trends Food Sci Technol 10:393–399, 1999.
65. T Vesa, P Pochart, P Marteau. Pharmacokinetics of *Lactobacillus plantarum* NCIMB 8826, *Lactobacillus fermentum* KLD, and *Lactococcus lactis* MG 1363 in the human gastrointestinal tract. Aliment Pharmacol Ther 14: 823–828, 2000.
66. S Salminen, E Salminen. Lactulose, lactic acid bacteria, intestinal microecology and mucosal protection. Scand J Gastroenterol 32(Suppl 222):45–48, 1997.
67. MW Oli, BW Petschow, RK Buddington. Evaluation of fructooligosaccharide supplementation of oral electrolyte solutions for treatment of diarrhea: recovery of the intestinal bacteria. Dig Dis Sci 43:138–147, 1998.
68. JO Hunter, Q Tuffnell, AJ Lee. Controlled trial of oligofructose in the management of irritable bowel syndrome. J Nutr 129:1451S–1453S, 1999.
69. G Jacobasch, D Schmiedl, M Kruschewski, K Schmehl. Dietary resistant starch and chronic inflammatory bowel diseases. Int J Colorectal Dis 14:201–211, 1999.
70. J van Loo, J Cummings, N Delzenne, H Englyst, A Franck, M Hopkins, N Kok, G Macfarlane, D Newton, M Quigley, M Roberfroid, T van Vliet, E van den Heuvel. Functional food properties of non-digestible oligosaccharides: A consensus report from the ENDO project (DGXII AIRII-CT94-1095). Br J Nutr 81:121–132, 1999.
71. EJ Schiffrin, F Rochat, H Link-Amster, JM Aeschlimann, A Donnet-Hughes. Immunomodulation of human blood cells following the ingestion of lactic acid bacteria. J Dairy Sci 78:491–497, 1995.
72. EJ Schiffrin, D Brassart, AL Servin, F Rochat, A Donnet-Hughes. Immune modulation of blood leukocytes in humans by lactic acid bacteria: criteria for strain selection. Am J Clin Nutr 66:515S–520S, 1997.
73. G Perdigon, E Vintini, S Alvarez, M Medina, M Medici. Study of the possible mechanisms involved in the mucosal immune system activation by lactic acid bacteria. J Dairy Sci 82:1108–1114, 1999.
74. M Kalliomäki, S Salminen, H Arvilommi, P Kero, P Koskinen, E Isolauri. Probiotics in primary prevention of atopic disease: a randomised placebo-controlled trial. Lancet 357:1076–1079, 2001.
75. EG van den Heuvel, G Schaafsma, T Muys, W van Dokkum. Nondigestible oligosaccharides do not interfere with calcium and nonheme-iron absorption in young, healthy men. Am J Clin Nutr 67:445–451, 1998.
76. U Teuri, R Korpela, M Saxelin, L Montonen, S Salminen. Increased fecal frequency and gastrointestinal symptoms following ingestion of galacto-oligosaccharide-containing yogurt. J Nutr Sci Vitaminol 44:465–471, 1998.
77. C Coudray, J Bellanger, C Castiglia-Delavaud, C Remesy, M Vermorel, Y Rayssignuier. Effect of soluble or partly soluble dietary fibres supplementation on absorption and balance of calcium, magnesium, iron and zinc in healthy young men. Eur J Clin Nutr 51:375–380, 1997.
78. EG van den Heuvel, T Muijs, W van Dokkum, G Schaafsma. Lactulose stim-

ulates calcium absorption in postmenopausal women. J Bone Miner Res 14: 1211–1216, 1999.
79. EG van den Heuvel, T Muys, W van Dokkum, G Schaafsma. Oligofructose stimulates calcium absorption in adolescents. Am J Clin Nutr 69:544–548, 1999.
80. W van Dokkum, B Wezendonk, TS Srikumar, EG van den Heuvel. Effect of nondigestible oligosaccharides on large-bowel functions, blood lipid concentrations and glucose absorption in young healthy male subjects. Eur J Clin Nutr 53:1–7, 1999.
81. KG Jackson, GR Taylor, AM Clohessy, CM Williams. The effect of the daily intake of inulin on fasting lipid, insulin and glucose concentrations in middle-aged men and women. Br J Nutr 82:23–30, 1999.
82. HP Kruse, B Kleessen, M Blaut. Effects of inulin on faecal bifidobacteria in human subjects. Br J Nutr 82:375–382, 1999.
83. F Brighenti, MC Casiraghi, F Canzi, A Ferrari. Effect of consumption of a ready-to-eat breakfast cereal containing inulin on the intestinal milieu and blood lipids in healthy male volunteers. Eur J Clin Nutr 53:726–733, 1999.
84. MS Alles, MN de Roos, JC Bakx, E van de Lisdonk, PL Zock, JGAC Hautvast. Consumption of fructo-oligosaccharides does not favourably affect blood glucose and serum lipids in non-insulin dependent diabetic patients. Am J Clin Nutr 69:64–69, 1999.
85. JE Prenosil, E Stucker, JR Bourne. Formation of oliogosaccharides during enzymatic lactose. Part 1. State of art. Biotechnol Bioeng 30:1019–1025, 1987.
86. KGI Nilsson, Enzymatic synthesis of oligosaccharides. Trends Biotechnol 6:256–264, 1988.
87. RA Rastall, C Bucke. Enzymatic synthesis of oligosaccharides. Biotechnol Gen Eng Rev 10:253–281, 1992.
88. JB Smart. Transferase reactions of β-galactosidases—new product opportunities. Bull IDF 289:16–22, 1993.
89. N Onishi, T Tanaka. Purification and properties of a novel thermostable galacto-oligosaccharide-producing β-galactosidase from *Sterigmatomyces elviae* CBS 8119. Appl Environ Microbiol 61:4026–4030, 1995.
90. K Matsumoto, Y Kobayashi, S Ueyama, T Watanabe, R Tanaka, T Kan, A Kuroda, Y Sumihara. Galactooligosaccharides. In: T Nakakuki, ed. Oligosaccharides: Production, Properties, and Applications. Japanese Technology Reviews, Vol. 3, No. 2. Tokyo: Gordon & Breach, 1993, pp 90–106.
91. M Playne. Galacto-oligosaccharides. In: H Roginski, PF Fox, JW Fuquay, eds. Encyclopaedia of Dairy Sciences. San Diego, CA: Academic Press, 2000, pp 1151–1158.
92. K Fujita, S Kitahata, K Hara, H Hashimoto. Production of lactosucrose and its properties. In: MA Clarke, ed. Carbohydrates in industrial synthesis. New York: Bartens, 1992, pp 68–76.
93. T Yatake. Anomalously linked oligosaccharides. In: T Nakakuki, ed. Oligosaccharides. Japanese Technology Reviews, Vol. 3, No. 2. Tokyo: Gordon & Breach, 1993, pp 79–89.

94. M Okado, T Nakakuki. Oligosaccharides—production, properties and applications. In: FW Schenck, RE Hebeda, eds. Starch Hydrolysis Products—Worldwide Technology, Production, and Applications. New York: VCH, 1994.
95. S Koizumi, T Endo, K Tabata, A Ozaki. Large-scale production of UDP-galactose and globotriose by coupling metabolically engineered bacteria. Nat Biotechnol 16:847–850, 1998.
96. T Mizota, Y Tamura, M Tomita, S Okonogi. Lactulose as a sugar with physiological significance. Bull IDF 212:69–76, 1987.
97. CJ Booy. Lactitol: "A new food ingredient." Bull IDF 212:62–68, 1987.
98. D Zopf, S Roth. Oligosaccharide anti-infective agents. Lancet 347:1017–1021, 1996.
99. T Idota, H Kawakama, I Nakajima. Growth-promoting effects of N-acetylneuraminic acid-containing substance on bifidobacteria. Biosci Biotechnol Biochem 58:1720–1722, 1994.
100. C Kunz, S Radloff. Strukturelle und funktionelle aspekte von oligosacchariden in frauenmilch. Z Ernahrungswiss. 35:22–31, 1996.
101. E Vaughan, B Mollet, W de Vos. Functionality of probiotics and intestinal lactobacilli: Light in the intestinal tract tunnel. Curr Opin Biotechnol 10:505–510, 1999.
102. RM Satokari, EE Vaughan, AD Akkermans, M Saarela, WM de Vos. Bifidobacterial diversity in human feces detected by genus-specific PCR and denaturing gradient gel electrophoresis. Appl Environ Microbiol 67:504–513, 2001.
103. T Nakakuki. Present status and future of functional oligosaccharide development in Japan. Pure Appl Chem 74:1245–1251, 2002.
104. MJ Playne. The health benefits of probiotics. Food Australia 54:71–72, 2002.
105. IR Rowland, CA Bearne, R Fischer, BL Pool-Zobel. The effect of lactulose on DNA damage induced by DMH in the colon of human flora-associated rats. Nutr Cancer 26: 37–47, 1996.
106. IR Rowland. Metabolic profiles of intestinal floras. In: T Hattori, ed. Recent Adv Microb Ecology pp. 510–514, 1990.
107. BS Reddy. Possible mechanisms by which pro- and prebiotics influence colon carcinogenesis and tumor growth. J Nutr 129:(75): 1478–1482, 1999.
108. S Kolida, K Tuohy, G Gibson. Prebiotic effects of inulin and oligofructose. Br J Nutr 87 (Suppl 2): S 193–197, 2002.

7

Dextran and Glucooligosaccharides

Pierre F. Monsan and Daniel Auriol
Centre de Bioingénierie Gilbert Durand, INSA, Toulouse, France

I. INTRODUCTION

Very recently, the European Commission authorized the use of a dextran preparation as a new food ingredient for bakery applications (1). This clearly illustrates the renewed interest in dextran as a food ingredient, in addition to the nutritional applications of glucooligosaccharides. In fact, dextran oligosaccharides and isomaltooligosaccharides, which are closely related in structure as they present a high proportion of α-1,6 glucosidic linkages, are partly or totally resistent to attack by humans' and animals' digestive enzymes. They are not absorbed in the small intestine. In the large intestine, they are then metabolized by the colonic bacterial flora and fermented into short-chain fatty acids (2,3). Such nondigestible glucooligosaccharides are regarded as prebiotics or colonic foods. A *prebiotic* is "a nondigestible food ingredient that beneficially affects the host by selectively stimulating the growth and/or activity of one or a limited number of bacteria in the colon, and thus improves host health" (4). On the other hand, a *colonic food* is a food ingredient enterring the colon and serving as substrate for the endogenous bacteria, thus indirectly providing the host with energy, metabolic substrates, and essential micronutrients.

II. DEXTRAN

Dextran is a D-glucosyl homopolysaccharide produced by lactic acid bacteria of the genera *Leuconostoc*, *Streptococcus*, *Lactococcus*, and *Lactobacillus* (5–7). It contains more than 50% α-1,6 glucosidic linkages, with different additional branching through α-1,2, α-1,3, and α-1,4 linkages (8). This branching is the basis of the classification of dextrans into three groups, A, B, and C, corresponding to α-1,2, α-1,3, and α-1,4 branching, respectively (9). Dextran biosynthesis does not involve the usual glycosyltransferase scheme with nucleotide–glucose substrate, as is the case for starch or cellulose biosynthesis, for example. In fact, it was demonstrated by Hehre that dextran is synthesized from sucrose by a cell-free filtrate (10). The corresponding extracellular enzyme was named *dextransucrase* (E.C. 2.4.1.5) by Hestrin, Averini-Shapiro, and Aschner (11).

Dextransucrase is able to use the energy of the glucose–fructose osidic bond of sucrose to catalyze the transfer of the D-glucosyl moiety through the formation of a covalent glucosyl-enzyme intermediate (12,13). A dextran polysaccharide is obtained, with a molecular weight ranging from 500 to 6,000 kda (14,15). D-fructose is released by the enzyme as a coproduct. In addition, in the presence of an efficient acceptor molecule, which is generally a carbohydrate, dextransucrase catalyzes the transfer of the D-glucosyl residue onto the acceptor to yield low-molecular-weight oligosaccharides (16–19). Maltose is among the most efficient acceptors tested (20). When D-fructose accumulates in the reaction medium, its pyranosyl form acts as an acceptor to yield the disaccharide leucrose (21).

The most widely known dextran is produced by the strain *Leuc. mesenteroides* NRRL B-512F. It contains 95% α-1,6 osidic linkages and 5% α-1,3 branching. The products obtained by controlled chemical hydrolysis are used for the production of chromatography supports (Sephadex) and blood plasma substitutes or iron carriers (22).

A. Dextransucrase Structure

Dextransucrases have a common structure, similar to the structure of streptococcal glucosyltransferases (13,23–25), which consists of the following (see Fig. 1):

An N-terminal signal peptide involved in protein excretion (26). The gene encoding the dextransucrase DSR-A from *Leuc. mesenteroides* NRRL B-1299 is the only known exception (27);

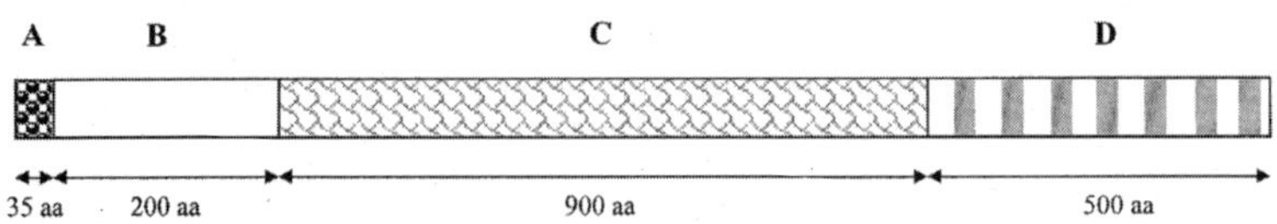

Figure 1 General structure of dextransucrases. A, N-terminal signal peptide; B, variable region; C, catalytic domain; D, glucan binding domain, containing repeated units. Average amino acid number in the different domains is given.

A variable region of unknown role (28), which is not present in DSR-A (27);

A highly conserved N-terminal catalytic domain, which contains the catalytic triad consisting of two aspartic acids and one glutamic acid residue (29,30) and presents a permutated $(\beta/\alpha)_8$ barrel structure (13);

A C-terminal glucan binding domain involved in both dextran and glucooligosaccharide synthesis in the case of the dextransucrase from *Leuc. mesenteroides* NRRL B-512F (31); this domain contains a series of repeating units (28,32,33).

The only exception to this general structure is the dextransucrase from *Leuc. mesenteroides* NRRL B-1299, which catalyzes the synthesis of α-1,2 osidic linkages: This enzyme contains an additional catalytic domain at the C-terminal end, which is specifically involved in the synthesis of such α-1,2 branching (34).

The availability of an increasing number of dextransucrase genes, and more broadly of glucansucrase genes, associated with a deeper characterization of the structure–function relationships of the corresponding enzymes at the molecular level, suggests the potential to design new enzymes with controlled specificity toward both substrates and acceptors, and controlled regioselectivity in D-glucopyranosyl unit transfer (35). In fact, it is possible to combine site-directed and random mutagenesis and/or molecular evolution approaches. The single substitution of the threonine amino acid residue at position 667 in the dextransucrase DSR-S from *Leuc. mesenteroides* NRRL B-512F by an arginine residue, for example, results in the synthesis of a dextran polymer that contains 13% α-1,3 glucosidic bonds, instead of 5% with the native enzyme (35). In the case of glucosyltransferase GTF-S from *Streptococcus mutans*, the substitution of the threonine residue 589 by an aspartic acid residue results in a 30% increase of the amount of α-1,3 osidic linkage in the synthesized glucan (36).

B. "Prebiotic Colonic Food" Dextran Oligosaccharides

The strain *Leuc. mesenteroides* NRRL B-1299 produces a dextran polymer containing 27% to 35% α-1,2 osidic branching linkages, besides a limited amount of α-1,3 osidic branching linkages (37–40). The dextransucrase activity is mostly associated with the cell fraction (41–46). Both cell-associated and soluble dextransucrase fractions catalyze the synthesis of α-1,2 branched dextran polymers. In addition, these enzyme preparations keep their selectivity in acceptor reactions, in the presence of maltose as D-glucopyranosyl residue acceptor, for example (47,48). This reaction yields a mixture of three families of glucooligosaccharides, with a maltose residue at the reducing end (49) and containing in addition (Fig. 2)

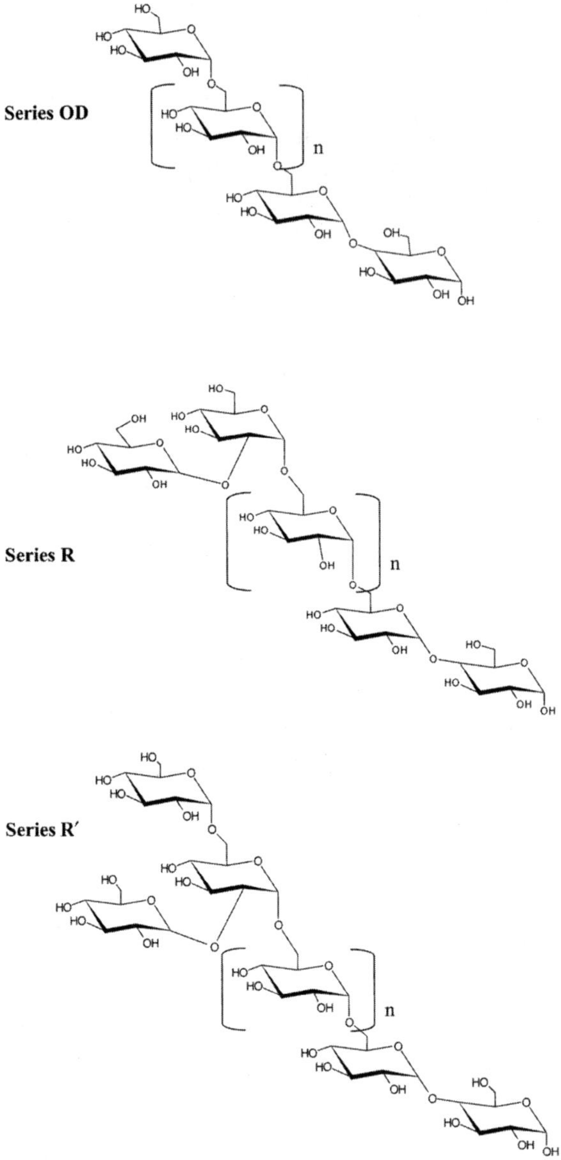

Figure 2 Structure of the glucooligosaccharides obtained by the acceptor reaction catalyzed by the dextransucrase from *Leuconostoc mesenteroides* NRRL B-1299 in the presence of sucrose (D-glucosyl donor) and maltose (D-glucosyl acceptor). Series OD, containing only α-1,6 osidic linkages in addition to the maltosyl residue located at the reducing end; Series R, containing an additional α-1,2 osidic linkage at the nonreducing end; Series R′, containing an additional α-1,2 osidic branching linkage on the penultimate glucosyl residue at the nonreducing end.

Only α-1,6 linkages (series OD), which are closely related in structure to isomaltooligosaccharides

α-1,6 Linkages and one α-1,2 linkage at the no-reducing end (series R)

α-1,6 Linkages and one α-1,2 linkage on the penultimate D-glucosyl residue (series R′)

Such α-1,2 osidic linkages present at or near the no-reducing end result in a very high resistance to hydrolysis by digestive enzymes in both humans and animals (50). In fact, these glucooligosaccharides were initially designed as low-calorie food-bulking agents to be used in combination with intense sweeteners (47). But these glucooligosaccharides are metabolized by certain species of beneficial intestinal flora, e.g., bifidobacteria, lactobacilli, and particularly bacteroides, and they are poorly metabolized by potentially detrimental strains (51,52). In fact, they are not significantly metabolized by germ-free rats that have no intestinal flora (50).

Such glucooligosaccharides have thus been developed as "prebiotic colonic foods" for both animal (52) and human applications (BioEcolians, BioEurope-Solabia). They are fermented by the intestinal flora to short-chain fatty acids (SCFAs). In heteroxenic rats inoculated with complex human intestinal flora, the short-chain fatty acid profile in the cecum changes (50), with a decrease in butyric, isobutyric, and isovaleric acid proportions ($p < 0.01$) and an increase in the proportion of caproic acid ($p < 0.05$). When compared to fructooligosaccharides and galactooligosaccharides, glucooligosaccharides are not bifidogenic, but they promote the growth of the cellulolytic intestinal flora. More important is the fact that they induce a broader range of glycolytic enzymes, without any significantly increased side production of gases, and thus without detrimental effects, as demonstrated by using rats inoculated with human cecal flora (53). These glucooligosaccharides are used in human food complement formulations, in combination with probiotics (beneficial living microorganisms) and B vitamins, to regulate the intestinal transit. Their indigestibility has been compared to that of several commercially available oligosaccharides and polysaccharides (54). Short-chain fatty acid production for glucooligosaccharides is similar to that of fructooligosaccharides, guar gum, guar hydrolysate, and gum arabic. Glucooligosaccharides and maltooligosaccharide-like oligosaccharides were added to an enteral formula control diet and fed to ileal-cannulated dogs. Ileal digestibility was lower for both products when compared to that of the control diet. Total fecal weights were higher and fecal concentration of bifidobacteria was numerically increased. These products can thus be regarded as dietary fiber–like ingredients (54).

The production of dextransucrase by fed-batch cultivation of *Leuc. mesenteroides* NRRL B-1299 was optimized, using sucrose as both carbon source and enzyme inducer (55,56). The addition of D-glucose to the sucrose fed-batch feed prevents the repressor effect of D-fructose released by dextransucrase activity (57). This results in a 100% increase in dextransucrase production, up to 9.7 U/mL culture medium. This dextransucrase activity is strongly associated with the insoluble phase comprising the bacterial cells and the surrounding dextran slime (41–46). It is thus easily recovered by centrifugation and immobilized by entrapment in alginate gel beads for developing a continuous process for glucooligosaccharide synthesis from a sucrose and maltose mixture (58). The optimal operational conditions have been determined, stressing the key effect of the sucrose/maltose concentration ratio on the glucooligosaccharide yield and size distribution (59).

The in vitro utilization by human gut microflora of (a) industrial-grade dextran (very probably produced by *Leuc. mesenteroides* NRRL B-512F dextransucrase), (b) oligodextran fractions produced by controlled enzymatic hydrolysis of dextran (60), and (c) maltodextrin was studied by using anaerobic batch culture fermenters (61). Glucose and fructooligosaccharides were used as reference carbohydrates. Fructooligosaccharides selectively increased numbers of bifidobacteria in the early stages of the fermentation. Dextran and oligodextran resulted in an enrichment of bacteria with high levels of persistence up to 48 hr, with production of elevated levels of butyrate ranging from 5 to 14.85 mmol/L. A three-stage continuous culture cascade system was used for more effective simulation of the conditions that prevail in different regions of the large intestine. A low-molecular-mass oligodextran fraction was then shown to be the best substrate for bifidobacteria and lactobacilli, when compared to dextran and maltodextrin. In addition, dextran and oligodextran stimulate butyrate production more efficiently, a characteristic that has been shown to present very interesting potential antineoplastic properties (61). These results underline the interest of oligodextrans as modulators of the gut microflora. As the dextran polymers produced by the dextransucrase from *Leuc. mesenteroides* NRRL B-512F contain about 95% α-1,6 glucosidic linkages (8), the oligodextran molecules resulting from their controlled enzymatic hydrolysis present a molecular structure very similar to that of isomaltooligosaccharides (Fig. 3).

III. ISOMALTOOLIGOSACCHARIDES

Isomaltooligosaccharides (Fig. 3) are produced on the industrial scale (approximately 10,000 tons per year) in Japan, which is the market leader in the field of nondigestible oligosaccharides (61). They are obtained by the enzymatic transformation of starch, using a combination of α-amylase, β-amy-

A

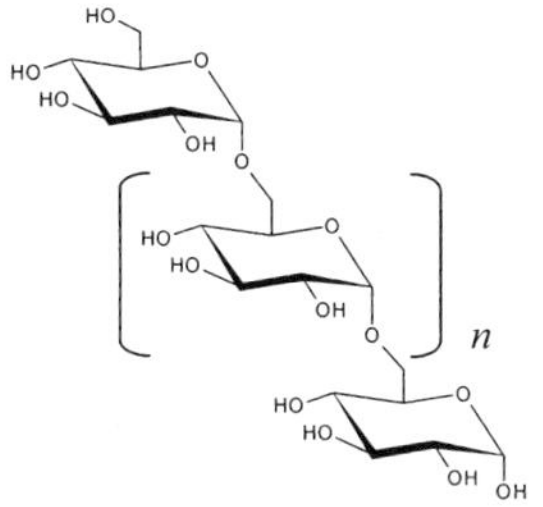

B

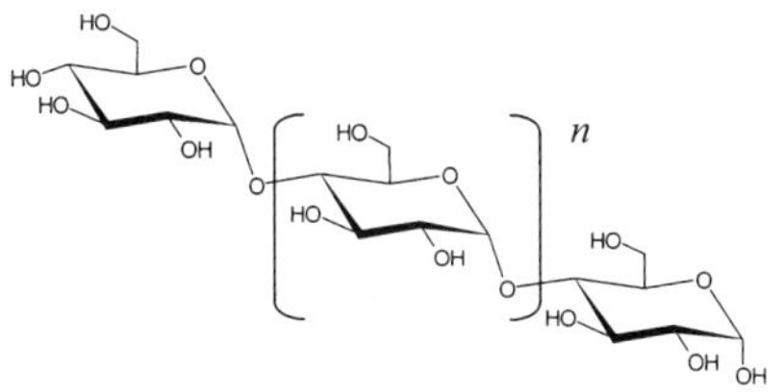

C

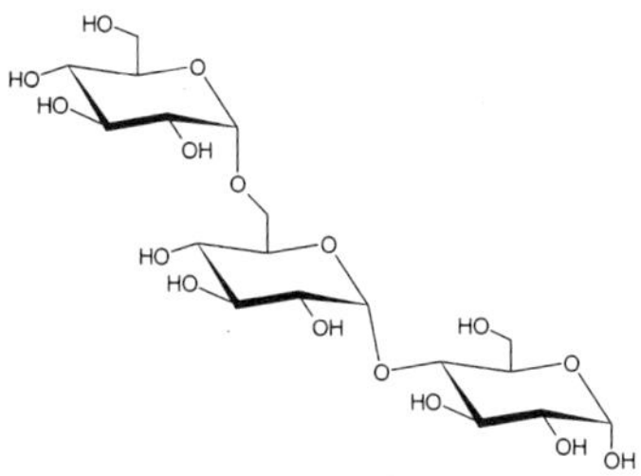

D

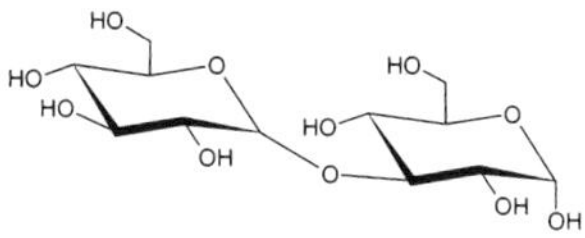

E

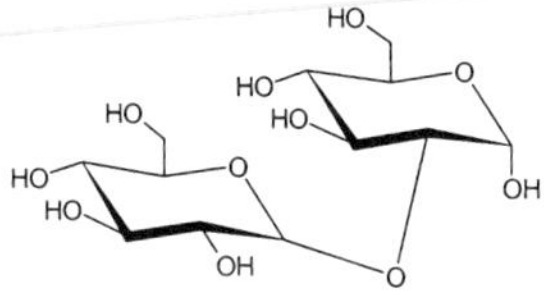

Figure 3 Structure of the glucooligosaccharides found in commercial isomaltooligosaccharides. A, Isomaltooligosaccharides, containing α-1,6 osidic linkages ($n = 0$, isomaltose; $n = 1$, isomaltotriose; $n = 2$, isomaltotetraose); B, maltooligosaccharides, containing α-1,4 osidic linkages ($n = 0$, maltose; $n = 1$, maltotriose; $n = 2$, maltotetraose); C, panose (α-D-Glc-1,6-maltose); D, nigerose (α-D-Glc-1,3-D-Glc); E, kojibiose (α-D-Glc-1,2-D-Glc).

lase, pullulanase, and α-glucosidase (62). Commercial preparations contain (60–65) a mixture of isomaltose (DP 2, 23%), isomaltotriose (DP 3, 17%), oligosaccharides (DP 4–6, 26%), and nonisomaltooligosaccharides (panose, maltose, maltotriose, nigerose, kojibiose, 30%).

But isomaltooligosaccharides are partly digestible in the small intestine. The degree of digestion decreases when the oligosaccharide DP is higher than 3 (64,65). At high dosages, they selectively promote the growth of bifidobacteria in the human intestine. A minimal dosage of isomaltooligosaccharides of 8–10 g per day was necessary to increase fecal bifidobacteria (62). By comparison, a minimal dosage of fructooligosaccharides of 1–2 g per day results in a similar effect. This difference is attributed to the fact that fructooligosaccharides are not digested in the small intestine, whereas isomaltooligosaccharides are partially digested (62). A commercially available mixture of isomaltooligosaccharides was fractionated by preparative high-performance liquid chromatography (HPLC). Two fractions were obtained, IMO2, containing mainly disaccharides (86.4%), and IMO3, containing trisaccharides and higher-DP oligosaccharides (89.9%). The administration of IMO2 or IMO3 to humans in amounts ranging from 5 to 20 g per day resulted in an increase in bifidobacteria in a dose-dependent manner: An IMO2 intake of 10 g per day and an IMO3 intake of 5 g per day both produced a significant increase of bifidobacterial number in feces and an enhanced ratio in fecal microflora within 12 days (64). The rat jejunal loop method was used to characterize the luminal clearance of isomaltooligosaccharides and their hydrogenated derivatives, as an indication of their digestibility (65). They were compared with isomaltooligosaccharide fractions IMO2 and IMO3 (see previous discussion) with typical digestible saccharides and with typical nondigestible saccharides. The clearance of isomaltooligosaccharides was significantly smaller than that of the IMO2 fraction and digestible saccharides. It was larger than that of the IMO3 fraction and significantly larger than that of nondigestible saccharides. Hydrogenated isomaltooligosaccharides presented a clearance similar to that of maltitol and significantly smaller than that of isomaltooligosaccharides. These results show that isomaltooligosaccharides are slowly digested in the rat jejunum. The components having higher molecular mass are less digestible, and hydrogenated isomaltooligosaccharides are nondigestible (65).

The metabolic fate of isomaltooligosaccharides in healthy men was investigated by using carbon-13-labeled products (66). The expiration rates of excess $^{13}CO_2$ and hydrogen of six men were measured while sedentary and while taking physical exercise after the carbon-13-labeled isomaltooligosaccharide intakes. Breath hydrogen excretion remained constant after isomaltooligosaccharide ingestion in the sedentary state and increased in the exercise test. Serum glucose and serum insulin increased 30 min after the oligosaccharide ingestion. The $^{13}CO_2$ recovery levels were 28.7% in the sedentary test and 60.9% in the exercise test, versus 70–80% in the case of maltose. This indicates that isomaltooligosaccharides were partially digested and partially fermented by the intestinal flora. The energy value of isomaltooligosaccharides was about 75% of that of maltose (66).

Closely related in structure to isomaltooligosaccharides are the branched maltooligosaccharides, which are oligomers of D-glucose linked primarily by α-1,4 osidic linkages but containing at least one α-1,6 bond, with a DP ranging from 2 to 5 (67): isomaltose, isomaltotriose, panose, and several higher-molecular-mass oligosaccharides. They are generally produced by serial reactions of starch with α-amylase and transglucosidase (67). They can also be obtained from a 15% starch suspension using a *Bacillus licheniformis* maltogenic amylase presenting both hydrolyzing and transglycosylation activities (68). The nonbranched carbohydrate content of the resulting products can be greatly decreased by successive fermentation with immobilized yeast cells (69). The functional properties of such branched oligosaccharide mixtures have been determined and compared with those of commercial isomaltooligosaccharide preparations, maltodextrin, and sucrose (70). In addition, branched oligosaccharides present interesting prebiotic properties (71–72).

IV. CONCLUSIONS

It is increasingly clear that the resident bacterial flora of the colon plays a major role in human nutrition and health, both directly, through metabolite production and occupation of epithelial adhesion sites, and indirectly, through the immunomodulatory response of the digestive tract. From this point of view, prebiotic compounds are of outstanding interest, as they are specifically fermented and thus allow the control and modulation of the large gut flora. Unfortunately, the individual roles of the roughly 500 different bacterial species present in the colon are not at all understood as yet. *Bifidobacterium* and *Lactobacillus* species are the two main bacterial groups identified as presenting health-promoting effects. Molecular biological tools now introduce the potential to identify precisely the microbial strains present at any site of the digestive tract, even if they cannot be grown on Petri dishes. This will allow a more precise determination of the composition of the intestinal microbial flora than the simple counting of colony-forming units from feces samples. In this context, it is essential to be able to design the broadest possible range of nondigestible oligosaccharides, built from simple and safe carbohydrate units. Such oligosaccharides will help to obtain precise control of the colonic flora, and, for example, to control the nature and the level of short-chain fatty acidsing produced in the colon. The main effects targeted are better resistance to pathogens, decreased gut tumor risks, and reduced levels of blood lipids.

From this point of view, glucooligosaccharides and glucansucrases are very good candidates. Several glucooligosaccharide structures have been proved to be partly or totally resistant to digestive enzymes, and particularly the structures containing α-1,2 osidic linkages produced by the dextransucrase from *Leuc. mesenteroides* NRRL B-1299. They present interesting properties, with very limited detrimental effects (abdominal distension, flatulence, diarrhea). In addition, the availability of several dextransucrase genes encoding enzymes catalyzing the synthesis of mostly α-1,6 osidic linkages, but with different branching selectivities (α-1,2, α-1,3, and α-1,4 osidic linkages), opens the way to the generation of a broad diversity of glucooligosaccharide structures, through the generation of enzyme variants by site-directed mutagenesis, random mutagenesis, and directed evolution [deoxyribonucleic acid (DNA) and family shuffling]. The corresponding glucooligosaccharides will have to be screened to identify the best functional candidates.

REFERENCES

1. Commission decision dated January 31, 2001. Official Journal of the European Communities, February 15, L 44/46-L 44/47, 2001.
2. GR Gibson. Dietary modulation of the human gut microflora using prebiotics. Br J Nutr 80:S209–S212, 1998.
3. LJ Fooks, R Fuller, GR Gibson. Prebiotics, probiotics and human gut microbiology. Int Dairy J 9:53–61, 1999.
4. RL Gibson, MB Robertfroid. Dietary modulation of the human colonic microbiota: Introducing the concept of prebiotics. J Nutrition 125:1401–1412, 1995.
5. RL Sidebotham. Dextrans. Adv Carbohydr Chem Biochem 30:371–444, 1974.
6. MD Hare, S Svensson, GJ Walker. Characterization of the extracellular, water-insoluble (α-*D*-glucans of oral streptococci by methylation analysis, and by enzymic synthesis and degradation. Carbohydr Res 66: 254–264, 1978.
7. GH Van Geel, EJ Faber, E Smit, K Bonting, MR Smith, B Ten Brink, JP Kamerling, JFG Vliegenhart, L Dijkhuisen. Biochemical and structural characterization of the glucan and fructan exopolysaccharides synthesized by the *Lactobacillus reuteri* wild-type strain and by mutant strains. Appl Environ Microbiol 65:3008–3014, 1999.
8. A Jeanes, WC Haynes, CA Williams, JC Rankin, EH Melvin, MJ Austin, JE Cluskev, BE Fischer, HM Tsuchiya, CE Rist. Characterization and classification of dextrans from ninety six strains of bacteria. J Am Chem Soc 76:5041–5052, 1954.

9. FR Seymour, RD Knapp. Structural analysis of dextrans from strains of *Leuconostoc* and related genera, that contain 3-O-α-*D*-glucosylated-*D*-glucopyranosyl residues at the branched points, or in consecutive, linear position. Carbohydr Res 81:105–129, 1980.
10. EJ Hehre. Production from sucrose of a serologically reactive polysaccharide by a sterile bacterial extract. Science 93:237–238, 1941.
11. S Hestrin, S Avireni-Shapiro, M Aschner. The enzymic production of levan. Biochem J 37:450–456, 1943.
12. JF Robyt. Mechanism in the glucansucrase synthesis of polysaccharides and oligosaccharides from sucrose. Adv Carbohydr Chem Biochem 51:133–168, 1995.
13. V Monchois, RM Willemot, P Monsan. Glucansucrases: Mechanism of action and structure–function relationships. FEMS Microbiol Rev 23:131–151, 1999.
14. AJ Groenwall, BGA Ingelman. Manufacture of infusion and injection fluids. US Patent 2,437,518, 1948.
15. KH Ebert, G Schenk. Mechanism of biopolymer growth: The formation of dextran and levan. Adv Enzymol 30:179–210, 1968.
16. H Koepsell, H Tsuchiya, N Hellman, A Kasenko, C Hoffman, E Sharpe, R Jackson. Enzymatic synthesis of dextran: Acceptor specificity and chain initiation. J Biol Chem 200:793–801, 1952.
17. JF Robyt, TF Walseth. The mechanism of acceptor reactions of *Leuconostoc mesenteroides* B-512 F dextransucrase. Carbohydr Res 61:433–445, 1978.
18. JF Robyt, SH Eklund. Stereochemistry involved in the mechanism of action of dextransucrase in the synthesis of dextran and the formation of acceptor products. Bioorg Chem 11:115–132, 1982.
19. KG Ebert, G Schenk, Kinetik und Mechanismus der enzymatischen Dextransynthese und Wirkung der Akzeptoren auf diese Reaktion. Z Naturforsch 23b: 788–798, 1968.
20. M Killey, RJ Dimler, JE Cluskey. Preparation of panose by the action of NRRL B-512 dextransucrase on a sucrose–maltose mixture. J Am Chem Soc 77:3315–3318, 1955.
21. FH Stodola, HJ Koepsell, ES Sharpe. The preparation, properties and structure of the disaccharide leucrose. J Am Chem Soc 78:2514–2518, 1956.
22. EI Garvie. Separation of species of the genus *Leuconostoc* and differentiation of the *Leuconostoc* from other lactic acid bacteria. In: T Bergan, ed. Methods in Microbiology. London: Academic Press,1984, pp 147–178.
23. RRB Russell. Molecular genetics of glucan metabolism in oral Streptococci. Arch Oral Biol 35:53S–58S, 1990.
24. CL Simpson, PM Giffard, NA Jacques. *Streptococcus salivarius* ATCC 25975 possesses at least two genes coding for primer independent glucosyltransferases. Infect Immun 63:609–621, 1995.
25. MM Vickerman, MC Sulavik, JD Nowak, NM Gardner, CW Jones, DB Clewell. Nucleotide sequence analysis of the *Streptococcus gordonii* glucosyltransferase gene, *gtfG*. DNA Seq 7:83–95, 1997.
26. V Monchois, M Remaud-Siméon, P Monsan, RM Willemot. Cloning and se-

quencing of an extracellular dextransucrase (DSR-B) from *Leuconostoc mesenteroides* NRRL B-1299 synthesizing only α(1-6) glucan. FEMS Microbiol Lett 159:307–315, 1998.
27. V Monchois, RM Willemot, M Remaud-Siméon, C Croux, P Monsan. Cloning and sequencing of a gene coding for a novel dextransucrase from *Leuconostoc mesenteroides* NRRL B-1299 synthesizing only α(1-6) and α(1-3) linkages. Gene 182:23–32, 1996.
28. H Abo, T Matsumura, T Kodama, H Ohta, K Fukui, K Kato, H Kagawa. Peptide sequences for sucrose splitting and glucan binding within *Streptococcus sobrinus* glucosyltransferase (water-insoluble glucan synthetase). J Bacteriol 173:989–996, 1991.
29. G Mooser, KR Iwakoa. Sucrose 6-α-*D*-glucosyltransferase from *Streptococcus sobrinus*: Characterisation of a glucosyl–enzyme complex. Biochemistry 28:443–449, 1989.
30. G Mooser, SA Hefta, RJ Paxton, JE Shively, TD Lee. Isolation and sequence of an active-site peptide containing a catalytic aspartic acid from two *Streptococcus sobrinus*-glucosyltransferases. J Biol Chem 266:8916–8922, 1991.
31. V Monchois, A Reverte, M Remaud-Siméon, P Monsan, RM Willemot. Effect of *Leuconostoc mesenteroides* NRRL B-512F dextransucrase carboxy-terminal deletions on dextran and oligosaccaride synthesis. Appl Environ Microbiol 64: 1644–1649, 1998.
32. JJ Ferretti, ML Gilpin, RRB Russell. Nucleotide sequence of a glucosyltransferase gene from *Streptococcus sobrinus* Mfe28. J Bacteriol 169:4271–4278, 1987.
33. KS Gilmore, RRB Russell, JJ Ferretti. Analysis of the *Streptococcus downei gtfS* gene, which specifies a glucosyltransferase that synthesizes soluble glucans. Infect Immun 58:2452–2458, 1990.
34. S Bozonnet, M Remaud-Siméon, RM Willemot, P Monsan. Molécules d'acides nucléiques codant une dextrane saccharase catalysant la synthèse de dextrane portant des ramifications de type alpha-1,2 osidiques. French Patent Appl no 0103631, 2001.
35. M Remaud-Siméon, RM Willemot, P Sarcabal, G Potocki de Montalk, P Monsan. Glucansucrases: Molecular engineering and oligosaccharide synthesis. J Mol Catal B: Enzymatic 10:117–128, 2000.
36. A Shimamura, YJ Nakano, H Musaka, HK Kuramitsu. Identification of amino acid residues in *Streptococcus mutans* glucosyltransferases influencing the structure of the glucan product. J Bacteriol 176:4845–4850, 1994.
37. M Kobayashi, K Shishido, T Kikushi, K Matsuda. Fractionation of the *Leuconostoc mesenteroides* NRRL B-1299 dextran and preliminary characterisation of the fractions. Agric Biol Chem 37:357–365, 1973.
38. M Kobayashi, K Shishido, T Kikushi, K Matsuda. Methylation analysis of fractions from the *Leuconostoc mesenteroides* NRRL B-1299 dextran. Agric Biol Chem 37:2763–2769, 1973.
39. M Kobayashi, K Matsuda. Structural characteristics of dextrans synthesized by dextransucrases from *Leuconostoc mesenteroides* NRRL B-1299. Agric Biol Chem 41:1931–1937, 1977.

40. M Kobayashi, Y Mitsuhishi, S Tagagi, K Matsuda. Enzymatic degradation of water-soluble dextran from *Leuconostoc mesenteroides* NRRL B-1299. Carbohydr Res 127:305–317, 1984.
41. EE Smith. Biosynthetic relation between the soluble and insoluble dextrans produced by *Leuconostoc mesenteroides* NRRL B-1299. FEBS Lett 12:33–37, 1970.
42. M Kobayashi, K Matsuda. The dextransucrase isoenzymes of *Leuconostoc mesenteroides* NRRL B-1299. Biochim Biophys Acta 370:441–449, 1974.
43. M Kobayashi, K Matsuda. Conversion of the extracellular dextransucrase isoenzymes from *Leuconostoc mesenteroides* NRRL B-1299. Agric Biol Chem 39: 2087–2088, 1975.
44. M Kobayashi, K Matsuda. Purification and characterisation of two activities of the intracellular dextransucrase from *Leuconostoc mesenteroides* NRRL B-1299. Biochim Biophys Acta 397:69–79, 1975.
45. M Kobayashi, K Matsuda. Purification and properties of the extracellular dextransucrase from *Leuconostoc mesenteroides* NRRL B-1299. J Biochem 79: 1301–1308, 1976.
46. M Dols, M Remaud-Siméon, RM Willemot, M Vignon, P Monsan. Characterization of dextransucrases from *Leuconostoc mesenteroides* NRRL B-1299. Appl Biochem Biotechnol 62:47–59, 1997.
47. F Paul, A López-Munguía, M Remaud, V Pelenc, P Monsan. Method for the production of α-1,2 oligodextrans using *Leuconostoc mesenteroides* NRRL B-1299. US Patent 5, 141, 858, 1992.
48. M Remaud-Siméon, A López-Munguía, V Pelenc, F Paul, P Monsan. Production and use of glucosyltransferases from *Leuconostoc mesenteroides* NRRL B-1299 for the synthesis of oligosaccharides containing α(1-2) linkages. Appl Biochem Biotechnol 44:101–117, 1994.
49. M Dols, M Remaud-Siméon, RM Willemot, M Vignon, P Monsan. Structural characterization of the maltose acceptor products synthesized by *Leuconostoc mesenteroides* NRRL B-1299 dextransucrase. Carbohydr Res 305:549–559, 1998.
50. P Valette, V Pelenc, Z Djouzi, C Andrieux, F Paul, P Monsan, O Szylit. Bioavailability of new synthesised glucooligosaccharides in the intestinal tract of gnotobiotic rats. J Sci Food Agric 62:121–127, 1993.
51. Z Djouzy, C Andrieux, V Pelenc, S Somarriba, F Popot, F Paul, P Monsan, O Szylit. Degradation and fermentation of α-gluco-oligosaccharides by bacterial strains from human colon: In vitro and in vivo studies in gnotobiotic rats. J Appl Bacteriol 79:117–127, 1995.
52. P Monsan, F Paul. Oligosaccharide feed additives. In: RJ Wallace, A Chesson, eds. Biotechnology in animal feeds and animal feeding. Weinheim: VCH, 1995, pp 233–245.
53. Z Djouzi, C Andrieux. Compared effects of three oligosaccharides on metabolism of intestinal microflora in rats inoculated with a human faecal flora. Br J Nutr 78:313–324, 1997.
54. EA Flickinger, BW Wolf, KA Garleb, JM Chow, GJ Leyer, PW Johns, GC Fahey Jr. Glucose-based oligosaccharides exhibit different in vitro fermentation

patterns and affect in vivo apparent nutrient digestibility and microbial populations in dogs. J Nutr 130:1267–1273, 2000.
55. M Dols, M Remaud-Siméon, P Monsan. Dextransucrase production by *Leuconostoc mesenteroides* NRRL B-1299: Comparison with *L. mesenteroides* NRRL B-521F. Enzyme Microb Technol 20:523–530, 1997.
56. M Dols, M Remaud-Siméon, P Monsan. Characterization of dextransucrases from *Leuconostoc mesenteroides* NRRL B-1299. Appl Biochem Biotechnol 62: 47–59, 1997.
57. M Dols, W Chraibi, M Remaud-Siméon, ND Lindley, P Monsan. Growth and energetics of *Leuconostoc mesenteroides* NRRL B-1299 during metabolism of various sugars and consequences on dextransucrase production. Appl Environ Microbiol 63:2159–2165, 1997.
58. M Dols-Lafargue, M Remaud-Siméon, RM Willemot, P Monsan. Factors affecting α-1,2 oligosaccharide synthesis by *Leuconostoc mesenteroides* NRRL B-1299 dextransucrase. Biotechnol Bioeng 74:498–504, 2001.
59. M Dols-Lafargue, M Remaud-Siméon, RM Willemot, P Monsan. Reactor optimization for α-1,2 glucooligosaccharide synthesis by immobilized dextransucrase. Biotechnol Bioeng 75:276–284, 2001.
60. KC Mountzouris, SG Gilmour, AS Grandison, RA Rastall. Modeling of oligodextran production in an ultrafiltration stirred-cell membrane reactor. Enzyme Microb Technol 24:75–85, 1999.
61. E Olano-Martin, KC Mountzouris, GR Gibson, RA Rastall. In vitro fermentability of dextran, oligodextran and maltodextrin by human gut bacteria. Br J Nutr 83:247–255, 2000.
62. T Kohmoto, F Fukui, H Takaku, T Mitsuoka. Dose–response test of isomaltooligosaccharides for increasing fecal bifidobacteria. Agric Biol Chem 55: 2157–2159, 1991.
63. MJ Playne, RG Crittenden. Commercially available oligosaccharides. Bull Int Dairy Fed 313:10–22, 1996.
64. T Kaneko, T Kohmoto, H Kikuchi, M Shiota, H Iino, T Mitsuoka. Effects of isomaltooligosaccharides with different degrees of polymerisation on human fecal bifidobacteria. Biosci Biotech Biochem 58:2288–2290, 1994.
65. T Kaneko, A Yokoyama, M Suzuki. Digestibility characteristics of isomaltooligosaccharides in comparison with several saccharides using the rat jejunum loop method. Biosci Biotechnol Biochem 59:1190–1194, 1995.
66. T Kohmoto, K Tsuji, T Kaneko, M Shiota, F Fukui, H Takaku, Y Nakagawa, T Ichikawa, S Kobayashi. Metabolism of ^{13}C-isomaltooligosaccharides in healthy men. Biosci Biotechnol Biochem 56:937–940, 1992.
67. H Takaku. Anomalously linked oligosaccharides mixture. In: The Amylase Research Society of Japan, ed. Handbook of Amylase and Related Enzymes: Their Sources, Isolation Methods, Properties and Applications. New York: Pergamon Press, 1988, pp 215–217.
68. IC Kim, SH Yoo, SJ Lee, BH Oh, JW Kim, KH Park. Synthesis of branched oligosaccharides from starch by two amylases cloned from *Bacillus licheniformis*. Biosci Biotechnol Biochem 58:416–418, 1994.

69. SH Yoo, MR Kweon, MJ Kim, JH Auh, DS Jung, JR Kim, C Yook, JW Kim, KH Park. Branched oligosaccharides concentrated by yeast fermentation and effectiveness as a low sweetness humectant. J Food Sci 60:516–519, 1995.
70. KS Kwon, JH Auh, GJ Kang, JW Kim, KH Park. Characterization of branched oligosaccharides produced by *Bacillus licheniformis* maltogenic amylase. J Food Chem 64:258–261, 1999.
71. JH Park, JY You, OH Shim, OH Shin, HK Shin, SH Lee, KH Park. Growth effect of branched oligosaccharides on principal intestinal bacteria. Korean J Appl Microbiol Biotechnol 20:237–242, 1992.
72. H Tomomatsu. Health effects of oligosaccharides. Food Technol 48:61–65, 1994.

8

Prebiotics and the Prophylactic Management of Gut Disorders: Mechanisms and Human Data

Robert A. Rastall and Glenn R. Gibson
The University of Reading, Reading, England

I. INTRODUCTION

In recent times, the activities of the human gut microbiota have been more fully elucidated. Although it is apparent that certain species may be involved in gut disorders such as ulcerative colitis, bowel cancer, acute enteritis, and pseudomembranous colitis (1), there has been much momentum for dietary approaches that modulate the gut flora composition toward improved health (2). Mainly historical associations have been made between probiotics and gastrointestinal improvements, but there is a very rapidly developing functional food sector in Europe based on food ingredients containing prebiotics. These are nonviable food components that are selectively fermented in the colon (by a beneficial and not detrimental flora). In many cases, literature on the effects of these ingredients in human studies is sparse, and our mechanistic understanding inadequate. However, the prebiotic approach may be a very straightforward route for preventative management of gut disease. The aim of this chapter is to take a critical view of the proposed mechanisms for the health promoting effects of prebiotics and an overview of recent human studies in the area. It is not our aim to provide a comprehensive review of the literature, but rather to focus on relevant recent research reports, concentrating on human trials and mechanisms of effect.

II. BACKGROUND TO THE PREBIOTIC CONCEPT

The human colon is an immensely complex microbial ecosystem containing several hundred bacterial species (3). Within this diversity it is possible to recognize bacteria with potentially positive or negative effects upon human health (Fig. 1). The concept that diet can shift the balance away from undesirable microorganisms toward physiologically positive species is not new (Fig. 1). To this end, there is a thriving functional food industry based on the concept of including lactic acid bacteria (probiotics) in foods such as yogurt and other fermented milks. There are, however, concerns about the survival of such microorganisms during processing, storage, and passage through the alimentary system. For example, gastric acid, bile salts, and pancreatic secretions are all barriers to long-term probiotic persistence in the gut. Moreover, in the colon they then face a huge indigenous microflora with which they have to compete effectively if any advantageous properties are to ensue.

Although certain probiotic strains may be robust enough to overcome some or all of these confrontations, they are bound to be few in number and

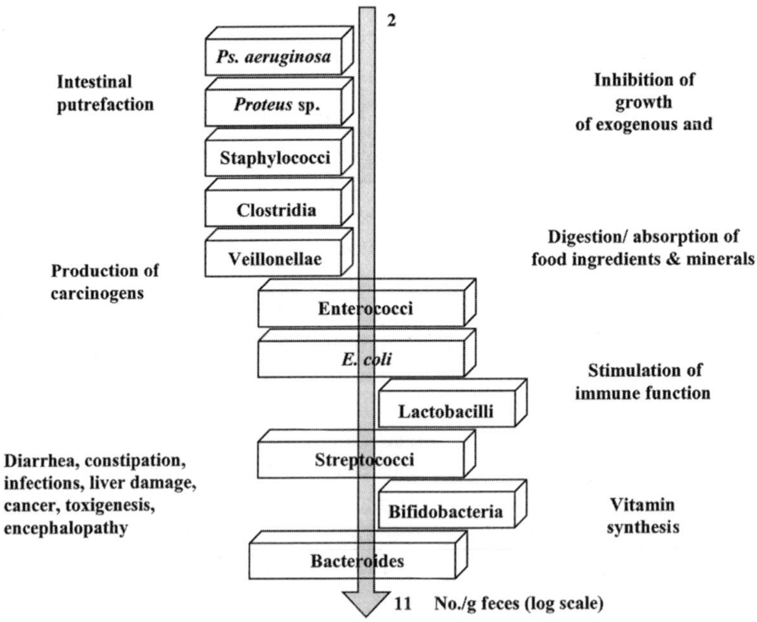

Figure 1 Properties of bacterial genera in the colon.

perhaps compromised in terms of activity. An alternative approach to improved microflora modulation is therefore to intake carbohydrates that resist digestion in the stomach and small intestine, reach the colon intact, and are then selectively metabolized by (beneficial) bifidobacteria and/or lactobacilli. This process is known as the *prebiotic approach* (4). In Europe, there are four prebiotics commonly used in foods: inulin, fructooligosaccharides (FOSs), lactulose, and galactooligosaccharides (GOSs). In Japan, the situation is more widespread with many oligosaccharides being incorporated into foods for their prebiotic effects (5). For a summary of in vitro and in vivo trials with prebiotics see Ref. (6). It can be said with some confidence that good prebiotics exist, and are being developed, that have selective effects on the gut microbiota composition. A key question remains about the health bonuses, however. Four promising areas of importance have arisen. They are discussed in Sections III–VI.

III. THE BARRIER EFFECT AGAINST GASTROINTESTINAL PATHOGENS

One of the most mechanistically strong health benefits proposed for prebiotics is the barrier function against invading gastrointestinal pathogens. If particular, this may be a successful way of prophylactically addressing the burden of microbial food safety through tackling problems such as those caused by campylobacters, salmonellae, and *Escherichia coli*.

A. Proposed Mechanisms

There are several prospective mechanisms for the inhibition of pathogens by bifidobacteria and lactobacilli (Fig. 2). Fermentation of carbohydrates results in the production of short-chain fatty acids (SCFAs) (4). These reduce luminal pH in the colon to reach levels below those at which pathogens such as *E. coli* can effectively multiply. In addition, increased populations of bifidobacteria and lactobacilli can compete with other organisms for nutrients and receptors on the gut wall (7). Probiotics (the target microorganisms for prebiotic intake) can also inhibit pathogens via a more direct mechanism. They are known to produce antimicrobial agents active against a range of pathogens (8). Although it is not known whether such metabolites function effectively in the human gut, their powerful effects have led to their widespread use as food preservatives (9).

Probiotics are also reputed to modulate the activities of the immune system, resulting in nonspecific enhancement of immune function. In particular, encouraging results have been obtained with both lactobacilli (10) and

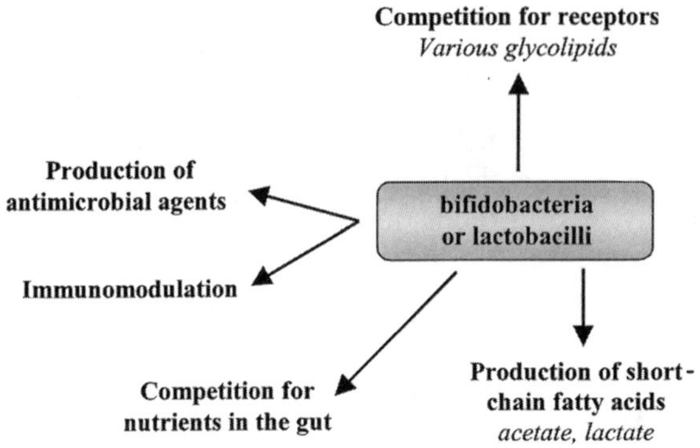

Figure 2 The probiotic barrier to gastrointestinal infections.

bifidobacteria (11) against rotaviral infection in infants. Fortification of indigenous probiotics, by efficient prebiotic use, should have similar effects.

B. Human Studies

It is, of course, impossible to collect human data on the probiotic barrier against infection with pathogens through challenge-type testing. Most fermentation studies in this area are alternatively carried out using in vitro models of the human gut (12) or animals (13). Both have limitations but do generate useful mechanistic data relevant to the situation for humans. Although bacterial pathogenesis in humans is difficult to predict, certain situations, such as antibiotic-associated diarrhea and gastrointestinal problems suffered by frequent travelers, seem good avenues for prebiotic use. Moreover, our own recent data have exploited the use of a rhesus monkey colony to infect animals with enteropathogenic *E. coli* (14). The experiments were carried out using placebo and blind control, with genotypic probes for the bacteriological characteristics. In essence, some protection against diarrhea was seen in the presence of bifidogenic substrates. These types of studies need to be taken further through human trials that apply sound genomic principles to the bacteriological characteristics (15).

IV. PROTECTION AGAINST COLON CANCER

A. Possible Mechanisms

The most likely means by which prebiotics could influence the development of bowel cancer is by modulation of the colonic flora. Prebiotics are fermented to organic acids, and in some cases this includes butyrate (4,16). Butyrate inhibits apoptosis (17) and is thought to be protective against colon cancer.

Many fecal microorganisms produce carcinogens and tumor promoters from dietary and other components entering the colon (Fig. 3). In addition, several enzymic activities, associated with fecal bacteria produce toxic or carcinogenic products from substrates entering the colon (Table 1). The microbial species responsible for these activities have not been unambiguously identified beyond certain *Bacteroides* spp. and *Clostridium* spp. It is known, however, that bifidobacteria and lactobacilli do not have such capabilities and a reduction in these activities can be demonstrated in feces from humans or rats fed lactic acid bacteria or prebiotics (18,19).

B. Human Studies

Many in vivo data on the protective effects of prebiotics on colon cancer are obtained from animal studies, in which, for example, inulin has been shown to inhibit formation of aberrant crypt foci (20). Human studies are few in number and tend to focus on fecal markers of carcinogenesis rather than being epidemiological in nature. Human data from healthy volunteers are summarized in Table 2. In three of the four trials, a decrease in several markers of carcinogenesis was seen. In two of these, significant increases in bifidobacteria were also observed; no microbiological analysis was carried out in the third.

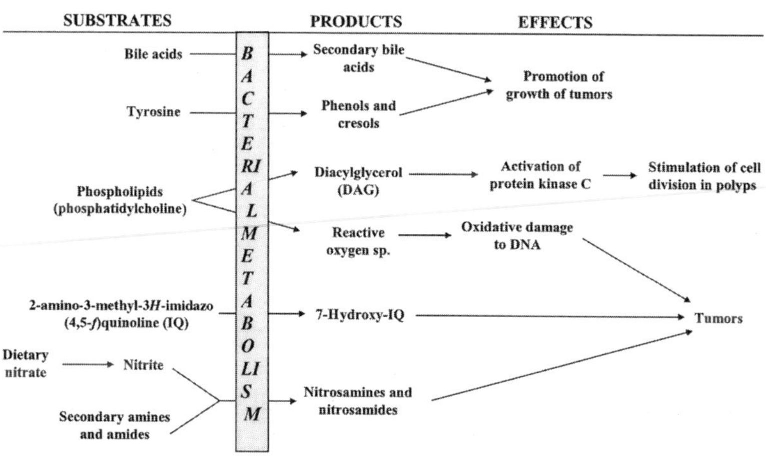

Figure 3 Production of carcinogens and tumor promoters by fecal bacterial metabolism.

Table 1 Fecal Bacterial Enzymes Producing Carcinogenic or Toxic Products

Enzyme	Substrate
β-Glucosidase	Plant glycosides (e.g., rutin, cycasin)
Azoreductase	Azocompounds (e.g., benzidines)
Nitroreductase	Nitrocompounds (e.g., nitrochrysene)
β-Glucuronidase	Biliary glucuronides (e.g., benzidine)
IQ hydratase-dehydrogenase	2-Amino-3-methyl-3*H*-imidazo (4,5-*f*) quinoline (IQ)
Nitrate/nitrite reductase	Nitrate, nitrite

A 1999 study, however (24), found no significant changes in bifidobacteria or in markers of carcinogenesis. These results might at first sight seem anomalous, as the test substrate (GOSs) has been found to be prebiotic in many studies in the past (25). However, starting populations of bifidobacteria in the volunteers were rather high (9.2–9.4 log). It has been noted (26,27) that the magnitude of the response to prebiotics by bifidobacteria depends upon the starting levels. It appears that there is a maximal level of bifidobacteria (about 10 log values) achievable in the human gut. If populations are already at or near this level, then little or no further increase in numbers is generally seen. This is an important point and is currently the subject of further research. The implication is that a healthy diet supporting a balanced gastrointestinal microflora may not further benefit from prebiotic functional foods (24).

V. IMPROVED CALCIUM ABSORPTION

There has been increasing interest in recent years in the possibility of increasing mineral (particularly calcium) absorption through the consumption of prebiotics. Although the small intestine is the principal site of calcium absorption in humans, significant amounts are also thought to be absorbed throughout the length of the gut; consequently, a maximizing of colonic effects is desirable.

Table 2 Human Studies Investigating the Anticancer Effects of Prebiotics[a]

Prebiotic	Subjects and study design	Dose, g/day	Increase in bifidobacteria	Markers of carcinogenesis	Reference
FOS	20, Placebo-controlled, 12 days	12.5	1.2 log	No change in bile acids, neutral sterols, nitroreductase, azoreductase, or β-glucuronidase	21
FOS	12, Controlled diet, control and treatment periods, 26 days	4	0.5 log	Significant decreases in β-glucuronidase (75%) and glycocholic acid hydroxylase (90%)	22
Resistant starch (RS)	12, Controlled diet, high- and low-RS periods, 28 days	55.2 ± 3.5 and 7.7 ± 0.3	Not investigated	Significant decreases in neutral sterols (30%), 4-cholesten-3-one (36%), total bile acids (30%), secondary bile acids (32%), and β-glucosidase activity (26%) in high-RS phase compared to low RS-phase.	23
GOS	40, Placebo-controlled, 21 days	7.5 or 15	No significant increase	No significant changes in SCFA, bile acids, ammonia, or skatoles	24

[a] FOS, fructooligosaccharide; RS, resistant starch; GOS, galactooligosaccharide; SCFA, short-chain fatty acid.

A. Possible Mechanisms

Several mechanisms have been postulated for increased calcium absorption as induced by prebiotics (28) (Fig. 4), although it is far from clear at the present time which (if any) actually operate in vivo.

1. Fermentation of prebiotics such as inulin results in a significant production of SCFA, leading to a reduction in luminal colonic pH. This is likely to increase calcium solubility and overall levels in the gut.

2. Phytate (myoinositol hexaphosphate) is a component of plants that reaches the colon largely intact (29). It also forms stable, insoluble complexes with divalent cations, such as calcium, rendering them unavailable for transport. Fermentation results in the bacterial metabolism of phytate, thereby liberating calcium.

3. It is postulated that a calcium exchange mechanism operates in the colon. In this system, SCFAs enter the colon in a protonated form and then dissociate in the intracellular environment. The liberated proton is then secreted into the lumen in exchange for a calcium ion.

B. Human Studies

Numerous animal studies have indicated that prebiotics increase absorption of calcium from the colon, thereby decreasing losses from bone tissue (30). Very few human studies have been carried out, however. In one such study, the feeding of 40 g inulin/day for 28 days to nine healthy subjects resulted in a significant increase in calcium absorption (31). A more realistic 15-g inulin

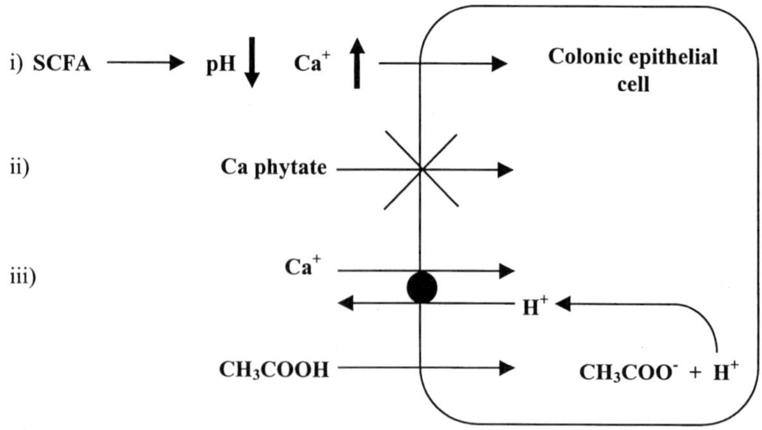

Figure 4 Possible mechanisms of enhanced calcium uptake as a result of prebiotics.

FOS or GOS/day, when fed to 12 healthy subjects for 21 days, resulted in no significant effect on the absorption of calcium or iron (32).

In a more recent study, 12 adolescent boys (aged 14–16) were fed 15 g FOS/day for 9 days in a placebo-controlled trial against sucrose (33). The data showed a 10.8% increase in calcium balance with no significant effect on urinary excretion.

VI. EFFECTS ON BLOOD LIPIDS

There is intense interest in the food industry in developing functional foods to modulate concentrations of blood lipids such as cholesterol and triglycerides.

A. Possible Mechanisms

The mechanisms suggested by which prebiotics may influence blood lipids are summarized in Fig. 5. There is evidence that FOSs decrease the de novo synthesis of triglycerides by the liver. The means by which this occurs is not

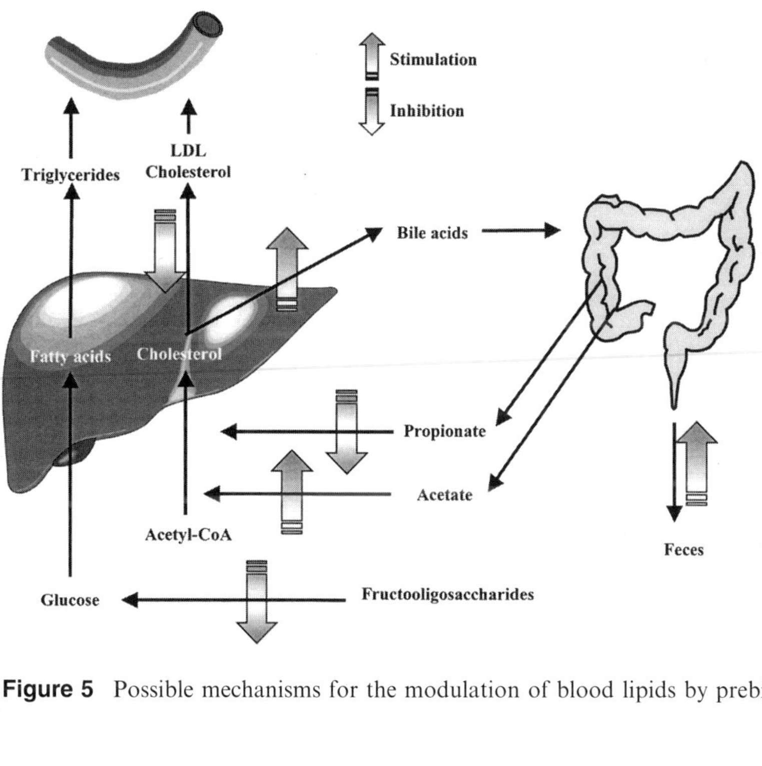

Figure 5 Possible mechanisms for the modulation of blood lipids by prebiotics.

fully understood, but the effect appears to be exerted at the transcriptional level. It is also possible that prebiotics (such as inulin) can modulate insulin-induced inhibition of triglyceride synthesis (34).

Effects on serum cholesterol levels have also been postulated for prebiotics, although the mechanism is more difficult to envisage. It has been suggested (for reviews, see 34 and 35) that propionate produced by the bacterial fermentation of prebiotics inhibits the formation of serum low-density lipoprotein (LDL) cholesterol. The difficulty with this hypothesis is that bacterial fermentation of prebiotics generally produces much more acetate than propionate. Moreover, acetate is a metabolic precursor of cholesterol and may therefore tend to increase, not decrease, serum levels. A more likely role for the gut microflora in the reduction of cholesterol is direct metabolism of cholesterol by colonic bacteria, or its conversion to other metabolites such as coprostanol (36). The evidence for this is not presently well clarified, however.

B. Human Studies

Human studies on the lipid lowering properties of prebiotics when consumed at a realistic (tolerable) dose are not clear-cut (37). These can be divided into trials carried out on subjects with hyperlipidemia and on normal subjects (Table 3). Hitherto, data indicated that there may be a significant effect on

Table 3 Effects of Prebiotics on Blood Lipids[a]

					Effect		
Reference	Date	Prebiotic	Dose	Duration	TG	LDL-Ch	Ch
Hyperlipidemic subjects							
38	1984	FOS	8 g/day	14 days	NS	−10%	−8%
39	1998	Inulin	18 g/day	6 weeks	NS	−14%	−8.7%
Normal subjects							
40	1995	Inulin	9 g/day	4 weeks	−27%	−7%	−5%
41	1996	FOS	20 g/day	4 weeks	NS	NS	NS
42	1997	Inulin	14 g/day	4 weeks	NS	NS	NS
43	1998	Inulin	10 g/day	8 weeks	−19%	NS	NS
44	1999	Inulin	15 g/day	3 weeks	NS	NS	NS
44	1999	FOS	15 g/day	3 weeks	NS	NS	NS
44	1999	GOS	15 g/day	3 weeks	NS	NS	NS

[a] NS, no significant; TG, triglyceride; LDL Ch, LDL cholesterol; Ch, cholesterol; FOS, fructooligosaccharide; NS, not significant; GOS, galactooligosaccharide.

blood triglycerides but not cholesterol in normal subjects, although three of the five studies did not show any effect on concentrations. For subjects with elevated lipid levels, however, there does seem to be a useful decrease in cholesterol levels. It would seem to be of high priority to carry out more research in this area—especially now that reliable methods for tracking microbiota changes through molecular procedures are available (45).

VII. CONCLUSIONS

Prebiotics have a long history of use in Japan, but the market for prebiotics as food ingredients in Europe is also now established (46). There has been a tendency in the past for unsubstantiated health claims to be made for functional foods, and it is essential, if we are to have confidence in the protective effects of prebiotics, for any claims to be based on rigorous science—preferably carried out in humans. To date, there have been a few well-designed volunteer studies, and these have sometimes yielded contradictory conclusions. One problem with evaluation of the effects of prebiotics lies is the difficulty of identifying fecal microorganisms by using conventional culture-based approaches. It is estimated that a large percentage of the total fecal microflora has yet to be described and is probably unculturable (47). The advent of molecular methods of bacterial identification has undoubtedly improved this situation (45).

Given the significance of the human gut microbiota and its activities [the colon is the body's most metabolically active organ (48), it seems a very reasonable approach to advocate dietary modulation by prebiotics. At present, we are at the stage at which efficacious forms exist and can be made to operate in the food matrix (49). The health benefits that have been suggested are varied but also very important. In addition to good human volunteer studies we need enhanced mechanistic understanding of the health effects of prebiotics. Progress is being made in this area, and it is to be expected that the prebiotic approach to prevention of disease will have a much stronger foundation. This will lead to better informed decisions by clinicians, nutritionists, and consumers.

REFERENCES

1. Gibson, G.R., Saavedra, J.M., Macfarlane, S. and Macfarlane, G.T. Gastrointestinal Microbial Disease. In: Fuller, R. editor. Probiotics. 2. Application and Practical Aspects. Chapman & Hall, Andover, 1997 pp. 10–39.

2. Salminen S., Bouley C., Boutron-Ruault M.C., Cummings J.H., Franck A., Gibson G.R., Isolauri E., Moreau M.C., Roberfroid M. and Rowland I.R. Functional food science and gastrointestinal physiology and function. Br. J. Nutr. 1998; 80:S147–S171.
3. Cummings, J.H. and Macfarlane, G.T. The control and consequences of bacterial fermentation in the human colon. J. Appl. Bacteriol. 1991; 70:443–459.
4. Gibson, G.R. and Roberfroid, M.B. Dietary modulation of the human colonic microbiota: Introducing the concept of prebiotics. J. Nutr. 1995; 125:1401–1412.
5. Japanscan. Functional Foods and Drinks in Japan. Leatherhead Food RA, Leatherhead, England, 1998.
6. Gibson, G.R., Berry Ottaway, P. and Rastall, R.A. Prebiotics: New Developments in Functional Foods. Chandos Publishing, Oxford, England, 2000.
7. Araya-Kojima, A., Yaeshima, T., Ishibashi, N., Shimamur, S. and Hayasawa, H. Inhibitory effects of *Bifidobacterium longum* BB536 on harmful intestinal bacteria. Bifid. Microflora 1995; 14:59–66.
8. Gibson, G.R. and Wang, X. Regulatory effects of bifidobacteria on the growth of other colonic bacteria. J. Appl. Bacteriol. 1994; 77:412–420.
9. de Vuyst, L. and Vandamme, E.J. Bacteriocins of lactic acid bacteria. Blackie Academic & Professional, Glasgow, Scotland, 1994.
10. Isolauri, E., Juntunen, M., Rautanen, T., Sillanaukee, P. and Koivula, T.A. Human *Lactobacillus* strain (*Lactobacillus* GG) promotes recovery from acute diarrhoea in children. Pediatrics 1991; 88:90–97.
11. Saavedra J.M., Bauman, N.A., Oung, I., Perman, J.A. and Yolken R.H. Feeding of *Bifidobacterium bifidum* and *Streptococcus thermophilus* to infants in hospital for prevention of diarrhoea and shedding of rotavirus. Lancet 1994; 344:1046–1049.
12. Macfarlane G.T., Macfarlane S. and Gibson G.R. Validation of a three-stage compound continuous culture system for investigating the effect of retention time on the ecology and metabolism of bacteria in the human colonic microbiota. Microb. Ecol. 1998; 35:180–187
13. Rowland, I.R. and Tanaka, R. The effects of transgalactosylated oligosaccharides on gut flora metabolism in rats associated with a human faecal microflora. J. Appl. Bacteriol. 1993; 74:667–674.
14. Brück, W.M., Kelleher, S.L., Lönnerdal, B., Nielsen, K.E., Chatterton, D. and Gibson, G.R. Fermentation studies on selected infant milk components using in vitro models of the human gut and rhesus monkeys. J. Pediatric Res, in press.
15. Tuohy, K.M., Kolida, S., Lustenberger, A. and Gibson, G.R. The prebiotic effects of biscuits containing partially hydrolyzed guar gum and fructooligosaccharides—a human volunteer study. Br. J. Nutr. 2001; 86:341–348.
16. Olano-Martin, E., Mountzouris, K.C., Gibson, G.R. and Rastall, R.A. In vitro fermentability of dextran, oligodextran and maltodextrin by human gut bacteria. Br. J. Nutr. 2000; 83: 247–255.
17. Hague, A., Elder, D.J.E., Hicks, D.J. and Paraskeva, C. Apotosis in colorectal tumour cells: Induction by the short chain fatty acids butyrate, propionate and acetate and by the bile salt deoxycholate. Int. J. Cancer 1995; 60:400–406.

18. Rowland, I.R. Metabolic interactions in the gut. In: Fuller, R. editor. Probiotics: The Scientific Basis. Andover, Chapman & Hall, 1992 pp. 29–53.
19. Rowland, I.R. and Tanaka, R. The effects of transgalactosylated oligosaccharides on gut flora metabolism in rats associated with a human faecal microflora. J. Appl. Bacteriol. 1993; 74:667–674.
20. Reddy, B.S., Hamid, R. and Rao, C.V. Effect of dietary oligofructose and inulin on colonic preneoplastic aberrant crypt foci inhibition. Carcinogenesis 1997; 18:1371–1374.
21. Bouhnik, Y., Flourie, B., Riottot, M., Bisetti, N., Gailing, M.F., Guibert, A., et al. Effects of fructo-oligosaccharides ingestion on faecal bifidobacteria and selected metabolic indexes of colon carcinogenesis in healthy humans. Nutr. Cancer 1996; 26:21–29.
22. Buddington, R.K., Williams, C.H., Chen, S.-C. and Witherly, S.A. Dietary supplementation of neosugar alters the fecal flora and decreases activities of some reductive enzymes in human subjects. Am. J. Clin. Nutr. 1996; 63:709–716.
23. Hylla, S. Gostner, A., Dusel, G., Anger, H., Bartram, H.-P., Christl, S.U., et al. Effects of resistant starch on the colon in healthy volunteers: possible implications for cancer prevention. Am. J. Clin. Nutr. 1998; 67:136–142.
24. Alles, M.S., Hartemink, R., Meyboom, S., Harryvan, J.L., van Laere, K.M.J., Nagengast, F.M., et al. Effect of transgalactooligosaccharides on the composition of the human intestinal microflora and on putative risk markers for colon cancer. Am. J. Clin. Nutr. 1999; 69:980–991.
25. Schoterman, H.C. and Timmermans, H.J.A.R. Galacto-oligosaccharides. In: Gibson, G.R. and Angus, F. editors. Prebiotics and Probiotics. LFRA Ingredients Handbook. Leatherhead Food RA, Leatherhead, England, 2000, pp. 19–46.
26. Roberfroid, M.B., Van Loo, J.A.E. and Gibson, G.R. The bifidogenic nature of inulin and its hydrolysis products. J. Nutr. 1998; 128:11–19.
27. Hidaka, H., Eida, T., Takizawa, T., Tokunaga, T. and Tashiro, Y. Effects of fructooligosaccharides on intestinal flora and human health. Bifid. Microflora 1986; 5:37–50.
28. Fairweather-Tait, S.J. and Johnson, I.T. Bioavailability of minerals. In: Gibson, G.R. and Roberfroid, M.B. editors. Colonic microbiota, nutrition and health. Kluwer, Dordrecht, 1999, pp. 233–244.
29. Cummings, J.H., Hill, M.J., Houston, H., Branch, W.J. and Jenkins, D.J.A. The effect of meat protein and dietary fibre on colonic function and metabolism. 1. Changes in bowel habit, bile acid excretion and calcium absorption. Am. J. Clin. Nutr. 1979; 32:2086–2093.
30. Greger, J.L. Nondigestible carbohydrates and mineral bioavailability. J. Nutr. 1999; 129:1434S–1435S.
31. Coudray, C., Bellanger, J., Castiglia-Delavaud, C., Rémésy, C., Vermorel, M. and Rayssignuier, Y. Effect of soluble or partly soluble dietary fibres supplementation on absorption and balance of calcium, magnesium, iron and zinc in healthy young men. Eur. J. Clin. Nutr. 1997; 51:375–380.
32. van den Heuvel, E., Schaafsma, G., Muys, T. and van Dokkum, W. Non-

digestible oligosaccharides do not interfere with calcium and non-haeme iron absorption in young, healthy men. Am. J. Clin. Nutr. 1998; 67:445–451.
33. van den Heuvel, E., Muys, T., van Dokkum, W. and Schaafsma, G. Oligofructose stimulates calcium absorption in adolescents. Am. J. Clin. Nutr. 1999; 69:544–548.
34. Delzenne, N.M. and Kok, N.N. Biochemical basis of oligofructose-induced hypolipidaemia in animal models. J. Nutr. 1999; 129:1467S–1470S.
35. Delzenne, N.M. and Williams, C.M. Actions of non-digestible carbohydrates on blood lipids in humans and animals. In: Gibson, G.R. and Roberfroid, M.B. editors. Colonic Microbiota, Nutrition and Health. Kluwer, Dordrecht; 1999, pp. 213–231.
36. Chezem, J., Furumoto, E. and Story, J. Effects of resistant potato starch on cholesterol metabolism and bile acid metabolism in the rat. Nutr. Res. 1997; 17:1671–1682
37. Williams, C.M. Effects of inulin on lipid parameters in humans. J. Nutr. 1999; 129:1471S–1473S.
38. Yamashita, K., Kawai, K. and Itakura, M. Effects of fructo-oligosaccharides on blood glucose and serum lipids in diabetic subjects. Nutr. Res. 1984; 4:961–966.
39. Davidson, M.H., Synecki, C., Maki, K.C. and Drennan, K.B. Effects of dietary inulin in serum lipids in men and women with hypercholesterolaemia. Nutr. Res. 1998; 3:503–517.
40. Canzi, E., Brighenti, F., Casiraghi, M.C., Del Puppo, E. and Ferrari, A. Prolonged consumption of inulin in ready to eat breakfast cereals: Effects on intestinal ecosystem, bowel habits and lipid metabolism. Cost 92. Workshop on Dietary Fibre and Fermentation in the Colon 15:17/04. Helsinki, 1995.
41. Luo, J., Rizkalla, S.W., Alamowitch, C., Boussairi, A., Blayo, A., Barry, J.-L., et al. Chronic consumption of short-chain fructo-oligosaccharides by healthy subjects decreased basal hepatic glucose production but had no effect on insulin-stimulated glucose metabolism. Am. J. Clin. Nutr. 1996; 63:939–945.
42. Pedersen, A., Sandstrom, B. and van Amelsvoort, J.M.M. The effect of ingestion of inulin on blood lipids and gastrointestinal symptoms in healthy females. Br. J. Nutr. 1997; 78:215–222.
43. Jackson, K.G., Taylor, G.R.J., Clohessy, A.M. and Williams, C.M. The effect of the daily intake of inulin on fasting lipid, insulin and glucose concentration in middle-aged men and women. Br. J. Nutr. 1999; 82:23–30.
44. van Dokkum, W., Wezendonk, B., Srikumar, T.S. and van den Heuvel, E.G.H.M. Effects of nondigestible oligosaccharides on large-bowel functions, blood lipid concentrations and glucose absorption in young healthy male subjects. Eur. J. Clin. Nutr. 1999; 53:1–7.
45. Steer, T., Carpenter, H., Tuohy, K. and Gibson, G.R. Perspectives on the role of the human gut microbiota and its modulation by pro- and prebiotics. Nutr. Res. Rev. 2000; 13:1–27.
46. Young, J. European market developments in prebiotic- and probiotic-containing foodstuffs. Br. J. Nutr. 1998; 80:S231–S233.
47. Suau, A., Bonnet, R., Sutren , M., Godon, J.J., Gibson, G.R., Collins, M.D. and

Dore, J. Direct analysis of genes encoding 16S rRNA from complex communities reveals many novel molecular species within the human gut. Appl. Environ. Microbiol. 1999; 65:4799–4807.

48. Tannock, G.W. Influences of the normal microbiota on the animal host. In: Mackie, R.I., White, B.A. and Isaacson, R.E. editors. Gastrointestinal Microbiology, Vol. 2. Chapman & Hall, New York, 1997, pp. 537–587.
49. Franck, A. Prebiotics in consumer products. In: Gibson, G.R. and Roberfroid, M.B. editors. Colonic Microbiota, Nutrition and Health. Kluwer, Dordrecht, 1999, pp. 291–300.

9

Proteins and Peptides

Yuriko Adkins and Bo Lönnerdal
University of California, Davis, Davis, California, U.S.A.

I. INTRODUCTION

Recombinant deoxyribonucleic acid (DNA) technology has made it possible for genes to be introduced into different expression systems ranging from microorganisms (bacteria, fungi, yeast) (Georgiou, 1988) to insect cells (Luckow and Summers, 1988), plants (rice, potatoes, tobacco leaves) (Kusnadi et al., 1997), and transgenic animals (cows, pigs, sheep, goat) (Lubon et al., 1996). Different expression systems have been used to obtain high levels of human proteins for therapeutic use, such as tissue plasminogen activator (Gordon et al., 1992), α_1-antitrypsin (Archibald et al., 1990), protein C (Velander et al., 1992), blood clotting factor VII (Niemann et al., 1999), and apolipoprotein C-II (Wang et al., 1996). Many proteins of pharmaceutical and/or commercial interest undergo complex posttranslational modifications that influence biological function. Therefore, not only should high-level protein expression be achieved by a specific system, but the system should also have the necessary machinery to produce correctly processed proteins when this is required. In addition, the choice of an expression system depends upon the amount of recombinant protein desired, cost-effectiveness, the time required for the isolation and purification of the recombinant protein, and its intended use. Current developments have focused on producing recombinant proteins that may be used as food ingredients. Expressing recombinant proteins in systems that introduce few or no contaminants to the final product or present little risk of an allergic response is paramount. Expression of bioactive recombinant proteins in edible plants that may be consumed uncooked, such as bananas or tomatoes, also offers an attractive delivery system.

Over the years, significant changes have been made in the composition of infant formulas to reflect the nutrient composition of human milk closely. Nutrient requirements for infant formulas are based upon the intakes of breast-fed infants plus an additional amount to compensate for the possibility of lower nutrient bioavailability. Even though the lipid, amino acid, carbohydrate, vitamin, and mineral profile of infant formulas can be manipulated to resemble the proportions present in human milk, there are some properties of human milk that cannot be attained in either cow milk– or soy-based formulas. Numerous studies have shown that human milk contains bioactive proteins that possess antimicrobial/anti-infective, immunomodulating, and nutrient utilization and absorption activities. Breast-fed infants not only have different growth patterns and nutritional status than formula-fed infants, but also have lower incidences of infections and when ill, experience faster recovery times. With the advances made in recombinant DNA technology, the potential exists for adding recombinant bioactive human milk proteins into infant formula so that formula-fed infants may enjoy some of the benefits so far only available to breast-fed infants. Many recombinant human milk

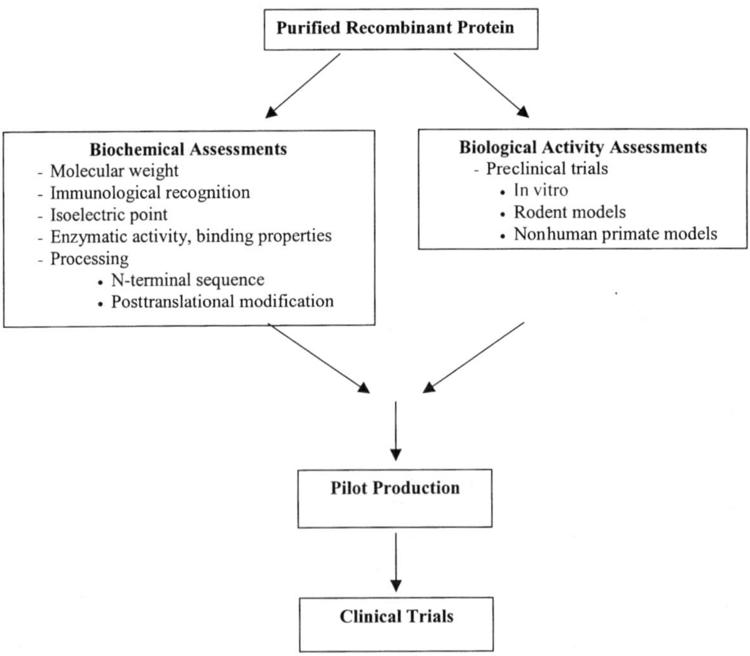

Figure 1 Strategies for evaluating recombinant proteins.

proteins have already been produced and secreted at high expression levels in various systems, and the addition of these bioactive milk proteins to infant formula is feasible if their biochemical, immunological, and physiological activities are similar to those of their native counterparts and their safety and efficacy can be proved. Since significant advances have been made in this area, and infant foods are more tightly regulated than other foods (Infant Formula Act) and therefore require extensive evidence of efficacy and safety, we have chosen to focus primarily on these proteins in this chapter (Fig. 1).

II. PRODUCTION SYSTEMS

A. Transgenic Animals

Expression of recombinant proteins in the milk of transgenic animals was introduced two decades ago. Transgenic mice have been used to produce high quantities of biologically active recombinant proteins in their milk (Archibald et al., 1990; Meade et al., 1990; Gordon et al., 1992); however, they are not a suitable expression system for large-scale production. Nevertheless, the transgenic mouse model set the stage for the expression of recombinant proteins in transgenic livestock such as cows, sheep, pigs, goats, and rabbits. The mammary gland of transgenic livestock has been viewed as a "bioreactor" for the production of proteins of pharmaceutical importance. Through appropriate gene constructs using mammary gland–specific promoter elements derived from genes encoding milk proteins in animals, high-level expression of recombinant proteins can be achieved. For example, bovine α_{S1}-casein promoter gene constructs have been used for the expression of recombinant proteins in transgenic rabbits (Riego et al., 1993) and cows (Krimpenfort et al., 1991), and the ovine β-lactoglobulin promoter has been used for recombinant protein expression in transgenic sheep (Niemann et al., 1999).

The technology for high-level expression of recombinant proteins in transgenic livestock is well established and straightforward. In addition, tissue-specific expression of human milk proteins in transgenic livestock can produce recombinant proteins with phosphorylation and glycosylation patterns very similar to those of their native counterparts that may not be achieved in other expression systems. However, a major disadvantage in using transgenic livestock is that their maintenance is expensive; periods of gestation and maturation are long, and even after the transgenic embryos are produced, the animals must be screened for the stable integration of the transgene, and finally, these selected animals must be raised to maturity and bred to create offspring with the same genetic background (Colman, 1996). The use of microinjection for in vitro embryo production and potent promoters has increased expression levels of lactoferrin in transgenic cows up to 3 g of protein/ L milk (van Berkel et al., 2002). With such expression levels, one cow with an annual milk output of ~8000 L could produce 24 kg of recombinant protein per year. Herds with a large number of such animals could supply thousands of kilograms annually, making this economically viable. The recent introduction of nuclear transfer techniques (instead of microinjection) as a means of producing transgenic cattle has proved more efficient and allows the sex of founders to be predetermined, allowing quick generation of herds, especially when lactation is induced with hormones in young cattle (Brink et al., 2000).

B. Microorganisms

Perhaps the best-studied microorganism expression system is *Escherichia coli.* Although the system is still widely used for large-scale production of certain proteins, bacteria lack the enzymatic machinery needed to perform complex posttranslational modifications. Many mammalian proteins undergo elaborate posttranslational modification processes, and without the proper glycosylation, hydroxylation, phosphorylation, cleavage of the polypeptide chains, or other alterations to the primary translation product, the biological activity of the protein may be compromised. Furthermore, *E. coli* may produce recombinant proteins that accumulate as undesirable inclusion bodies, which require additional purification processes.

Yeasts have been used to produce large quantities of foreign proteins for both research and pharmaceutical applications. They have the capability to perform eukaryotic-specific posttranslational modifications, such as folding, glycosylation, and disulfide bridge formation (Eckart and Bussineau, 1996). For this reason, the yeast system is preferred to the bacterial system. However, more sophisticated modifications, such as amidation, hydroxylation, and some types of phosphorylation, are lacking in certain strains of yeast (Cregg and Higgins, 1995). Depending upon the protein of interest, high yields can be achieved in yeast; yeast is therefore regarded as an economical alternative to transgenic livestock (Cereghino and Cregg, 1999).

The use of filamentous fungi, such as *Aspergillus* species, for large-scale, cost-effective fermentation expression of glycoproteins is also well established. Certain *Aspergillus* spp. strains have been genetically well characterized, and recombinant proteins can be produced under the control of inducible promoters; cell cultures can be grown to high cell density before induction (Sanders et al., 1989). By controlling the induction of the recombinant protein to a specific stage in the growth cycle of *Aspergillus* spp., not only is the exposure of the cells to the recombinant protein decreased, but protein degradation by proteases secreted into the growth medium by *Aspergillus* spp. is also minimized. Another advantage is that specific industrial strains of fungi can secrete large quantities of the protein into the growth medium, thus allowing for an easily purifiable protein (Ward et al., 1992). For example, when using *Aspergillus niger* var. *awamori*, expression levels of 5 g of human lactoferrin/L fermentation medium have been achieved (Headon, 2002).

C. Plants

There is an increased demand for safe recombinant proteins. The production of recombinant proteins with potential commercial application in transgenic plants has become an alternative to other expression systems such as transgenic animals, insect cells, and microorganisms because of its inexpensive large-scale production of highly purifiable, functional proteins. Transgenic plant expression of recombinant human proteins has several distinct advantages. Because genetically engineered proteins produced in transgenic microorganisms or animals have the potential to elicit an immune response from contaminants or to be copurified with human or animal pathogens that may cause illness, recombinant proteins produced in certain plants may be viewed more favorably by consumers. Furthermore, when the plant cell culture is ultimately scaled up to a level at which transgenic plants are grown in fields, well-established agricultural technologies are available for cultivation and harvesting.

Transgenic potatoes have become the edible plant of choice for the production of vaccine antigens (Haq et al., 1995; Arakawa et al., 1997; Tacket et al., 1998). A gene encoding a subunit of a pathogen cloned into an expression cassette that contains plant regulatory sequences was introduced into potato plants by using *Agrobacterium tumefaciens*–mediated leaf disk transformation methods. Furthermore, similar transformation techniques have been used to produce recombinant human milk proteins in transgenic potatoes (Chong et al., 1997; Chong and Langridge, 2000). The potential for using edible plants to produce recombinant proteins for commercial applications is large—the need for purification of the protein is minimal or nonexistent if the plant is to be eaten raw. High-level protein expression in transgenic tomatoes (Ruf et al., 2001) and bananas (Sagi et al., 1995) was achieved in the 1990s. If raw consumption is not possible, then freeze-drying the transgenic plant could be a plausible process. Not only would the amount of the recombinant protein expressed by the plant be concentrated, but each dose delivered to the recipient (vaccines) would be consistent.

Many pharmaceutical proteins, e.g., growth hormones (Leite et al., 2000), erythropoietin (Matsumoto et al., 1995), and serum albumin (Sijmons et al., 1990), have been expressed in tobacco plants. An advantage in using tobacco as the expression system of choice is that the transformation of the plants by *Agrobacterium tumefaciens*–based gene transfer is well established and high-level expression of recombinant proteins can be accomplished by using the 35S promoter from cauliflower mosaic virus. In addition, the recombinant proteins can easily be tested in the tobacco leaves within 3 to 4 months after the first transformation (Theisen, 1999). Obvious disadvantages of using tobacco plants are the presence of nicotine and other toxic alkaloids that must be eliminated in the purification procedure and potential negative response by consumers.

Finally, the use of rice may be the most efficient expression system for the production of recombinant proteins because (a) it is cost-effective, with regard to materials and time; (b) rice cereal is often the first baby food recommended by pediatricians to be introduced to infants, because of its low allergenicity; and (c) rice has high nutritional value. In addition, protein regulation and secretion can be controlled, thus providing the capacity to achieve industrial levels. For example, Simmons and colleagues (1991) have shown that the rice *Amy3D* gene encoding the α-amylase isozyme is expressed in response to sugar deprivation by either medium change or metabolism. Therefore, this transcriptional control allows the induction of recombinant proteins to be expressed at high levels in sucrose-free media. Recombinant proteins expressed by rice cell culture can be evaluated in as little as 48 hr versus a year or longer for stable transgenic rice grown in open fields (Terashima et al., 1999). Depending upon the intended use of the recombinant protein, a highly purified form may have to be isolated from the crude medium; however, purification of the recombinant protein expressed in seeds of rice may not be necessary if it is to be used in rice-based infant formulas or cereals.

III. RECOMBINANT MILK PROTEINS

Human milk contains proteins that participate in defense against infection, nutrient digestion, and absorption and may also influence enterocyte growth differentiation, and function. Many human milk proteins have been expressed in various systems (Table 1). Achieving high-level protein expression allows not only further structural and functional characterization of the protein but also evaluation of the advantages and disadvantages of the expression system used to produce the recombinant proteins that will ultimately be added to foods.

A. Caseins

Human casein, which accounts for ~40% of the total protein in human milk, has been shown to provide a limited amount of phosphorus, calcium, and amino acids to the newborn infant. On the basis of their electrophoretic mo-

Table 1 Summary of Recombinant Milk Proteins Expressed in Various Expression Systems

Recombinant human milk proteins	Production system	Biological function	Reference
α-Casein	Bacteria	Inhibitor of angiotensin-converting enzyme	Kim et al., 1997
β-Casein	Bacteria, potato plants	Opioid activity; facilitation of absorption of calcium	Hansson et al., 1993a Chong et al., 1997
κ-Casein	Rice	Antibacterial properties	Personal communication, Ning Huang, 2002
α-Lactalbumin	Bacteria, tobacco plants	Lactose synthesis	Kumagai et al., 1990 Takase and Hagiwara, 1998
Lactoferrin	Fungi, tobacco and potato plants, rice, cows, mice	Iron-binding protein; antibacterial, antiviral, and antifungal properties; growth factor(?)	Ward et al., 1992a Ward et al., 1992b Nuijens et al., 1997 Salmon et al., 1998 Chong and Langridge, 2000 Nandi et al., 2002 Suzuki et al., 2002 van Berkel et al., 2002
Lysozyme	Yeast, rice	Bactericidal properties	Maullu et al., 1999 Huang et al., 2002
Secretory immunoglobulin A	Insect and mammalian cells, transgenic mice	Antibodies; source of amino acids	Morton et al., 1993 Carayannopoulos et al., 1994 Maga et al., 1994
Lactoperoxidase	Insect cells	Bactericidal properties	Shin et al., 2000
α_1-Antitrypsin	Mice, rice	Protection of milk proteins during digestion	Archibald et al., 1990 Terashima et al., 1999 Huang et al., 2001
Bile salt–stimulated lipase	Mice, yeast, mammalian cells	Lipid digestion	Hansson et al., 1993b Strömqvist et al., 1996 Sahasrabudhe et al., 1998 Murasugi et al., 2001
Insulin-like growth factor I	Rabbits	Growth factor	Wolf et al., 1997

bility, casein subunits are classified into α-, β-, and κ-casein (Jenness, 1985). Current interest in caseins has largely focused on the biologically active peptides resulting from the proteolysis of casein subunits.

In human milk, the major casein subunit is β-casein, a phosphorylated protein with a molecular weight of 24 kd. Peptides resulting from the proteolytic degradation of human β-casein have exhibited biological effects. Some peptides have been shown to possess opioid activity—the peptides have an affinity for opiate receptors and exhibit opiate-like effects (Brantl, 1984). Other peptides resulting from the proteolysis of β-casein have been shown to consist of N-terminal fragments containing phosphorylated amino acid residues that have been shown not only to keep calcium in soluble form, but also to facilitate calcium uptake in the intestinal tract of rats (Sato et al., 1986). Recombinant human β-casein has been expressed in *E. coli* (Hansson et al., 1993a). Assessment of the molecular weight and phosphate content of the purified recombinant human β-casein showed that only one isoform of β-casein (zero phosphate groups) was produced in *E. coli*; however, there are up to seven isoforms (zero to five phosphate groups) present in human milk. This finding further confirms the lack of sophisticated posttranslational machinery in *E. coli* that is necessary for the expression of complex mammalian proteins. Nevertheless, the recombinant β-casein isoform exhibited similar micelle formation in aqueous solution and decreased solubility in water after lyophylization as native human milk β-casein. Potato (*Solanum tuberosum*) plants have also been transformed with complementary deoxyribonucleic acid (cDNA) encoding human β-casein (Chong et al., 1997). The molecular weight of the recombinant β-casein was slightly lower than that of human milk β-casein; the difference may be attributable to differences in posttranslational modifications and/or the removal of a 15-amino-acid leader sequence.

κ-Casein, a minor casein subunit in human milk, is a glycoprotein with charged sialic acid residues. It has a molecular weight of 37 kd, of which about 19 kd is carbohydrate (Brignon et al., 1985). κ-Casein has been shown to prevent the attachment of bacteria to the mucosal lining by acting as a receptor analog (Newburg, 1997). The glycan of the surface-exposed carbohydrates of cells in the gastrointestinal tract is similar to that of κ-casein, and therefore, it may serve as a soluble "decoy" for pathogens. Strömqvist and associates (1995) demonstrated that native human milk κ-casein inhibited the adhesion of *Helicobacter pylori* to human gastric mucosa in vitro. Thus, κ-casein may be a component of the defense against infection provided by breast milk. To our knowledge, the only expression of recombinant human κ-casein has been in transgenic rice suspension culture that is currently in progress. We have cloned the human κ-casein gene into a plasmid containing the *RAmy3D* promoter, signal peptide, and terminator and introduced the gene into calli of rice (*Oryza sativa*) cultivar Taipei 309 by particle bombardment–mediated transformation (personal communication, Ning Huang, 2002).

Another minor casein subunit is α-casein. α-Casein is present in human milk in very small amounts; it has been found to account for only 0.06% of the total protein content in human milk (Cavaletto et al., 1994). *Escherichia coli* has been used to express recombinant human α_{S1}-casein (Kim et al., 1999). Three peptides resulting from the tryptic digest of recombinant human α_{S1}-casein were found to exhibit angiotensin I–converting enzyme inhibitory activity in a dose-dependent manner, but this finding has not yet led to concrete applications.

B. α-Lactalbumin

α-Lactalbumin is a major whey protein in the milk of most species. In human milk, α-lactalbumin accounts for 10%–20% of total protein, thus supplying a significant amount of amino acids to the breast-fed infant. It is the regulatory subunit of lactose synthase, a Golgi enzyme complex that catalyzes the formation of lactose in the mammary gland (Brew and Hill, 1975). In vitro pancreatic enzyme digestion of bovine α-lactalbumin has been shown to produce three polypeptide fragments with antibacterial activity against *E. coli*, *Klebsiella pneumoniae*, *Staphylococcus aureus*, *S. epidermidis*, *Streptococcus* spp., and *Candida albicans* (Pelligrini et al., 1999). Recombinant α-lactalbumin has been produced in *E. coli*; however, the protein accumulated as undesirable inclusion bodies in an insoluble fraction (Kumagai et al., 1990). On the other hand, expression in transgenic tobacco plants, under the control of the cauliflower mosaic virus 35S promoter, produced a soluble form of α-lactalbumin that was biochemically indistinguishable from human milk α-lactalbumin (Takase and Hagiwara, 1998). In this expression system, the signal peptide of the recombinant human α-lactalbumin was correctly processed to produce a mature form of the protein, and its biological activity, as assessed by the synthesis of lactose when combined with lactose synthase, was similar to that of human milk α-lactalbumin. In 2000 Svensson and colleagues found that human milk α-lactalbumin, in the presence of oleic acid (C18:1), partially unfolded to a conformational state that induced apoptosis in several human and murine tumor cell lines. They also successfully converted recombinant human α-lactalbumin expressed in *E. coli* to its active apoptosis-inducing form from its inactive, native conformation. Thus, the possibility exists that the conformation of recombinant human α-lactalbumin expressed in other systems may also be converted to this apoptosis-inducing form by using similar methods.

C. Lactoferrin

Lactoferrin, an 80 kd monomeric glycoprotein in human milk, has a high affinity for iron; thus, a role in iron absorption in the newborn infant has been proposed (Cox et al., 1979). Other functions for lactoferrin include antibacterial (Arnold et al., 1980) as well as antiviral (Harmsen et al., 1995) activities, production of cytokines (Crouch et al., 1992), and activation of natural killer cells (Nishiya and Horwitz, 1982). Because of the variety of functions it possesses—hence its potential for pharmaceutical applications—recombinant human milk lactoferrin has been produced in various expression systems, from microorganisms to plants to animals.

Lactoferrin has been expressed in filamentous fungi (*Aspergillus oryzae*) under the control of the *A. oryzae* α-amylase promoter (Ward et al., 1992a) and in *A. nidulans* under the control of the strong ethanol-inducible alcohol dehydrogenase (*alcA*) promoter (Ward et al., 1992b). Because lactoferrin also has antifungal properties (Andersson et al., 2000; Kuipers et al., 2002-a; Kuipers et al., 2002-b), long exposure of the recombinant protein to *Aspergillus* spp. cells could be deleterious; however, the ability to induce protein expression at a specific stage in the growth cycle in *Aspergillus* spp. minimizes exposure of the fungal cells to recombinant lactoferrin secreted into the media. In both *Aspergillus* spp. strains, the secreted recombinant human lactoferrin was identical to native human milk lactoferrin with regard to molecular weight, immunoreactivity, as well as iron-binding capacity. In addition, the recombinant human lactoferrin produced in *A. oryzae* had types of glycans that were similar to those of the native human lactoferrin, as determined by the comigration of both proteins on sodium dodecyl sulfate-polyacrylamide gel electrophoresis (SDS-PAGE) after treatment of the proteins with glycosidase. The amounts of recombinant lactoferrin secreted in both *Aspergillus* spp. systems were relatively high (up to 25 mg/L), but the potential exists to increase these levels considerably by optimizing growth conditions. More recently, expression levels of 5 g/L of human lactoferrin have been achieved in *A. niger* var. *awamori* (Headon, 2002).

Recombinant human milk lactoferrin has also been successfully produced in tobacco plants (*Nicotiana tabacum xanthi NC*) under the control of the 35S promoter from cauliflower mosaic virus by *Agrobacterium tumefaciens* (Salmon et al., 1998). N-terminal sequencing of the recombinant human lactoferrin showed a sequence identical to that of native human milk lactoferrin, indicating that signal peptides were correctly processed. Mass spectrometry analysis of the purified recombinant human lactoferrin showed two major species that may be attributable to differences in the carbohydrate moiety. Carbohydrate analysis of the recombinant human lactoferrin showed a lower amount of galactose and *N*-acetylglucosamine residues, presence of xylose, and absence of sialic acid residues when compared to native lactoferrin. Nevertheless, the slight difference in carbohydrate groups did not hinder binding of the recombinant human lactoferrin to lactoferrin receptors found on Jurkat human lymphoblastic T cells or the human colon carcinoma cells HT29-18C1.

More recently, full-length bioactive human lactoferrin, under the control of the auxin-inducible manopine synthase P2 promoter and the cauliflower mosaic virus 35S tandem promoter, was expressed in transgenic potato plants (Chong and Langridge, 2000). An advantage of an expression system such as this is the introduction of an antimicrobial human milk protein into an edible plant. Although the likelihood of consuming raw potatoes may be minimal, as mentioned earlier, freeze-drying of transgenic potatoes, possibly with partial purification, may be a possibility if lactoferrin is to be added to baby foods. Therefore, expression of recombinant human milk lactoferrin in transgenic rice provides another "edible" alternative—rice expressing recombinant human milk lactoferrin may be used in baby cereals. We have expressed recombinant human lactoferrin in transgenic rice (Suzuki et al., 2002). We have cloned the human lactoferrin gene into a plasmid containing the *RAmy3D* promoter, signal peptide, and terminator and introduced the gene into calli of rice (*Oryza sativa*) cultivar Taipei 309 by particle bombardment–mediated transformation. Primary assessment of the secreted recombinant human lactoferrin showed similar biochemical as well as functional characteristics to those of native human milk lactoferrin.

Recombinant human lactoferrin has also been secreted in the milk of transgenic mice (Nuijens et al., 1997) and other animals (Krimpenfort, 1993). Although the posttranslational modification machinery in the mammary epithelial cells of transgenic animals may more closely resemble that of humans, the degree of glycosylation was different in the recombinant protein expressed in mice milk (Nuijens et al., 1997). Nevertheless, iron-binding and -releasing properties as well as binding to a range of ligands, such as heparin, lipopolysaccharide, and lectins, were similar to those of native lactoferrin. Studies in 2002 on recombinant lactoferrin expressed at very high levels in transgenic cows showed that deglycosylated lactoferrin had a molecular mass identical to that of native lactoferrin but that the mass of the glycan was 1–2 kd smaller (van Berkel et al., 2002). Monosaccharide analysis showed that recombinant human lactoferrin contained relatively more mannose and less sialic acid and fucose than the native form. The types of N-linked glycans (oligomannose-, hybrid-type, complex type) also differed, and there was a substitution of galactose with *N*-acetylgalactosamine. These differences, however, did not result in any differences with regard to the biological properties tested.

D. Lysozyme

Lysozyme is a 15-kd protein present in human milk at concentrations higher than in milk of most other species (Chandan et al., 1968). Lysozyme destroys the cell walls of bacteria by catalyzing the hydrolysis of β-1,4 linkages of *N*-

acetylmuramic acid acid and 2-acetylamino-2-deoxy-D-glucose residues (Chipman and Sharon, 1969). Lysozyme alone can kill several gram-positive bacteria; however, it has also been shown that lysozyme and lactoferrin can work synergistically to produce bactericidal effects on gram-negative bacteria—lactoferrin disrupts the outer bacterial membrane by binding to lipopolysaccharide (LPS), thereby allowing lysozyme access to the inner proteoglycan matrix (Ellison and Giehl, 1991). High levels of recombinant human lysozyme have been achieved in yeast, *Kluyveromyces lactis* K7, grown in cottage cheese whey (Maullu et al., 1999). The recombinant lysozyme was secreted into the culture medium in a soluble form and possessed biological activity as assessed by a bacteriolytic assay using *Micrococcus luteus*. Recombinant human lysozyme has also been expressed in transgenic rice cell cultures (Huang et al., 2002b). The human lysozyme gene, cloned into a plasmid containing the *RAmy3D* promoter, signal peptide, and terminator, was introduced into calli of rice (*Oryza sativa*) cultivar Taipei 309 by particle bombardment–mediated transformation. N-terminal amino acid sequencing of the recombinant human lysozyme revealed that rice cells correctly processed the signal peptide, and functional assays demonstrated bactericidal activity. Human lysozyme has subsequently been expressed in rice grains (Huang et al., 2002a). Expression levels were very high, 45% of soluble protein or 0.6% of the brown rice weight, and they were maintained over six generations. Biochemical, biophysical, and functional comparisons of native and recombinant human lysozyme showed identical N-terminal sequence, molecular weight, pI, and specific activity. Similar thermal and pH stability was also demonstrated as well as bactericidal activity against *E. coli*.

E. Secretory Immunoglobulin A

Approximately 90% of immunoglobulins in human milk is attributable to secretory immunoglobulin A (sIgA). When the infant receives sIgA from breast milk, the infant receives a portion of the maternal immune defense system. Furthermore, since sIgA is resistant to proteolytic degradation (Lindh, 1985), it survives the passage through the immunologically immature gastrointestinal tract of the breast-fed infant (Davidson and Lönnerdal, 1987). Once in the gut of the infant, the sIgA antibodies offer protection of the mucosal lining against colonization and invasion by pathogenic microorganisms. The IgA antibodies have been used for passive immunization in animals; studies in mice have shown that oral administration of IgA antibodies conferred protection against various bacteria such as *Salmonella* spp. (Michetti et al., 1992), *Vibrio cholerae* (Winner et al., 1991; Apter et al., 1993), and *Helicobacter pylori* (Blanchard et al., 1995). In humans, oral IgA administration appeared to protect against *Clostridium difficile* (Tjellström et al., 1993). Because sIgA has four different polypeptide chains and is highly glycosylated with both N- and O- linked glycans, any expression system that is used will have its advantages and limitations in producing recombinant sIgA. Nevertheless, recombinant sIgA has been produced in heterologous expression systems such as insect cells (Carayannopoulos et al., 1994) and mammalian cells (Morton et al., 1993). Since sIgA is the product of two different cell types in mammals, transgenic plants may be a suitable expression system for large-scale production of sIgA for mucosal immunotherapy because plants have the ability to produce the antibodies in single cells (Ma et al., 1995). Because of the complex nature of sIgA, structural and biochemical assessments and in vivo biological function assays of recombinant sIgA produced in any expression system must be conducted.

F. Lactoperoxidase

Lactoperoxidase, in the presence of hydrogen peroxide, catalyzes the oxidation of thiocyanate to produce reactive products with antibacterial properties capable of destroying both gram-positive (Steele and Morrisons, 1969) and gram-negative (Björck et al., 1975) bacteria. Catalytically active recombinant human lactoperoxidase has been expressed in the culture supernatant of *Tricoplusia ni* cells infected with recombinant baculovirus (Shin et al., 2000). Since lactoperoxidase is a heme peroxidase, addition of δ-aminolevulinic acid, a heme precursor, into the insect cell culture medium increased expression levels in this insect cell culture expression system. Currently, transgenic rice cell culture is being used for the expression of recombinant human lactoperoxidase. Methods similar to those used for the insect cell culture system can be implemented in the rice cell culture system in order to enhance protein expression levels.

G. α_1-Antitrypsin

The exact physiological significance of α_1-antitrypsin in human milk is not clear. α_1-Antitrypsin may act in the mammary gland, where it may protect the gland from proteolytic activity of leukocytic and lysosomal proteases during stages of differentiation and/or inhibit the degradation of other enzymes and proteins in milk during mastitis (Hamosh, 1995). Since the gastric pH of newborns is relatively high, human milk α_1-antitrypsin remains immunologically intact and biologically active in the infant intestine (Davidson and Lönnerdal, 1990). Thus, another function of α_1-antitrypsin may be that it offers protection to some milk proteins from hydrolysis in the infant gastrointestinal tract (Lindberg, 1979; Hamosh, 1995). High levels of bioactive α_1-antitrypsin, as assessed by trypsin-inhibition assay, have been expressed in the milk of transgenic mice (Archibald et al., 1990). However, the electrophoretic pattern of recombinant human α_1-antitrypsin was distinct from that of the native human α_1-antitrypsin, possibly because of differences in posttranslational modifications. Recombinant α_1-antitrypsin has also been expressed and secreted from transgenic rice cell suspension culture under the control of the promoter, signal peptide, and terminator from the rice α-amylase gene *Amy3D* (Terashima et al., 1999; Huang et al., 2001). Because the expression of *Amy3D* gene is regulated by the sugar concentration in the culture media, expression of recombinant α_1-antitrypsin was inducible. From this system, ~5 mg of recombinant α_1-antitrypsin per gram dry cells was obtained. The correctly processed recombinant α_1-antitrypsin demonstrated the same kinetic constant and thermostability and exhibited biological activity similar to that of the native human α_1-antitrypsin, although slight differences in glycan structures may be present.

H. Bile Salt–Stimulated Lipase

The presence of a lipase in human milk that is stimulated by bile salts has been found to be part of the reason for the high digestibility of lipids in breast-fed infants (Fredrikzon et al., 1978; Hernell and Bläckberg, 1983). The high efficiency of bile salt–stimulated lipase (BSSL) is due in part to its wide substrate specificity, it hydrolyzes mono-, di-, and triacylglycerols; cholesterol esters; retinyl esters; diacylphosphatidylglycerols; and micellar and water-soluble substrates (Hernell et al., 1989). In addition, in vitro analysis suggests that BSSL remains active in the gastrointestinal tract of infants, thus contributing significantly to lipid digestion (Hamosh, 1982).

Recombinant human BSSL has been successfully produced in transgenic mice (Strömqvist et al., 1996). Under the control of regulatory elements derived from murine whey acidic protein, recombinant human BSSL was expressed in milk at levels of ~1 mg/mL. Although the recombinant human BSSL exhibited a lower degree of O-glycosylation, the recombinant BSSL demonstrated similar lipase activity, pH stability, and thermal sensitivity to those of the native enzyme.

Stable expression of BSSL has also been achieved in a murine mammary tumor–derived cell line, C127 (Hansson et al., 1993b). Enzyme activity as well as bile salt dependence were found to be similar to those of the native milk enzyme. In addition, BSSL has been produced in a yeast expression system. Sahasrabudhe and associates (1998) recovered 0.3 g of recombinant human BSSL expressed by *Pichia pastoris* per liter of fermentation medium. The purified recombinant BSSL was found to be similar in its biochemical properties to native BSSL. Murasugi and associates (2001) further optimized the culture conditions of *Pichia pastoris* and successfully expressed ~1 g recombinant BSSL/L. Assessment of enzyme activity after varying storage times showed that recombinant BSSL maintained most of its lipase activity even after 200 days, an element of great importance when considering commercial applications for a bioactive protein.

I. Growth Factors

Insulin-like growth factors I and II (IGF-I and IGF-II) are factors present in human milk that have been shown to stimulate DNA synthesis and to promote the growth of various cells in culture (Klagsburn, 1978; Corps et al., 1987; Corps et al., 1988). In addition, these factors promote the development of the neonatal gastrointestinal tract (Berseth et al., 1983; Read et al., 1984; Corps and Brown, 1987). Human IGF-I has been expressed in the milk of transgenic rabbits under the transcriptional control of regulatory elements of the bovine α_{S1}-casein gene (Wolf et al., 1997). High levels of recombinant human IGF-I (50–300 μg/mL) were expressed throughout lactation, processed correctly, and demonstrated biological activity by binding to IGF-I receptors and stimulating DNA synthesis in growth-arrested IGF-I-responsive human osteosarcoma cells.

IV. BIOCHEMICAL ASSESSMENT OF RECOMBINANT MILK PROTEINS

Assessment of the structure and function of recombinant proteins begins with the isolation and purification from the expression source, i.e., culture medium, leaves, seeds, or milk. This is usually accomplished by affinity, ion exchange, reverse-phase, and/or size-exclusion column chromatography or by membrane filtration technology. Even though high levels of the recombinant protein may be achieved in an expression system, if the protein purification step is labor-intensive and many ex vivo manipulations must be performed, then achieving high quantities of purified protein at a reasonable cost may be difficult. For example, purification of a soluble protein from the milk of transgenic cows requires multiple steps, which include the removal of the fat and casein fractions, before column chromatography. In contrast, purification of a secreted recombinant protein from cell culture medium may require only one step. Or, depending upon the expression system, it is possible that the recombinant protein may not need any purification. Expression of recombinant human milk proteins in rice that will ultimately be added to baby cereals or to rice-based infant formulas may require only partial purification or no purification at all. Once the recombinant protein is successfully isolated, the molecular weight should be determined by SDS-PAGE. Western blotting techniques using IgG specifically directed against the protein can subsequently be used to identify the recombinant protein positively. Further evaluation of the purity of the isolated recombinant milk protein, as assessed by mass spectrometry, as well as other biochemical parameters, such as stoichiometry and isoelectric point, is often useful in determining similarities to the native protein.

As mentioned previously, it is important not only to achieve high expression levels of the recombinant protein, but also to obtain high amounts of correctly processed protein. Attainment of correctly processed full-length, mature recombinant proteins is paramount for optimal biological function. Electrophoretic mobility of the recombinant and native protein can initially show whether or not the recombinant protein underwent correct proteolytic processing of precursors during synthesis and secretion from cells, i.e., removal of the signal peptide. Furthermore, N-terminal amino acid sequencing and peptide mass mapping by mass spectrometry can be performed for additional information. In some cases, amino acid analysis may also be needed. For instance, if several recombinant human milk proteins would have to be added to infant formula, then the amino acid composition of those proteins must be known so that the amino acid profile of the infant formula could be adjusted according to the amino acid requirements of infants.

Many human proteins undergo other posttranslational modifications that significantly influence their biological activity. In a simple assay to determine the extent of glycosylation of the recombinant protein, both the native and the recombinant protein can be treated with *N*-glycosidase F, resolved on SDS-PAGE, transferred to a nitrocellulose membrane, and probed by using IgG specifically directed against the protein. Because *N*-glycosidase F hydrolyzes the glycosylamine linkage that generates a carbohydrate-free peptide that is lower in molecular weight, comigration of the native and recombinant protein would suggest similar N-linked glycosylation. For further glycan characterization, the monosaccharide composition can be analyzed by using gas–liquid chromatography. Data from mass spectrometry calculations and monosaccharide composition can be used to assign specific glycans to recombinant proteins (Salmon et al., 1998).

Proper phosphorylation of the recombinant protein may be equally important. Although the role of phosphates in proteins for structure–function relationship needs further investigation, it is possible that biological function of a protein may be affected by conformational changes resulting from varying phosphate content. For example, the number of phosphates attached to recombinant human β-casein does not change the apparent molecular weight; however, attachment of the first phosphate group changes the conformation of the protein and adds ~2 kd to the apparent molecular weight (Hansson et al., 1993a). Recombinant proteins can be analyzed for different phosphorylation isoforms by ion-exchange chromatography (elution order) followed by phosphate staining on SDS-PAGE, as described by Green and colleagues (1973). More accurate determination of the number of phosphate groups attached to the recombinant protein can be achieved by mass spectrometry. The presence of other posttranslational modifications, such as hydroxylation or sulfation, should be assessed if the native proteins possess such characteristics.

V. FUNCTIONAL EVALUATION OF RECOMBINANT MILK PROTEINS

Recombinant milk proteins produced in various expression systems should be biochemically, immunologically, and physiologically similar to their native counterparts. Differences in glycosylation or phosphorylation patterns may affect the biological activity of the protein. The physiological activity of the recombinant milk protein is of importance if its intended use is as a food additive. The biological activity of the recombinant milk protein must be evaluated both in a physiologically neutral buffer and in the matrix of the food to which it will be added, e.g., cow milk–based infant formula, since aggregation and/or precipitation of the recombinant proteins with existing food components may occur. Determination of the concentration of the recombinant milk protein to be added to the food for optimal biological activity is also important. If more than one recombinant protein would have to be added to the food, then synergistic as well as counteractive effects of the proteins should be determined. It should also be noted that recombinant milk proteins should be evaluated for structural integrity and biological function after purification from the expression source as well as after the proteins are subjected to varying degrees of heat (to simulate sterilization procedures), and to various pH conditions and digestive enzymes (to simulate digestion in the gastrointestinal tract).

Recombinant milk proteins with enzymatic activity can easily be evaluated in vitro. For example, the bacteriolytic activity of recombinant human milk lysozyme can be assessed by its reduction in turbidity of a suspension of *Micrococcus lysodeikticus* (Shugar, 1952). Substrate binding and specificity of the recombinant enzyme can be determined as in the case of recombinant human milk BSSL (Hansson et al., 1993b). The hydrolyzing capacity of recombinant BSSL can be determined by incubating the enzyme with substrates, i.e., mono-, di-, and triacylglycerol. Furthermore, monitoring the synthesis of lactose when combined with lactose synthase can assess the biological activity of recombinant α-lactalbumin (Brew and Hill, 1975).

Some human milk proteins have the capacity to enhance nutrient absorption. To determine whether or not the recombinant milk proteins have similar function, the extent of binding and release of the nutrient can be assessed. For example, the iron-binding capacity and pH-dependent iron-releasing properties of recombinant lactoferrin can be studied in vitro. Cell culture experiments can also help evaluate binding properties of the recombinant human milk protein to specific sites with which they interact. Since human milk lactoferrin binds to lactoferrin receptors in the infant gastrointestinal tract (Kawakami and Lönnerdal, 1991), interactions of recombinant human lactoferrin with various cells that express lactoferrin receptors can be determined. In these cell binding and uptake experiments, dissociation constants and binding curves should be similar to those of the native protein (Suzuki et al., 2002). Although it is possible that the recombinant and the native milk protein have slight biochemical differences, i.e., glycosylation patterns, little or no difference in binding parameters may be observed.

Human milk lactoferrin, lysozyme, and lactoperoxidase are known to have antimicrobial functions. Some pathogens that have been shown to infect infants include enteropathogenic and enterotoxigenic *E. coli*, *Helicobacter pylori*, *Campylobacter jejuni*, *Yersinia enterocolitica*, *Salmonella* spp., and *Clostridium* spp. Determination of bacteriostatic or bactericidal activity by recombinant milk proteins can be performed by mixing a bacterial cell suspension with the recombinant milk protein, plating the mixture on agar plates, then incubating the mixture at 37°C for subsequent enumeration of viable colonies. However, not all microorganisms in the infant gastrointestinal tract are pathogenic. It has been shown that the fecal flora of breast-fed infants is dominated by gram-positive, non-spore-forming rods whereas formula-fed infants have potentially pathogenic gram-negative anaerobic rods and *E. coli* (Kleesen et al., 1995). Host-friendly bacteria, such as *Lactobacillus* and *Bifidobacterium* spp., found primarily in breast-fed infants as compared to formula-fed infants, produce acetic and lactic acids from lactose fermentation that may provide an unfavorable environment for pathogenic bacteria. Therefore, assessment of recombinant milk proteins that stimulate the growth of host-friendly bacteria should also be performed. Another human milk protein, κ-casein, has been shown to prevent the adhesion of *Helicobacter pylori* to gastric mucosa in vitro; therefore, the capacity of recombinant κ-casein to prevent this adhesion can be evaluated.

Finally, peptides derived from human milk proteins have been shown to have physiological activities. Casein-derived peptides have been shown to exhibit opiate, antithrombotic, antihypertensive, and immunomodulatory activities (Brantl, 1984; Fiat et al., 1993; Schlimme and Meisel, 1995). Peptides derived from lactoferrin (Tomita et al., 1991) and α-lactalbumin (Pelligrini et al., 1999) have also been shown to possess biological functions. Therefore, full-length recombinant α_{S1}, β-, and κ-casein as well as lactoferrin and α-lactalbumin can be enzymatically hydrolyzed and the peptides tested for biological activity.

VI. IN VITRO ASSESSMENT OF RECOMBINANT MILK PROTEIN DIGESTIBILITY

The digestive system of the newborn infant is not fully developed; therefore, some human milk proteins are resistant to degradation (Davidson and Lönnerdal, 1987). In the neonatal stomach, the degree of protein hydrolysis is limited. The pH of the gastric contents has been found to be around 4.0–4.5 (Agunod et al., 1969; Hamosh, 1996), and pepsin secretion has been found to be less developed than in adults (Adamson et al., 1988; DiPalma et al., 1991). In addition, since the pH of the infant stomach is above the optimum for pepsin activity (pH 2), pepsin can only act to a limited extent. Furthermore, the development of a functional response by the pancreas is not complete at birth (Lebenthal and Lee, 1980), and the activity of intestinal enterokinase, the enzyme that converts the proteolytic enzyme trypsinogen to trypsin, which in turn activates other zymogens, is low at birth but increases with age (Antonowicz and Lebenthal, 1977). In addition, the transit time through the infant gastrointestinal tract is rapid during the neonatal period, thus limiting the time proteins are exposed to enzymatic degradation (Lönnerdal, 1994).

Even though it may be known that a native human milk protein can resist degradation by proteases encountered in the digestive tract of infants, it is still necessary to evaluate the recombinant milk protein in vitro for structural and biological stability after treatment with varying pH and proteases. A pH stability assay, which mimics the acidic conditions of the stomach (Huang et al., 2002b), can be used to subject the recombinant milk proteins to varying pH (pH 2–7) for various periods (30 min–2 hr). An in vitro protein digestibility assay, designed to represent the conditions of the infant gastrointestinal tract, can be used to predict the structural, biochemical, and functional fate of the ingested recombinant milk protein (Rudloff and Lönnerdal, 1992). First, the conditions of the neonatal stomach are simulated, the pH of the recombinant milk proteins suspended in either physiologically neutral buffer or infant formula is reduced to pH 2, 3, or 4 with HCl, followed by the addition of pepsin and incubation with moderate shaking at 37°C for 30 min to 1 hr. Second, the conditions of the infant small intestine is simulated, the pH of the protein mixture is raised to 7 with bicarbonate, followed by the addition of pancreatin and incubation at 37°C for 1 to 2 hr with moderate shaking. After enzyme inactivation, the recombinant protein is assessed for structural, biochemical, and functional integrity, which is compared to that of its native counterpart.

VII. IN VIVO EVALUATION OF RECOMBINANT MILK PROTEINS

Many recombinant human milk proteins have been expressed in various systems and their biological activity evaluated in vitro; however, their in vivo efficacy has yet to be determined. In addition, it is important to assess whether or not recombinant milk proteins can exert their biological activity in a diet other than human milk in vivo and whether or not anticipated synergistic effects by multiple milk proteins can be exhibited in vivo. Rodent models can be used to evaluate the activity of the recombinant milk protein before using nonhuman primates or humans in clinical trials. The choice of the animal model is important since the data that are obtained from these studies must be relevant to human infants. The use of rat pups may be inappropriate to study the binding capacity of recombinant human lactoferrin to lactoferrin receptors, since rat small intestine has no lactoferrin receptors, but rather receptors for transferrin, the main iron-binding protein in rat milk (Kawakami et al., 1990). However, rat pups can be used to study the protective effects of recombinant human lactoferrin against intestinal infection from *E. coli* (Edde et al., 2001). We are currently examining the in vivo activity of recombinant human lactoferrin and lysozyme produced in transgenic rice in rat pups infected with enteropathogenic *E. coli*, the most common bacteria known to cause diarrhea in infants in developing countries. The antimicrobial properties of the two proteins are tested not only alone, but together, to determine possible synergistic effects. A dose–response relationship of the recombinant proteins on infection has also been determined.

Infant rhesus monkeys can be used as an animal model for preclinical trials to better determine the in vivo effects of the recombinant milk proteins. Safety and efficacy of the recombinant proteins are evaluated in infant monkeys before the clinical trials in human infants. The use of infant rhesus monkeys has advantages over use of human infants, in that the studies can be performed in a controlled environment; also, ethical standard dictate that invasive procedures, e.g., dissection of tissues and use of radioisotopes, are feasible only in monkeys. In addition, the composition of monkey milk is very similar to that of human milk (Lönnerdal et al., 1984), and the neonatal gastrointestinal tract of infant monkeys is similar to that of human infants (Lindberg et al., 1997). Therefore, the physiological fate of recombinant human milk proteins in infant monkeys may be similar to that in human infants.

Human infants should be used in the final evaluation of the recombinant milk proteins. Data obtained from in vitro studies and animal experiments should offer information with regard to the optimal dose to be added to the infant formulas on the basis of potency/strength, half-life (stability), and affinity/specificity (function) of the recombinant milk proteins.

After obtaining data on the efficacy of addition of a novel food ingredient, such as a recombinant human milk protein, it is essential to acquire data on the safety of the added protein. The efficacy data, however, are essential for proceeding with the regulatory processes, in particular for infant formulas. Regulatory agencies, e.g., the U.S. Food and Drug Administration (FDA) (in the United States), have strict definitions provided by the Infant Formula Act, a federal legal document, which closely defines what is allowed. This law stipulates that no novel component can be added unless a significant beneficial outcome can be correlated with this component. Thus, "cosmetic" additions of novel compounds, i.e., addition of small quantities of a novel compound without documented beneficial effect (under rigorous conditions) for marketing purposes are not allowed. Many other countries or organizations [e.g., the European Union (EU)] have similar regulations.

Safety data will always be required before approval of novel food components, whether for adults generally regarded as safe (GRAS) or for infants (Infant Formula Act). Such studies include rat bioassays, growth and acceptance studies in human infants or adults, allergenicity tests, and clinical trials.

VIII. DETERMINATION OF STABILITY AFTER FOOD PROCESSING

Sterilization of infant formula requires elevated temperatures that can potentially destroy the biological activity of the recombinant milk protein. The effect of various heat treatments on the biological function of milk proteins is likely to vary, depending on their biochemical properties. Compact, stable proteins, such as lactoferrin, may tolerate higher temperatures than enzymes, for which a small change in tertiary structure (denaturation) may prove irreversible and affect activity negatively. However, their susceptibility to heat varies greatly among enzymes, and for some the food matrix may actually help the enzyme to retain its activity. Both the temperature and the duration of heat treatment are likely to affect protein bioactivity. Naturally, a combination of high temperature and long heat exposure, such as so-called in-can or retort sterilization, is likely to have more negative impact than lower temperatures for longer periods (e.g., spray-drying) or high temperature for a very brief time [e.g., ultra high temperature (UHT)]. This is illustrated by the fact that protein digestibility, indirectly a measure of heat denaturation, was lowest for retort sterilized infant formula, and highest for UHT formula, with powdered (spray-dried) formula being intermediate (Rudloff and Lönnerdal, 1992). If all current processing methods affect the bioactivity of the milk proteins negatively, other alternatives, such as aseptic packaging, may be used.

IX. CONCLUSIONS: PRINCIPLES FOR ADDING RECOMBINANT MILK PROTEINS INTO INFANT FORMULA AND CEREALS

Safety, allergenicity, potency, and, ultimately, cost are factors that will determine novel applications for recombinant proteins. As the addition of these proteins in most countries will not be allowed without any proven benefit, efficacy trials are going to drive progress in this field. Studies in vitro and in animal models will provide support for this, to be followed by safety studies in rodents as well as higher species, such as nonhuman primates. Then, clinical studies in humans will provide information on efficacy as well as preliminary safety data before launching of larger acceptability and safety studies.

REFERENCES

Adamson, I., Esangbedo, A., Okolo, A.A. and Omene, J.A. (1988). Pepsin and its multiple forms in early life. Biol Neonate 53: 267–273.

Agunod, M., Yamaguchi, N., Lopez, R., Luhby, A.L. and Glass, G.B.J. (1969). Correlative study of hydrochloric acid, pepsin, and intrinsic factor secretion in newborns and infants. Am J Dig Dis 14: 401–414.

Andersson, Y., Lindquist, S., Lagerqvist, C. and Hernell, O. (2000). Lactoferrin is responsible for the fungistatic effect of human milk. Early Hum Dev 59: 95–105.

Antonowicz, I. and Lebenthal, E. (1977). Developmental pattern of small intestinal enterokinase and disaccharidase activities in the human fetus. Gastroenterology 72: 1299–1303.

Apter, F.M., Michetti, P., Winner, L.S., 3rd, Mack, J.A., Mekalanos, J.J. and Neutra, M.R. (1993). Analysis of the roles of antilipopolysaccharide and anti-cholera toxin immunoglobulin A (IgA) antibodies in protection against Vibrio cholerae and cholera toxin by use of monoclonal IgA antibodies in vivo. Infect Immun 61: 5279–5285.

Arakawa, T., Chong, D.K., Merritt, J.L. and Langridge, W.H. (1997). Expression of cholera toxin B subunit oligomers in transgenic potato plants. Transgenic Res 6: 403–413.

Archibald, A.L., McClenaghan, M., Hornsey, V., Simons, J.P. and Clark, A.J. (1990). High-level expression of biologically active human alpha 1-antitrypsin in the milk of transgenic mice. Proc Natl Acad Sci USA 87: 5178–5182.

Arnold, R.R., Brewer, M. and Gauthier, J.J. (1980). Bactericidal activity of human lactoferrin: Sensitivity of a variety of microorganisms. Infect Immun 28: 893–898.

Berseth, C.L., Lichtenberger, L.M. and Morriss, F.H. (1983). Comparison of the gastrointestinal growth-promoting effects of rat colostrum and mature milk in newborn rats in vivo. Am J Clin Nutr 37: 52–60.

Björck, L., Rosen, C.G., Marshall, V. and Reiter, B. (1975). Antibacterial activity of lactoperoxidase system in milk against pseudomonas and other gram-negative bacteria. Appl Microbiol 30: 199–204.

Blanchard, T.G., Czinn, S.J., Maurer, R., Thomas, W.D., Soman, G. and Nedrud, J.G. (1995). Urease-specific monoclonal antibodies prevent Helicobacter felis infection in mice. Infect Immun 63: 1394–1399.

Brantl, V. (1984). Novel opioid peptides derived from human β-casein. Eur J Pharmacol 106: 213–214.

Brew, K. and Hill, R.L. (1975). Lactose biosynthesis. Rev Physiol Biochem Pharmacol 72: 105–157.

Brignon, G., Chtourou, A. and Ribadeau-Dumas, B. (1985). Preparation and amino acid sequence of human κ-casein. FEBS Lett 188: 48–54.

Brink, M.F., Bishop, M.D. and Pieper, F.R. (2000). Developing efficient strategies for the generation of transgenic cattle which produce biopharmaceuticals in milk. Theriogenology 53: 139–148.

Carayannopoulos, L., Max, E.E. and Capra, J.D. (1994). Recombinant human IgA expressed in insect cells. Proc Natl Acad Sci USA 91:8348-8352.

Cavaletto, M., Cantisani, A., Giuffrida, G., Napolitano, L. and Conti, A. (1994). Human α_{S1}-casein like protein: Purification and *N*-terminal sequence determination. Biol Chem Hoppe-Seyler 375: 149–151.

Cereghino, G.P.L. and Cregg, J.M. (1999). Applications of yeast in biotechnology: Protein production and genetic analysis. Curr Opin Biotechnol 10: 422–427.

Chandan, R.C., Parry, R.M. and Shahani, K.M. (1968). Lysozyme, lipase, and ribonuclease in milk from various species. J Dairy Sci 51: 606–607.

Chipman, D.M. and Sharon, N. (1969). Mechanism of lysozyme action. Science 165: 454–465.

Chong, D.K. and Langridge, W.H. (2000). Expression of full-length bioactive antimicrobial human lactoferrin in potato plants. Transgenic Res 9: 71–78.

Chong, D.K.X., Roberts, W., Arakawa, T., Illes, K., Bagi, G., Slattery, C.W. and Langridge, W.H.R. (1997). Expression of the human milk protein β-casein in transgenic potato plants. Transgenic Res 6: 289–296.

Colman, A. (1996). Production of proteins in the milk of transgenic livestock: Problems, solutions, and successes. Am J Clin Nutr 63: 639S–645S.

Corps, A.N., Blakeley, D.M., Carr, J., Rees, L.H. and Brown, K.D. (1987). Synergistic stimulation of Swiss mouse 3T3 fibroblasts by epidermal growth factor and other factors in human milk. J Endocrinol 112: 151–159.

Corps, A.N. and Brown, K.D. (1987). Stimulation of intestinal cell proliferation in culture by growth factors in human and ruminant mammary secretions. J Endocrinol 113: 285–290.

Corps, A.N., Brown, K.D., Rees, L.H., Carr, J. and Prosser, C.G. (1988). The insulin-like growth factor I content in human milk increases between early and full lactation. J Clin Endocrinol Metab 67: 25–29.

Cox, T.M., Mazurier, J., Spik, G., Montreuil, J. and Peters, T.J. (1979). Iron binding proteins and influx of iron across the duodenal brush border: Evidence for specific lactotransferrin receptors in the human intestine. Biochim Biophys Acta 588: 120–128.

Cregg, J.M. and Higgins, D.R. (1995). Production of foreign proteins in the yeast *Pichia pastoris*. Can J Bot 73(suppl 1): 891–897.

Crouch, S.P., Slater, M.K.J. and Fletcher, J. (1992). Regulation of cytokine release from mononuclear cells by the iron-binding protein lactoferrin. Blood 80: 235–240.

Davidson, L.A. and Lönnerdal, B. (1987). Persistence of human milk proteins in the breast-fed infant. Acta Paediatr Scand 76: 733–740.

Davidson, L.A. and Lönnerdal, B. (1990). Fecal alpha$_1$-antitrypsin in breast-fed infants is derived from human milk and is not indicative of enteric protein loss. Acta Paediatr Scand 79: 137–141.

DiPalma, J.S., Kirk, C. and Hamosh, M. (1991). Lipase and pepsin activity in the gastric mucosa of infants, children, and adults. Gastroenterology 101: 116–121.

Eckart, M.R. and Bussineau, C.M. (1996). Quality and authenticity of heterologous proteins synthesized in yeast. Curr Opin Biotechnol 7: 525–530.

Edde, L., Hipolito, R.B., Hwang, F.F., Headon, D.R., Shalwitz, R.A. and Sherman, M.P. (2001). Lactoferrin protects neonatal rats from gut-related systemic infection. Am J Physiol Gastrointest Liver Physiol 281: G1140–G1150.

Ellison, R.T.J. and Giehl, T.J. (1991). Killing of Gram-negative bacteria by lactoferrin and lysozyme. J Clin Invest 88: 1080–1091.

Fiat, A.M., Migliore-Samour, D., Jolles, P., Drouet, L., Sollier, C.B. and Caen, J. (1993). Biologically active peptides from milk proteins with emphasis on two examples concerning antithrombotic and immunomodulating activities. J Dairy Sci 76: 301–310.

Fredrikzon, B., Hernell, O., Bläckberg, L. and Olivecrona, T. (1978). Bile salt-stimulated lipase in human milk: Evidence of activity in vivo and of a role in the digestion of milk retinol esters. Pediatr Res 12: 1048–1052.

Georgiou, G. (1988). Optimizing the production of recombinant proteins in microorganisms. Am Inst Chem Eng J 34: 1233–1248.

Gordon, K., Lee, E., Vitale, J.A., Smith, A.E., Westphal, H. and Henninghausen, L. (1992). Produciton of human plasminogen activator in transgenic mouse milk. Biotechnology 24: 425–428.

Green, M.R., Pastewka, J.V. and Peacock, A.C. (1973). Differential staining of phosphoproteins on polyacrylamide gels with a cationic carbocyanine dye. Anal Biochem 56: 43–51.

Hamosh, M. (1982). Lingual and breast milk lipases. Adv Pediatr 29: 33–67.

Hamosh, M. (1995). Enzymes in human milk. In: R.G. Jensen, ed. Handbook of Milk Composition. San Diego, Academic Press: 388–427.

Hamosh, M. (1996). Digestion in the newborn. Clin Perinatol 23: 191–209.

Hansson, L., Bergström, S., Hernell, O., Lönnerdal, B., Nilsson, A.K. and Strömqvist, M. (1993a). Expression of human milk β-casein in *Escherichia coli*: Comparison of recombinant protein with native isoforms. Protein Expr Purif 4: 373–381.

Hansson, L., Bläckberg, L., Edlund, M., Lundberg, L., Strömqvist, M. and Hernell, O. (1993b). Recombinant human milk bile salt-stimulated lipase: Catalytic activity is retained in the absence of glycosylation and the unique proline-rich repeats. J Biol Chem 268: 26692–26698.

Haq, T.A., Mason, H.S., Clements, J.D. and Arntzen, C.J. (1995). Oral immunization with a recombinant bacterial antigen produced in transgenic plants. Science 268: 714–716.

Harmsen, M.C., Swart, P.J., Debethune, M.P., Pauwels, R., Declercq, E., The, T.H. and Meijer, D.K.F. (1995). Antiviral effects of plasma and milk proteins: Lactoferrin shows potent activity against both human immunodeficiency virus and human cytomegalovirus replication in vitro. J Infect Dis 172: 380–388.

Headon, D. (2002). Recombinant human lactoferrin produced in *Aspergillus*. Biochem Cell Biol 80: 142.

Hernell, O. and Bläckberg, L. (1983). Digestion of human milk lipids: Physiologic significance of sn-2-monoacylglycerol hydrolysis by bile salt-stimulated lipase. Pediatr Res 16: 882–885.

Hernell, O., Bläckberg, L. and Lindberg, T. (1989). Human milk enzymes with emphasis on the lipases. In: Textbook of Gastroenterology and Nutrition in Infancy. Ed. Lebenthal, E. New York, Raven Press: 209–217.

Huang, J., Nandi, S., Wu, L., Yalda, D., Bartley, G., Rodriguez, R., Lönnerdal, B. and Huang, N. (2002a). Expression of natural antimicrobial human lysozyme in rice grains. Mol Breeding 10: 83–94.

Huang, J., Sutliff, T.D., Wu, L., Nandi, S., Benge, K., Terashima, M., Ralston, A.H., Drohan, W., Huang, N. and Rodriguez, R.L. (2001). Expression and purification of functional human alpha-1-antitrypsin from cultured plant cells. Biotechnol Prog 17: 126–133.

Huang, J., Wu, L., Yalda, D., Adkins, Y., Kelleher, S., Crane, M., Lönnerdal, B. and Huang, N. (2002b). Expression of functional recombinant human lysozyme in transgenic rice cell culture. Transgenic Res 11: 229–239.

Jenness, R. (1985). Biochemical and nutritional aspects of milk and colostrum. In: Lactation. Ed. Larson, B.L. Ames, University of Iowa Press: 78–94.

Kawakami, H., Dosako, S. and Lönnerdal, B. (1990). Iron uptake from transferrin and lactoferrin by rat intestinal brush-border membrane vesicles. Am J Physiol 258: G535–G541.

Kawakami, H. and Lönnerdal, B. (1991). Isolation and function of a receptor for human lactoferrin in human fetal intestinal brush-border membranes. Am J Physiol 261: G841–G846.

Kim, Y.-K., Yu, D.-Y., Lönnerdal, B. and Chung, B.-H. (1999). Novel angiotensin-I-converting enzyme inhibitory peptides derived from recombinant human α_{S1}-casein expressed in *Escherichia coli*. J Dairy Res 66: 431–439.

Klagsburn, M. (1978). Human milk stimulates DNA synthesis and cellular proliferation in cultured fibroblasts. Proc Natl Acad Sci USA 75: 5057–5061.

Kleesen, B., Bunke, H., Tovar, K., Noack, J. and Sawatzki, G. (1995). Influence of two infant formulas and human milk on the development of the fecal flora in newborn infants. Acta Paediatr 84: 1347–1356.

Krimpenfort, P. (1993). The production of human lactoferrin in the milk of transgenic animals. Cancer Detect Prev 17: 301–305.

Krimpenfort, P., Rademakers, A., Eyestone, W., van der Schans, A., van den Broek, S., Kooiman, P., Kootwijk, E., Platenburg, G., Pieper, F., Strijker, R. and De Boer, H. (1991). Generation of transgenic cattle using 'in vitro' embryo production. Biotechnology 9: 844–847.

Kuipers, M.E., Beljaars, L., van Beek, N., de Vries, H.G., Heegsma, J., van den Berg, J.J., Meijer, D.K. and Swart, P.J. (2002a). Conditions influencing the in vitro antifungal activity of lactoferrin combined with antimycotics against clinical isolates of Candida: Impact on the development of buccal preparations of lactoferrin. APMIS 110: 290–298.

Kuipers, M.E., Heegsma, J., Bakker, H.I., Meijer, D.K., Swart, P.J., Frijlink, E.W., Eissens, A.C., De Vries-Hospers, H.G. and van den Berg, J.J. (2002b). Design and fungicidal activity of mucoadhesive lactoferrin tablets for the treatment of oropharyngeal candidosis. Drug Deliv 9: 31–38.

Kumagai, I., Takeda, S., Hibino, T. and Miura, K. (1990). Expression of goat α-lactalbumin in *Escherichia coli* and its refolding to biologically active protein. Protein Eng 3: 449–452.

Kusnadi, A.R., Nikolov, Z.L. and Howard, J.A. (1997). Production of recombinant proteins in transgenic plants: Practical consideration. Biotechnol Bioeng 56: 473–484.

Lebenthal, E. and Lee, P.C. (1980). Development of functional response in human exocrine pancreas. Pediatrics 66: 556–560.

Leite, A., Kemper, E.L., Da Silva, M.J., Luchessi, A.D., Siloto, R.M.P., Bonaccorsi, E.D., El-Dorry, H.F. and Arruda, P. (2000). Expression of correctly processed human growth hormone in seeds of transgenic plants. Mol Breeding 6: 47–53.

Lindberg, T. (1979). Protease inhibitors in human milk. Pediatr Res 13: 969–972.

Lindberg, T., Engberg, S., Jakobsson, I. and Lönnerdal, B. (1997). Digestion of proteins in human milk, human milk fortifier, and preterm formula in infant rhesus monkeys. J Pediatr Gastroenterol Nutr 24: 537–543.

Lindh, E. (1985). Increased resistance of immunoglobulin dimers to proteolytic degradation after binding of secretory component. J Immunol 113: 284–288.

Lönnerdal, B. (1994). Digestibility and absorption of protein in infants. In: N.C.R. Räihä, ed. Protein Metabolism During Infancy, Vol. 33. New York, Raven Press: 53–65.

Lönnerdal, B., Keen, C.L., Glazier, C.E. and Anderson, J. (1984). A longitudinal study of rhesus monkeys (*Macaca mulatta*) milk composition: Trace elements, minerals, protein, carbohydrate and fat. Pediatr Res 18: 911–914.

Lubon, H., Paleyanda, R.K., Velander, W.H. and Deohan, W.N. (1996). Blood proteins from transgenic animal bioreactor. Med Rev 10: 131–143.

Luckow, V.A. and Summers, M.D. (1988). Trends in the development of baculovirus expression vectors. Biotechnology 6: 47–55.

Ma, J.K., Hiatt, A., M. Hein, Vine, N.D., Wang, F., Stabila, P., van Dolleweerd, C., Mostov, K. and Lehner, T. (1995). Generation and assembly of secretory antibodies in plants. Science 268(5211): 716–719.

Maga, E.A., Anderson, G.B., Huang, M.C., and Murray, J.D. (1994). Expression of human lysozyme mRNA in the mammary gland of transgenic mice. Transgenic Res 3: 36–42.

Matsumoto, S., Ikura, K., Ueda, M. and Sasaki, R. (1995). Characterization of a human glycoprotein (erythropoietin) produced in cultured tobacco cells. Plant Mol Biol 27: 1163–1172.

Maullu, C., Lampis, G., Desogus, A., Ingianni, A., Rossolini, G.M. and Pompei, R. (1999). High-level production of heterologous protien by engineered yeasts grown in cottage cheese whey. Appl Environ Microbiol 65: 2745–2747.

Meade, H., Gates, L., Lacy, E. and Lonberg, N. (1990). Bovine α_{S1}-casein gene sequences direct high level expression of active human urokinase in mouse milk. Biotechnology 8: 443–446.

Michetti, P., Mahan, M.J., Slauch, J.M., Mekalanos, J.J. and Neutra, M.R. (1992). Monoclonal secretory immunoglobulin A protects mice against oral challenge with the invasive pathogen Salmonella typhimurium. Infect Immun 60: 1786–1792.

Morton, H.C., Atkin, J.D., Owens, R.J. and Woof, J.M. (1993). Purification and characterization of chimeric human IgA1 and IgA2 expressed in COS and Chinese hamster ovary cells. J Immunol 151: 4743–4752.

Murasugi, A., Asami, Y. and Mera-Kikuchi, Y. (2001). Production of recombinant human bile salt-stimulated lipase in *Pichia pastoris*. Protein Expr Purif 23: 282–288.

Nandi, S., Suzuki, Y.A., Huang, J., Yalda, D., Pham, P., Wu, L., Bartley, G., Huang, N., and Lönnerdal, B. (2002). Expression of human lactoferrin in transgenic rice grains for the application in infant formula. Plant Sci 163: 713–722.

Newburg, D.S. (1997). Do the binding properties of oligosaccharides in milk protect human infants from gastrointestinal bacteria? J Nutr 127: 980S–984S.

Niemann, H., Halter, R., Carnwath, J.W., Herrmann, D., Lemme, E. and Paul, D. (1999). Expression of human blood clotting factor VIII in the mammary gland of transgenic sheep. Transgenic Res 8: 237–247.

Nishiya, K. and Horwitz, D.A. (1982). Contrasting effects of lactoferrin on human lymphocytes and monocytes natural killer activity and antibody dependent cell mediated cytotoxicity. J Immunol 129: 2519–2523.

Nuijens, J.H., van Berkel, P.H.C., Geerts, M.E.J., Hartevelt, P.P., de Boer, H.A., van Veen, H.A. and Pieper, F.R. (1997). Characterization of recombinant human lactoferrin secreted in milk of transgenic mice. J Biol Chem 272: 8802–8807.

Pelligrini, A., Thomas, U., Bramaz, N., Hunziker, P. and Von Fellenberg, R. (1999).

Isolation and identification of three bactericidal domains in the bovine α-lactalbumin molecule. Biochim Biophys Acta 1426: 439–448.

Read, L.C., Upton, F.M., Francis, G.L., Wallace, J.C., Dahlenburg, G.W. and Ballard, F.J. (1984). Changes in the growth-promoting activity of human milk during lactation. Pediatr Res 18: 133–139.

Riego, E., Limonta, J., Aguilar, A., Perez, A., Armas, R.D., Solano, R., Ramos, B., Castro, F. and De la Fuente, J. (1993). Production of transgenic mice and rabbits that carry and express the human tissue plasminogen activator cDNA under the control of a bovine alpha S1 casein promoter. Theriogenology 39: 1173–1185.

Rudloff, S. and Lönnerdal, B. (1992). Solubility and digestibility of milk proteins in infant formulas exposed to different heat treatments. J Pediatr Gastroenterol Nutr 15: 25–33.

Ruf, S., Hermann, M., Berger, I.J., Carrer, H. and Bock, R. (2001). Stable genetic transformation of tomato plastids and expression of a foreign protein in fruit. Nat Biotechnol 19: 870–875.

Sagi, L., Panis, B., Remy, S., Schoofs, H., Smet, K.D., Swennen, R. and Cammue, B.P. (1995). Genetic transformation of banana and plantain (Musa spp.) via particle bombardment. Biotechnology 13: 481–485.

Sahasrabudhe, A.V., Solapure, S.M., Khurana, R., Suryanarayan, V., Ravishankar, S., deSousa, S.M. and Das, G. (1998). Production of recombinant human bile salt stimulated lipase and its variant in *Pichia pastoris*. Protein Expr Purif 14: 425–433.

Salmon, V., Legrand, D., Slomianny, M.-C., ElYazidi, I., Spik, G., Gruber, V., Bournat, P., Olagnier, B., Mison, D., Theisen, M. and Mérot, B. (1998). Production of human lactoferrin in transgenic tobacco plants. Protein Expr Purif 13: 127–135.

Sanders, G., Picknett, T.M., Tuite, M.F. and Ward, M. (1989). Heterologous gene expresssion in filamentous fungi. Trends Biotechnol 7: 283–287.

Sato, R., Noguchi, T. and Naito, H. (1986). Casein phosphopeptide (CPP) enhances calcium absorption from the ligated segment of rat small intestine. J Nutri Sci Vitaminol 32: 67–76.

Schlimme, E. and Meisel, H. (1995). Bioactive peptides derived from milk proteins. Structural, physiological and analytical aspects. Nahrung 39: 1–20.

Shin, K., Hayasawa, H. and Lönnerdal, B. (2000). PCR cloning and baculovirus expression of human lactoperoxidase and myeloperoxidase. Biochem Biophys Res Commun 271: 831–836.

Shugar, D. (1952). Measurement of lysozyme activity and the ultraviolet inactivation of lysozyme. Biochim Biophys Acta 8: 302.

Sijmons, P.C., B.M. Dekker, Schrammeijer, B., Verwoerd, T.C., van Den Elzen, P.J. and Hoekema, A. (1990). Production of correctly processed human serum albumin in transgenic plants. Biotechnology 8: 217–221.

Simmons, C.R., Huang, N., Cao, Y. and Rodriguez, R.L. (1991). Synthesis and secretion of α-amylase by rice callus: Evidence for differential gene expression. Biotechnol Bioeng 38: 545–551.

Steele, W.F. and Morrisons, M. (1969). Antistreptococcal activity of lactoperoxidase. J Bacteriol 97: 635–639.

Strömqvist, M., Falk, P., Bergström, S., Hansson, L., Lönnerdal, B., Normark, S. and Hernell, O. (1995). Human milk kappa-casein and inhibition of *Helicobacter pylori* adhesion to human gastric mucosa. J Pediatr Gastroenterol Nutr 21: 288–296.

Strömqvist, M., Tornell, J., Edlund, M., Edlund, A., Johansson, T., Lindgren, K., Lundberg, L. and Hansson, L. (1996). Recombinant human bile salt-stimulated lipase: An example of defective O-glycosylation of a protein produced in milk of transgenic mice. Transgenic Res 5: 475–485.

Suzuki, Y.A., Kelleher, S.L., Yalda, D., L.Wu, Huang, J., Huang, N. and Lönnerdal, B. (2003). Expression, characterization and biological activity of recombinant human lactoferrin in rice. J Pediatr Gastroenterol Nutr 36: 190–199.

Svensson, M., Håkansson, A., Mossberg, A.-K., Linse, S. and Svanborg, C. (2000). Conversion of α-lactalbumin to a protein inducing apoptosis. Proc Natl Acad Sci USA 97: 4221–4226.

Tacket, C.O., Mason, H.S., Losonsky, G., Clements, J.D., Levine, M.M. and Arntzen, C.J. (1998). Immunogenicity in humans of a recombinant bacterial antigen delivered in a transgenic potato. Nat Med 4: 607–609.

Takase, K. and Hagiwara, K. (1998). Expression of human α-lactalbumin in transgenic tobacco. J Biochem 123: 440–444.

Terashima, M., Murai, Y., Kawamura, M., Nakanishi, S., Stoltz, T., Chen, L., Drohan, W., Rodriguez, R.L. and Katoh, S. (1999). Production of functional human alpha 1-antitrypsin by plant cell culture. Appl Microbiol Biotechnol 52: 516–523.

Theisen, M. (1999). Production of recombinant blood factors in transgenic plants. Adv Exp Med Biol 464: 211–220.

Tjellström, B., Stenhammar, L., S. Eriksson and Magnusson, K.E. (1993). Oral immunoglobulin A supplement in treatment of Clostridium difficile enteritis. Lancet 341: 701–702.

Tomita, M., Bellamy, W., Takase, M., Yamauchi, K., Wakabayashi, H. and Kawase, K. (1991). Potent antibacterial peptides generated by pepsin digestion of bovine lactoferrin. J Dairy Sci 74: 4137–4142.

van Berkel, P.H., Welling, M.M., Geerts, M., van Veen, H.A., Ravensbergen, B., Salaheddine, M., Pauwels, E.K., Pieper, F., Nuijens, J.H. and Nibbering, P.H. (2002). Large scale production of recombinant human lactoferrin in the milk of transgenic cows. Nat Biotechnol 20: 484–487.

Velander, W.H., Page, R.L., Morcol, T., Russell, C.G., Cajseco, R., Young, J.M., Drohan, W.N., Gwaz-Daukas, F.C., Wilkins, T.D. and Hohnson, J.L. (1992). Production of biologically active human protein C in the milk of transgenic mice. Ann NY Acad Sci 665: 391–403.

Wang, C.S., Downs, D., Dashti, A. and Jackson, K.W. (1996). Isolation and characterization of recombinant human apolipoprotein C-II expressed in *Escherichia coli*. Biochim Biophys Acta 1302: 224–230.

Ward, P.P., Lo, J.-Y., Duke, M., May, G.S., Headon, D.R. and Conneely, O.M.

(1992a). Production of biologically active recombinant human lactoferrin in *Aspergillus oryzae*. Biotechnology 10: 784–789.

Ward, P.P., May, G.S., Headon, D.R. and Conneely, O.M. (1992b). An inducible expression system for the production of human lactoferrin in *Aspergillus nidulans*. Gene 122: 219–233.

Winner, L., Mack, J., Weltzin, R., Mekalanos, J.J., Kraehenbuhl, J.P. and Neutra, M.R. (1991). New model for analysis of mucosal immunity: Intestinal secretion of specific monoclonal immunoglobulin A from hybridoma tumors protects against Vibrio cholerae infection. Infect Immun 59: 977–982.

Wolf, E., Jehle, P.M., Weber, M.M., Sauerwein, H., Daxenberger, A., Breier, B.H., Besenfelder, U., Frenyo, L. and Brem, G. (1997). Human insulin-like growth factor I (IGF-I) produced in the mammary glands of transgenic rabbits: Yield, receptor binding, mitogenic activity, and effects on IGF-binding proteins. Endocrinology 138: 307–313.

10
Enzymes

Jun Ogawa and Sakayu Shimizu
Kyoto University, Kyoto, Japan

I. INTRODUCTION

Enzymes have a long history of safe use in human foods, not only for processing but as ingredients. Recently the use of enzymes themselves as functional foods is rising together with the spreading concept of dietary supplements. Although there are only a few examples of practical applications now, the use of enzymes for dietary purposes will increase along with increasing attention to health maintenance and disease prevention. Here, an overview of the present status of enzymes as functional foods and their future prospects is briefly summarized together with examples of production technology and safety guidelines.

II. TRADITIONAL USE OF ENZYMES AS FOOD INGREDIENTS

In the 1970s, several enzymes, listed in Table 1, were used as optional ingredients in some foods for various purposes (1). The enzymes derived from well-known sources were generally recognized as safe. The examples of animal enzymes included catalase (from bovine liver), animal lipase, pepsin, rennin, and trypsin. Vegetable-derived enzymes included bromelain, papain, and various malts. Microbial enzymes included preparations from *Aspergillus niger* (lipase, catalase, carbohydrases, cellulase, pectinase, and glucose oxidase), *A. oryzae* (lipase, carbohydrases, and proteases), *Bacillus subtilis* or *B. licheniforms* (carbohydrases and proteases), and *Saccharomyces* species (carbohydrases, invertase). These enzymes were used to improve and/or modify

Table 1 Enzymes Used as Food Ingredients and Their Purposes

Food	Enzyme and source	Purpose
Flour	Malted wheat flour Malted barley flour α-Amylase (*A. oryzae*)	Provide sugar, modify dough properties
Farina	Papain, pepsin	Accelerate cooking
Bread	Wheat malt Barely malt Malt extract α-Amylase/protease (*A. oryzae*) Papain Bromelain Lipoxygenase (soy)	Provide sugar, modify dough properties
Cheese	Catalase Rennin Microbial rennin	Remove H_2O_2 from milk, precipitate casein
Soda water	Glucose oxidase Catalase	Stabilize drinks
Dried whole egg, dried egg white	Glucose oxidase Catalase	Stabilize

the properties of the foods and to facilitate the food processing. The other examples of enzyme preparations that are used in foods today are summarized in Table 2. They have future potential for use as functional foods, dietary supplements, and nutraceuticals (see Sec. III).

III. ENZYMES AS FUNCTIONAL FOODS, DIETARY SUPPLEMENTS, AND NUTRACEUTICALS

Recently use of enzymes as functional foods (2) and dietary supplements (3), in which they offer many benefits to the consumer, has been increasing. Initially, the term *dietary supplements* referred to products made of one or more of the essential nutrients, such as vitamins, minerals, and proteins. Today, they include, with some exceptions, any product intended for ingestion as a supplement to the diet. This includes vitamins, minerals, herbs, botanicals, and other plant-derived substances, amino acids, peptides, proteins, and enzyme. The enzymes in use before 1995 for dietary supplements are listed in Table 3. Most of them are used as digestive enzymes. Popular

Table 2 Enzyme Preparations Used in Foods Today

Enzyme (origin)	Function and purpose
From microorganisms	
α-Amylase	Flour processing
(*Aspergillus oryzae*) (*Bacillus stearothermophilus*)	Hydrolysis of edible starch to produce maltodextrin and nutritive carbohydrate sweeteners
Amyloglucosidase (*Rhizopus niveus*)	Degradation of gelatinized starch into constituent sugars
Aminopeptidase (*Lactococcus lactis*)	Optional ingredient for flavor development in manufacture of cheddar cheese
Carbohydrase and cellulase (*Aspergillus niger*)	Clam and shrimp processing
Carbohydrase (*Rhizopus oryzae*)	Production of dextrose from starch
Catalase (*Micrococcus lysodeikticus*)	Manufacture of cheese
Esterase-lipase (*Mucor miehei* var. *Coony et Emarson*)	Flavor enhancer in cheeses, fats and oils, and milk products
α-D-Galactosidase (*Mortierella vinaceae* var. *raffinoseutilizer*)	Production of sucrose from sugar beets
Glucose isomerase (*Streptomyces rubiginosus, Actinoplane missouriensis, Streptomyces olivaceus, Streptomyces olivochromogenes, Bacillus coagulans*)	Conversion of glucose syrups (pure culture fermantation that produces no antibiotic)
Lactase (*Candida pseudotropicalis, Kluyveromyces lactis*)	Hydrolysis of lactose in milk to glucose and galactose
Lipase (*Rhizopus niveus*)	Interesterification of fats and oils
Milk-clotting enzymes (*Endothia parasitica, Bacillus cereus, Mucor pusillus* Lindt, *Mucor miehei* var. *Cooney et Emerson, Aspergillus oryzae*/containing the gene for aspartic proteinase from *Rhizomucor miehei* var *Cooney et Emerson*)	Production of cheese
Mixed carbohydrase and protease (*Bacillus licheniformis*)	Hydrolysis of proteins and carbohydrates in preparation of alcoholic beverages, candy, nutritive sweeteners, and protein hydrolysates

Table 2 Continued

Enzyme (origin)	Function and purpose
Rennin and chymosin (animal-derived preparation from *Escherichia coli* K-12, *Kluyveromyces marxianus* var. *lactis*, or *Aspergillus niger* var. *awamori*)	Coagulation of milk in cheeses and other dairy products
Urease (*Lactobacillus fermentum*)	Production of wine
From plants	
Bromelain (pineapples/*Ananas comosus*, *Ananas bracteatus*)	Hydrolysis of proteins and polypeptides
Ficin/peptide hydrolase (*Ficus* spp.)	Hydrolysis of proteins and polypeptides
Malt/α-amylase and β-amylase (barley)	Hydrolysis of starch
Papain (papaya, *Carica papaya*)	Hydrolysis of proteins or polypeptides
From animals	
Catalase (bovine liver)	Decomposition of hydrogen peroxide
Lipase (edible forestomach of calves, kids, or lambs)	Hydrolysis of fatty acid glycerides
Pancreatin/peptide hydrolase (porcine or bovine pancreatic tissue)	Hydrolysis of proteins or polypeptides
Pepsin/peptide hydrolase (hog stomach)	Hydrolysis of proteins
Rennin (abomasums of the sucking calf)	Coagulation of milk in cheeses and other dairy products
Trypsin/peptide hydrolase (porcine or bovine pancreas)	Hydrolysis of proteins

examples of the enzymes already used as dietary supplements are described in the following discussion.

A. Lactase

Lactase (β-D-galactosidase) is used for persons with lactose intolerance who are unable to digest significant amounts of lactose because of a genetically inadequate expression of the enzyme lactase. Common symptoms of lactose intolerance include abdominal pain and bloating, excessive flatus, and watery stool after the ingestion of foods containing lactose. Lactase deficiency is present in up to 15% of persons of northern European descent, up to 80% of blacks and Latinos, and up to 100% of American Indians and Asians. Treatment consists primarily of avoiding lactose-containing foods. Lactase enzyme supplements may be helpful. Lactase from *Aspergillus oryzae* (4) is used as a dietary supplement.

Table 3 Enzymes Used as Dietary Supplements and Their Origins

Enzyme	Origin
Amylase	*Bacillus subtilis, Bacillus amyloliquefaciens, Aspergillus oryzae, Aspergillus niger,* malt (malt diastase)
Amyloglucosidase	*Aspergillus niger, Rhizopus oryzae*
β-Amylase	wheat, *Bacillus* spp.
Catalase	*Aspergillus niger*
Cellulase	*Aspergillus niger, Trichoderma longibrachiatum* (*reesei*)
α-D-Galactosidase	*Aspergillus niger*
Glucose oxidase	*Aspergillus niger*
Hemicellulase	*Aspergillus niger, Trichoderma longibrachiatum* (*reesei*), *Aspergillus oryzae, Bacillus subtilis*
Invertase	*Saccharomyces cerevisiae*
Lactase	*Aspergillus oryzae, Kluyveromyces lactis*
Lipase	*Aspergillus niger, Arthrobacter ureafaciens, Candida cylindracea, Rhizomucor miehei, Rhizopus oryzae, Rhizopus delemar*
Lysophospholipase	*Aspergillus niger*
Lysozyme	Egg white
Pancreatin	Porcine pancreas
Pancrelipase	Bovine and porcine pancreas
Pectinase	*Aspergillus niger, Aspergillus japonicus, Rhizopus oryzae*
Phospholipase	Porcine
Phytase	*Aspergillus niger*
Protease (microbial)	*Aspergillus oryzae, Aspergillus niger, Aspergillus melleus, Bacillus licheniformis, Bacillus subtilis, Bacillus thermoproteolyticus, Rhizopus niveus*
Protease (botanical)	Papaya latex (papain), pineapple, root and stem (bromelain)
Protease (animal)	Bovine or porcine (trypsin), bovine or porcine (chymotrypsin), bovine or porcine (pepsin), rennin
Superoxide dismutase	*Bacillus* spp.

B. α-D-Galactosidase

Foods such as broccoli, cabbage, and beans are famous for causing gas because they contain complex sugars (oligosaccharides), which some people cannot easily digest. These sugars ferment in the large intestine, causing gas and distress. α -D-Galactosidase breaks down raffinose and stachyose in gassy foods, making them more digestible (5). α-D-Galactosidase is made from a safe food-grade mold, *Aspergillus niger* (6,7).

C. Proteases

Especially in a person of advanced age, the absorption of amino acids decreases with decreasing activity of the digestive enzymes, resulting in a low level of serum albumin concentration. Serum albumin transports various biomaterials and pharmaceuticals, so its decrease causes various disorders. Proteases increase amino acid absorption and are used for maintaining the health of aged persons.

Dietary supplements are taken as part of a healthy lifestyle. But some of them are expected to be effective for clinical purposes. Some enzymes are used to reduce the risk of, or to modulate the risk factors for, specific chronic diseases such as heart disease and cancer and certain birth defects. Furthermore, they are used for short-term benefits such as sleep management and enhanced physical performance. Dietary supplements with such uses are called *nutraceuticals*. A good example of a nutraceutical enzyme is bromelain (8). Bromelain has been used successfully as a digestive enzyme after pancreatectomy, in cases of exocrine pancreas insufficiency, and in other intestinal disorders. The combination of ox bile, pancreatin, and bromelain is effective in lowering stool fat excretion in patients with pancreatic steatorrhea, resulting in a symptomatic improvement in pain, flatulence, and stool frequency. Moreover, bromelain is reported to have beneficial effects on tumorigenesis, immune modulation, potentiation of antibiotics, and other processes. The therapeutic applications of bromelain have included administration as a synergistic drug to potentiate the action of antibiotics, as a mucolytic to assist with the clearing of sputum from the respiratory tract, as a proteolytic digestive enzyme, and as a nutraceutical to prevent the symptoms of angina.

IV. GUIDELINES FOR THE PRODUCTION OF FOOD-GRADE ENZYMES

In general, enzymes are unremarkable in terms of their toxicological profiles: That is, they possess a low order of toxicity. Some types of enzymes, e.g., proteases, are irritating to skin and eyes, particularly at high concentrations or on prolonged contact. The main safety concern associated with enzymes is the induction of respiratory allergies among workers in the manufacturing environment, where repeated exposure to a foreign protein may induce sensitivity. Enzymes are proteins, and as can many other proteins foreign to the human body, they can stimulate the body's immune system if inhaled at sufficient concentrations. The safety of enzymes commonly used in foods and food processing is well established. However, for future expanded use of enzymes in foods, several guidelines were established. In 1990, the U.S. Food and Drug Administration (FDA) published Enzyme Preparations: Chemistry Recommendation for Food Additive and GRAS (Generally Recognized As Safe) Affirmation Petitions. The European Science Committee of Food published Guidelines for the Presentation of Data on Food Enzymes in 1991. These guidelines deal with (a) the toxicological properties of the enzyme preparation; (b) the quantity of enzyme consumed; (c) the allergies and irritations caused by the enzyme activity in the final product; (d) the unintended reaction products in the food, caused by enzymatic reactions in the final foodstuffs; and (e) the safety of the source organism. In 1994, the U.S. FDA announced the Dietary Supplement Health and Education Act (http://vm.cfsan.fda.gov/~dms/dietsupp.html), which clearly regulated the use of enzymes as dietary supplements. Any enzyme that was grandfathered as a dietary supplement before October 15, 1994 (see Table 3), any enzyme that is generally recognized as safe (GRAS), or any enzyme that meets the criteria of a food additive petition may not require additional toxicological testing. In addition, the U.S. Enzyme Technical Association has published Industry Guidelines for the Use of Enzymes in Dietary Supplements (http://www.enzymetechnicalassoc.org/dietary.html), which refers to the safe handling of enzymes during production processes and recommends that the following testing be performed on the enzyme preparation (test material):

1. Subchronic toxicity study of 90 days in a rodent species. The no observable adverse effect level (NOAEL) should be sufficiently high to assure safety.
2. Mutagenicity studies using gene mutation in *Salmonella typhimurium* (9,10).
3. Chromosome aberration test.

V. PRODUCTION OF ENZYMES FOR DIETARY PURPOSES

General concepts for the production of enzymes for dietary purposes are basically the same as those for industrial purposes, with the addition of the special safety considerations described.

A. Sources of Enzymes

Enzymes are produced from animals, plants, and microorganisms (11). Animals have traditionally produced some enzymes for food and medical use. The best-known food enzyme of animal origin is rennin. In general, however, animals are a poor source of enzymes as they are slow-growing and expensive. Extraction of enzymes from animal tissues can also be difficult, further adding to the production cost of the enzymes. Plants grow more quickly than most animals and can be produced in quantity on an annual basis. Again, this time scale is too long for enzyme manufacturers, and the only commercially important plant enzymes are proteases obtained from crops such as pineapple (bromelain) and papaya (papain). For these reasons, enzyme production from microorganisms is preferred as they are fast growing, can be easily controlled during growth, and produce enzymes that are easy to extract. In some cases, microorganisms produce extracellular enzymes, making extraction and purification even simpler. The microorganisms listed in Table 2 and 3 are representative examples used in the production of enzymes generally recognized as safe.

B. Microbial Production of Food-Grade Enzymes

The production and use of microbial enzymes have been reviewed (12). To produce food-grade enzymes of microbial origin, the materials for the cultivation medium should be food-grade. As a general principle, pathogenic microorganisms should not be used. The equipment for cultivation, extraction, and purification should be as clean as that used for food processing. Microbial production processes can be summarized as follows (13):

1. Maintenance of the master or seed culture
2. Subculturing of the inoculation phase
3. Volume fermentation and harvesting of organisms
4. Cell removal, extraction, purification, and packing

The procedures for cultivation (1 to 3) are general and common. Adequate purification processes should be used to prevent contamination of harmful materials. Usually, alcohol precipitation, salting-out, ultrafiltration, and size-exclusion techniques are used for purification. Column chromatography is cost-ineffective but is required to produce enzymes for clinical purposes.

ACKNOWLEDGMENT

The authors wish to thank Dr. Satoshi Koikeda of Amano Enzyme Inc. for his help in gathering valuable information.

REFERENCES

1. G Leed. Health and legal aspects of the use of enzymes. In: G Leed, ed. Enzymes in Food Processing. 2nd ed. New York: Academic Press, 1975, pp 549–554.

2. S Ross. Functional foods: The food and drug administration perspective. Am J Clin Nutr 71: 1735S–1738S, 2000.
3. J Hathcock. Dietary supplements: How they are used and regulated. J Nutr 131: 1114S–1117S, 2001.
4. S Ogushi, T Yoshimoto, D Tsuru. Purification and comparison of two types of β-galactosidase from *Aspergillus oryzae*. J Ferment Technol 58: 115–122, 1980.
5. M Kagaya, M Iwata, Y Tosa, Y Nakase, T Kondo. Circadian rhythm of breath hydrogen in young women. J Gastroenterology 33: 472–476, 1998.
6. R Kaneko, I Kusakabe, K Murakami. Substrate specificity of α-galactosidase from *Aspergillus niger* 5-16. Agric Biol Chem 55: 109–115, 1991.
7. P Manzanares, LH de Graaff, J Visser. Characterization of galactosidases from *Aspergillus niger*: Purification of a novel α-galactosidase activity. Enzyme Microb Technol 22: 383–390, 1998.
8. Bromelaine. Altern Med Rev 3: 302–305, 1998.
9. BN Ames, J McCann, E Yamasaki. Methods for detecting carcinogens and mutagens with the *Salmonella*/Mammalian-microsome mutagenicity test. Mutat Res 31: 347–364, 1975.
10. D Maron, B Ames. Revised methods for the *Salmonella* mutagenicity test. Mutat Res 113:173–215, 1983.
11. AJ Taylor. Enzymes in the food industry. In: GA Tucker, LFJ Woods eds. Enzymes in Food Processing. London: Blackie, 1991, pp 22–34.
12. WM Fogarty. Microbial Enzymes and Biotechnology. London: Applied Science, 1983.
13. RI Farrow. Enzymes: Health and safety considerations. In: GG Birch, N Blakebrough, KJ Parker eds. Enzyme and Food Processing, London: Applied Science, 1981, pp 239–259.

11

Chemical Analysis and Health Benefits of Isoflavones

Shaw Watanabe, Sayo Uesugi, and Ryota Haba
Tokyo University of Agriculture, Tokyo, Japan

I. INTRODUCTION

Health benefits of soya isoflavones (IFs) for prevention of certain cancers, osteoporosis, and climacteric symptoms at menopause have been recognized by many epidemiological studies. A recent recommendation to use soy protein or IFcontaining tablets, however, does not stand on firm evidence that clarifies safety and toxicity potential in humans. Various mechanisms of IF action have been found, in addition to classical action as phytoestrogens. Absorption, distribution, and excretion of IF have been clarified, but internal metabolism and interaction with macromolecules in the body need further study. Recent research on IF, including analytical methods, is reviewed in this chapter.

II. PRESENCE AND SYNTHESIS OF ISOFLAVONES IN NATURAL PRODUCTS

Soybeans are consumed in many Asian countries, and they are a major source of isoflavones (1–7). Isoflavonoids are divided into seven categories: isoflavone, isoflavane, isoflavanone, coumestane, pterocarpane, rotenoid, and coumaronochromone (2) (Fig. 1). Isoflavonoids are distributed in various gymnosperms and angiosperms, and the only IFs naturally occur in animals as 5,7,2′,4′-tetrahydroxyisoflavone and 5-hydroxy-7,2′,4′ trimethoxy isoflavone in the mollusk *Nerita albicilla*. Some IF metabolites (such as equol in the

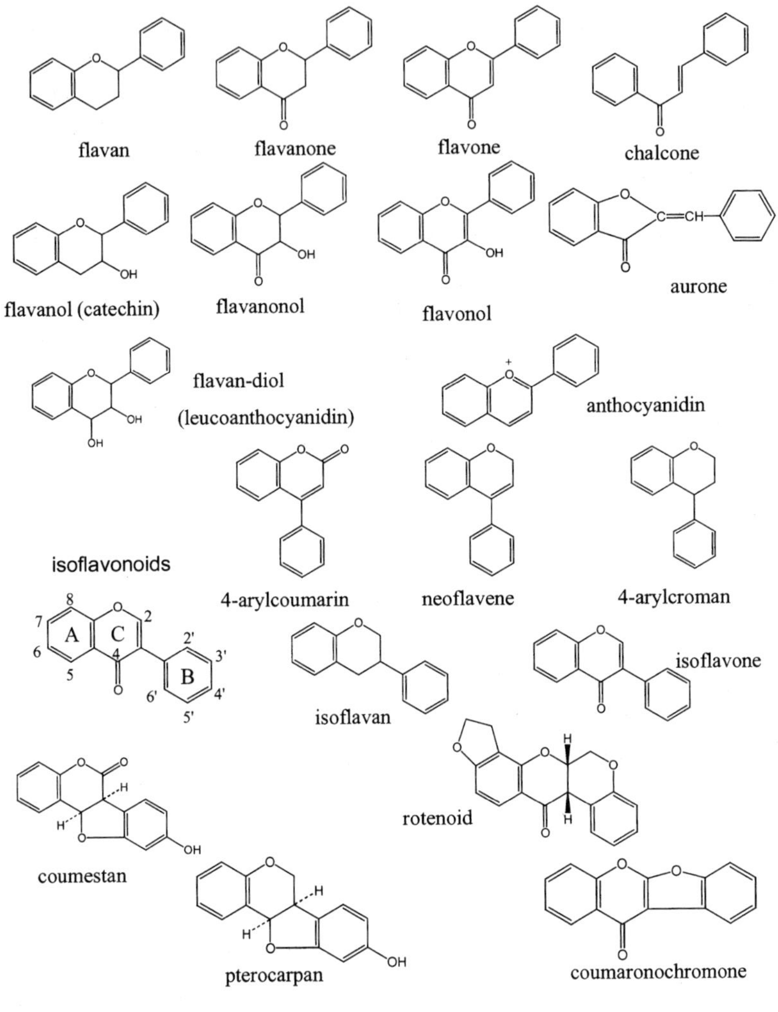

Figure 1 Basic structure of the flavonoid family.

human body) are considered animal IF. Hegnauer and Grayer-Barkmeijier (8) reviewed IFs in Leguminosae and found that 489 of 517 isoflavonoid aglycones were present in beans: 89% isoflavones, 100% isoflavanes, 100% isoflavonones, 100% coumestanes, 100% pterocarpanes, and 95% rotenoids (Table 1). In plants, almost all isoflavonoids are esterified with glucosides. In beans IFs usually take a glycoside or malonyl glycoside form. Major IFs in soybeans, such as genistein, daidzein, and glycitein, are present in the form of

Table 1 Proportion of Flavonoids in Legumes Among All Plants

Compound class	Aglicones in all plants, total no.	Aglicones in Leguminosae, no.	Legume compounds compared to the total, percentage
Anthocyanidins	17	6	35
Flavones	287	40	14
Flavonols	357	63	18
Chalcones	164	82	50
Aurone	16	8	50
Dihydrochalcones	59	13	22
Flavanones	244	83	34
Dehydroflavonols	75	27	36
Flavans	24	2	8
Flavane-3-ols(catechins)	26	10	38
Flavane-3,4-diol	31	23	74
Flavan-4-ols	7	1	14
Peltogynoids	9	7	78
Total flavonoid aglycones	1346	366	28
Isoflavones	223	198	89
Isoflavanones	50	50	100
Pterocarpans	114	114	100
Isoflavans	42	42	100
Rotenoids	56	53	95
Caumestans	32	32	100
Total isoflavonoid aglycones	517	489	95

Source: Ref. 8.

7-*O*-glycosides and further esterified between malonic acid and the 6 positions of glucose. Raw soybeans are rich in malonyl glycosides, and acetyl glycoside concentration increase by decarboxylation of malonyl residue during roast (9).

Soybeans are well known to be rich in genistein and daidzein (3–8). Their 6″-*O*-malonylgenistin and 6″-*O*-malonyldaidzein concentration increase through ripening of the seeds (9) (Fig. 2). Soybeans grown at low temperature contained more IF in beans when compared to those grown at higher temperature (Table 2), but this difference was smaller inside the hypocotyls (10). Glycitin and its malonyl derivatives are only present in hypocotyls as well as soyasaponin A and soyasaponin α-g. The synthetic pathway of IF is shown in Fig. 3. A key enzyme is phenylalanin-lyase (PAL), which transforms

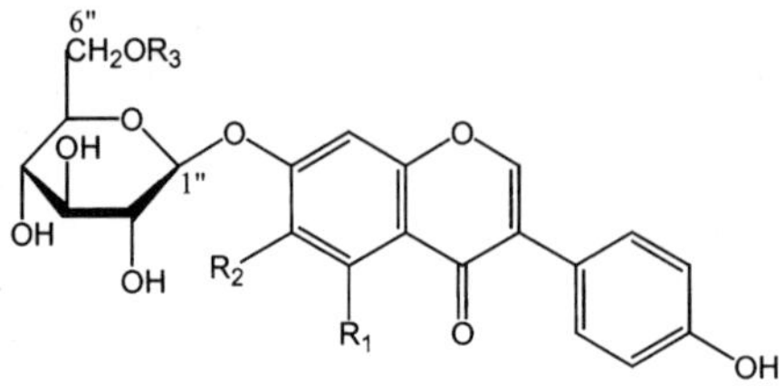

	R_1	R_2	R_3
Daidzin	H	H	H
glycitin	H	OCH_3	
genistin	OH	H	H
6″-*O* -acetyldaidzin	H	H	$COCH_3$
6″-*O* -acetylglycitin	H	OCH_3	$COCH_3$
6″-*O* -acetylgenistin	OH	H	$COCH_3$
6″-*O* -malonyldaidzin	H	H	$COCH_3$
6″-*O* -malonylglycitin	H	OCH_3	$COCH_3$
6″-*O* -malonylgenistin	OH	H	$COCH_3$

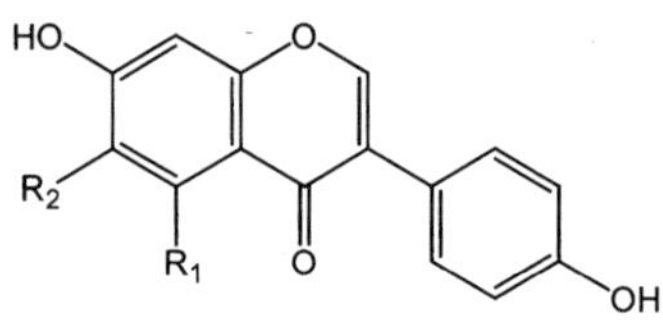

	R_1	R_2
daidzein	**H**	**H**
glycitein	**H**	**OCH_3**
genistein	**OH**	**H**

Figure 2 Isoflavones in the soybean.

phenylalanine to cinnamic acid. Three molecules of malonyl-coenzyme A (CoA) condense to chalcone via an intermittent molecule. Arylmigration of chalcone from C2 to C3 yields daidzein, and genistein is made by arylmigration from naringenin, which is made from chalcone by chalcone-flavone isomerase. Coumestrol is synthesized from daidzein but is only detected in the sprout.

The physiological function of IF in plants is considered to be related to node formation (11,12). But antistress molecules, phytoalexin of soybean

Table 2 Effects of Temperature on Soy (Kogane) Isoflavones and DDP Binding Saponins During Ripening

Temperature	Chemicals	High temperature	Low temperature
Cotyledon	Daidzein	0.9 ± 0.6	10.2 ± 1.8
	Genistein	1.1 ± 0.6	15.0 ± 1.5
	Malonyldaidzein	3.5 ± 0.5	69.0 ± 10.3
	Malonylgenistein	5.2 ± 1.4	100.7 ± 5.1
	Total	10.7 ± 3.1	194.9 ± 15.1
Hypocotyl	Daidzein	14.5 ± 4.0	48.3 ± 13.0
	Glycitein	4.3 ± 1.8	39.8 ± 14.6
	Genistein	9.1 ± 0.9	45.9 ± 10.2
	Malonyldaidzein	84.3 ± 19.1	428.6 ± 69.5
	Malonylglycitein	13.2 ± 5.1	155.5 ± 31.2
	Malonylgenistein	10.2 ± 3.9	119.9 ± 24.1
	Total	135.6 ± 33.1	838.0 ± 161.4
DDMP saponin	Soyasaponin αa	11.7 ± 1.0	9.9 ± 0.9
	Soyasaponin βg	165.6 ± 6.3	137.8 ± 11.5
	Soyasaponin βa	71.8 ± 3.1	89.3 ± 10.7
	Total	249.1 ± 7.5	237.0 ± 23.0

glyceollin I, glyceollin II, and glyceollin III, are also produced from daidzein (13,14).

III. MEASUREMENT OF ISOFLAVONOIDS IN BIOLOGICAL SPECIMENS

Development of separation techniques such as (HPLC), associated with development of detectors such as (^{1}H-NMR), ^{13}C-NMR, (2D)-NMR, made it possible to identify many IFs without hydrolysis and/or derivative production (9,10,15,16). Usually IF is extracted from samples by 60%–80% ethanol or methanol. Acetonitrile with 0.1N HCl has also been used. Heating during extraction caused deacylation of malonyl residues, and glycoside forms increased (9). Highest concentration of aglycon was obtained after 2-hr hydrolysis with 1M HCl at 98–100°C.

Absorption of IF in the alimentary tract mostly occurred in the form of aglycone (17–21). It has been suggested that bacterial hydrolysis of the glycosyl bond of glycosidic forms of isoflavonoids must precede absorption from the intestinal tract. In humans, absorption occurs for the unconjugated form

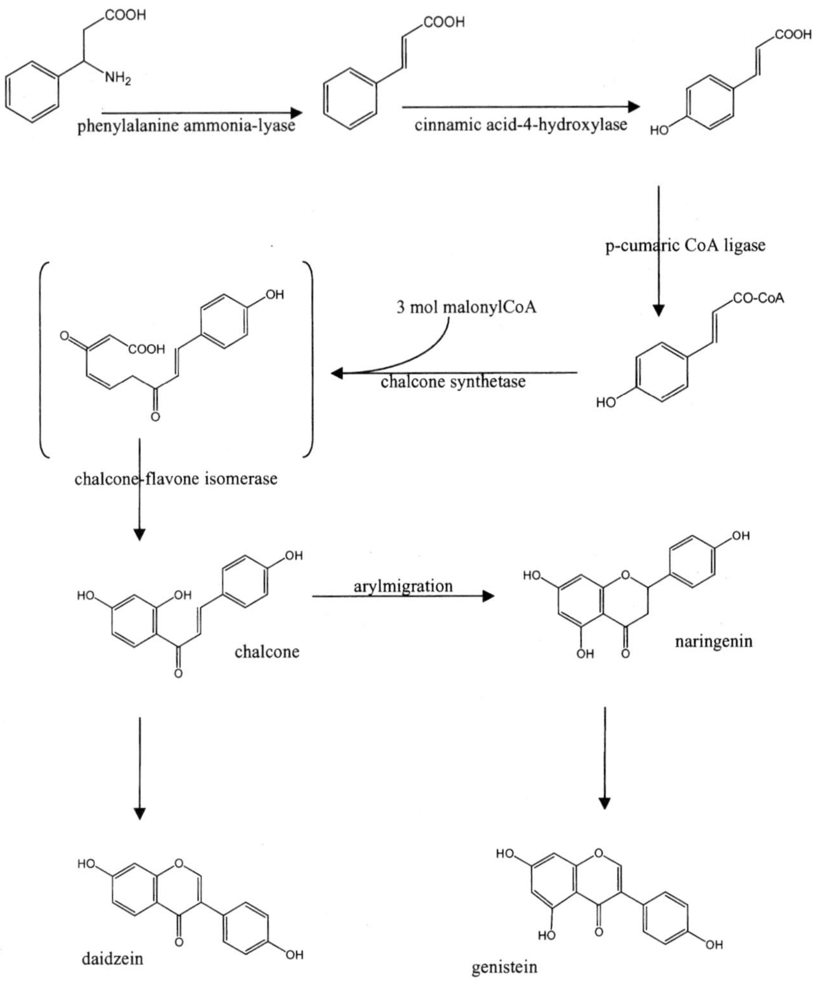

Figure 3 The synthetic pathway of daidzein and genistein.

of IF in the small intestine and large intestine. Enterohepatic circulation of glucuronized IFs has also been considered (17). Fecal excretion of intact isoflavonoids varied from 0.4% to 8%. This difference arises from different compositions of intestinal bacterial flora. Isoflavonoids that possess a 5-OH group, such as genistein, are much more susceptible to C-ring cleavage by rat intestinal bacteria (22). Certain strains of *Clostridium* spp. in the human colon can cleave the C-ring and produce monophenolic compounds anaerobically (23).

O-DMA and equol, the main metabolites of daidzein, are probably produced in the colon (20,21). Hutchins and associates (24) compared urinary excretion after consumption of fermented and unfermented soy foods and found that the early absorption was obtained by fermented soy, such as shoyu and miso (fermented soy bean paste).

Measurement of plasma isoflavonoid level is difficult, because of the small amounts present in the blood and the necessity for pretreatment to get active forms (15). The commonly used HPLC untraviolet (UV) detector does not have enough sensitivity (usually 100–1000 ng/mL). A high-performance liquid chromatography electrochemical detector (ECD) that detects 10–100 ng/mL has become available. Gas chromatography–mass spectrometry (GC-MS) is so far the most sensitive method to detect 1 ng/mL, but it requires a skillful technique. A recently developed time-resolved fluoroimmunoassay provides a different sensitive method, especially for the large number of specimens from an epidemiological study. It can measure below 5 nmol/L (25).

To measure the urinary isoflavonoids, vitamin C (final concentration is about 1%) and NaN_3 (final concentration is 0.003M) in storage bottles prevent denaturation of phytoestrogens and bacterial contamination. We compared the IF concentration in plasma, urine, and feces after a single ingestion of baked soy powder (*kinako*) and further analyzed the kinetic properties of IF (18,26). Eight kinds of phytoestrogens, daidzein, genistein, O-DMA, equol, enterolacton, enterodiol, matairesinol, and secoisolacriresinol, were measured. Ingested kinako cakes contained 103.23 μmol daidzein and 111.89 μmol genistein, on average, after complete hydrolysis to aglycone. From the calculation of the highest level of plasma, at least 15 μmol (3000 nmol/L × 5 L: about 3.76 mg) should be present in the whole blood in the body. Recovery of daidzein and genistein from urine was 35.8% and 17.6%, respectively, and 4.4% of ingested daidzein and 4% of genistein were recovered in feces. The plasma half-life of daidzein was 6.31 hr and that of genistein was 8.95 hr. King and Bursill (27) found the same values. In addition, equol excretion was dependent upon the composition of intestinal microflora, such as *Streptococcus intermedius* ssp., *Bacteroides ovatus* ssp., and *Ruminococcus* ssp. (personal communication, Uchiyama, 2001). Genistein seems to be absorbed more efficiently from the intestinal tract and to remain in circulation longer.

IV. BIOAVAILABILITY OF ISOFLAVONES

Estrogenic effects of isoflavones have been extensively investigated (28–31) (Table 3). Kuiper and colleagues (32) found that genistein and daidzein strongly bind ER-beta, more than ER-alpha (Table 4). They are beneficial to

Table 3 Estrogenic Activities of Isoflavones

Chemical	Recombinant yeast	MCF 7 cell
Estradiol-17β[a]	reference	reference
Equol	2.3×10^{-2}	1.3×10^{-2}
Genistein	4.5×10^{-3}	2.6×10^{-2}
Daidzein	2.8×10^{-1}	1.1×10^{-2}

[a] The activity of 17β E2 is taken as unity.
Source: Ref. 32.

prevention of osteoporosis, because bone formation is stimulated by estrogens through ER-beta.

Equol showed much greater affinity for binding to estrogen receptors than the other isoflavones. Duncan and coworkers (33) reported that equol excretors generally had lower concentrations of estrone, estrone sulfate, testosterone, androstendione, dehydroepiandrosterone (DHEA), DHEA sulfate, and cortisol and higher concentrations of sex hormone–binding globulin and midluteal progesterone, a hormonal pattern generally consistent with lowered breast cancer risk. We also found that equal excretors showed a stronger antioxidant effect (34).

Table 4 Soy-Isoflavone Elements Affected by Temperature of Extraction

	Room temperature mg %		80°C mg %	
	Hypocotyls	Cotyledon	Hypocotyls	Cotyledon
Daidzin	320	45	838	145
Glycitin	485	ND	1004	ND
Genistin	118	80	246	210
6″-*O*-Malonyldaidzin	423	70	8	3
6″-*O*-Malonylglycitin	445	ND	11	ND
6″-*O*-Malonylgenistin	144	117	4	ND
6″-*O*-Acetyldaidzin	2	2	57	8
6″-*O*-Acetylglycitin	6	ND	89	ND
6″-*O*-Acetylgenistin	105	1	39	1
Daidzein	102	33	35	11
Glycitein	ND	ND	15	ND
Genistein	35	48	16	14
Total	2185	396	2362	392

ND, not detected.

Administration of isoflavones showed decreased estradiol concentration and a prolonged menstruation cycle, as well as a decrease in luteinizing hormone (LH) and follicle stimulating hormone (FSH) concentration. On the contrary, Smith and colleagues (35) found a 15% increase in mean estradiol concentration through administration of 86 mg/day isoflavones in tablet form for two menstrual cycles, followed by a placebo for an equivalent period. Steroid hormones, SHBG, total cholesterol, and triacylglycerol did not change, however.

Administration of 40-mg IF tablets per day for 4 weeks showed that the inhibition of ovarian steroid synthesis did not appear to be mediated by gonadotrophins, since IF tablet intake increased LH and FSH in two of three subjects (36). Lu and associates (37) reported that high levels of isoflavones did not affect gonadotrophins but did increase prolactin levels. These data suggest that soy isoflavones may have selective effects on hypothalamic receptors that regulate prolactin but not on gonadotrophin production.

Most IFs in blood are glucuronide conjugates, and small amounts are sulfated or unconjugated free molecules. Isoflavone stimulates SHBG production in the liver, so increased SHBG diminishes free active IF.

Free radicals are considered to be among the most potent carcinogens in the body. They also act on lipid peroxidation, which is related to the aging process (38). Suppressed production of peroxides, such as PCOOH and PEOOH, is thought to decrease the mutations leading to cancer. Inhibitory action of IF on protein kinases was studied by Akiyama and colleagues (39) (Table 5). Genistein selectively showed strong inhibition on the EGF receptor.

In our animal experiments, treatment of SD rats that drank soy hypocotyl tea in place of ordinary water showed decreased blood PCOOH and PEOOH levels, as well as decreased ^{8}OHdG in the liver and kidneys (39a). Under iron-deficient conditions, which significantly increased peroxide level in the body, the effects were more clearly observed.

^{8}OHdG is also considered an indicator of deoxyribonucleic acid (DNA) damage (40–42). The excision repair of DNA followed by excretion in urine suggests dose-dependent oxidative damage to DNA, decreased urinary excretion of 80HdG would reflect the amount of DNA damage in the body (42). Our preliminary study on SD rats fed a mixture of soy hypocotyl showed marked inhibition of breast cancer induced by PHIP treatment (unpublished).

Adlercreutz and coworkers (30) and Murkies and associates (43) found that urinary excretion of phytoestrogens was significantly lower in the breast cancer patients than in controls.

The 1997 case–control study of Ingram and colleagues (44) clearly showed a significantly reduced odds ratio of 0.27 [95% confidence interval

Table 5 Effect of Flavonoids on Protein Kinases[a]

	EGF receptor kinase	cAMP-dependent protein kinase
Genistein	0.7	>100
Genistin	>100	NA
Prunetin	4.2	NA
Daidzein	>100	NA
Biochanin A	26.0	>100
Apigenin	25.0	>100
Acacetin	40.0	NA
Flavone	50.0	>100
Kaempherol	3.2	>100
Quercetin	5.0	>100

[a] EGF, epidermal growth factor; cAMP, cyclic adenosine monophosphate; NA, not applicable.
Source: Ref. 39.

(CI): 0.10–0.69] for breast cancer patients with >185 nmol/day excretion of equol, compared to <70 nmol/day.

Functions of IF are summarized in Table 6 (45–47). The low prostatic cancer incidence among Japanese has been attributed to high IF level (48). Carcinogenesis requires accumulation of complex events in target cells, so various pharmacological functions should act collaboratively for the prevention of cancer. In the case of breast cancer, transfer of IF from mother to fetus could effect a preventive effect on the mammary gland after birth (49). The parallel phenomenon for rats and humans would support the possibility that

Table 6 Multiple Isoflavone Functions[a]

Estogenic effect (Messina, 1991)
Inhibition of
- Tyrosine protein kinase (Akiyama et al., 1987)
- DNA topoisomerase (Naim, 1987)
- Ornithine decarboxylase, ribosomal S6 kinase (Linassier, 1990)
- Aromatase (Adlercreutz, 1986)

Induction of
- Specific P450, (Sariaslani, 1986)
- Glutathione transferase, catalase
- SHBG

[a]DNA, deoxyribonucleic acid; SHBG, steroid hormone binding globulin.

the higher intake of soy-derived food among Japanese would lower the incidence of breast cancer.

V. EPIDEMIOLOGY AND INTERVENTION TRIALS BY ISOFLAVONES

Traditional Japanese women with their high intake of soy—a rich source of phytoestrogens—have a low incidence of breast and other estrogen-related cancers and few menopausal symptoms (31). The U.S. FDA has proposed allowing some soy products to carry a label indicating that a reduction of heart disease risk may result from ingestion of 25 g/day soy protein as a source of 60 mg of soy phytoestrogens. The average Japanese consumes about 30 mg of isoflavones per day (1,50,51). Three of four case–control studies since 1991 have suggested that soy intake reduces breast-cancer risk; one recent case–control study showed a decreased risk in women who had high amounts of urinary phytoestrogens (44,52) (Table 7). Whereas circumstantial evidence for the beneficial effects of phytoestrogens is increasing, results of clinical trials are insufficient (53–59). Studies on soy have been limited by several factors: the amount of phytoestrogens in soy varies and is rarely measured; other factors in soy protein could not be neglected; and trials have been short. Previous intervention studies used various types of soy proteins, including soy milk and soy-rich foods, so the results were often controversial. It is necessary to determine the most effective dose that has no adverse effects. This approach may serve to maximize the efficacy of phytoestrogens in clinical studies.

Table 7 Case–Control Study on the Relationship Between Urinary Isoflavones and Breast Cancer

	Urine excretion nmol/day	Case	Control	(95% CI)[a]	Odds ratio
Daidzein	≦600	51	31	1.00	
	>600 ~ ≦900		29	36	0.60 (0.27, 1.33)
	>900 ~ ≦1300		29	35	0.80 (0.36, 1.80)
	>1300		24	32	0.47 (0.17, 1.33) $p < 0.241$
Equol	≦70	47	35	1.00	
	>70 ~ ≦110		37	37	0.45 (0.40, 1.02)
	110 ~ ≦185		35	36	0.52 (0.23, 1.17)
	>185		24	36	0.27 (0.10, 0.69) $p < 0.009$

[a] CI, confidence interval.
Source: Ref. 44.

There have been seven intervention studies to observe hormonal effects of isoflavone intake (36). Most of them administered the isoflavones by soy protein powder, soy milk, or isoflavone-rich food, so the composition and content of isoflavones varied. The ratio of daidzein, genistein, and glycitin in soy protein is 4:5:1; that of a hypocotyl tablet is 7:1:4 (data not shown). This difference would yield some controversial results.

Daily drinking of soy hypocotyl tea at 12–15 mg/day showed antioxidative activity in the body and low urinary excretion of ^{8}OHdG compared to those of controls (34). Total consumption, including that in the diet, was estimated to be 32–35 mg/day. Japanese consumed 20–30 mg isoflavone per day on average. Such a daily dosage would have bioactivity at this concentration. An upper limit that may have adverse effects has not been determined for humans. We examined the health effects on young women by adding 20 mg, 40 mg, or 50 mg of IF in tablets to about 10-mg daily consumption in foods for 1 month. The concern was that excessive isoflavone in the diet might have negative effects, yet there is a paucity of information on the pharmacokinetics of chronic exposure to isoflavones.

Chemoprevention trials of soy isoflavones are now being carried out in Phase II/III Cancer Prevention Trials of the National Cancer Institute (NCI) (55). Isoflavone-rich tablets can reduce circulating levels of ovarian steroid hormones. Ovarian steroids (estradiol, progesterone, and estron sulfate) regulate breast cell proliferation, so these reductions may lead to a lowered risk of breast cancer.

The observed decreases in ovarian steroid levels after soy consumption are consistent with epidemiological data suggesting that Asians consuming a diet rich in soy have lower levels of ovarian steroids and lower incidence of breast cancer than populations consuming diets that contain less soy.

VI. CONCLUSIONS

Estrogenic and antiestrogenic effects of IF should be studied more in relation to the nuclear protein. A 2001 study with GeneChips may show unexpected action of IF on multiple genes. Interaction of IF metabolites with macromolecules in the body should also be clarified. Quercetin metabolites, for example, showed stronger biological activity (60).

Regulation of the menstrual cycle occurs as a result of a number of interactions between the hormones of the hypothalamus–pituitary–gonadal axis. We found that a depressed FSH level in preovulatory phase was prolonged by IF tablet intake, resulting in a prolongation of LH surge. A shortened menstruation cycle resulted from a shortened follicular phase with a normal 14-day luteal phase. This means that isoflavones affect the hypothalamopituitary level through a feedback-like mechanism. Changes in androstendione, testosterone, and thyroxin, caused by IF tablet intake, would also be caused by altered function of the hypothalamus and hypophysis. Further study of the hypothalamus–pituitary–gonadal axis is necessary. Metabolites of isoflavones, such as equol, could doubly influence the endocrine functions in this case.

Increased natural killer (NK) cell activity has newly suggested the influence of IF on immune functions. We found in 1999 that isoflavones can be transferred to the human fetus from maternal blood (49). That means that intrauterine development may be modulated by dietary factors.

In 2001 supplements of IF became very popular (61). Further multipotential effects of phytoestrogens should be clarified before their extensive use is recommended for disease prevention.

ACKNOWLEDGMENTS

A part of this study was supported by the research fund from the Ministry of Education, Science Technology, Culture and Sport. The authors thank Fuji Protein Research Foundation for providing us with IF tablets,

REFERENCES

1. M Kimira, Y Arai, K Shimoi, S Watanabe. Japanese intake of flavonoids and isoflavonoids from foods. J Epidemiol 8:168–175,1998.
2. JB Harborne, TJ Mabry. The flavonoids. London, Chapman & Hall, 1988.
3. H-J Wang, PQ Murphy. Isoflavone content in commercial soybean foods. J Agric Food Chem 42: 1666–1673, 1994.
4. K Okubo, F Yamauchi. Science of soy bean. Tokyo, Asakura, 1996.
5. KR Price, GR Feuwick. Naturally ocurring oestrogens in foods—a review. Food Addict Contam 2:73–106, 1985.
6. L Bravo. Polyphenols: Chemistry, dietary sources, metabolism, and nutritional significance. Nutr Rev 56: 317–333, 1998.
7. S Barnes, M Kirk, L Coward. Isoflavones and their conjugates in soy foods: Extraction and analysis by HPLC-mass spectrometry. J Agric Food Chem 42: 2464–2474, 1994.
8. R Hegnauer, RJ Grayer-Barkmeijer. Relevance of seed plysaccharides and flavonoids for the classification of the leguminosae: A chemotaxonomic approach. Phytochemistry 34:3–16, 1993.
9. S Kudou, Y Fleury, D Welti, D Magnolato, T Uchida, K Kitamura, K Okubo. Malonyl isoflavone glycosides in soybean seads (Glycine max MERRILL). Agric Biol Chem 55: 2227–2233, 1991.

10. C Tsukamoto, S Shimada, K Igita, S Kudou, M Kokubun, K Okubo, K Kitamura. Factors affecting isoflavone content in soybean seeds: Changes in isoflavones, saponins, and composition of fatty acids at different temperatures during seed development. J Agric Food Chem 43: 1184–1192, 1995.
11. G Smit, V Puvanesarajah, RW Carlson, WM Barbour, G Stacey. Bradyrhizobium japonicum nodD1 can be specifically induced by soy-bean flavonoids that do not induce the nodYABCSUIJ operon. J Biol Chem 267: 310–318, 1992.
12. ME Baker. Evolution of regulation of steroid-mediated intercellular communication in vertebrates: Insights from flavonoids, signals that mediate plant-rhizobia symbiosis. J Steroid Biochem Mol Biol 41: 301–308, 1992.
13. M Pariniske, B Ahiborn, D. Werner. Isoflavonoid-inducible resistance to the phytoalexin glyceollin in soybean rhizobia. J Bacteriol 173: 3432–3439, 1991.
14. M Naim, B Gestetner, S Zilkah, Y Birk, A Bondi. Soybean isoflavones, characterization, determination, and antifungal activity. J Agric Food Chem 22:806–810, 1974.
15. H Adlercreutz, T Fotisis, KJ Lampe, K Wahala, T Makela, G Brunow, T Hase. Quantitative determination of lignans and isoflavonoids in plasma of omnivorous and vegetarian women by isotope dilution gas chromatography-mass spectrometry. Scand J Clin Lab Invest 53 (suppl 215): 5–18, 1993.
16. L Coward, H Kirk, N Albin, S Barnes. Analysis of plasma isoflavones by reversed-phase HPLC-multiple reaction ison monitoring-mass spectrometry. Clin Chem Acta 247: 121–142, 1996.
17. J Sfakianos, L Coward, M Kirk, S Barnes. Intestinal uptake and biliary excretion of the isoflavone genistein in rats. J Nutr 127: 1260–1268, 1997.
18. S Watanabe, M Yamaguchi, T Sobue, T Takahashi, Y Miura, Y Arai, W Mazur, K Wahaha, H Adlercreutz. Pharmacokinetics of soybean isoflavonoids in plasma, urine and feces of men after ingestion of 60g baked soybean powder (kinako). J Nutr 128: 1710–1715,1998.
19. GE Kelly, C Nelson, MA Waring, GE Joannou, AY Reeder. Metabolites of dietary (soya) isoflavones in human urine. Clin Chim Acta 223: 9–22, 1993.
20. JW Lampe, SC Karr, AM Hutchins, JL Slavin. Urinary equol excretion with a soy challenge: Influence of habitual diet. Proc Soc Exp Biol Med 217: 335–339, 1998.
21. X Xu, KS Harris, HJ Wang, PA Murphy, S Hendrich. Bioavailability of soybean isoflavones depends upon gut microflora in women. J Nutr 125: 2307–2315, 1995.
22. LA Griffiths, A Barrow. Metabolism of lavonoid compounds in germ-free rats. Biochem J 130: 1161–1162, 1972.
23. J Winter, LH Moore, VR Dowell Jr, VD Bokkenheuser. C-ring cleavage of flavonoids by human intestinal bacteria. J App Envion Microbiol 55: 1203–1208, 1989.
24. AM Hutchins, JL Slavin, JW Lampe. Urinary isoflavonoid phytoestrogen and lignan excretion after consumption of fermented and unfermented soy products. J Am Diet Assoc 95: 545–551, 1995.

25. M Uehara, O Lapcik, R Hampl, N Al-Maharik, T Makela, K Wahala, H Mikola, H Adlercreutz. Rapid analysis of phytoestrogens in human urine by time-resolved fluoroimmunoassay. J Steroid Biochem Mol Biol 72: 273–282, 2000.
26. S Watanabe, H Adlercreutz. Pharmacokinetics of soy phytoestrogens in humans. ACS Symp Series 702: 198–208,1998.
27. RA King, DB Bursill. Plasma and urinary kinetics of the isoflavones daidzein and genistein after a single soy meal in humans. Am J Clin Nutr 67: 867–672, 1998.
28. S Watanabe, S Koessel. Colon cancer: An approach from molecular epidemiology. J Epidemiol 3: 47–61,1993.
29. E Finkel. Phyto-oestrogens: The way to postmenopausal health? Lancet 352: 1762, 1998.
30. H Adlercreutz, BR Godlin, SL Gorbach, KAV Hockerstedt, S Watanabe, EK Hamalalainen, HH Markkanen, TH Makela, KT Wahala, TA Hase, T Fotsis. Soybean phytoestrogen intake and cancer risk. Nutr Cancer 125: 757S–770S, 1995.
31. C Nagata, N Takatsuka, N Kawakami, H Shimizu. Soy product intake and hot flashes in Japanese women: Results from a community-based prospective study. Am J Epidemiol 15: 790–792, 2001.
32. GG Kuiper, JG Lemmen, B Calsson, JC Corton, SH Safe, PT van der Saag, B van der Burg, JA Gustafsson. Interaction of estrogenic chemicals and phytoestrogens with estrogen receptor beta. Endocrinology 139: 4252–4263, 1998.
33. AM Duncan, BE Merz-Demlow, X Xu, WR Phipps, MS Kurzer. Premenopausal equol excretors show plasma hormone prolifes associated with lowered risk of breast cancer. 3rd International Symposium on the Role of Soy in Preventing and Treating Chronic Diseases, p 41 (1999), Washington DC.
34. S Watanabe, R Haba, K Tarashima, Y Arai, T Miura, D Chiba, M Takamatsu. Antioxidant activity of isoflavones in humans. BioMarker 10: 12:227–232, 2000.
35. SJ Smith, GSM Chan, M Chan, S Samman, PM Lyon-Wall. Effects of an isoflavone supplement on risk factors for breast cancer and coronary heart disease in premenopausal women. 3rd International Symposium on the Role of Soy in Preventing and Treating Chronic Diseases, p 49 (1999), Washington DC.
36. S Watanabe, K Terashima, Y Sato, S Arai, A Eboshida. Effects of isoflavone supplement on healthy women. Biofactors 12: 233–241, 2000.
37. LJ Lu, KE Anderson, JJ Grady, M Nagamani. Effects of soya consumption for one month on steroid hormones in premenopausal women: Implications for breast cancer risk reduction. Cancer Epidemiol Biomarkers Prev 5: 63–70, 1996.
38. CG Fraga, MK Shigenaga, J-W Park, P Degan, BN Ames. Oxidative damage to DNA during aging: 8-hydroxy-2′-deoxyguanosine in rat organ DNA and urine. Proc Natl Acad Sci USA 87: 4533–4537, 1990.
39. T Akiyama, J Ishida, S Nakagawa, H Ogawara, S Watanabe, N Itoh, M Shibuya, Y Fukami. Genistein, a specific inhibitor of tyrosine-specific protein kinases. J Biol Chem 262: 5592–5595, 1987.
39a. R Haba, S Watanabe, Y Arai, H Chiba, T Miora. Suppression of lipid hydro-

peroxide and DNA-adduct formation by isoflavone-containing soy hypocotyl tea in rat. Env Health Prev Med 7:64–73, 2002.

40. H Kasai. Analysis of a form of oxidative DNA damage, 8-hydroxy-2′-deoxyguanosine, as a marker of cellular oxidative stress during carcinogenesis. Mutat Res 387: 147–163. 1997.
41. H Kasai, PF Crain, Y Kuchino, S Nishimura, A Ootsuyama, H Tanooka. Formation of 8-hydroxyguanine moiety in cellular DNA by agents producing oxygen radicals and evidence for its repair. Carcinogenesis 7: 1849–1851, 1986.
42. MG Simic. Urinary biomarkers and the rate of DNA damage in carcinogenesis and anticarcinogenesis. Mutat Res 267: 277-290, 1992.
43. AL Murkies, FS Dalais, EM Briganti, DL Healy, HG Urger, ML Wahlqvist, SR Davis. Phytoestrogen and androgen levels in Australian. 3rd International Symposium on the Role of Soy in Preventing and Treating Chronic Diseases, p. 38 (1999), Washington DC.
44. D Ingram, K Sanders, M Kolybaba, D Lopez. Case-control study of phyto-oestrogens and breast cancer. Lancet 350: 990–994, 1997.
45. X Xu, AM Duncan, BE Merz, MS Kurzer. Effects of estrogen and phytoestrogen metabolism in premenopausal women. Cancer Epidemiol Biomarkers Prev 7: 1101–1108, 1998.
46. Y Zhang, TT Song, JE Cunnick, PA Murphy, S Hendrich. Daidzein and genistein glucuronides in vitro are weakly estrogenic and activate human natural killer cells at nutritionally relevant concentration. J Nutr 129: 399–405. 1999.
47. YC Kao, C Zhou, M Sherman, CA Laughton, S Chen. Molecular basis of the inhibition of human aromatase (estrogen synthetase) by flavone and isoflavone phytoestrogens: A site-directed mutagenesis study. Environ Health Perspect 106: 85–92, 1998.
48. H Adlercreutz, H Markkanen, S. Watanabe. Plasma concetnrations of phyto-oestrogens in Japanese men. Lancet 342: 1209–1210, 1993.
49. H Adlercreutz, T Yamada, T Wahala, S Watanabe. Maternal and neonate phytoestrogens in Japanese women during birth. Am J Gynecol Obst 180: 737–743, 1999.
50. Y Arai, M Uehara, Y Sato, M Kimira, A Eboshida, H Adlercreutz, S Watanabe. Comparison of isoflavone among dietary intake, plasma concentration and urinary excretion for accurate estimation of phytoestrogens intake. J Epidemiol 10: 127–135, 2000.
51. Y Arai, S Watanabe, M Kimira, K Shimoi, R Mochizuki, N Kinae. Dietary intake of flavonols, flavone and isoflavones in Japanese women: The quercetin intake inversely correlated with the plasma LDL-cholesterol concentration. J Nutr.
52. LJ Lu, KE Anderson. Sex and long-term soy diets affect the metabolism and excretion of soy isoflavones in humans. Am J Clin Nutr 68: 1500S-1504S, 1998.
53. S Barnes, TG Peterson, L Coward. Rationale for the use of genistein-containing soy matrices in chemoprevention trials for breast and prostate cancer. J Cell Biochem Suppl 22: 181–187, 1995.
54. DD Baird, DM Umbach, L Lansdell, CL Hughes, KD Setchell, CR Weinberg, AF Haney, AJ Wilcox, JA Mclachlan. Dietary intervention study to assess

estrogenicity of dietary soy among postmenopausal women. J Clin Endocrinol Metab 5: 1685–1690, 1995.

55. A Cassidy, S Bingham, K Setchell. Biological effects of isoflavones in young women: Importance of the chemical composition of soybean products. Br J Nutr 74: 587–601, 1995.
56. A Cassidy, S Bingham, KD Setchell. Biological effects of a diet of soy protein rich in isoflavones on the menstrual cycle of premenopausal women. Am J Clin Nutr 60: 333–340, 1994.
57. AM Duncan, BE Merz, X Xu, TC Nagel, WR Phipp, MS Kurzer. Soy isoflavones exert modest hormonal effects in premenopausal women. J Clin Endocrinol Metab 84: 192–197, 1999.
58. AM Duncan, KE Underhill, X Xu, J Lavalleur, WR Phipps, MS Kurzer. Modest hormonal effects of soy isoflavones in postmenopausal women. J Clin Endocrinol Metab 10: 3479–3484, 1999.
59. W L-J Lu, KE Anderson, M Nagamani. Reproductive endocrine effects of a soya diet in premenopausal women. 3rd International Symposium on the Role of Soy in Preventing and Treating Chronic Diseases, p. 22 (1999), Washington DC.
60. R Binsack, BJ Boersma, RP Patel, M Kirk, CR White, V Darley-Usmar, S Barnes, F Zhou, DA Parks. Enhanced antioxidant activity after chlorination of quercetin by hypochlorous acid. Alcohol Clin Exp Res 25: 434–443, 2001.
61. KD Setchell, NM Brown, P Desai, L Zimmer-Nechemias, BE Wolfe, WT Brashear, AS Kirschner, A Cassidy, JE Heubi. Bioavailability of pure isoflavones in healthy humans and analysis of commercial soy isoflavone supplements. J Nutr 131 (suppl 4): 1362S–1375S, 2001.

12
Anandamides and Diet: A New Pot of Nutritional Research Is Simmering

A. Berger and G. Crozier Willi
Nestlé Research Center, Lausanne, Switzerland

V. Di Marzo
Istituto per la Chimica di Molecole di Interesse Biologico, Naples, Italy

I. INTRODUCTION

A. Discovery of *N*-Acylethanolamines

Although Δ^9(-)-tetrahydrocannabinol (THC) is not naturally found in the body, receptors for this compound in the brain have been described since 1988 (Devane et al., 1988). These receptors were later named *cannabinoid 1* (CB1), and a receptor subtype (CB2) has been found in immune tissues (Munro et al., 1993). An endogenous ligand for these receptors has more recently been discovered: It is the *N*-ethanolamine form of arachidonic acid and was named *anandamide* from the Sanskrit word *ananda*, which means "bliss". When anandamide [20:4n6 *N*-acylethanolamine (20:4n6 NAE)] was injected into animals, hypomobility, hypothermia, analgesia, and catalepsy were observed (Fride and Mechoulam, 1993). These are the classic behavioral effects of THC administration. Since this discovery, other endogenous metabolites that have been shown to be functional agonists of these receptors have been described: several polyunsaturated fatty acid derivatives, some of which are 22:6n3 NAE, 20:4n3 NAE, 20:3n6 NAE (Barg et al., 1995), 20:3n9 NAE (Pillar et al., 1995), and 2-arachidonoylglycerol (2-AG) (Mechoulam et al., 1995; Sugiura et al., 1995) (Fig. 1). These compounds are collectively termed *endocannabinoids* and have been shown to bind to brain CB receptors with an IC_{50} value equal to or 1.5-fold that of 20:4n6 NAE (Felder et al., 1995; Hillard and

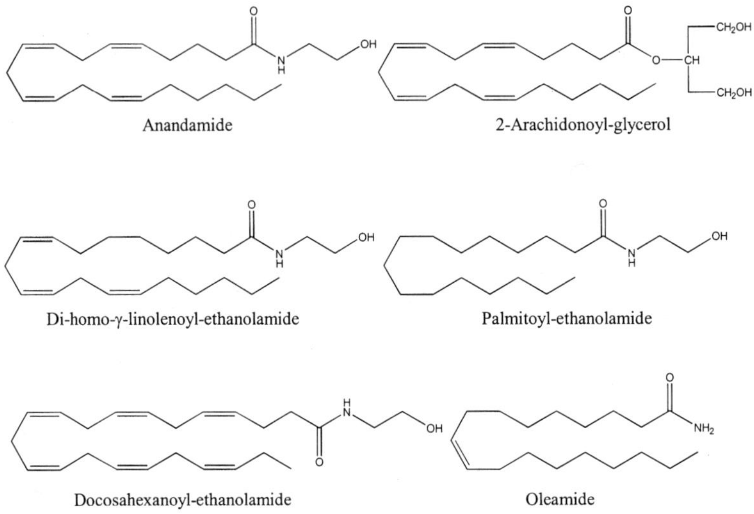

Figure 1 Chemical structures of endocannabinoids and other cannabimimetic fatty acid derivatives.

Campbell, 1997). The presence of CB receptors in diverse invertebrates evidences that the CB signaling system has been conserved for at least 500 million years (Stefano et al., 1996; De Petrocellis et al., 1999).

B. Tissue Distribution of *N*-Acylethanolamines

The CB1 receptor is widely distributed in several brain regions, including the hippocampus, cortex and striatum, cervical ganglia, and peripheral autonomic fibers (Pertwee, 1997). It is also present in nonnerve tissues such as prostate, uterus, testes, bladder, small intestine, spleen, and lymphocytes. The CB2 receptor is found in immune tissues such as the spleen, thymus, and tonsils and in immunocompetent blood cells with highest concentration in B cells, natural killer cells, mast cells, and monocytes.

The distribution of endocannabinoids in the brain is not completely consonant with the regional distribution of their binding sites. This suggests either that they may not necessarily be produced near their targets or that they play functional roles other than CB receptor activation (Bisogno et al., 1999a). The highest concentrations of 20:4n6 NAE found so far in animal studies occur in uterine tissue (Schmid et al., 1997). Down regulation of 20:4n6 NAE levels is associated with uterine receptivity for implantation, whereas up regulation is associated with uterine refractoriness. A largely unexplored hypothesis is that 20:4n6 NAE will prove to play an important role in reproduction.

C. Specific Types of Endocannabinoids

20:4n6 *N*-acylethanolamine has been isolated from the human hippocampus, striatum, and cerebellum—brain areas known to express high levels of CB1-type CB receptors—together with its putative biosynthetic precursor, *N*-arachidonoylphosphatidylethanolamine (Bisogno et al., 1999a). Both 22:6n3 and 20:3n6 NAE bind to rat brain CB1 receptor and show similar potency in IC_{50} values for inhibition of adenylate cyclase. When injected into animals these two compounds exerted effects on behavior similar to those of THC and 20:4n6 NAE, indicating that the two ethanolamides are highly potent endocannabinoids (Barg et al., 1995). The first NAE discovered in the 1950s 16:0 NAE was found to be the active anti-inflammatory component in fractions of peanut oil, soybean lecithin, and egg yolk with demonstrated activity in animal models (Kuehl et al., 1957; Ganley et al., 1958). More recently 16:0 NAE has been shown to act by inhibiting synthesis of interleukins -4, -6, and -8 (Berdyshev et al., 1997) and diminishing inflammatory response in a number of experimental models. Although, 16:0 NAE has low affinity for CB1 or CB2 receptors, data have been recently presented indicating that it competes for the binding of a high-affinity ligand to membranes from rat basophilic leukemia cells (Facci et al., 1995), increasing the likelihood that fatty acid ethanolamides represent a family of transmitters, each with a distinct receptor (Calignano et al., 1998). The significance of this finding was brought out in a study of pain reception in skin in which pain responses were reduced 100-fold more potently when both 20:4n6 NAE and 16:0 NAE were administered simultaneously. Synergism and "entourage" effects are highly important concepts in the understanding of endocannabinoid action (Ben Shabat et al., 1998). Other naturally occurring NAEs do not bind to CB1 and CB2 receptors with high affinity (Mechoulam et al., 1994; Mechoulam et al., 1996; Di Marzo, 1998; Di Marzo and Deutsch, 1998; Di Marzo et al., 1998a), and their biological significance is not clear. These compounds, however, can inhibit the breakdown of the more biologically potent NAEs, such as 20:4n6 NAE, by acting as alternative substrates for fatty acid amide hydrolase (FAAH), the enzyme that degrades 20:4n6 NAE.

Oleamide (*cis*-9-octadecenoamide) is a major primary amide found in mammals. It was first identified from the cerebral spinal fluid of a sleep-deprived cat and when injected into a rat induced rapid physiological sleep (Cravatt et al., 1995). It can be synthesized from oleic acid and ammonia in brain microsomes (Sugiura et al., 1996b). Oleamide does not bind to CB receptors at physiological concentrations (Boring et al., 1996), but similarly to linoleoylethanolamide, oleoylethanolamide, and 2-arachidonoylglycerol (discussed later), it inhibits the breakdown of 20:4n6 NAE by acting as an alternative substrate for fatty acid amide hydrolase (Maurelli et al., 1995; Cravatt et al., 1996). In addition, oleamide has been observed to enhance binding of 20:4n6 NAE to CB1 receptors (Mechoulam et al., 1997) and potentiate 20:4n6 NAE antiproliferative action on human breast cancer cells (Bisogno et al., 1998).

sn-2-Arachidonoylglycerol (2-AG) resembles 20:4n6 NAE in its molecular conformation and has an even greater efficacy acting at the CB1 receptor than 20:4n6 NAE itself. It may play an even more important biological role than 20:4n6 NAE since it was shown to be present at approximately 200- (Sugiura et al., 1995) to 800-fold (Sugiura et al., 1996d) greater concentration than 20:4n6 NAE in some regions of the rodent brain. It has been found also in spleen, pancreas, and intestinal cells of rats. 2-Agrchidionylglycerol is a CB1 receptor agonist in neuroblastoma cells (Sugiura et al., 1999) and inhibits the breakdown of 20:4n6 NAE (Di Marzo and Deutsch, 1998). The two other *sn*-2-acylglycerol compounds, *sn*-2-linoleylglycerol and *sn*-2-palmitoylglycerol, have been shown to potentiate the activity of 2-AG with respect to both its binding to CB receptors and its capacity to inhibit adenylyl cyclase, as well as in its effects on behavior (Ben Shabat et al., 1998). Free arachidonic acid, 2-palmitoylglycerol, 2-oleylglycerol, 2-linoleylglycerol, and 2-docosahexaenoylglycerol are inactive or weakly active on CB1 receptors in neuroblastoma cells (Sugiura et al., 1999).

D. Synthesis and Degradation of *N*-Acylethanolamines

The NAEs are phospholipase D metabolites of membrane-bound *N*-acyl phosphatidylethanolamines (NAPEs), which are formed from calcium-dependent transacylase reactions utilizing the *sn*-1 position of phosphatidyethanolamine, phosphatidylcholine, or cardiolipin as the acyl group donor (Schmid et al., 1990). This pathway has been shown to apply also to 20:4n6 NAE (Di Marzo et al., 1994). Typically, saturated or monounsaturated fatty acids predominate on the *sn*-1 position; thus a selective pool of *sn*-1 polyunsaturated phospholipid molecules must be used in the transacylase reaction (Di Marzo, 1998; Di Marzo and Deutsch, 1998; Di Marzo et al., 1998a).

Degradation occurs by the action of fatty acid amide amidohydrolase (FAAH) (Cravatt et al., 1996), an enzyme with broad substrate specificity found in many tissues, including small intestine, liver, eye, and brain (Katayama et al., 1997; Matsuda et al., 1997; Maccarrone et al., 1998). The lack of specificity of this enzyme may be important in physiological regulation since the corelease of varying substrates that compete for its hydrolytic activity may potentiate the actions of others. This interaction was demonstrated in 1997 for oleamide (Mechoulam et al., 1997). Although FAAH is capable of synthesis and degradation in vitro (Kurahashi et al., 1997), degradation is probably more important under physiological conditions.

The NAEs are also substrates for lipoxygenases and form new eicosanoid products, some of which have been shown to be bioactive (Ueda et al., 1995; Hillard and Campbell, 1997; Edgemond et al., 1998).

E. Mechanisms of Action of *N*-Acylethanolamines

Cannabinoid receptor–triggered intracellular neuronal signaling leads to downregulation of adenylate cyclase (AC) and reduction of cyclic adenosine manophosphate (cAMP), ultimately leading to various physiological sequelae, including changes in appetite, peristalsis, mood and reward, motor control, heart rate variability, and blood pressure (described later). In the striatum, activation of CB receptors leads to decreased glutamate release from afferent presynaptic terminals, acting after presynaptic action potentials and calcium channels (Gerdeman and Lovinger, 2001). In the nucleus accumbens, activation of CB receptors on presynaptic excitatory fibers contacting γ-aminobufyric acid–ergic (GABAergic) neurons leads to decreased sodium channel conductance, to decreased glutamergic transmission, to decreased excitation, to decreased firing by postsynaptic inhibitory cells, to increased activation of the mesolimbic dopamine neurons, finally leading to increased dopamine release, which could in turn lead to increased drug dependency (Robbe et al., 2001; Vaughan, 2001).

There has been a 10-year search for a signaling molecule that fine-tunes signals released by neurons [depolarization-induced suppression of inhibition (DSI)].For example, DSI is important in the hippocampus, the amygdala (emotional memory), and the cerebral cortex (movements). The DSI signaling molecule is released from depolarized postsynaptic neurons when internal calcium level is increased. The CB1 receptor antagonists have been found to block DSI in hippocampus, and NAEs are now known to be the previously unknown DSI molecules. Cannabinoids also turn down *excitatory* neuronal inputs (DSE). Overall, endocannabinoids are now thought to prime hippocampal neurons for long-term potentiation important for learning and memory (Barinaga, 2001; Wilson and Nicoll, 2001).

II. FUNCTIONS OF ENDOCANNABINOIDS

Cannabis (marijuana) has been used medicinally for more than 4000 years for the treatment of disorders that include migraine, anorexia, asthma, muscle spasms, seizures, glaucoma, neuralgia, pain, diarrhea, and nausea. When 20:4n6 NAE is injected into rodents, the classic behavioral effects induced by marijuana or its active ingredient, THC, namely, hypomobility, hypothermia, analgesia, and catalepsy, are reproduced (Crawley et al., 1993; Fride and Mechoulam, 1993; Mechoulam et al.,1998b). In general the effects occur with rapid onset and short duration (Smith et al., 1994).

A. Memory

There are a number of animal studies that suggest that 20:4n6 NAE and potentially other CB1 receptor agonists induce forgetfulness, a critical "sorting" process necessary in the processing and retention of important memories. Evidence from rodent trials has shown that 20:4n6 NAE impairs working memory (Mallet and Beninger, 1998) and memory consolidation (Castellano et al., 1997), and Derkinderen and associates have postulated a role of 20:4n6 NAE in making and breaking short neural connections (Derkinderen et al., 1996).

Intracerebroventricular administration of 20:4n6 NAE strikingly increased sleep—both slow-wave and rapid eye movement (REM) types—at the expense of wakefulness, and deterioration of memory consolidation was confirmed (Murillo-Rodriguez et al., 1998). These observations suggested that the brain CB system participates in the modulation of both vigilance states and mnemonic processes. In concert with these observations, further rodent studies have shown that the CB1 receptor antagonist SR141716A improved memory (Terranova et al., 1996; Mallet and Beninger, 1998) and facilitated short-term olfactory memory in the social recognition test (Terranova et al., 1996). These results strongly support the concept of a role for endocannabinoids in forgetting and lead to the tantalizing suggestion that CB1 receptor blockers might aid the establishment of effective memory.

B. Sleep Induction

The presence of CB1 receptors and 20:4n6 NAE and 2-AG in the hypothalamus implies that endocannabinoids may play a role in the tuning of sleep–wake cycles. For example, the CB1 antagonist SR141716A increased waking time in rodents (Santucci et al., 1996; Mechoulam et al., 1997). Since oleamide has been recognized as the sleep inducing agent, there may be hitherto unrecognized links between this bioactive compound and the CB1 receptor system (Mechoulam et al., 1997).

Cannabinoids can also affect other hypothalamic functions such as thermoregulation and food intake, which affect overall sedativity. Indeed, the physiological responses of hypothermia and hypomobility are found with

20:4n6 NAE or THC injection (Smith et al., 1994; Adams et al., 1998). One of the mechanisms by which this occurs has been shown to be through interference with dopaminergic and GABAergic transmission (Romero et al., 1995).

C. Vocalizations

Ultrasonic vocalizations in rat pups separated momentarily from their mothers is a recognized sign of anxiety and is typically activated by a decrease in body temperature. This behavior was decreased by administration of the CB receptor agonist CP 55,940 and increased by the CB receptor antagonist SR141716A (McGregor et al., 1996).

D. Appetite

The control of appetite is partially directed by the central CB system. Williams and associates observed overeating in mice given any dose of 20:4n6 NAE from 0.5 to 10 mg/kg, an effect that was attenuated by the previous administration, in a dose-dependent manner, of the CB receptor antagonist SR141716A (Williams and Kirkham, 1999). In 2001 ob/ob genetically obese mice, which are known to have huge appetites, were reported to have increased hypothalamic 20:4n6 NAE levels (Di Marzo et al., 2001). These same mice have defective leptin signaling. Acute leptin injection was found to decrease levels of hypothalamic 20:4n6 NAE and 2-AG and to decrease appetite. Together, these observations suggest that the leptin signaling cascade to decrease appetite involves endocannabinoids (20:4n6 NAE and 2-AG). In the future, it may be possible to block brain CB actions for antiobesity treatments (in combination with other molecular targets to decrease appetite) and, conversely, to administer cannabinoids to prevent weight loss in persons with loss of appetite [e.g., cancer and, (AIDS) patients].

E. Pain

TΔ^9(-)-Tetrahydrocannabinol has well-known antinociceptive properties. Peripheral administration of 20:4n6 NAE has been shown to inhibit the induction of hyperalgesia by capsaicin (Richardson et al., 1998). Calignano and associates (1998) have shown highly effective decrease in skin pain (100-fold) with administration of 16:0 NAE and 20:4n6 NAE. The precise mechanism of action of NAEs in modulating pain is a focus of current investigation. For example, Monhemius and associates (2001) have shown in an animal model of neuropathic pain that CB receptors that mediate analgesia are located on the nucelus reticularis gigantocellularis pars alpha.

F. Gut Motility

The effects of endocannabinoids on gut peristalsis were reviewed in 2000 and 2001 (Izzo et al., 2000; Pertwee, 2001). There is evidence that endocannabinoids decrease peristalsis (delayed gastric emptying, decreased transit of nonabsorbable markers) in rodents, guinea pigs, dogs, and humans. For example, in mice, 20:4n6 NAE injection inhibited the passage of a charcoal meal, an effect that was reversed by the administration of the CB receptor antagonist SR141716A (Calignano et al., 1997b). In intestinal strip models, cannabinoids decreased evoked acetylcholine release, peristalsis, and cholinergic and nonadrenergic noncholinergic contractions of longitudinal smooth muscle. The effects of cannabinoids on gastric emptying and intestinal transport are mediated by CB receptors in both the brain and the intestine. Gastric acid secretion is also decreased after CB receptor activation. In humans, CB receptor agonists also delay gastric emptying (and may decrease gastric acid secretion).

G. Hypotension and Modulation of Vascular Tension

20:4n6 *N*-Acylethanolamine influences vascular tension through vasorelaxant and neuromodulatory effects, as shown by studies in renal arterial segments (Deutsch et al., 1997). Administration of 20:4n6 NAE was shown to decrease systemic blood pressure in rats (Lake et al., 1997) and in anaesthetized guinea pigs, in which the effect was prolonged if an inhibitor of 20:4n6 NAE transport was administered (Calignano et al., 1997a). The hypotension and bradycardia that have been observed with 20:4n6 NAE administration seem likely to be mediated by action on presynaptic CB1 receptors in sympathetic nerves (Varga et al., 1995; Varga et al., 1996). 20:4n6 *N*-Acylethanolamine has also been shown to activate vanilloid receptors on perivascular sensory nerves and to induce vasodilatation (Zygmunt et al., 1997).

The NAEs may affect arterial pressure via effects on platelets and endothelial cells. Platelets contribute to mean arterial pressure through formation of 2-AG selectively over 20:4n6 NAE (Varga et al., 1998) or via disposal of endocannabinoids through reuptake and hydrolysis processes (Maccarrone et al., 1999). Accordingly, 2-AG has been shown to induce vasodilatation in vivo and in vitro (Mechoulam et al., 1998a; Varga et al., 1998; Járai et al.,1999). Endothelial cells from the vascular bed also produce and inactivate endocannabinoids (Mechoulam et al., 1998a; Sugiura et al., 1998; Maccarrone et al., 2000a). Endocannabinoids derived from macrophages, but not from platelets, are the cause of the hypotensive state that arises during hemorrhagic shock (Wagner et al., 1997). The hypotensive effect of 20:4n6 NAE is not limited to the systemic vasculature—it also plays a role in ocular pressure (Pate et al., 1995; Pate et al., 1996; Mikawa et al., 1997), leading to the development of new strategies for treatment of eye disease such as glaucoma. In the lungs, 20:4 NAE synthesized after calcium stimulation has been shown to have dual effects (dilatation, bronchospasm) via binding to CB receptors in axonal terminals of airway nerves (Calignano et al., 2000).

H. Immune System

There is evidence for the occurrence of anabolic and catabolic reactions involving 20:4n6 NAE and 2-AG in murine macrophages, mast cells, basophils, neutrophils, lymphocytes, platelets, and endothelial cells. Lipopolysaccharides (LPSs) of bacterial origin stimulate 20:4n6 NAE and 2-AG formation in macrophages (Di Marzo et al., 1999; Parolaro, 1999). Thus, bacterial infections may trigger release of endocannabinoids, which in turn have immune-modulatory actions. Endocannabinoids may have a role in immune hyperreactivity, as evidenced by the fact that the immortalized mast cell–like line leukemia RBL-2H3 cells respond to immunoglobulin E (IgE) challenge (a stimulus leading to cell degranulation and histamine and serotonin release) by producing 20:4n6 NAE and its congeners (Bisogno et al., 1997a). The RBL-2H3 cells and human mast cells also inactivate 20:4n6 NAE and/or 2-AG via several mechanisms [(Bisogno et al., 1997a); Di Marzo, 1999 #1601; Maccarrone, 2000 #1803]. In sum, endocannabinoids variously affect immune cell development, proliferation, and functionality (Parolaro, 1999) and vascular tone and cardiac output (Kunos et al., 2000). It is thus likely that endocannabinoids participate in the control of immune and vascular functions, particularly during uncontrolled immune reactions (allergies, septic shock, autoimmune diseases) or hemorrhage.

I. Cell Protection

Noxious agents such as ethanol and sodium azide, exposure to Ultraviolet B (UVB) radiation, and glutamate receptor stimulation all lead to an increased intracellular calcium concentration and an increased synthesis of 20:4n6 NAE and other NAEs (Basavarajappa and Hungund, 1999; Berdyshev et al., 2000). This relationship is the basis for the speculation that these compounds may have cell protective roles. However, in the nervous system, direct evidence for a neuroprotective action for 20:4n6 NAE and 2-AG was reported in 2000 not to involve CB receptors (Sinor et al., 2000). This report is surprising because 20:4n6 NAE (as well as 2-AG) should be able to inhibit pathological conditions arising from high intracellular calcium concentrations by acting at CB1 receptors to inhibit neuronal calcium influx through voltage-operated calcium channels (Mackie et al., 1993) and by counteracting membrane permeability to calcium via NMDA receptor-coupled channels (Hampson et al., 1998).

J. Apoptosis

The NAEs may affect cell death (apoptosis) via effects on ceramide. Ceramide is generally an activator of apoptosis. *N*-oleoylethanolamine (NOE) was reported to inhibit ceramidase (deacylation of ceramide), leading to increased levels of ceramide and increased apoptosis (via the ERK cascade); it also decreased glucosylation of GlcCer, leading to lesser amounts of higher glycoslylated compounds in neuroepithelioma cells (Spinedi et al., 1999). Δ^9(-)-Tetrahydrocannabinol is known to increase ceramide in the short term via activation of neutral sphingomyelinase and FAN adapter protein, and in the long term via activation of serine palmitoyl transferase.

K. Reproduction

20:4n6 *N*-Acylethanotamine, 2-AG, and cannabinoids inhibit mouse embryo development and blastocyst implantation in the uterus via CB1, but not CB2, receptors (Paria et al., 1998). This finding, together with the observation that 20:4n6 NAE levels or the expression of FAAH are correlated with uterine receptivity to embryo implantation (Paria et al., 1999), suggests endocannabinoids play a role in pregnancy. In particular, it was found that uterine 20:4n6 NAE levels are lower when the uterus is receptive, and higher when the organ is refractory, to embryo implantation. Activity of FAAH is highly expressed in the peri-implantation period; that expression may possibly explain the low levels of 20:4n6 NAE found in the uterus at this stage. On the basis of these studies, 20:4n6 NAE may have a role in directing the timing and site of embryo implantation. One can speculate that disorders such as spontaneous termination of pregnancy may be linked to a malfunctioning uterine endocannabinoid system. Indeed, evidence in 2000 showed decreased FAAH levels and activity in lymphocytes, a condition that is likely to determine high levels of endocannabinoids in the uterus, predicts spontaneous abortion in women (Maccarrone et al., 2000b). On the basis of studies carried out in sea urchin sperm (Schuel et al., 1994; Berdyshev, 1999), endocannabinoids may also interfere with sperm cell fertilizing capability. These findings were extended in 1998 to human sperm, with preliminary evidence for CB receptor expression (Schuel et al., 1998). A role for the endocannabinoid system in reproduction is also suggested by the capability of 20:4n6 NAE and 2-AG to inhibit electrically stimulated mouse vas deferens contractions (Devane et al., 1992; Mechoulam et al., 1995) via CB receptor activation (Pertwee and Fernando, 1996). Cannabinoids can possibly affect sexual behavior via phosphorylation of the progesterone receptor, via two hypothetical pathways. First, when THC binds to the CB1 receptor, this binding leads to dopamine release, binding to the dopamine D1B receptor, cAMP induction of protein kinase A, phosphorylation of DARP-32 protein, and effects on the progesterone receptor. Second, binding to the dopamine D1B receptor may lead to MAP kinase mediated phosphorylation of the progesterone receptor (Mani et al., 2001).

L. *N*-Acylethanolamines in Demyelinating and Other Disease States

Multiple sclerosis (MS) patients are known to smoke marijuana to decrease spasticity. In a mouse model of MS (CREAE mice), intravenous injection of either 20:4 NAE, 16:0 NAE, or 2-AG led to decreased spasticity. The three compounds had different kinetics properties, based on hind limb flexion resistance. Effects were mediated via binding to CB1/CB2 receptors (Baker et al., 2000; Di Marzo et al., 2000b).

In Parkinson's disease, NAE levels are reported to be increased in the basal ganglia, which have a role in motor activity. Levels of 20:4n6 NAE were three-fold higher in the globus pallidus and the substantia nigra relative to those of other brain regions. In reserpine-treated rats, locomotion is restored with a dopamine D2 agonist and with a CB1 receptor antagonist (Di Marzo et al., 2000c).

In schizophrenics, 20:4n6 NAE levels are increased in the cerebrospinal fluid as a result of perturbed dopamine D2 receptor activation, which causes increased 20:4n6 NAE release. Thus, the effects of blocking the CB1 receptor in schizophrenics might be therepeutically important (Leweke et al., 1999).

Endocannabinoids may have a role in regulating psychomotor activity (Beltramo et al., 2000). The anandamide transport inhibitor AM404 reduced dopamine D2 receptor mediated stimulation of motor activity in rats (utilizing the D2 receptor agonist quinpirole). Also AM404 decreased hyperactivity in juvenille spontaneously hypertensive rats, a putative model for attention deficit disorder.

III. ENDOCANNABINOIDS IN FOODS AND ORAL ADMINISTRATION

A. Sources of Endocannabinoids in Foods

Although the amounts of 20:4n6 NAE in typical animal tissues (0.35–1.75 μg/kg tissue) are well below those necessary to elicit a cannabimimetic response after oral administration, 16:0 NAE and 2-AG amounts in these tissues can be one or two to three orders of magnitude higher than those of 20:4n6 NAE, respectively. For example, edible mussels may contain up to 55 μg/kg of 16:0 NAE (Sepe et al., 1998).

B. Lecithin

A source of NAEs could be lecithin, a widely used additive in many food products. Although one is used to thinking of soy lecithin as a crude source of PC, it actually contains NAPEs and lysoNAPEs up to 20% by weight. Soybean lecithin also contains 16:0 NAE (Ganley et al., 1958), but it is not clear whether it is artefactually generated from the corresponding NAPE during the preparation of the lecithin. Plants have very active phospholipase D enzymes, which, depending on how the lecithin is obtained, could also be present in lecithin crude preparations.

C. Chocolate

In 1996, a study that claimed that 20:4n6 NAE had been discovered in chocolate (di Tomaso et al., 1996), a rather surprising finding as it is known that higher-order plants do not synthesize arachidonic acid, was published. Nevertheless the press and some members of the scientific community (Bruinsma and Taren, 1999) seized on this finding as the explanation for the desirability of chocolate and the feelings of well-being associated with its consumption. Whether the quantitative aspects of this finding were realistic was also discussed. Some calculations indicating that 25 lbs of chocolate would have to be *injected* in order to create marijuana-like effects were reported in the press.

D. Coffee Cherries and Beans, Unfermented Cocoa Beans, Fermented Unroasted and Fermented Roasted, Cocoa Powders, Dark Chocolate, Peanuts, Hazelnuts, Walnuts, Soybeans, Oatmeal, Millet, Olives, and Milks—Bovine Milk and Colostrum and Early and Mature Human Milk

After the finding that chocolate might contain 20:4n6 NAE and other NAEs discussed earlier (di Tomaso et al., 1996), we analyzed all of the food products indicated for MAGs, primary amides, and NAEs (Di Marzo et al., 1998b) and later determined oral bioactivity of some endocannabinoids (discussed later).

E. Analysis of Food Products for Endocannabinoids

Materials were extracted with $CHCl_3$/MeOH 2:1 (v:v) containing $[^2H]_8$-20:4n6 NAE, $[^2H]_4$-16:0, -18:0, -18:1n9, -18:2n3, and -18:3n3 NAEs, and $[^2H]_8$-2-AG as internal standards. In order to purify and characterize the NAEs and 2-AG, the organic phase was brought to dryness; submitted to a series of chromatographic steps, including SiO_2 open bed chromatography and normal phase high-pressure liquid chromatography (NP-HPLC) (Berrendero et al., 1999); and NAEs and 2-AG quantified by gas chromatography–electron impact mass spectrometry (GC-EIMS) of the corresponding TMSE derivatives (Di Marzo et al., 1998b). Oleamide was analyzed in the total ion current mode (Di Marzo et al., 1998b).

F. 20:4n6 *N*-Acylethanolamine Levels in Food Products

Results (Table 1) show that 20:4n6 NAE was very low or not present in cocoa beans, unfermented or fermented, unroasted, or roasted; 20:4n6 NAE was also not present in cocoa powder or dark chocolate. Thus, di Tomaso and colleagues' reported finding of 20:4n6 NAE in chocolate (di Tomaso et al., 1996) may be due to an artefact, or to microbial contamination, and 20:4n6 NAE is not likely to account for chocolate's purported pleasurable properties. In the present analyses, 20:4n6 NAE was also not found in caturra coffee cherries, green beans, peanuts, hazelnuts, walnuts, soybeans, oatmeal, millet, or olives. On the other hand, human breast milk was found to contain small amounts of 20:4n6 NAE (21 ng/g total extracted lipid), and the possible biological significance of this finding is not known.

G. Other Fatty Acid Amides and 2-*sn*-2-Arachidonoylglycerol in Food Products

In contrast to the absence of 20:4n6 NAE in plant materials, other NAEs and oleamide were found to variable degrees in the materials studied. Oleoyl and linoleoyl ethanolamides were the most important ethanolamides found in the plant materials. Some 16:0 NAE was also present, and this congener was found in milk as well. All materials contained oleamide, and milk proved to be the richest source of this fatty acid amide of all materials tested. Human milk contained higher 2-AG levels compared to bovine milk.

H. Physiological Effects of Orally Administered Endocannabinoids

Although numerous investigators have studied the effects of ingested endocannabinoids on a plethora of physiological functions, few if any researchers have examined the bioactivity of *orally* delivered endocannabinoids. We utilized a series of tests developed for CBs (Martin et al., 1991; Fride and Mechoulam, 1993; Sulcova et al., 1998). Sabra female mice received food and water ad libitum and were maintained at 20°C–22°C on a 12-hr light–dark

Table 1 *N*-Acylethanolamines, Oleamide, and Arachidonylglycerol in Food Products and Human Milks[a]

Material	NAEs					
	Starting material, ng/g					
Fatty acyl moiety	16:0	18:1n9	18:2n6	20:4n6	Oleamide	2-AG
Coffee						
Caturra coffee cherries with skin, Ecuador		54 ± 35	228 ± 75		720 ± 207	
Coffee green beans, Arabica, Colombia		63 ± 22	17 ± 6		42 ± 10	
Cocoa						
Cocoa beans, unfermented, unroasted, unhulled, Amelonado, Ivory Coast	10 ± 4	148.8 ± 87	108 ± 35		170 ± 43	
Cocoa beans, fermented, unroasted, unhulled		121 ± 53			268 ± 77	
Cocoa roast, fermented, roasted, hulled	20 ± 18	214 ± 144	41 ± 9		5781 ± 1633	
Cocoa powder	1464 ± 401	2172 ± 695	5844 ± 1515	3 ± 2 (or ND)	3687 ± 1237	
Dark chocolate, 70% cocoa	14 ± 3.9	435 ± 147	224 ± 125		5990 ± 4035	
Nuts, soy, grains, olives						
Peanuts, with salt	77 ± 27	273 ± 31	260 ± 98		620 ± 489	

Hazelnuts	58 ± 22	1055 ± 399	247 ± 102		1476 ± 828	
Walnuts	13 ± 4	21 ± 4	76 ± 27		90 ± 31	
Soybeans, white, whole, dehulled, dried	126 ± 43	302 ± 101	805 ± 249		2289 ± 987	
Oatmeal large flakes		890 ± 298	2750 ± 977		170 ± 93	
Millet	44 ± 12	9 ± 4	431 ± 133		221 ± 29	
Olives, green, with salt, spice water, acidifiant	11 ± 4	95 ± 27	47 ± 20		33 ± 9	
			Total extracted lipid, ng/g			
Milks						
Bovine milk, early, fresh frozen	112 ± 37	111 ± 41		4.2 ± 0.6	24200 ± 15600	1100 ± 300
Bovine milk, mature, fresh frozen	271 ± 53	65 ± 39	405 ± 198	ND	8500 ± 6100	2400 ± 1000
Bovine milk, pasteurized, Italy	140 ± 29	452 ± 210	117 ± 31	94 ± 6	400 ± 30	1800 ± 300
Human milk early, pooled, frozen				ND	4000 ± 1000	6400 ± 1900
Human milk, mature, pooled, frozen	156 ± 117	227 ± 7	38 ± 15	11 ± 6	1500 ± 300	8700 ± 2800
Goat milk, Italy commercial	528 ± 7	57 ± 5		9 ± 4	34,500 ± 17,500	8300 ± 300

[a] Data expressed as nanograms per gram starting material (in plant foods) or as nanograms per gram total extracted lipid (in milks, in which lipid concentration is about 36 g/L) and are means ± SD of at least three separate determinations. NAE, *N*-acylethanolamine; 2-AG, *sn*-2-arachidonoylglycerol; NO, not detected.

cycle. Compounds were dissolved in olive oil and administered orally (100 μL/10-g mouse) with intubation needles inserted 3–3.5 cm into the esophagus. Pups were tested 15 min after gavaging, since 20:4n6 NAE effects can begin10 min after oral administration (Fride and Mechoulam, 1993).

Horizontal ambulatory motor activity was measured in a 20- by 30-cm *open field* divided into 12 squares of equal size, over 8 min, for the numbers of squares crossed. Vertical *rearing* was assessed as the number of rears in an open field. R*ing test* immobility (catalepsy) was then assessed as time spent motionless on a 5.5-cm diameter ring, for up to 4 min (Pertwee, 1972). Rectal body temperature was measured immediately before food administration and again after the ring test. Hot plate *analgesia* was also investigated (Ankier, 1974). Finally, the number of fecal pellets accumulating during open field testing was a measure of *intestinal motility* (Chesher and Jackson, 1974).

I. Results of Behavioral Testing

20:4n6 *N*-Acylethanolamine (300 mg/kg), oleamide (200 mg/kg), and 2-AG (400 mg/kg) had demonstrated effects on three of the four types of behavior tested and on body temperature but had no influence on analgesia or intestinal motility (Figs. 2–4). Lower doses were inactive in all tests. Intraperitoneally injected doses of these compounds have been shown to be effective at much lower doses (Fride and Mechoulam, 1993; Mechoulam et al., 1998b).

What is the potential for these compounds to influence behavior when consumed in a normal diet? On the basis of the present study, calculations show that unreasonably large amounts of a food would have to be consumed to achieve levels of ingestion of fatty acid amides such as were shown to be effective in the mouse. However, there are several factors to consider. First, it is known from in vitro studies that some of the amide species, although having lesser affinity to the receptor themselves, inhibit the breakdown of other, more active species. Therefore, a mixture of the compounds may have a synergistic effect that is difficult to predict. Second, in the present study, the animal testing was based on clearly evident responses to stimuli: very strong, discernible, and pharmacological. The question remains whether lower levels result in a much attenuated effect, although not discerned and perhaps not discernible with the present methods used.

J. Fatty Acids and Monoglyceride Effects in the Gut?

Other digestive processes starting from fatty foods could, in principle, also lead to the formation of cannabimimetic fatty acids in the gut of animals and humans. The amount of free arachidonic acid in the diet before digestion would be minute because all would normally be esterified to phospholipids

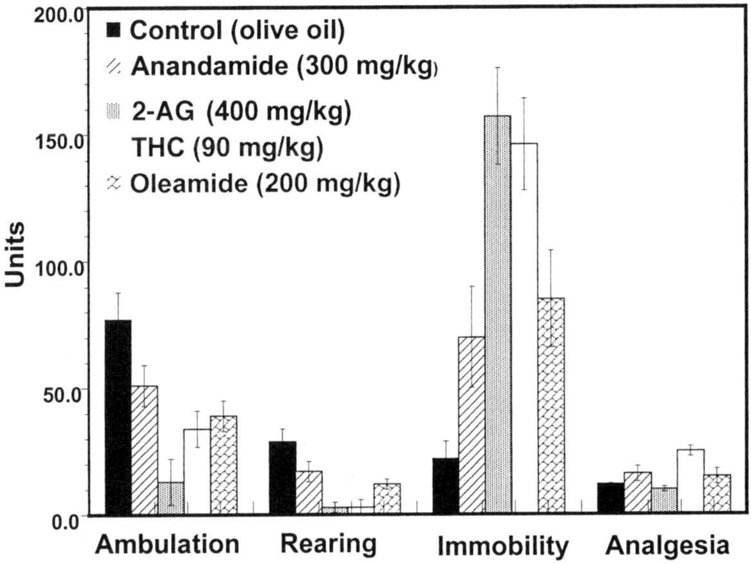

Figure 2 MAG and NAE had calming properties when orally administered. MAG, monoacylglycerol; NAE, *N*-acylethanolamine; 2-AG, *sn*-2-arachidonoylglycerol; THC, Δ^9(-)-tetrahydrocannabinol.

(Carrie et al., 2000) or, more importantly, triacylglycerols. During the digestive process, however, phospholipids containing arachidonate are converted via gut phospholipase A_2 to 1-acyllysophospholipids and free arachidonic acid. Triacylglycerols containing arachidonate (derived from fish, meats, fungal, and algal sources, all of which are common foods today) can contain the fatty acid on all 3 positions. During digestion by pancreatic, lingual, and other lipases this leads to 2-AG and the free fatty acids formerly on the *sn*-1 and *sn-3* positions, including arachidonate. The *sn*-2-acylglycerols are better absorbed than free fatty acids because the latter metabolites can form insoluble calcium soaps that are poorly absorbed and, consequently, are likely to stay in the gut for a longer period. If a man consumes 600 mg of arachidonic acid per day (this fatty acid is present in high amounts in foods of animal origin), this means about 200 mg maximum per serving, and up to two-thirds of this (120 mg, formerly on the 1- and 3- positions of triacylglycerols) could be converted to free arachidonate in the gut. Thus, it is possible that the arachidonic acid and other free fatty acids are converted in the gut to the corresponding NAEs by FAAH, since Katayama and associates (1997) have shown that because of the presence of lipid inhibiting the hydrolysis reaction in the gastrointestinal (G.I.) tract, the enzyme can catalyze the condensation

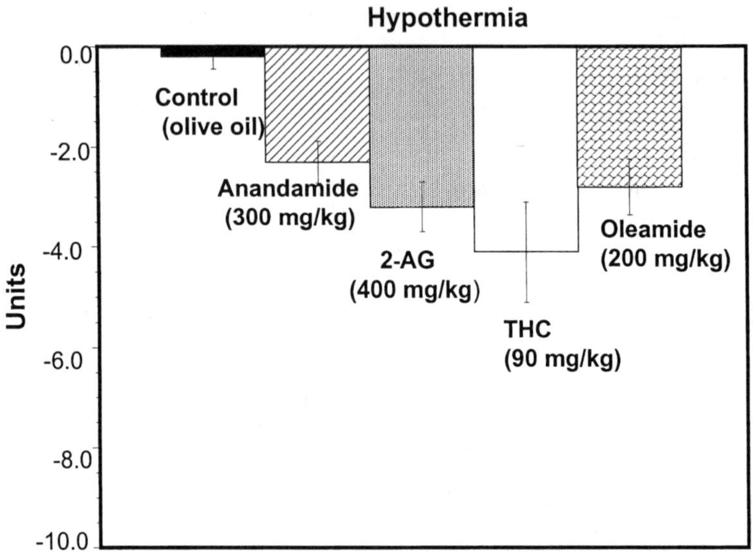

Figure 3 All compounds reduced body temperature vs. control body temperature. 2-AG, *sn*-2-arachidonoylglycerol; THC, Δ^9(-)-tetrahydrocannabinol.

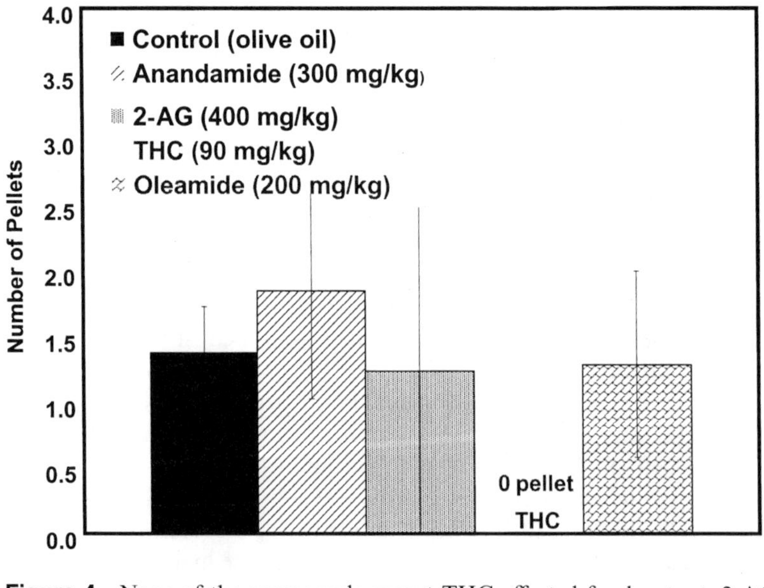

Figure 4 None of the compounds except THC affected fecal output. 2-AG, *sn*-2-arachidonoylglycerol; THC, Δ^9(-)-tetrahydrocannabinol.

reaction (i.e., the formation of NAEs and 20:4n6 NAE) under these conditions. This hypothesis needs to be tested experimentally by feeding animals diets with and without arachidonic acid, then aspirating the stomach and gut contents, and measuring the levels of NAEs in the different gut regions.

IV. DOES DIETARY LONG-CHAIN POLYUNSATURATED FATTY ACID CONSUMPTION LEAD TO HIGHER CELLULAR LEVELS OF CORRESPONDING *N*-ACYLETHANOLAMINES, MONOACYLGLYCEROLS, AND PRIMARY AMIDES BY INFLUENCING *N*-ACYLETHANOLAMINE PRECURSOR LIPID POOLS OF 20:4n6 AND 22:6n3 IN WHOLE BRAIN AND BRAIN REGIONS?

As a next step, the hypothesis related to LCPUFA consumption and its influence on NAE precursor lipid pools was tested in piglets that were fed diets with and without 20:4n6 and 22:6n3 at levels similar to those found in porcine milk, and brain levels of corresponding NAEs were assessed (Berger et al., 2001).

A. Animals and Diets

Male piglets (>1 kg at birth, <12 hr old) were fed liquid formulas based on the macro- and micronutrient composition of pig milk (n = 6 piglets per group; see Table 2). Piglets were bottle-fed by hand until 18 days of age (Innis and Dyer, 1997), and littermates assigned to different diets. In a separate feeding study, piglets were fed identical diets, stored at 80°C under N_2, and individual brain sections further analyzed for NAEs (n = 6 per group). Formulas contained 8.30 en% 18:2n6 and 0.80 en% 18:3n3 and were supplemented optionally with 0.20 en% 20:4n6 and 0.16 en % 22:6n3 (0.30–0.40 g/100 g of total fatty acids; see Table 3). The 20:4n6 and 22:6n3 were from single-celled-organism triacylglycerols. Formulas contained 57.9 g of fat/L, 4.143 MJ/L.

B. Sample Collection

Piglets were anesthetized at 18 days age, 3–4 hr after the last feeding (Innis and Dyer, 1997), and intracardiac blood obtained. Brains were removed and weighed, and the entire frontal cortex, striatum, hippocampus, and hypothalamus were removed for neurotransmitter analysis (de la Presa Owens and Innis, 1999, 2000). Remaining brain tissue was homogenized in buffer, frozen, and later analyzed for phospholipids [de la Presa Owens and Innis, 1999] and

Table 2 Fat Components of Formulas Varying in 20:4n6 and 22:6n3 Content[a]

Type of oil	Formula C−	Formula C+
Milk fat	1.70	1.70
MCT	8.70	8.70
Palm olein	4.90	4.40
Trisun oil	8.00	8.30
Soybean oil	6.50	6.30
Single-cell AA	0.00	0.30
Single-cell DHA	0.00	0.20

[a] Grams per 100 g oil formula. C−, control diet without 20:4n6 and 22:6n3; C+, control diet with added 20:4n6 and 22:6n3; AA, arachidonic acid; DHA, docosahexanoic acid: MCT, medium-chain triacylglycerol, single-cell, single-cell organism triacylglycerol source.

Table 3 Major Fatty Acid Components of Formulas Varying in 20:4n6 and 22:6n3 Content[a]

Fatty acid, g FA/100 g total FA	Formula C−	Formula C+	Formula Sow milk
8:0	17.20	14.90	ND
10:0	13.50	13.00	ND
12:0	1.00	0.30	0.10
14:0	0.80	0.60	2.40
16:0	11.30	10.90	28.10
18:0	3.20	3.30	5.60
16:1n7	0.10	0.20	4.70
18:1n9	33.30	35.10	32.60
18:2n6	15.60	16.40	20.40
20:2n6	ND	ND	0.40
18:3n3	1.50	1.60	ND
18:3n6	ND	0.10	0.20
20:3n6	ND	ND	0.20
20:4n6	ND	0.40	0.70
22:4n6	ND	ND	0.10
20:5n3	ND	ND	0.10
22:5n3	ND	ND	0.60
22:6n3	ND	0.30	0.10

[a] C−, control diet without 20:4n6 and 22:6n3; C+, control diet with added 20:4n6 and 22:6n3; FA, fatty acid; ND, not detected.

primary amides, NAEs, and momoacylglycerols (MAGs) (Berger et al., 2001).

C. *N*-Acylethanolamines, Primary Amide, and Monoacylglycerol Analysis

Brain samples were extracted with $CHCl_3$/MeOH 2:1 (vol/vol) containing [$^2H_8$]-20:4n6-NAE and [$^2H_8$]-2-AG as internal standards. To purify and characterize NAEs and MAGs, the organic phase was dried and then was submitted to chromatographic steps, including SiO_2 open bed chromatography and normal-phase high-performance liquid chromatography (HPLC), then GC-EIMS on TMSE derivatives (Di Marzo et al., 1996a; Bisogno et al., 1997b; Di Marzo et al., 2000a; Berger et al., 2001).

D. Polyunsaturated *N*-Acylethanolamines

Supplementation of control diets with 20:4n6 and 22:6n3 led to brain homogenate increases of 20:4n6 (3.9-fold), 22:4n6 (1.6-fold), 20:5n3 (5.2-fold), 22:5n3 (9.4-fold), and 22:6n3 NAEs (9.5-fold; see Table 4). The NAE levels were statistically similar for piglets fed diets supplemented with 20:4n6 and 22:6n3 and those fed sow's milk. The increases in 22:4n6, 20:5n3, and 22:5n3 can be attributed to the facts that 22:4n6 is the two-carbon elongation product of 20:4n6, and 20:5n3 and 22:5n3 are retroconversion products of 22:6n3 (Sprecher, 1999). The increase in the indicated NAEs could be biologically important, since NAEs with fatty acids having 20 carbons and three double bonds bind CB1 receptors (Barg et al., 1995; Fride et al., 1995; Priller et al., 1995; Sugiura et al., 1996a). 20:4n6 *N*-Acylethanolamine and synthetic 22:6n3 NAE have similar cannabimimetic activities when injected into mice (Fride et al., 1995), but in vitro studies showed poor affinity of 22:6n3 NAE for CB1 receptors (Felder et al., 1993; Sheskin et al., 1997). Higher precursor substrate levels of 22:6n3 (phosphatidylethanolamine and phosphatidylcholine with 22:6n3 on both *sn*-1 and *sn*-2 positions, and *N*-docosahexaenoyl phosphatidylethanolamine) similarly lead to higher levels of 22:6n3 NAE in bovine retina (Bisogno et al., 1999b). Despite differences in 20:4n6 and 22:6n3 content between sow's milk and our 20:4n6- and 22:6n3-supplemented diet (Table 2) (de la Presa Owens et al., 1998; Devlin et al., 1998; Berger et al., 2000), levels of unsaturated NAEs were similar, suggesting that a threshold level of dietary 20:4n6 and 22:6n3 is needed to maintain "normal" NAE levels. Preliminary evidence (Berger et al., 2001) showed that the dietary precursors 18:2n6 and 18:3n3 added to pig formula were not able to elevate 20:4n6 and 22:6n3 brain NAE levels. Triacylglycerol positional distribution of fatty acids (affecting fatty acid absorption), cholesterol, and hormones in

Table 4 Diet-Induced Changes in Polyunsaturated Brain (Monoacylglycerols) and *N*-Acylethanolamines[a]

Diet	Formula		
	C−	C+	Sow milk
MAG, 4–6 double bonds			
20:4n6	66.00 ± 9.38[b]	44.40 ± 16.13	44.10 ± 21.73
22:4n6	3.53 ± 0.65[b]	6.23 ± 1.82	6.13 ± 1.42
22:6n3	3.87 ± 0.03[ab]	5.93 ± 1.37	6.67 ± 1.92
NAE, 4–6 double bonds			
20:4n6	1.02 ± 0.34[ab]	3.96 ± 0.91	3.33 ± 0.80
22:4n6	0.74 ± 0.03[ab]	1.15 ± 0.02	1.46 ± 0.48
20:5n3	0.33 ± 0.12[ab]	1.74 ± 0.47	1.66 ± 0.13
22:5n3	0.18 ± 0.09[ab]	1.69 ± 0.49	1.40 ± 0.52
22:6n3	0.95 ± 0.21[ab]	9.03 ± 3.60	3.94 ± 1.24

[a] Picomoles per milligram brain lipid extract. Values represent means ± standard errors. Pairwise comparisons using an unpaired, one-sided student *t*-test ($p < 0.05$, $n = 3$/group) are inxdicated as follows: a, C− different from C+; b, C− different from sow-fed group. C+ group was not statistically significant different from sow-fed group ($p > 0.10$);

[b] Statistically different at $0.05 < p < 0.10$. C−, control diet without 20:4n6 and 22:6n3; C+, control diet with added 20:4n6 and 22:6n3. 18:3n3 NAE was not detected and is omitted from Table 4. MAG, monoacylglycerol; NAE, *N*-acylethanolamine.

sow milk are examples of other factors in sow milk that could affect NAE metabolism (Berger et al., 2000).

In a separate feeding experiment (Berger et al., 2001), 20:4n6 and 22:6n3 NAEs were measured in distinct piglet brain regions, focusing on regions with higher amounts of CB receptors and NAEs (Table 5). Levels of NAEs in hippocampus were resistant to dietary modulation. Relative to piglets fed control diet, in piglets fed diets supplemented with 20:4n6 and 22:6n3, there were increases in 22:6n3 in auditory cortex (5.5-fold), brain stem (7.0-fold, statistical trend), and striatum (2.0-fold). In contrast to results with brain homogenates (Berger et al., 2001), there were no statistical differences in 20:4n6 NAE between supplemented- and control-diet groups for the specific brain regions examined. However, supporting the observation that 20:4n6 and 22:6n3 can increase corresponding NAEs is that when compared to sow's milk–fed piglets, piglets fed diets lacking 20:4n6 and 22:6n3 had reduced amounts of 20:4n6 NAE in brain stem, auditory cortex, and cerebellum and reduced amounts of 22:6n3 NAE in brain stem, auditory cortex, and striatum. In the same regions, there were not significant differences in NAEs between sow's milk–and supplemented diet–fed piglets.

In a mouse experiment (Berger et al., 2001), mice received fat-free diets supplemented with 0.4% milk fat, 1.2% palm olein, 1.9% sunflower oil, 1.5%

Table 5 Diet-Induced Changes in *N*-acylethanolamines from Specific Brain Regions[a]

N-Acylethanolamines Brain sections	Formula C−	Formula C+	Formula Sow milk
Brainstem			
20:4n6	1.10 ± 0.09[b]	1.60 ± 0.23	2.50 ± 0.35
22:6n3	0.70 ± 0.06[ab]	4.90 ± 1.79	9.90 ± 1.39
Auditory cortex			
20:4n6	3.00 ± 0.23[b]	4.00 ± 0.69	7.30 ± 1.33
22:6n3	0.80 ± 0.23[ab]	4.40 ± 0.87	8.00 ± 1.62
Visual cortex			
20:4n6	3.50 ± 0.35	—	4.00 ± 0.58
22:6n3	3.30 ± 0.17	—	2.10 ± 0.35
Cerebellum			
20:4n6	0.50 ± 0.06[b]	—	0.80 ± 0.06
22:6n3	0.50 ± 0.17	—	0.40 ± 0.06
Striatum			
20:4n6	1.80 ± 0.02[b]	1.50 ± 0.06	1.10 ± 0.17
22:6n3	1.00 ± 0.12[ab]	2.00 ± 0.29	3.20 ± 0.17
Hippocampus			
20:4n6	2.20 ± 0.46	2.20 ± 0.17	2.00 ± 0.17
22:6n3	1.20 ± 0.12	0.90 ± 0.06	1.40 ± 0.17

[a] Picomoles per milligram brain lipid extract. Values represent means ± standard errors. Pairwise comparisons using an unpaired, one-sided student t-test ($p < 0.05$, $n = 3$/group) are indicated as follows: a, C− different from C+; b, C− different from sow-fed group. C+ group was not statistically significant different from sow-fed group ($p > 0.10$);

[b] Statistically different at $0.05 < p < 0.10$; C−, control diet without 20:4n6 and 22:6n3; C+, control diet with added 20:4n6 and 22:6n3;—, not determined for technical reasons; ND, not detected.

soybean oil, and 5.1% medium-chain triacylglycerol (MCT) for 58 days. The MCT was partly replaced with 1.1% single-celled algal oil, providing 0.5% 20:4n6 in a 20:4n6-supplemented diet. Similarly to pig results, supplementation with 20:4n6 increased 20:4n6 NAE 5.8-fold (21.8–125.8 pmol/g wet weight of brain tissue).

E. How Dietary Long-Chain Polyunsaturated Fatty Acid Can Modulate Brain *N*-Acylethanolamine Levels

Dietary 20:4n6 and 22:6n3 can modulate NAE levels by influencing levels of NAE precursors such as *N*-acylphosphatidylethanolamines in neuronal membranes (Di Marzo et al., 1994) or by providing direct fatty acid substrate for de novo synthesis (Devane and Axelrod, 1994). Additionally, the phospholipid acyl donor during *N*-acylphosphatidylethanolamine synthesis is unusual in being an *sn*-1-arachidonoyl phospholipid species (Schmid et al., 1990; Sugiura et al., 1996c). The extent to which the *sn*-1-polyunsaturated phospholipid pool is modifiable by diet is not known. Dietary LCPUFA could affect NAE metabolism by a number of other means, including physical effects on membranes such as the blood–brain barrier (Yehuda et al., 1999), which can influence receptor number and kinetics (Zimmer et al., 2000); modulation of intracellular calcium levels in some cell types (Desai et al., 1993); and modulation of protein kinase C (Raygada et al., 1998) and hormones.

F. Physiological Effects of Increasing Brain Levels of Polyunsaturated *N*-Acylethanolamines

As decribed in Sec. I, activated CB1 and CB2 receptors mediate inhibition of adenylate cyclase and *N*- and P/Q-type calcium channels, and activation of potassium channels and mitogen-activated protein kinase and affect a wide variety of physiological processes. The NAEs also have non-CB1/non-CB2 receptor-mediated activities (Di Marzo et al., 1998a; Howlett et al., 1999). High doses of injected 20:4n6 NAE or 20:4n6 NAE analogues in mice (10–100 mg/kg of body weight) cause typical in vivo cannabimimetic inhibitory effects (motor activity, rearing activity, ring catalepsy, hypothermia, analgesia, and agonistic behavior) and inhibition of leukocyte phagocytosis. In contrast, low doses of injected 20:4n6 NAE (0.01 mg/kg of body weight) stimulate activities in open field and ring, increase aggressive behavior, and stimulate phagocytosis (Sulcova et al., 1998). Thus, in the present nutritional studies, it is difficult to predict whether LCPUFA-induced increases in NAEs will increase or decrease these behavioral effects and other physiological sequelae observed after direct NAE injection or consumption.

G. Monoarachidonoylglycerols

sn-2-Arachidonoylglyceral resembles 20:4n6 NAE in molecular conformation, and *sn*-1-, *sn*-2-, and *sn3*-arachidonoyl MAGs have CB1 receptor affinity (Stella et al., 1997; Sugiura et al., 1999). The 2-AG concentration can be 200- to 800-fold greater than that of 20:4n6 NAE in some rodent brain regions (Sugiura et al., 1996d). In addition, 2-AG is found in spleen, pancreas, and intestinal cells of rats and inhibits 20:4n6 NAE breakdown (Di Marzo and Deutsch, 1998). Such MAGs as *sn*-2-linoleoylglycerol and *sn*-2-palmitoylglycerol could also be important, because they can potentiate CB receptor binding of 2-AG (Di Marzo and Deutsch, 1998). Free 20:4n6 and *sn*-2-

palmitoylglycerol, 2-oleoylglycerol, 2-linoleylglycerol, and 2-docosahexaenoylglycerol are inactive or weakly active on CB1 receptors in neuroblastoma cells (Sugiura et al., 1999). Berger and associates (2001) found levels of 20:4n6 MAG (1- and 2-isomers combined) that did not increase in response to dietary 20:4n6 and 22:6n3 supplementation, whereas levels of 22:6n3 MAG did increase. Kinetic studies with labeled fatty acids, with particular focus on *sn*-2-arachidonoyl-containing phospholipids (Di Marzo et al., 1996b; Bisogno et al., 1999c) and triacylglycerols, are needed to elucidate the proceeding finding.

H. Diet-Induced Changes in Phospholipids Compared with *N*-Acylethanolamines

After feeding of the experimental diets described earlier to pigs and preparation of an identical brain homogenate (Berger et al., 2001), relative to that of control, supplementation with 20:4n6 and 22:6n3 led to a small increase in brain 22:6n3 level in only phosphatidylcholine, and no change in 20:4n6 in any phospholipid classes examined [de la Presa Owens and Innis, 1999]. No distinction was, however, made between *sn*-1 and *sn*-2 position of phospholipids, and 20:4n6 and 22:6n3 NAEs are derived from phospholipids with LCPUFA on the *sn*-1 position (Di Marzo et al., 1998a). The lack of change in 20:4n6 level in brain phospholipids after feeding different diets exemplifies the importance of examining additional nonphospholipid lipid pools, such as the NAE pool, to predict functional sequelae and to use in follow-up experimentation in human and animal fatty acid feeding trials.

V. CONCLUSIONS

We have demonstrated that dietary 20:4n6 and 22:6n3 can modulate levels of bioactive brain NAEs (see Fig. 5 for summary). This intriguing observation has raised a number of important questions. What is the optimal dietary dose of LCPUFA (arachidonic acid and docosahexaenoic acid, added separately and together) for increasing levels of brain NAEs and what are the physiologic and behavioral implications? What are the pathways and kinetics for formation of brain NAEs and MAGs from dietary LCPUFA? Does the same increase in 20:4n6 NAE and 22:6n3 NAE observed in piglets occur in brain and other organs from other animals, human infants, clinical populations, and elderly persons after 20:4n6 and 22:6n3 supplementation. What could be the clinical applications of these findings?

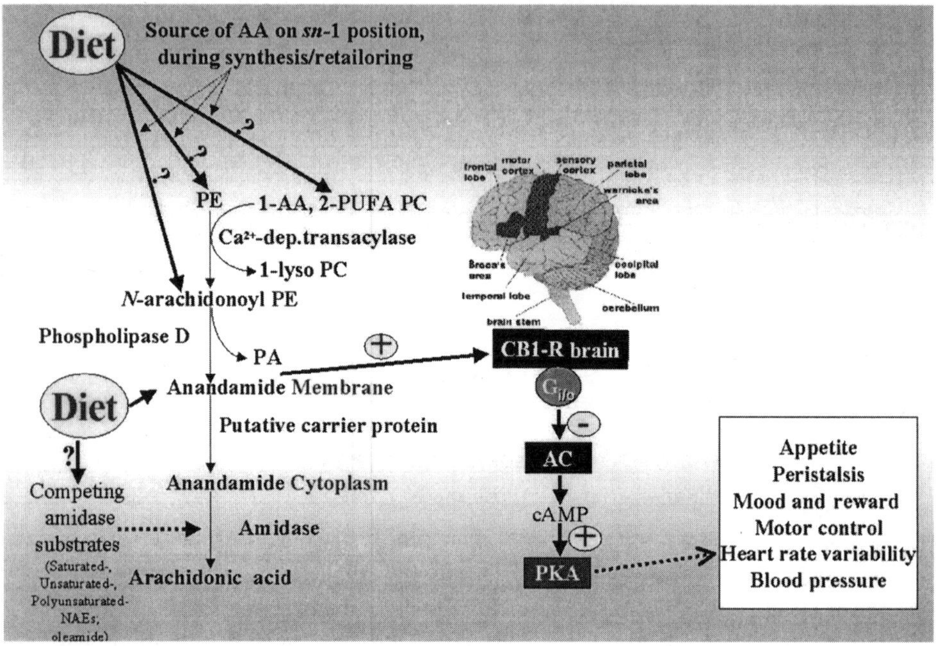

Figure 5 Hypothetical pathways for dietary AA modulation of *20:4n6 NAE* (anandamide) metabolism and the subsequent possible physiological downstream sequelae. A unique pool of phospholipids [one candidate is phosphatidylcholine (PC)] with polyunsaturated fatty acids (PUFAs) including arachidonic acid (AA) on the *sn*-1 position, rather than the usual *sn*-2 position, are believed to be precursors in a transacylase reaction leading to formation of *N*-acylphosphatidylethanolamines (NAPEs). *N*-arachidonylphosphatidylethanolamine represents one NAPE species. Dietary AA could enrich the *sn*-1-arachidonyl 2-PUFA acyl PC pool of lipids by affecting direct synthesis or retailoring, as depicted. Likewise, dietary fatty acids could also lead to alteration of the acyl chains of phosphatidylethanolamine (PE), a donor in the transacylase reaction, which subsequently could affect kinetics of the transacylase reaction. Dietary lipids could also influence retailoring of *N*-arachidonylphosphatidylethanolamine after its formation. Thereafter, *N*-arachidonylphosphatidylethanolamine is converted to anandamide in a phospholipase D–mediated reaction. Dietary fatty acids could influence levels of anandamide at this step by acting as alterative substrates for amidase, an enzyme that degrades anandamide. Anandamide released from membranes of neuronal cells can bind to CB receptors (CB1-R) in the brain (and other organs). CB1-R-triggered intracellular signaling, e.g., down regulation of adenylate cyclase (AC), can ultimately lead to various physiological sequelae, including changes in appetite, peristalsis, mood and reward, motor control, heart rate variability, and blood pressure.

REFERENCES

Adams, I. B., Compton, D. R., & Martin, B. R. (1998) Assessment of anandamide interaction with the cannabinoid brain receptor: SR 141716A antagonism studies in mice and autoradiographic analysis of receptor binding in rat brain. J Pharmacol Exp Ther 284: 1209–1217.

Ankier, S. I. (1974) New hot plate tests to quantify antinociceptive and narcotic antagonist activities. Eur J Pharmacol 27: 1–4.

Baker, D., Pryce, G., Croxford, J. L., Brown, P., Pertwee, R. G., Huffman, J. W., & Layward, L. (2000) Cannabinoids control spasticity and tremor in a multiple sclerosis model. Nature 404: 84–87.

Barg, J., Fride, E., Hanus, L., Levy, R., Matus-Leibovitch, N., Heldman, E., Bayewitch, M., Mechoulam, R., & Vogel, Z. (1995) Cannabinomimetic behavioral effects of and adenylate cyclase inhibition by two new endogenous anandamides. Eur J Pharmacol 287: 145–152.

Barinaga, M. (2001) How cannabinoids work in the brain. Science 291: 2530–2531.

Basavarajappa, B. S., & Hungund, B. L. (1999) Chronic ethanol increases the cannabinoid receptor agonist anandamide and its precursor N-arachidonoylphosphatidylethano lamine in SK-N-SH cells. J Neurochem 72: 522–528.

Beltramo, M., Fonseca, F. R., Navarro, M., Calignano, A., Gorriti, M. A., Grammatikopoulos, G., Sadile, A. G., Guiffrida, A., & Piomelli, D. (2000) Reversal of dopamine D(2) receptor responses by an andamide transport inhibitior. J Neurosci 2: 3401–3407.

Ben Shabat, S., Fride, E., Sheskin, T., Tamiri, T., Rhee, M. H., Vogel, Z., Bisogno, T., De Petrocellis, L., Di Marzo, V., & Mechoulam, R. (1998) An entourage effect: Inactive endogenous fatty acid glycerol esters enhance 2-arachidonoyl-glycerol cannabinoid activity. Eur J Pharmacol 353: 23–31.

Berdyshev, E. V. (1999) Inhibition of sea urchin fertilization by fatty acid ethanolamides and cannabinoids. Comp Biochem Physiol C Pharmacol Toxicol Endocrinol 122: 327–330.

Berdyshev, E. V., Boichot, E., Germain, N., Allain, N., Anger, J. P., & Lagente, V. (1997) Influence of fatty acid ethanolamides and delta9-tetrahydrocannabinol on cytokine and arachidonate release by mononuclear cells. Eur J Pharmacol 330: 231–240.

Berdyshev, E. V., Schmid, P. C., Dong, Z., & Schmid, H. H. (2000) Stress-induced generation of N-acylethanolamines in mouse epidermal JB6 P+ cells. Biochem J 346 (Pt 2): 369–374.

Berger, A., Crozier, G., Bisogno, T., Cavaliere, P., Innis, S., & Di Marzo, V. (2001) Anandamide and diet: Inclusion of dietary arachidonate and docosahexaenoate leads to increased brain levels of the corresponding N-acylethanolamines in piglets. Proc Natl Acad Sci USA 98: 6402–6406.

Berger, A., Fleith, M., & Crozier, G. (2000) Nutritional implications of replacing bovine milk fat with vegetable oil in infant formulas (Invited review). J Pediatr Gastroenterol Nutr 30: 115–130.

Berrendero, F., Sepe, N., Ramos, J. A., Di Marzo, V., & Fernández-Ruiz, J. J. (1999)

Analysis of cannabinoid receptor binding and mRNA expression and endogenous cannabinoid contents in the developing rat brain during late gestation and early postnatal period. Synapse 33: 181–191.

Bisogno, T., Berrendero, F., Ambrosino, G., Cebeira, M., Ramos, J. A., Fernandez-Ruiz, J. J., & Di Marzo, V. (1999a) Brain regional distribution of endocannabinoids: Implications for their biosynthesis and biological function. Biochem Biophys Res Commun 256: 377–380.

Bisogno, T., Delton-Vandenbroucke, I., Milone, A., Lagarde, M., & Di Marzo, V. (1999b) Biosynthesis and inactivation of N-arachidonoylethanolamine (anandamide) and N-docosahexaenoylethanolamine in bovine retina. Arch Biochem Biophys 370: 300–307.

Bisogno, T., Katayama, K., Melck, D., Ueda, N., De Petrocellis, L., Yamamoto, S., & Di Marzo, V. (1998) Biosynthesis and degradation of bioactive fatty acid amides in human breast cancer and rat pheochromocytoma cells1/Mimplications for cell proliferation and differentiation. Eur J Biochem 254: 634–642.

Bisogno, T., Maurelli, S., Melck, D., De Petrocellis, L., & Di Marzo, V. (1997a) Biosynthesis, uptake, and degradation of anandamide and palmitoylethanolamide in leukocytes. J Biol Chem 272: 3315–3323.

Bisogno, T., Melck, D., De Petrocellis, L., & Di Marzo, V. (1999c) Phosphatidic acid as the biosynthetic precursor of the endocannabinoid 2-arachidonoylglycerol in intact mouse neuroblastoma cells stimulated with ionomycin. J Neurochem 72: 2113–2119.

Bisogno, T., Sepe, N., Melck, D., Maurelli, S., De Petrocellis, L., & Di Marzo, V. (1997b) Biosynthesis, release and degradation of the novel endogenous cannabimimetic metabolite 2-arachidonoylglycerol in mouse neuroblastoma cells. Biochem J 322: 671–677.

Boring, D. L., Berglund, B. A., & Howlett, A. C. (1996) Cerebrodiene, arachidonylethanolamide, and hybrid structures: Potential for interaction with brain cannabinoid receptors. Prostaglandins Leukot Essent Fatty Acids 55: 207–210.

Bruinsma, K., & Taren, D. L. (1999) Chocolate: Food or drug? J Am Diet Assoc 99: 1249–1256.

Calignano, A., Katona, I., Desarnaud, F., Giuffrida, A., La Rana, G., Mackie, K., Freund, T. F., & Piomelli, D. (2000) Bidriectional control of airway responsiveness by endogenous cannabinoids. Nature 408: 96–101.

Calignano, A., La Rana, G., Beltramo, M., Makriyannis, A., & Piomelli, D. (1997a) Potentiation of anandamide hypotension by the transport inhibitor, AM404. Eur J Pharmacol 337: R1–2.

Calignano, A., La Rana, G., Giuffrida, A., & Piomelli, D. (1998) Control of pain initiation by endogenous cannabinoids. Nature 394: 277–281.

Calignano, A., La Rana, G., Makriyannis, A., Lin, S. Y., Beltramo, M., & Piomelli, D. (1997b) Inhibition of intestinal motility by anandamide, an endogenous cannabinoid. Eur J Pharmacol 340: R7–8.

Carrie, I., Clement, M., de Javel, D., Frances, H., & Bourre, J. M. (2000) Phospholipid supplementation reverses behavioral and biochemical alterations induced by n-3 polyunsaturated fatty acid deficiency in mice. J Lipid Res 41: 473–480.

Castellano, C., Cabib, S., Palmisano, A., Di Marzo, V., & Puglisi Allegra, S. (1997) The effects of anandamide on memory consolidation in mice involve both D1 and D2 dopamine receptors. Behav Pharmacol 8: 707–712.

Chesher, G. B., & Jackson, D. M. (1974) Anticonvulsant effects of cannabinoids in mice: Drug interactions within cannabinoids and cannabinoid interactions with phenytoin. Psychopharmacologia 37: 255–264.

Cravatt, B. F., Giang, D. K., Mayfield, S. P., Boger, D. L., Lerner, R. A., & Gilula, N. B. (1996) Molecular characterization of an enzyme that degrades neuromodulatory fatty-acid amides. Nature 384: 83–87.

Cravatt, B. F., Prospero-Garcia, O., Siuzdak, G., Gilula, N. B., Henriksen, S. J., Boger, D. L., & Lerner, R. A. (1995) Chemical characterization of a family of brain lipids that induce sleep. Science 268: 1506–1509.

Crawley, J. N., Corwin, R. L., Robinson, J. K., Felder, C. C., Devane, W. A., & Axelrod, J. (1993) Anandamide, an endogenous ligand of the cannabinoid receptor, induces hypomotility and hypothermia in vivo in rodents. Pharmacol Biochem Behav 46: 967–972.

de la Presa Owens, S., & Innis, S. M. (1999) Docosahexaenoic and arachidonic acid prevent a decrease in dopaminergic and serotoninergic neurotransmitters in frontal cortex caused by a linoleic and α-linolenic acid deficient diet in formula-fed piglets. J Nutr 129: 2088–2093.

de la Presa Owens, S., & Innis, S. M. (2000) Diverse, region-specific effects of addition of arachidonic and docosahexanoic acids to formula with low or adequate linoleic and alpha-linolenic acids on piglet brain monoaminergic neurotransmitters. Pediatric Res 48: 125–130.

de la Presa-Owens, S., Innis, S. M., & Rioux, F. M. (1998) Addition of triglycerides with arachidonic acid or docosahexaenoic acid to infant formula has tissue- and lipid class-specific effects on fatty acids and hepatic desaturase activities in formula-fed piglets. J Nutr 128: 1376–1384.

De Petrocellis, L., Melck, D., Bisogno, T., Milone, A., & Di Marzo, V. (1999) Finding of the endocannabinoid signalling system in Hydra, a very primitive organism: Possible role in the feeding response. Neuroscience 92: 377–387.

Derkinderen, P., Toutant, M., Burgaya, F., Le Bert, M., Siciliano, J. C., de Franciscis, V., Gelman, M., & Girault, J. A. (1996) Regulation of a neuronal form of focal adhesion kinase by anandamide. Science 273: 1719–1722.

Desai, N. N., Carlson, R. O., Mattie, M. E., Olivera, A., Buckley, N. E., Seki, T., Brooker, G., & Spiegel, S. (1993) Signaling pathways for sphingosylphosphorylcholine-mediated mitogenesis in Swiss 3T3 fibroblasts. J Cell Biol 121: 1385–1395.

Deutsch, D. G., Goligorsky, M. S., Schmid, P. C., Krebsbach, R. J., Schmid, H. H., Das, S. K., Dey, S. K., Arreaza, G., Thorup, C., Stefano, G., & Moore, L. C. (1997) Production and physiological actions of anandamide in the vasculature of the rat kidney. J Clin Invest 100: 1538–1546.

Devane, W. A., & Axelrod, J. (1994) Enzymatic synthesis of anandamide, an endogenous ligand for the cannabinoid receptor, by brain membranes. Proc Natl Acad Sci USA 91: 6698–6701.

Devane, W. A., Dysarz, F. A. D., Johnson, M. R., Melvin, L. S., & Howlett, A. C. (1988) Determination and characterization of a cannabinoid receptor in rat brain. Mol Pharmacol 34: 605–613.

Devane, W. A., Hanus, L., Breuer, A., Pertwee, R. G., Stevenson, L. A., Griffin, G., Gibson, D., Mandelbaum, A., Etinger, A., & Mechoulam, R. (1992) Isolation and structure of a brain constituent that binds to the cannabinoid receptor (see comments). Science 258: 1946–1949.

Devlin, A. M., Innis, S. M., Shukin, R., & Rioux, M. F. (1998) Early diet influences hepatic hydroxymethyl glutaryl coenzyme A reductase and 7alpha-hydroxylase mRNA but not low-density lipoprotein receptor mRNA during development. Metabolism 47: 20–26.

Di Marzo, V. (1998) "Endocannabinoids" and other fatty acid derivatives with cannabimimetic properties: Biochemistry and possible physiopathological relevance. Biochim Biophys Acta 1392: 153–175.

Di Marzo, V., Berrendero, F., Bisogno, T., González, S., Cavaliere, P., Romero, J., Cebeira, M., Ramos, J. A., & Fernández Ruiz, J. J. (2000a) Enhancement of anandamide formation in the limbic forebrain and reduction of endocannabinoid contents in the striatum of delta9-tetrahydrocannabinol-tolerant rats. J Neurochem 74: 1627–1635.

Di Marzo, V., Bifulco, M., & De Petrocellis, L. (2000b) Endocannabinoids and multiple sclerosis: A blessing from the 'inner bliss'? Trends Pharmacol Sci 21: 195–197.

Di Marzo, V., Bisogno, T., De Petrocellis, L., Melck, D., Orlando, P., Wagner, J. A., & Kunos, G. (1999) Biosynthesis and inactivation of the endocannabinoid 2-arachidonoylglycerol in circulating and tumoral macrophages. Eur J Biochem 264: 258–267.

Di Marzo, V., De Petrocellis, L., Sepe, N., & Buono, A. (1996a) Biosynthesis of anandamide and related acylethanolamides in mouse J774 macrophages and N18 neuroblastoma cells. Biochem J 316: 977–984.

Di Marzo, V., De Petrocellis, L., Sugiura, T., & Waku, K. (1996b) Potential biosynthetic connections between the two cannabimimetic eicosanoids, anandamide and 2-arachidonoyl-glycerol, in mouse neuroblastoma cells. Biochem Biophys Res Commun 227: 281–288.

Di Marzo, V., & Deutsch, D. G. (1998) Biochemistry of the endogenous ligands of cannabinoid receptors. Neurobiol Dis 5: 386–404.

Di Marzo, V., Fontana, A., Cadas, H., Schinelli, S., Cimino, G., Schwartz, J. C., & Piomelli, D. (1994) Formation and inactivation of endogenous cannabinoid anandamide in central neurons (see comments). Nature 372: 686–691.

Di Marzo, V., Goparaju, S. K., Wang, L., Liu, W., Bátkai, S., Járai, Z., Fezza, F., Miura, G. I., Palmiter, R. D., Sugiura, T., & Kunos, G. (2001) Leptin-regulated endocannabinoids are involved in maintaining food intake. Nature 410: 822–825.

Di Marzo, V., Hill, M. P., Bisogno, T., Crossman, A. R., & Brotchie, J. M. (2000c) Enhanced levels of endogenous cannabinoids in the globus pallidus are associated with a reduction in movement in an animal model of Parkinson's disease. FASEB J 14: 1432–1438.

Di Marzo, V., Melck, D., Bisogno, T., & De Petrocellis, L. (1998a) Endocannabinoids: Endogenous cannabinoid receptor ligands with neuromodulatory action. Trends Neurosci 21: 521–528.

Di Marzo, V., Sepe, N., De Petrocellis, L., Berger, A., Crozier, G., Fride, E., & Mechoulam, R. (1998b) Trick or treat from food endocannabinoids? (letter). Nature 396: 636–637.

di Tomaso, E., Beltramo, M., & Piomelli, D. (1996) Brain cannabinoids in chocolate (letter). Nature 382: 677–678.

Edgemond, W. S., Hillard, C. J., Falck, J. R., Kearn, C. S., & Campbell, W. B. (1998) Human platelets and polymorphonuclear leukocytes synthesize oxygenated derivatives of arachidonylethanolamide (anandamide): Their affinities for cannabinoid receptors and pathways of inactivation. Mol Pharmacol 54: 180–188.

Facci, L., Dal Toso, R., Romanello, S., Buriani, A., Skaper, S. D., & Leon, A. (1995) Mast cells express a peripheral cannabinoid receptor with differential sensitivity to anandamide and palmitoylethanolamide. Proc Natl Acad Sci USA 92: 3376–3380.

Felder, C. C., Briley, E. M., Axelrod, J., Simpson, J. T., Mackie, K., & Devane, W. A. (1993) Anandamide, an endogenous cannabimimetic eicosanoid, binds to the cloned human cannabinoid receptor and stimulates receptor-mediated signal transduction. Proc Natl Acad Sci USA 90: 7656–7660.

Felder, C. C., Joyce, K. E., Briley, E. M., Mansouri, J., Mackie, K., Blond, O., Lai, Y., Ma, A. L., & Mitchell, R. L. (1995) Comparison of the pharmacology and signal transduction of the human cannabinoid CB1 and CB2 receptors. Mol Pharmacol 48: 443–450.

Fride, E., Barg, J., Levy, R., Saya, D., Heldman, E., Mechoulam, R., & Vogel, Z. (1995) Low doses of anandamides inhibit pharmacological effects of delta 9-tetrahydrocannabinol. J Pharmacol Exp Ther 272: 699–707.

Fride, E., & Mechoulam, R. (1993) Pharmacological activity of the cannabinoid receptor agonist, anandamide, a brain constituent. Eur J Pharmacol 231: 313–314.

Ganley, O. H., Graessle, O., E., & Rathway, H. J. (1958) Anti-inflammatory activity of compounds obtained from egg yolks, peanut oil and soybean lecithin. J Lab Clin Med 51: 709–714.

Gerdeman, G., & Lovinger, D. M. (2001) CB1 cannabinoid receptor inhibits synaptic release of glutamate in the dorsolateral striatum. J Neurophysiol 85: 468–471.

Hampson, A. J., Bornheim, L. M., Scanziani, M., Yost, C. S., Gray, A. T., Hansen, B. M., Leonoudakis, D. J., & Bickler, P. E. (1998) Dual effects of anandamide on NMDA receptor-mediated responses and neurotransmission. J Neurochem 70: 671–676.

Hillard, C. J., & Campbell, W. B. (1997) Biochemistry and pharmacology of arachidonylethanolamide, a putative endogenous cannabinoid. J Lipid Res 38: 2383–2398.

Howlett, A. C., Mukhopadhyay, S., Shim, J. Y., & Welsh, W. J. (1999) Signal transduction of eicosanoid CB1 receptor ligands. Life Sci 65: 617–625.

Innis, S. M., & Dyer, R. (1997) Dietary triacylglycerols with palmitic acid (16:0) in the 2-position increase 16:0 in the 2-position of plasma and chylomicron triacylglycerols, but reduce phospholipid arachidonic and docosahexaenoic acids, and alter cholesteryl ester metabolism in formula-fed piglets. J Nutr 127: 1311–1319.

Izzo, A. A., Mascolo, N., & Capasso, F. (2000) Endocannabinoid system in the gut. Trends Pharmacol Sci 21: 372.

Járai, Z., Wagner, J., Varga, K., Lake, K., Compton, D., Martin, B., Zimmer, A., Bonner, T., Buckley, N., Mezey, E., Razdan, R., Zimmer, A., & Kunos, G. (1999) Cannabinoid-induced mesenteric vasodilatation through an endothelial site distinct from CB1 or CB2 receptors. Proc Natl Acad Sci USA 96: 14136–14141.

Katayama, K., Ueda, N., Kurahashi, Y., Suzuki, H., Yamamoto, S., & Kato, I. (1997) Distribution of anandamide amidohydrolase in rat tissues with special reference to small intestine. Biochim Biophys Acta 1347: 212–218.

Kuehl, F. A. J., Jacob, T. A., Ganley, O. H., Ormond, R. E., & Meisinger, M. A. P. (1957) The identification of an anti-inflammatory agent present in egg yolk, soy bean lecithin and peanut oil. J Am Chem Soc 79: 5577–5578.

Kunos, G., Járai, Z., Varga, K., Liu, J., Wang, L., & Wagner, J. A. (2000) Cardiovascular effects of endocannabinoids-the plot thickens. Prostaglandins Other Lipid Mediat 61: 71–84.

Kurahashi, Y., Ueda, N., Suzuki, H., Suzuki, M., & Yamamoto, S. (1997) Reversible hydrolysis and synthesis of anandamide demonstrated by recombinant rat fatty-acid amide hydrolase. Biochem Biophys Res Commun 237: 512–515.

Lake, K. D., Martin, B. R., Kunos, G., & Varga, K. (1997) Cardiovascular effects of anandamide in anesthetized and conscious normotensive and hypertensive rats. Hypertension 29: 1204–1210.

Leweke, F. M., Giuffrida, A., Wurster, U., Emrich, H. M., & Piomelli, D. (1999) Elevated endogenous cannabinoids in schizophrenia. Neuroreport 10: 1665–1669.

Maccarrone, M., Bari, M., Lorenzon, T., Bisogno, T., Di Marzo, V., & Finazzi-Agro, A. (2000a) Anandamide uptake by human endothelial cells and its regulation by nitric oxide. J Biol Chem 275: 13484–13492.

Maccarrone, M., Bari, M., Menichelli, A., Del Principe, D., & Agrò, A. F. (1999) Anandamide activates human platelets through a pathway independent of the arachidonate cascade. FEBS Lett 447: 277–282.

Maccarrone, M., Fiorucci, L., Erba, F., Bari, M., Finazzi Agrò, A., & Ascoli, F. (2000) Human mast cells take up and hydrolyze anandamide under the control of 5-lipoxygenase and do not express cannabinoid receptors. FEBS Lett 468: 176–180.

Maccarrone, M., Valensise, H., Bari, M., Lazzarin, N., Romanini, C., & Finazzi Agrò, A. (2000b) Relation between decreased anandamide hydrolase concentrations in human lymphocytes and miscarriage (see comments). Lancet 355: 1326–1329.

Maccarrone, M., van der Stelt, M., Rossi, A., Veldink, G. A., Vliegenthart, J. F., &

Agrò, A. F. (1998) Anandamide hydrolysis by human cells in culture and brain. J Biol Chem 273: 32332–32339.

Mackie, K., Devane, W. A., & Hille, B. (1993) Anandamide, an endogenous cannabinoid, inhibits calcium currents as a partial agonist in N18 neuroblastoma cells. Mol Pharmacol 44: 498–503.

Mallet, P. E., & Beninger, R. J. (1998) The cannabinoid CB1 receptor antagonist SR141716A attenuates the memory impairment produced by delta9-tetrahydrocannabin ol or anandamide. Psychopharmacology (Berl) 140: 11–19.

Mani, S. K., Mitchell, A., & Malley, B. W. (2001) Progesterone receptor and dopamine receptors are required in delta 9-tetrahydrocannabinol modulation of sexual receptivity in female rats. Proc Natl Acad Sci USA 98: 1249–1254.

Martin, B. R., Compton, D. R., Thomas, B. F., Prescott, W. R., Little, P. J., Razdan, R. K., Johnson, M. R., Melvin, L. S., Mechoulam, R., & Ward, S. J. (1991) Behavioral, biochemical, and molecular modeling evaluations of cannabinoid analogs. Pharmacol Biochem Behav 40: 471–478.

Matsuda, S., Kanemitsu, N., Nakamura, A., Mimura, Y., Ueda, N., Kurahashi, Y., & Yamamoto, S. (1997) Metabolism of anandamide, an endogenous cannabinoid receptor ligand, in porcine ocular tissues. Exp Eye Res 64: 707–711.

Maurelli, S., Bisogno, T., De Petrocellis, L., Di Luccia, A., Marino, G., & Di Marzo, V. (1995) Two novel classes of neuroactive fatty acid amides are substrates for mouse neuroblastoma "anandamide amidohydrolase". FEBS Lett 377: 82–86.

McGregor, I. S., Dastur, F. N., McLellan, R. A., & Brown, R. E. (1996) Cannabinoid modulation of rat pup ultrasonic vocalizations. Eur J Pharmacol 313: 43–49.

Mechoulam, R., Ben Shabat, S., Hanus, L., Fride, E., Vogel, Z., Bayewitch, M., & Sulcova, A. E. (1996) Endogenous cannabinoid ligands1/Mchemical and biological studies. J Lipid Mediat Cell Signal 14: 45–49.

Mechoulam, R., Ben Shabat, S., Hanus, L., Ligumsky, M., Kaminski, N. E., Schatz, A. R., Gopher, A., Almog, S., Martin, B. R., Compton, D. R., et al. (1995) Identification of an endogenous 2-monoglyceride, present in canine gut, that binds to cannabinoid receptors. Biochem Pharmacol 50: 83–90.

Mechoulam, R., Fride, E., Ben-Shabat, S., Meiri, U., & Horowitz, M. (1998a) Carbachol, an acetylcholine receptor agonist, enhances production in rat aorta of 2-arachidonoyl glycerol, a hypotensive endocannabinoid. Eur J Pharmacol 362: R1–R3.

Mechoulam, R., Fride, E., & Di Marzo, V. (1998b) Endocannabinoids. Eur J Pharmacol 359: 1–18.

Mechoulam, R., Fride, E., Hanus, L., Sheskin, T., Bisogno, T., Di Marzo, V., Bayewitch, M., & Vogel, Z. (1997) Anandamide may mediate sleep induction (letter). Nature 389: 25–26.

Mechoulam, R., Hanus, L., & Martin, B. R. (1994) Search for endogenous ligands of the cannabinoid receptor. Biochem Pharmacol 48: 1537–1544.

Mikawa, Y., Matsuda, S., Kanagawa, T., Tajika, T., Ueda, N., & Mimura, Y. (1997) Ocular activity of topically administered anandamide in the rabbit. Jpn J Ophthalmol 41: 217–220.

Monhemius, R., Azami, J., Green, D. L., & Roberts, M. H. T. (2001) CB1 receptor mediated analgesia from the Nucelus Reticularis Gigantocellularis pars alpha is activated in an animal model of neuropathic pain. Brain Res 908: 67–74.

Munro, S., Thomas, K. L., & Abu Shaar, M. (1993) Molecular characterization of a peripheral receptor for cannabinoids (see comments). Nature 365: 61–65.

Murillo-Rodríguez, E., Sánchez-Alavez, M., Navarro, L., Martínez-González, D., Drucker-Colín, R., & Prospéro-García, O. (1998) Anandamide modulates sleep and memory in rats. Brain Res 812: 270–274.

Paria, B. C., Ma, W., Andrenyak, D. M., Schmid, P. C., Schmid, H. H., Moody, D. E., Deng, H., Makriyannis, A., & Dey, S. K. (1998) Effects of cannabinoids on preimplantation mouse embryo development and implantation are mediated by brain-type cannabinoid receptors. Biol Reprod 58: 1490–1495.

Paria, B. C., Zhao, X., Wang, J., Das, S. K., & Dey, S. K. (1999) Fatty-acid amide hydrolase is expressed in the mouse uterus and embryo during the periimplantation period. Biol Reprod 60: 1151–1157.

Parolaro, D. (1999) Presence and funtional regulation of cannabinoid receptors in immune cells. Life Sci 65: 637–644.

Pate, D. W., Järvinen, K., Urtti, A., Jarho, P., Fich, M., Mahadevan, V., & Järvinen, T. (1996) Effects of topical anandamides on intraocular pressure in normotensive rabbits. Life Sci 58: 1849–1860.

Pate, D. W., Järvinen, K., Urtti, A., Jarho, P., & Järvinen, T. (1995) Ophthalmic arachidonylethanolamide decreases intraocular pressure in normotensive rabbits. Curr Eye Res 14: 791–797.

Pertwee, R. G. (1972) The ring test: A quantitative method for assessing the 'cataleptic' effect of cannabis in mice. Br J Pharmacol 46: 753–763.

Pertwee, R. G. (1997) Pharmacology of cannabinoid CB1 and CB2 receptors. Pharmacol Ther 74: 129–180.

Pertwee, R. G. (2001) Cannabinoids and the gastrointestinal tract. Gut 48: 859–867.

Pertwee, R. G., & Fernando, S. R. (1996) Evidence for the presence of cannabinoid CB1 receptors in mouse urinary bladder. Br J Pharmacol 118: 2053–2058.

Priller, J., Briley, E. M., Mansouri, J., Devane, W. A., Mackie, K., & Felder, C. C. (1995) Mead ethanolamide, a novel eicosanoid, is an agonist for the central (CB1) and peripheral (CB2) cannabinoid receptors. Mol Pharmacol 48: 288–292.

Raygada, M., Cho, E., & Hilakivi-Clarke, L. (1998) High maternal intake of polyunsaturated fatty acids during pregnancy in mice alters offsprings' aggressive behavior, immobility in the swim test, locomotor activity and brain protein kinase C activity. J Nutr 128: 2505–2511.

Richardson, J. D., Kilo, S., & Hargreaves, K. M. (1998) Cannabinoids reduce hyperalgesia and inflammation via interaction with peripheral CB1 receptors. Pain 75: 111–119.

Robbe, D., Alonso, G., Duchamp, F., Bockaert, J., & Manzoni, O. J. (2001) Localization and mechanisms of action of cannabinoid receptors at the glutamatergic synapses of the mouse nucleus accumbens. J Neurosci 21: 109–116.

Romero, J., García, L., Cebeira, M., Zadrozny, D., Fernández-Ruiz, J. J., & Ramos, J. A. (1995) The endogenous cannabinoid receptor ligand, anandamide, inhibits the motor behavior: Role of nigrostriatal dopaminergic neurons. Life Sci 56: 2033–2040.

Santucci, V., Storme, J. J., Soubrié, P., & Le Fur, G. (1996) Arousal-enhancing properties of the CB1 cannabinoid receptor antagonist SR 141716A in rats as assessed by electroencephalographic spectral and sleep-waking cycle analysis. Life Sci 58: 103–110.

Schmid, H. H., Schmid, P. C., & Natarajan, V. (1990) N-acylated glycerophospholipids and their derivatives. Prog Lipid Res 29: 1–43.

Schmid, P. C., Paria, B. C., Krebsbach, R. J., Schmid, H. H., & Dey, S. K. (1997) Changes in anandamide levels in mouse uterus are associated with uterine receptivity for embryo implantation. Proc Natl Acad Sci USA 94: 4188–4192.

Schuel, H., Burkman, L. J., Picone, R. P., & Makriyannis, A. (1998) Functional cannabinoid receptors in human sperm. ICRS Symposium 59.

Schuel, H., Goldstein, E., Mechoulam, R., Zimmerman, A. M., & Zimmerman, S. (1994) Anandamide (arachidonylethanolamide), a brain cannabinoid receptor agonist, reduces sperm fertilizing capacity in sea urchins by inhibiting the acrosome reaction. Proc Natl Acad Sci USA 91: 7678–7682.

Sepe, N., De Petrocellis, L., Montanaro, F., Cimino, G., & Di Marzo, V. (1998) Bioactive long chain N-acylethanolamines in five species of edible bivalve molluscs: Possible implications for mollusc physiology and sea food industry. Biochim Biophys Acta 1389: 101–111.

Sheskin, T., Hanus, L., Slager, J., Vogel, Z., & Mechoulam, R. (1997) Structural requirements for binding of anandamide-type compounds to the brain cannabinoid receptor. J Med Chem 40: 659–667.

Sinor, A. D., Irvin, S. M., & Greenberg, D. A. (2000) Endocannabinoids protect cerebral cortical neurons from in vitro ischemia in rats. Neurosci Lett 278: 157–160.

Smith, P. B., Compton, D. R., Welch, S. P., Razdan, R. K., Mechoulam, R., & Martin, B. R. (1994) The pharmacological activity of anandamide, a putative endogenous cannabinoid, in mice. J Pharmacol Exp Ther 270: 219–227.

Spinedi, A., Di Bartolomeo, S., & Piacentini, M. (1999) N-oleoylethanolamine inhibits glucosylation of natural ceramides in CHP-100 neuroepithelioma cells: Possible implications for apoptosis. Biochem Biophys Res Commun 255: 456–459.

Sprecher, H. (1999) An update on the pathways of polyunsaturated fatty acid metabolism. Curr Opin Clin Nutr Metab Care 2: 135–138.

Stefano, G. B., Liu, Y., & Goligorsky, M. S. (1996) Cannabinoid receptors are coupled to nitric oxide release in invertebrate immunocytes, microglia, and human monocytes. J Biol Chem 271: 19238–19242.

Stella, N., Schweitzer, P., & Piomelli, D. (1997) A second endogenous cannabinoid that modulates long-term potentiation. Nature 388: 773–778.

Sugiura, T., Kodaka, T., Kondo, S., Tonegawa, T., Nakane, S., Kishimoto, S., Yamashita, A., & Waku, K. (1996a) 2-Arachidonoylglycerol, a putative endogenous cannabinoid receptor ligand, induces rapid, transient elevation of intracellular free $Ca2^{+}$ neuroblastoma x glioma hybrid NG108-15 cells. Biochem Biophys Res Commun 229: 58–64.

Sugiura, T., Kodaka, T., Nakane, S., Kishimoto, S., Kondo, S., & Waku, K. (1998) Detection of an endogenous cannabimimetic molecule, 2-arachidonoylglycerol, and cannabinoid CB1 receptor mRNA in human vascular cells: is 2-arachidonoylglycerol a possible vasomodulator? Biochem Biophys Res Commun 243: 838–843.

Sugiura, T., Kodaka, T., Nakane, S., Miyashita, T., Kondo, S., Suhara, Y., Takayama, H., Waku, K., Seki, C., Baba, N., & Ishima, Y. (1999) Evidence that the cannabinoid CB1 receptor is a 2-arachidonoylglycerol receptor. Structure-activity relationship of 2-arachidonoylglycerol, ether-linked analogues, and related compounds. J Biol Chem 274: 2794–2801.

Sugiura, T., Kondo, S., Kodaka, T., Tonegawa, T., Nakane, S., Yamashita, A., Ishima, Y., & Waku, K. (1996b) Enzymatic synthesis of oleamide (cis-9, 10-octadecenoamide), an endogenous sleep-inducing lipid, by rat brain microsomes. Biochem Mol Biol Int 40: 931–938.

Sugiura, T., Kondo, S., Sukagawa, A., Nakane, S., Shinoda, A., Itoh, K., Yamashita, A., & Waku, K. (1995) 2-Arachidonoylglycerol: A possible endogenous cannabinoid receptor ligand in brain. Biochem Biophys Res Commun 215: 89–97.

Sugiura, T., Kondo, S., Sukagawa, A., Tonegawa, T., Nakane, S., Yamashita, A., Ishima, Y., & Waku, K. (1996c) Transacylase-mediated and phosphodiesterase-mediated synthesis of N-arachidonoylethanolamine, an endogenous cannabinoid-receptor ligand, in rat brain microsomes: Comparison with synthesis from free arachidonic acid and ethanolamine. Eur J Biochem 240: 53–62.

Sugiura, T., Kondo, S., Sukagawa, A., Tonegawa, T., Nakane, S., Yamashita, A., & Waku, K. (1996d) N-arachidonoylethanolamine (anandamide), an endogenous cannabinoid receptor ligand, and related lipid molecules in the nervous tissues. J Lipid Mediat Cell Signal 14: 51–56.

Sulcova, E., Mechoulam, R., & Fride, E. (1998) Biphasic effects of anandamide. Pharmacol Biochem Behav 59: 347–352.

Terranova, J. P., Storme, J. J., Lafon, N., Pério, A., Rinaldi-Carmona, M., Le Fur, G., & Soubrié, P. (1996) Improvement of memory in rodents by the selective CB1 cannabinoid receptor antagonist, SR 141716. Psychopharmacology (Berl) 126: 165–172.

Ueda, N., Yamamoto, K., Yamamoto, S., Tokunaga, T., Shirakawa, E., Shinkai, H., Ogawa, M., Sato, T., Kudo, I., Inoue, K., et al. (1995) Lipoxygenase-catalyzed oxygenation of arachidonylethanolamide, a cannabinoid receptor agonist. Biochim Biophys Acta 1254: 127–134.

Varga, K., Lake, K., Martin, B. R., & Kunos, G. (1995) Novel antagonist implicates the CB1 cannabinoid receptor in the hypotensive action of anandamide. Eur J Pharmacol 278: 279–283.

Varga, K., Lake, K. D., Huangfu, D., Guyenet, P. G., & Kunos, G. (1996) Mechanism of the hypotensive action of anandamide in anesthetized rats. Hypertension 28: 682–686.

Varga, K., Wagner, J. A., Bridgen, D. T., & Kunos, G. (1998) Platelet- and macrophage-derived endogenous cannabinoids are involved in endotoxin-induced hypotension. FASEB J 12: 1035–1044.

Vaughan, C. (2001) Cannabis: A pot of cellular actions in the mesolimbic system. Trends Pharmacol Sci 22: 110–111.

Wagner, J. A., Varga, K., Ellis, E. F., Rzigalinski, B. A., Martin, B. R., & Kunos, G. (1997) Activation of peripheral CB1 cannabinoid receptors in haemorrhagic shock. Nature 390: 518–521.

Williams, C. M., & Kirkham, T. C. (1999) Anandamide induces overeating: mediation by central cannabinoid (CB1) receptors. Psychopharmacology 143: 315–317.

Wilson, R. I., & Nicoll, R. A. (2001) Endogenous cannabinoids mediate retrograde signalling at hippocampal synapses. Nature 410: 588–591.

Yehuda, S., Rabinovitz, S., & Mostofsky, D. I. (1999) Essential fatty acids are mediators of brain biochemistry and cognitive functions. J Neurosci Res 56: 565–570.

Zimmer, L., Delion-Vancassel, S., Durand, G., Guilloteau, D., Bodard, S., Besnard, J.-C., & Chalon, S. (2000) Modification of dopamine neurotransmission in the nucleus accumbens of rats deficient in n-3 polyunsaturated fatty acids. J Lipid Res 41: 32–40.

Zygmunt, P. M., Högestätt, E.D., Waldeck, K., Edwards, G., Kirkup, A. J., & Weston, A. H. (1997) Studies on the effects of anandamide in rat hepatic artery. Br J Pharmacol 122: 1679–1686.

13

Obesity and Energy Regulation

Kevin J. Acheson
Nestlé Research Center, Lausanne, Switzerland

Luc Tappy
Physiology Institute, Lausanne, Switzerland

I. INTRODUCTION

There is little doubt that the prevalence of obesity is increasing worldwide and recent reports emphasize that it is increasing at an alarming rate. In the United States the prevalence of obesity increased from 12% to 18% during the period 1991 to 1998 (1), an increase of 50%! In Europe the prevalence of obesity is also high (2,3), with values of 19% in the United Kingdom (4,5), which is almost double that observed in the Netherlands and France (3,6). Although it could be argued that this is a consequence of the aging of the population in Europe and America, obesity in childhood and adolescence is also increasing and is now of growing concern (7–9). In fact, obesity is considered to be the most prevalent disease/metabolic disorder of children and adolescents in the United States (10). Unfortunately this dramatic increase in the prevalence of obesity is not confined to industrialized countries but is pandemic, with evidence of obesity in urban areas of many developing countries experiencing rapid economic growth (11–15).

Obesity has not always been associated with disease and a diminished quality of life. In the past, some societies considered obesity or an ample girth to be favorable, indicating affluence, a plentiful food supply, and even health (16). However, with changing fashion and, perhaps more importantly, the realization that obesity is a strong risk factor for a variety of diseases that can substantially decrease the length and quality of life (Table 1), individuals are now searching for ways to lose excess body fat or prevent its accumulation.

Table 1 Adverse Health Consequences of Obesity (Comorbidities)

Cardiovascular disease
Impaired glucose tolerance
Hyperinsulinemia
Type II diabetes
Hyperlipidemia
Hypertension
Gallbladder disease
Musculoskeletal disorders, osteoarthritis
Some cancers
Sleep apnea

It is now generally accepted that obesity is classified according to an individual's body mass index (BMI), which can be calculated easily from body weight and body height (17). However, this index takes no account of an individual's actual adiposity (18,19) and should occasionally be used with caution, particularly when some athletes or ethnic groups, with smaller body frames than Caucasians (20), are being assessed.

Epidemiological studies relating BMI to mortality have shown that the risk follows a J-shaped curve (21). The risk of mortality is moderate to high,

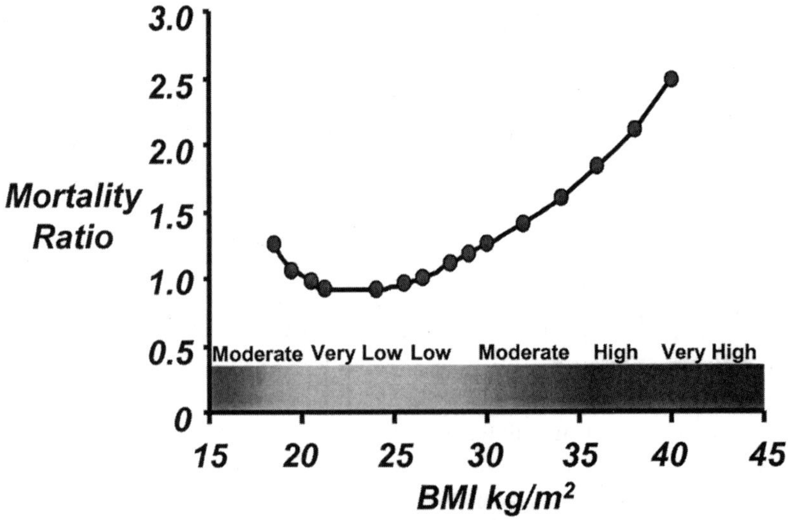

Figure 1 Relationship between body mass index and mortality risk.

below 19 kg/m^2, as a result of undernutrition and illness; is relatively low in the range between 20 kg/m^2 and 28 kg/m^2 (Fig. 1); but rises progressively once the BMI exceeds 30 kg/m^2, not only as a result of obesity itself but also the contribution of comorbidities listed in Table 1.

This is supported by recent data that have estimated that approximately 30,000, or 6%, of deaths registered in England in 1998 (5) and 13% of deaths registered in the United States in 1991 could be attributed to obesity (22). The direct and indirect costs of obesity have also been estimated to be in the region of £0.5 billion and £2 billion per year in the United Kingdom (5) and $52 billion and $99 billion, respectively, in 1995 in the United States (23).

With this background in mind it is no wonder that considerable resources are being invested in the search to prevent, palliate, as well as find remedies for obesity and its associated diseases.

II. CLASSICAL TREATMENTS OF OBESITY

The remedies proposed over the years for losing body weight, or more precisely body fat, are exhaustive and have essentially concentrated on traditional methods aimed at manipulating food intake and/or increasing physical activity. Many dietary regimens have been proposed with various permutations on the three major energy-providing components of the diet (24), and although they may work for a few individuals, the fact that overweight and obesity continue to rise is evidence that they do not work for the vast majority who wish to lose and maintain their lost weight. Increasing physical activity increases energy expenditure and provides other health benefits; however, when one calculates the amount of exercise that must be performed to lose a few grams of fat it must appear a daunting task for someone who has 20 to 50 kg of excess fat to lose (25). In reality it is far easier, if one has the strength of character, to resist consuming a 300-kcal snack than to exercise for 30 min to 1 hr, to "burn off" the same amount of energy. The observations that most weight-reducing programs have a low rate of success indicate that both diet and exercise have limited efficiency. Failing, or in addition to, dietary management and physical exercise, an individual can also use pharmaceutical aids to help control body weight. However, the number of drugs that are presently approved for the treatment of obesity are relatively few (26). In the past a variety of drugs that influence appetite, energy expenditure, or both were used. These included thyroid hormones (27,28), amphetamines (26), dinitrophenol (29), and, more recently, fenfluramine and dexfenfluramine (30); however, all have been found to cause serious side effects and are considered inappropriate for the pharmacological treatment of obesity. At the present time Sibutramine, a serotonin and norepinephrine uptake inhibitor, which reduces appetite (28) and stimulates thermogenesis (31), and Orlistat, an intestinal lipase inhibitor that inhibits intestinal fat digestion, are in use (28,32). However, a 2002 metaanalysis of clinical trials using antiobesity treatments concluded that no drug or class of drugs demonstrated clear superiority as an obesity medication (33). Although new drugs are being investigated (32), it may be many years before they are proved effective and considered safe. As a final resort, usually reserved for massively obese patients for whom all other treatments have failed, some surgical options have been proposed, such as jaw wiring (34,35), gastric bypass (36), stapled gastroplasty (37), and gastric banding (38). These interventions have been reported to be successful as far as weight loss and health-related quality of life are concerned (39). However, what might be considered more invasive procedures such as coagulation of the feeding centers in the brain (40) have failed to produce sustained weight loss!

III. ENERGY BALANCE

Humans and animals are capable of maintaining a stable body weight over very long periods, and for this reason many believe that body weight stability is under precise regulatory control. Discussions on the cause of obesity often resort to the energy balance equation, which reiterates the first law of thermodynamics, that energy cannot be created or destroyed, and states that changes in energy balance of the body are the result of differences between energy intake and energy expenditure. Although in essence this is true, it has led to some confusion inasmuch as investigators have perhaps concentrated too much on energy, as a general term, rather than the specific energy providing components of our food and the energetic substrates of our body (41). It is now known that there are biomarkers that reflect the state of the different stores in the body and that they are under neural and hormonal feedback control. Proof of this was provided several years ago with the discovery of leptin (42), a protein secreted by the obese gene, which has been demonstrated to have effects on both appetite control (43) and energy expenditure (44), as well as a variety of other functions.

Chemoreceptors sensitive to circulating molecules released from the body's energy stores signal the brain to modify food intake, energy expenditure, and the composition of substrate oxidation (45). A variety of signals such as insulin (46), blood glucose (47), free fatty acids that reflect the body fat stores (45), or amino acid concentrations (48) have been proposed to influence food intake and satiety (49).

Most animals, including humans, have evolved in conditions of periodic food shortages and plenty that follow the seasons, droughts, and floods. For this reason the body has become well adapted to conserving energy during starvation and laying down stores rapidly in times of plenty. Today, with food security established in many countries, the body no longer has to deal with food shortages and starvation but must now adapt to conditions in which a plentiful food supply is always available. In this "obesigenic" environment (50) the human body is ill equipped to expend more energy than it consumes, unless an individual makes a conscious effort to do so. In spite of this plethora of food many individuals do manage to remain slim, but unfortunately a large proportion do not.

Although the pharmaceutical industry is investing in research for antiobesity drugs it will be many years before safe and effective treatments are available. Consequently few options are available at the present time, other than prevention or reduction of obesity by healthy eating and lifestyle changes.

IV. ARE FUNCTIONAL FOODS FOR WEIGHT CONTROL AVAILABLE?

To date there are no effective functional foods for the prevention and treatment of obesity and long-term maintenance of reduced body weight. However, there are various steps in energy assimilation and metabolism that are potential targets for present and future functional foods.

V. NUTRIENT INTAKE

It is known that energy intake is controlled by many peripheral and central factors, and it has been shown that animals are able to regulate food intake very well in their natural environment. Humans, on the other hand, experience numerous social and psychological cues that can override normal physiological signals of satiety and satiation. Nutrient intake is regulated by short-term satiety factors from the gastrointestinal tract such as gastric distension, amino acids, and peptide hormones, as well as those, such as leptin, that are involved in more long-term appetite regulation. Gut peptides, cholecystokinin, and GLP1 are known to be involved in food intake control, whereas leptin, which is secreted by adipocytes, influences body weight via its effects on both appetite and cellular thermogenesis (51). Plasma leptin concentrations are proportional to body fat mass (52); thus any increase in fat mass increases leptin concentrations and serves as a negative feedback signal to decrease appetite, by inhibition of NPY, and to increase cellular thermogenesis. Unfortunately obese individuals have lower leptin concentrations in cerebrospinal fluid than the lean (53); thus its uptake or appearance in cerebrospinal fluid would appear to be defective in obesity and may contribute to the decreased sensitivity of the brain to leptin in this condition (53,54). Leptin gene expression is stimulated by insulin, which itself signals adiposity to the brain (55); therefore, any foods that increase insulin, such as carbohydrates and proteins, influence feeding behavior via the effect of insulin per se and its effect on leptin. However, it is not known whether specific nutrients regulate leptin secretion and/or its sensitivity in the brain.

VI. NUTRIENT DIGESTION AND ABSORPTION

Numerous low-energy sweeteners and artificial fat substitutes have been designed and produced. Although some reports demonstrate modest weight losses and improved plasma lipid profiles (56–59), the substitutes have not, so far, proved effective in the battle against obesity. Olestra, a nondigestible fat, is used in foods to provide the taste and texture of fat while denying the body its high-energy value. To date Olestra has only been approved for use in savory snacks, and although there were questions about its effect on absorption of fat-soluble vitamins and anal leakage, these concerns have been overcome. Since Olestra does not represent a large proportion of dietary fat intake, its effect on fat-soluble vitamins is limited, and during development its formulation has been modified to prevent the the problem of anal leakage.

Salatrim is a reduced-fat product that is not just one fat, but a family of triglycerides, which are produced by assembling different combinations of the short-chain, acetic, propionic, and butyric fatty acids with the long-chain stearic acid on the glycerol moiety. With these combinations it is possible to decrease the digestible energy value of fat from 9 to 5 kcal/g (60). Recently, another oil, Econa oil, was introduced in Japan and the United States. It contains a large proportion of diacylglycerol (DAG) that is processed from canola and soy oils. It is claimed that regular use of DAG causes weight loss (61,62), since it is directly oxidized rather than being stored in adipose tissue, and that it also reduces blood cholesterol concentrations.

Slowing or reducing the digestibility of carbohydrates by using slowly digested carbohydrates and fiber-containing foods has demonstrated functional properties. Slowly digested carbohydrates reduce plasma glucose and insulin concentrations (63–67), improve insulin sensitivity (68), and have been repoted to have satiating effects (69–72). Dietary fibers provide bulk, slow the digestion and absorption of carbohydrates, and at the same time provide less energy than well digested carbohydrates. In addition to their positive effects on glucose metabolism, many fibers are fermented in the large intestine to produce short-chain fatty acids, which provide energy and beneficial effects for the gastric mucosa (69). Although dietary guidelines advocate eating a high-fiber diet and increasing the consumption of fruits and vegetables, there is no indication that these recommendations have been widely accepted by the general population.

VII. THERMOGENESIS

Although much work has been done on ingredients and foods that increase thermogenesis, it should be realized that none stimulates energy expenditure sufficiently and for long enough periods to have a significant effect on energy balance without also producing side effects. As mentioned, it is perhaps far easier and safer to reduce energy intake by 300 kcal per day than either to exercise or to stimulate resting energy expenditure by 12% to 15% for 24 hr.

Coffee (73) and green tea (74) are habitually consumed drinks that have been reported to increase thermogenesis over relatively short periods. Much of the thermic effect of coffee is due to its caffeine content, which has been known since the beginning of the 20th century to increase cellular thermogenesis (75,76). However, with green tea, it has been proposed that the catechin-polyphenols it contains interact synergistically with caffeine to increase and prolong sympathetic stimulation of thermogenesis and lipolysis (77). Others have reported that epigallocatechin gallate, extracted from green tea, decreased body weight and waist circumference of moderately obese patients during a 3-month trial (78) and that this effect was due to inhibition of gastric and pancreatic lipases as well as stimulation of thermogenesis.

Potential synergistic actions of plant-derived active molecules have already been tested for weight control. Caffeine in combination with ephedrine, a sympathomimetic derived from the plant *Ephedra sinica*, has been shown to be effective in improving (24 weeks) and maintaining weight loss (a further 24 weeks) of obese patients (79). Although indiscriminate use without medical supervision can be potentially dangerous, a 2002 multicenter trial confirmed the efficacy of herbal ephedra/caffeine in promoting weight loss, fat loss, and improvement in blood lipid concentrations without significant adverse events (80).

VIII. LIPID OXIDATION

Pragmatically, since obesity is defined as an excess of body fat, an obvious target would be the adipose tissue stores of the body. Although increased lipid mobilization and oxidation of any mobilized lipid are advantageous, it has been known for several years that different fat depots in the body are more detrimental to health than others. Accumulation of fat in the visceral abdominal region, rather than the gluteal, femoral depots, is associated with increased risk for ischemic heart disease, stroke, and the metabolic syndrome (81,82). Differences in body fat distribution are thought to be due to an imbalance of steroid hormones and pertubations of the hypothalamic-pituitary-adrenal and pituitary-gonadal axes, which are amplified by stress conditions (83). However, it is not known whether foods or food ingredients can influence these endocrine changes or their consequences.

IX. TRIACYLGLYCEROL STORAGE AND MOBILIZATION

The major steps involved in triacylglycerol (TAG) storage and mobilization are illustrated in Fig. 2. It is well known that the sympathetic nervous system has a profound effect on substrate utilization by the body. The hormones epinephrine and norepinephrine not only stimulate glycogen breakdown but also increase lipolysis and energy expenditure. In adipose tissue, stimulation of the β-adrenergic receptor by norepinephrine (which is released from sympathetic nerve terminals and has the main lipolytic effect) or epinephrine (from the systemic circulation, which has limited lipolytic effect at physio-

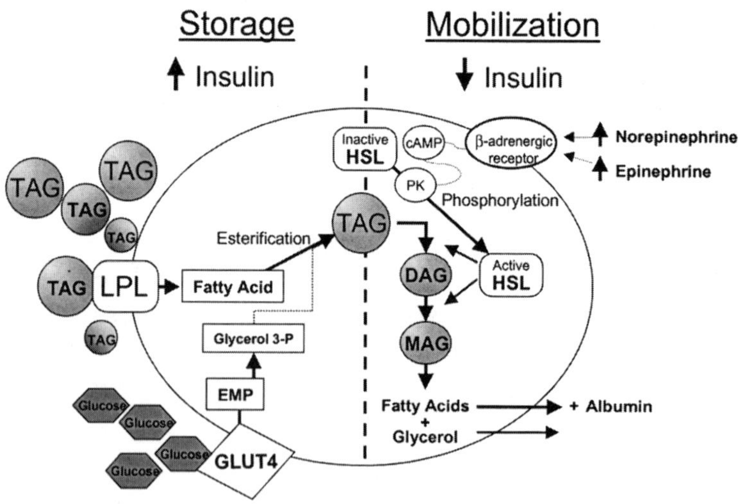

Figure 2 Major steps involved in the storage and mobilization of triacylglycerol in adipose tissue. TAG, triacylglycerol; LPL, lipoprotein lipase; GLUT 4, glucose transporter; EMP, Embden-Meyerhof pathway; glycerol 3-P, glycerol-3-phosphate; HSL, hormone-sensitive lipase; cAMP, cyclic adenosine monophosphate; PK, pyruvate kinase; DAG, diacylglycerol; MAG, monoacylglycerol.

logical concentrations) increases intracellular cyclic adenosine monophosphate (cAMP) concentrations. This in turn stimulates protein kinase (PK) to phosphorylate inactive hormone-sensitive lipase (HSL) into its active, phosphorylated form. Active HSL, together with monoacylglycerol lipase, hydrolyze triacylglycerol to produce three fatty acids and a glycerol moiety, which are released from the adipocyte. The fatty acids bind to albumin and are transported to other tissues, where they are subsequently oxidized or transported back to adipose tissue, where they are taken up and reesterified. However, the lipolytic effect of catecholamines only occurs in the presence of low insulin concentrations. In the presence of high insulin concentrations, any fatty acids produced intracellularly are taken up, without leaving the adipocyte, and are reesterified with glycerol 3-phosphate to TAG. High insulin concentrations also stimulate intracellular phosphodiesterase, which converts cyclic AMP to AMP and prevents the production of active HSL. Thus insulin not only prevents intracellular lipid mobilization but also encourages reesterification of any fatty acids present in the cell to the stable TAG storage form. This brief description and Fig. 2 suggest that it is theoretically possible to intervene at a number of sites of lipid metabolism to reduce lipid storage and encourage increased lipid mobilization. However, whether the mobilized lipids are subsequently oxidized depends very much on the energy requirements of the body at that moment in time.

Lipolysis is very sensitive to plasma insulin concentrations and is inhibited by much lower concentrations of insulin than are required to stimulate glucose transport (84); thus any foods that elicit a small insulin response limit its antilipolytic effects and facillitate lipid mobilization and oxidation. Formation of intracellular cyclic AMP is inhibited by stimulation of α_2-adrenergic and adenosine receptors and is broken down by phosphodiesterase. Therefore, inhibition of these receptors and phosphodiesterase activity allows lipolysis to proceed. Caffeine can increase lipolysis by stimulating the sympathetic nervous system (85) and inhibiting both adenosine and phosphodiesterase. More recently it has been proposed that dietary calcium regulates energy metabolism and lipolysis (86) by modulating intracellular phosphodiesterase and fatty acid synthase activities (87,88). High intracellular calcium concentrations, as a result of low-calcium diets and raised concentrations of 1,25 dihydroxyvitamin D (dihydroxycholecalciferol), increases adipocyte phosphodiesterase activity. This in turn reduces cAMP, decreases active HSL, and inhibits lipolysis. At the same time fatty acid synthase expression and activity are increased. Together, these two mechanisms favor lipid storage. Conversely, high-calcium diets decrease intra-adipocyte calcium flux, inhibiting fat synthesis and stimulating lipolysis (88). Whether dietary calcium also influences thermogenesis is not known; however, Zemel and associates observed increases in core temperature of mice receiving high-calcium diets, which they interpreted as a shift in efficiency from energy storage to thermogenesis (88). On the other hand, it has also been established that high dietary calcium intakes decrease the digestibility of dietary lipids, because of the formation of indigestible calcium soaps in the gastrointestinal tract, and increase fecal fat excretion (89,90). Addition of relatively small amounts of calcium to the diet also resulted in significant reductions of total cholesterol (89) and low-density lipoprotein (LDL) cholesterol (89,90). Regardless of whether dietary calcium influences lipolysis, thermogenesis, lipid absorption, or combinations of the three, recent epidemiological studies tend to support the positive effects of calcium and dairy products: Children with higher intakes of calcium and monounsaturated fat and more servings of dairy products had lower body fat (91), and the CARDIA study, investigating risk factors associated with type 2 diabetes and the insulin resistance syndrome (IRS), noted that there was a strong inverse association between increased dairy consumption and IRS among overweight adults (92).

The fatty acids mobilized from adipose tissue can be taken up by various organs or tissues to be oxidized as energetic fuels or be reesterified to triglyceride in the liver, to be released as very-low-density lipoprotein TG (VLDL-TG). Mobilization of fatty acid from adipose tissue is unlikely to lead to substantial weight loss if it is subsequently reesterified, and hence it is likely that factors promoting cellular fatty acid oxidation are required. Recent evidence indicates that transport of fatty acids inside the cell is regulated by several fatty acid binding proteins (FAT/CD36, FATP) and intracellular fatty acid binding proteins (FABPs). Once fatty acids are inside the cells, their oxidation is regulated at the leve of the entrance of fatty acyl coenzyme A (CoA) into the mitochondria, a process operated by CPT-1. Malonyl CoA is a key regulator of this process: When glucose metabolism is high, malonyl CoA concentrations increase and inhibit CPT-1 and lipid oxidation. Inside the cell, there are proteins that are able to sense the energy status: AMPK is a protein kinase activated when AMP concentrations are high, as is the case when the cell suffers a shortage of energy. This protein phosphorylates and deactivates acetyl Co A carboxylase (ACC), thus suppressing the biosynthesis of malonyl CoA and promoting fat oxidation (93).

Apart from its effects on food intake, leptin is recognized to stimulate lipid oxidation by activating AMPK and inhibiting ACC in several cell types, including skeletal muscle, and may contribute therefore to weight loss by promoting fat oxidation (94). It is also recognized that activation of PPARα by pharmacological ligands, such as fibrates, promotes lower plasma triglyceride concentrations by stimulating lipid oxidation in liver cells, and possibly in other cell types as well.

There is presently little information available regarding the effects of nutrients on leptin synthesis and intracellular ACC activity. Since ACC is primarily regulated by insulin, it can be expected, however, that nutrients that keep insulin levels low tend to decrease ACC activity and malonyl-CoA formation and hence stimulate lipid oxidation. Whether some types of nutrients modulate AMPK activity remains to be investigated.

Although dietary fat is believed to be the major culprit for weight gain and obesity, not all fats are detrimental to health. Literature is accumulating about the advantages of consuming the ω3 fatty acids eicosapentanoic acid (EPA 22:5ω3) and docosahexaenoic acid (DHA 22:6ω3), which protect against cardiovascular disease (95) and obesity (96) by favoring fatty acid oxidation and inhibiting fat storage (97). The metabolic effects of ω3 fatty acids appear to be mediated directly, by inhibiting arachadonic acid metabolism and replacing it as an eicosanoid substrate, and indirectly, by influencing transcription factors such as nuclear factor kappa B (NFκB) and peroxisome proliferator-activated receptors (PPARs) (98–100). Long-chain fatty acids or their fatty acid derivatives are endogenous ligands for PPARα, which is stimulated by ω3 long-chain fatty acids (101), and hence may promote weight loss by increasing fatty acid oxidation. A systematic study of the effect of various fat sources on lipid oxidation is, however, awaited.

X. CONCLUSIONS

Throughout history an ample body weight has been associated with fertility and, in some societies, has been considered an indication of wealth. However, with the change in interest to a more healthy and prolonged life, for which obesity is not an advantage, we have been searching for ways, with little success so far, to avoid, prevent, or reduce its deleterious effects. Although obesity was initially treated more as a cosmetic problem, the financial burden of treating obesity and its related disorders today has pushed it to the forefront of government and public attention. Many strategies have been proposed to prevent or correct obesity, but all, except gastroplastry, have met with with very modest or limited success. It is hoped that with a better understanding of the complex mechanisms involved in energy homeostasis, provided by continued research and gene technology, more successful treatments will be available in the future.

REFERENCES

1. AH Mokdad, MK Serdula, WH Dietz, BA Bowman, JS Marks, JP Koplan. The spread of the obesity epidemic in the United States, 1991–1998. JAMA 282:1519–1522, 1999.

2. BL Heitmann. Ten-year trends in overweight and obesity among Danish men and women aged 30–60 years. Int J Obes Relat Metab Disord 24:1347–1352, 2000.
3. JC Seidell. Obesity, insuln resistance and diabetes—a worldwide epidemic. Br J Nutr 83:S5–S8, 2000.
4. Joint Health Surveys Unit on behalf of the Department of Health. Health Survey of England: Cardiovascular Disease '98. London: The Stationery Office, 1999.
5. Report by the Comptroller and Auditor General. Tackling Obesity in England. London: The Stationery Office, 2001.
6. A Basdevant. Obesity: Epidemiology and public health. Ann Endocrinol (Paris) 61(Suppl6):6–11, 2000.
7. RP Troiano, KM Flegal. Overweight children and adolescents: Description, epidemiology, and demographics. Pediatrics 101:497–504, 1998.
8. G Frühbeck. Childhood obesity. BMJ 320:1401–1401, 2000.
9. RP Troiano, KM Flegal, RJ Kuczmarski, SM Campbell, CL Johnson. Overweight prevalence and trends for children and adolescents: The National Health and Nutrition Examination Surveys, 1963 to 1991. Arch Pediatr Adolesc Med 149:1085–1091, 1995.
10. WH Dietz. Health consequences of obesity in youth: Childhood predictors of adult disease. Pediatrics 101:518–525, 1998.
11. JC Seidell, A Rissanen. Time trends in the worldwide prevalence of obesity. In: GA Bray, C Bouchard, WPT James, eds. Handbook of Obesity, New York: Marcel Dekker, 1998, pp 79–79.
12. RB Singh, S Ghosh, R Beegom, AS Mehta, AK De, M Haque, GK Dube, GS Wander, S Kundu, S Roy, A Krishnan, H Simhadri, NB Paranjpe, N Agarwal, RH Kalikar, SS Rastogi, AS Thakur. Prevalence and determinants of central obesity and age-specific waist:hip ratio of people in five cities: The Indian Women's Health Study. J Cardiovasc Risk 5:73–77, 1998.
13. Y Wong, YC Huang. Obesity concerns, weight satisfaction and characteristics of female dieters: A study on female Taiwanese college students. J Am Coll Nutr 18:194–200, 1999.
14. AA Musaiger, HM Radwan. Social and dietary factors associated with obesity in university female students in United Arab Emirates. J R Soc Health 115:96–99, 1995.
15. F al-Mahroos, K al-Roomi. Overweight and obesity in the Arabian Peninsula: an overview. J R Soc Health 119:251–253, 1999.
16. L Grivetti. Psychology and cultural aspects of energy. Nutr Rev 59:S5–S12, 2001.
17. JS Garrow, J Webster. Quetelet's index (W/H2) as a measure of fatnesss. Int J Obes 9:147–153, 1985.
18. WM Chumlea, SS Guo. Assessment and prevalence of obesity: Application of new methods to a major problem. Endocrine 13:135–142, 2000.
19. ME Lean. Pathophysiology of obesity. Proc Nutr Soc 59:331–336, 2000.
20. GT Ko, J Tang, JC Chan, R Sung MM, Wu HP Wai, R Chen. Lower BMI

cut-off value to define obesity in Hong Kong Chinese: An analysis based on body fat assessment by bioelectrical impedance. Br J Nutr 85: 239–242, 2000.
21. EA Lew, L Garfinkel. Variations in mortality by weight among 750,000 men and women. J Chronic Dis 32:563–576 1979.
22. DB Allison, KR Fontaine, JE Manson, J Stevens, TB Vanltallie. Annual deaths attributable to obesity in the United States. JAMA 282:1530–1538, 1999.
23. AM Wolf, GA Colditz. Current estimates of the economic cost of obesity in the United States. Obes Res 6:97–106, 1998.
24. J Dwyer. Sixteen popular diets: Brief nutritional analyses. In AJ Stunkard, ed. Obesity. Philadelphia: WB Saunders, 1980, pp 276–291.
25. JS Garrow. Exercise in the treatment of obesity: A marginal contribution. Int J Obes Relat Metab Disord 19(Suppl 4):126–129, 1995.
26. GA Bray. A concise review on the therapeutics of obesity. Nutrition 16:953–960, 2000.
27. JS Garrow. Energy balance and obesity in man. Amsterdam:North-Holland, 1974, pp 335.
28. GA Bray, DA Fisher, IJ Chopra. Relation of thyroid hormones to body weight. Lancet 1:1206–1208, 1976.
29. M Wangness. Pharmacological treatment of obesity: Past, present and future. Minn Med 83:21–26, 2000.
30. JM Gardin, D Schumacher, G Constantine, KD Davis, C Leung, CL Reid. Valvular abnormalities and cardiovascular status following exposure to dexfenfluramine or phentermine/fenfluramine. JAMA 283:1730–1709, 2000.
31. A Astrup. Thermogenic drugs as a strategy for treatment of obesity. Endocrine 13:207–212, 2000.
32. J Proietto, BC Fam, DA Ainslie, AW Thorburn. Novel anti-obesity drugs. Expert Opin Invest Drugs 9:1317–1326, 2000.
33. CK Haddock, WSC Poston, PL Dill, JP Foreyt, M Ericsson. Pharmacotherapy for obesity: A quantitative analysis of four decades of published randomized clinical trials. Int J Obes 26:262–273, 2002.
34. JS Garrow. Dental splinting in the treatment of hyperphagic obesity. Proc Nutr Soc 33:29A, 1974.
35. S Rodgers, R Burnet, A Goss, P Phillips, R Goldney, C Kimber, D Thomas, P Harding, P Wise. Jaw wiring in treatment of obesity. Lancet 1(8024):1221–1222, 1977.
36. JG Kral. Surgical interventions for obesity. In: KD Brownell, CG Fairburn, eds. Eating Disorders and Obesity. New York: Guilford Press, 1995, pp 510–515.
37. EE Mason. Vertical banded gastroplasty for morbid obesity. Arch Surg 117: 701–706, 1982.
38. M Belachew, M Legrand. Update on laparoscopic surgery for morbid obesity. In: B Guy-Grand, G Ailhaud, eds. Progress in Obesity Research, Vol. 8. London: John Libbey, 1999, pp 789–793.
39. J Karlsson, L Sjöstöm, M Sullivan. Swedish obese subjects (SOS)—an inter-

vention study of obesity: Two year follow-up of health-related quality of life (HRQL) and eating behaviour after gastric surgery for severe obesity. Int J Obes Relat Metab Disord 22:113–126, 1998.

40. F Quaade, K Vaernet, S Larsson. Stereotaxic stimulation and electrocoagulation of the lateral hypothalamus in obese humans. Acta Neurochiurgica 30:111–117, 1974.
41. JP Flatt. How not to approach the obesity problem. Obes Res 5:632–633, 1997.
42. Y Zhang, P Proenca, M Maffei, M Barone, L Leopold, JM Friedman. Positional cloning of the mouse obese gene and its homologue. Nature 372:425–432, 1994.
43. JC Halford, JE Blundell. Separate systems for serotonin and leptin in appetite control. Ann Med 32:222–232, 2000.
44. AM Mistry, AG Swick, DR Romsos. Leptin rapidly lowers food intake and elevates metabolic rates in lean and ob/ob mice. J Nutr 127:2065–2072, 1997.
45. GC Kennedy. The role of depot fat in the hypothalamic control of food intake in the rat. Proc R Soc Lond 140:579–592, 1953.
46. SC Woods, EC Lotter, LD McKay, D Porte Jr. Chronic intracerebroventricular infusion of insulin reduces food intake and body weight of baboons. Nature 282:503–505 1979.
47. J Mayer. Glucostatic mechanism of regulation of food intake. New Engl J Med 249:13–16, 1953.
48. S Mellinkoff, S Frankland, D Boyle, M Greipel. Relationship between serum amino acid concentration and fluctuations in appetite. J Appl Physiol 8:535–538, 1956.
49. W Langhans, E Scharrer. Metabolic control of eating. World Rev Nutr Diet 70:1–67, 1992.
50. E Ravussin, C Bouchard. Human genomics and obesity: Finding appropriate drug targets. Eur J Pharmacol 410:131–145, 2000.
51. LA Campfield, FJ Smith. Overview: Neurobiology of OB protein (leptin), Proc Nutr Soc 57:429–440, 1998.
52. RV Considine, MK Sinha, ML Heiman, A Kriauciunas, TW Stephens, MR Nyce, JP Ohannesian, CC Marco, LJ McKee, TL Bauer, JF Caro. Serum immunoreactive-leptin concentrations in normal-weight and obese humans. N Engl J Med 334:292–295, 1996.
53. MW Schwartz, E Peskind, M Raskind, EJ Boyko, D Porte Jr. Cerebrospinal fluid leptin levels: Relationship to plasma levels and to adiposity in humans. Nat Med 2:589–593, 1996.
54. JF Caro, JW Kolaczynski, MR Nyce, JP Ohannesian, I Opentanova, WH Goldman, RB Lynn, PL Zhang, MK Sinha, RV Considine. Decreased cerebrospinal-fluid/serum leptin ratio in obesity: A possible mechanism for leptin resistance. Lancet 348:159–161, 1996.
55. MW Schwartz, DG Baskin, KJ Kaiyala, SC Woods. Model for the regulation of energy balance and adiposity by the central nervous system. Am J Clin Nutr 69:584–596, 1999.
56. AL Eldridge, DA Cooper, JC Peters. A role for olestra in body weight management. Obes Rev 3:17–25, 2002.

57. HJ Roy, MM Most, A Sparti, JC Lovejoy, J Volaufova, JC Peters, GA Bray. Original research effect on body weight of replacing dietary fat with olestra for two or ten weeks in healthy men and women. J Am Coll Nutr 21:259–267, 2002.
58. Stubbs RJ. The effect of ingesting olestra-based foods on feeding behavior and energy balance in humans. Crit Rev Food Sci Nutr 41:363–386, 2001.
59. RE Patterson, AR Kristal, JC Peters, ML Neuhouser, CL Rock, LJ Cheskin D, Neumark-Sztainer, MD Thornquist. Changes in diet, weight, and serum lipid levels associated with olestra consumption. Arch Intern Med 160:2600–2604, 2000.
60. JW Finley, LP Klemann, GA Leveille, MS Otterburn, CG Walchak. Caloric availability of salatrim in rats and humans. J Agric Food Chem 42:495–499, 1994.
61. T Nagao, H Watanabe, N Goto, K Onizawa, H Taguchi, N Matsuo, T Yasukawa, R Tsushima, H Shimasaki, H Itakura. Dietary diacylglycerol suppresses accumulation of body fat compared to triacylglycerol in men in a double-blind controlled trial. J Nutr 130:792–797, 2000.
62. T Murase, M Aoki, T Wakisaka, T Hase, I Tokimitsu. Anti-obesity effect of dietary diacylglycerol in C57BL/6J mice: Dietary diacylglycerol stimulates intestinal lipid metabolism. J Lipid Res 43(8):1312–1319, 2002.
63. M Chandalia, A Garg, D Lutjohann, K von Bergmann, SM Grundy, LJ Brinkley. Beneficial effects of high dietary fiber intake in patients with type 2 diabetes mellitus. N Engl J Med. 342:1392–1398, 2000.
64. M Rodriguez-Moran, F Guerrero-Romero, G Lazcano-Burciaga. Lipid- and glucose-lowering efficacy of Plantago Psyllium in type II diabetes. J Diabetes Complications 12:273–278, 1998.
65. ME Pick, ZJ Hawrysh, MI Gee, E Toth, ML Garg, RT Hardin. Oat bran concentrate bread products improve long-term control of diabetes: A pilot study. J Am Diet Assoc 96:1254–1261, 1996.
66. L Tappy, E Gugolz, P Wursch. Effects of breakfast cereals containing various amounts of beta-glucan fibers on plasma glucose and insulin responses in NIDDM subjects. Diabetes Care 19:831–834, 1996.
67. HG Liljeberg, YE Granfeldt, IM Bjorck. Products based on a high fiber barley genotype, but not on common barley or oats, lower postprandial glucose and insulin responses in healthy humans. 126:458–466, 1996.
68. NK Fukagawa, JW Anderson, G Hageman, VR Young, KL Minaker. High-carbohydrate, high-fiber diets increase peripheral insulin sensitivity in healthy young and old adults. Am J Clin Nutr 52:524–528, 1990.
69. JA Marlett, MI McBurney, JL Slavin. Position of the American Dietetic Association: Health implications of dietary fiber. J Am Diet Assoc 102:993–1000, 2002.
70. MA Pereira, DS Ludwig. Dietary fiber and body-weight regulation: Observations and mechanisms. Pediatr Clin North Am 48:969–980, 2001.
71. SH Holt, HJ Delargy, CL Lawton, JE Blundell. The effects of high-carbohydrate vs high-fat breakfasts on feelings of fullness and alertness, and subsequent food intake. Int J Food Sci Nutr 50:13–28, 1999.

72. WJ Pasman, WH Saris, MA Wauters, MS Westerterp-Plantenga. Effect of one week of fibre supplementation on hunger and satiety ratings and energy intake. Appetite 29:77–87, 1997.
73. KJ Acheson, B Zahorska-Markiewicz, P Pittet, K Anantharaman, E Jequier. Caffeine and coffee: Their influence on metabolic rate and substrate utilization in normal weight and obese individuals. Am J Clin Nutr 33:989–997, 1980.
74. AG Dulloo, C Duret, D Rohrer, L Girardier, N Mensi, M Fathi, P Chantre, J Vandermander. Efficacy of a green tea extract rich in catechin polyphenols and caffeine in increasing 24-h energy expenditure and fat oxidation in humans. Am J Clin Nutr 70:1040–1045, 1999.
75. WM Boothby, LG Rowntree. Drugs and basal metabolism. J Pharmacol Exp Ther 22:99–108, 1924.
76. K Horst, RJ Wilson, RG Smith. The effect of coffee and decaffeinated coffee on oxygen consumption, pulse rate and blood pressure. J Pharmacol Exp Ther 58:294–304, 1936.
77. AG Dulloo, J Seydoux, L Girardier, P Chantre, J Vandermander. Green tea and thermogenesis: Interactions between catechin-polyphenols, caffeine and sympathetic activity. Int J Obes Relat Metab Disord 24:252–258, 2000.
78. P Chantre, D Lairon. Recent findings of green tea extract AR25 (Exolise) and its activity for the treatment of obesity. Phytomedicine 9:3–8, 2002.
79. S Toubro, AV Astrup, L Breum, F Quaade. Safety and efficacy of long-term treatment with ephedrine, caffeine and an ephedrine/caffeine mixture. Int J Obes Relat Metab Disord 17(Suppl 1):S69–S72, 1993.
80. CN Boozer, PA Daly, P Homel, JL Solomon, D Blanchard, JA Nasser, R Strauss, T Meredith. Herbal ephedra/caffeine for weight loss: A 6-month randomized safety and efficacy trial. Int J Obes Relat Metab Disord 26:593–604, 2002.
81. P Bjorntorp. Regional patterns of fat distribution. Ann Intern Med 103:994–995, 1985.
82. P Bjorntorp. Abdominal fat distribution and the metabolic syndrome. J Cardiovasc Pharmacol 20(Suppl 8):S26–S28, 1992.
83. P Bjorntorp. The origins and consequences of obesity. Diabetes. Ciba Found Symp 201:68–80, 1996.
84. BB Kahn, JS Flier. Obesity and insulin resistance. J Clin Invest 106:473–481, 2000.
85. NL Benowitz, P Jacob, H Mayan, C Denaro. Sympathomimetic effects of paraxanthine and caffeine in humans. Clin Pharmacol Ther 58:684–691, 1995.
86. M Zemel. Regulation of adiposity and obesity risk by dietary calcium: Mechanisms and implications. J Am Coll Nutr 21:146S–151S, 2002.
87. B Xue, AG Greenberg, FB Kraemer, MB Zemel. Mechanism of intracellular calcium ([Ca2+]i) inhibition of lipolysis in human adipocytes. FASEB J 15: 2527–2529, 2001.
88. MB Zemel, H Shi, B Greer, D Dirienzo, PC Zemel. Regulation of adiposity by dietary calcium. FASEB J 2000:1132–1138, 14.
89. MA Denke, MM Fox, MC Schulte. Short-term dietary calcium fortification

increases fecal saturated fat content and reduces serum lipids in men. J Nutr 123:1047–1053, 1993.

90. Y Shahkhalili, C Murset, I Meirim, E Duruz, S Guinchard, C Cavadini, K Acheson. Calcium supplementation of chocolate: Effect on cocoa butter digestibility and blood lipids in humans. Am J Clin Nutr 73:246–252, 2001.
91. BR Carruth, JD Skinner. The role of dietary calcium and other nutrients in moderating body fat in preschool children. Int J Obes Relat Metab Disord 25:559–566, 2001.
92. MA Pereira, DR Jacobs Jr, L Van Horn, ML Slattery, AI Kartashov, DS Ludwig. Dairy consumption, obesity, and the insulin resistance syndrome in young adults: The CARDIA Study. JAMA 287(16):2081–2089, 2002.
93. WW Winder. Energy-sensing and signaling by AMP-activated protein kinase in skeletal muscle. J Appl Physiol 91:1017–1028, 2001.
94. Y Minokoshi, YB Kim, OD Peroni, LG Fryer, C Muller, D Carling, BB Kahn. Leptin stimulates fatty-acid oxidation by activating AMP-activated protein kinase. Nature 15:339–343, 2002.
95. WE Connor Importance of n-3 fatty acids in health and disease. Am J Clin Nutr 71:171S–175S, 2000.
96. H Wang, LH Storlien, XF Huang. Effects of dietary fat types on body fatness, leptin, and ARC leptin receptor, NPY, and AgRP mRNA expression. Am J Physiol Endocrinol Metab 282:E1352–E1359, 2002.
97. SD Clarke. Polyunsaturated fatty acid regulation of gene transcription: A mechanism to improve energy balance and insulin resistance. Br J Nutr 83 (Suppl 1):S59–S66, 2000.
98. PC Calder. Dietary modification of inflammation with lipids. Proc Nutr Soc 61:345–358, 2002.
99. S Kersten, B Desvergne, W Wahli. Roles of PPARs in health and disease. Nature 405:421–424, 2000.
100. R Walczak, P Tontonoz. PPARadigms and PPARadoxes: Expanding roles for PPARgamma in the control of lipid metabolism. J Lipid Res 43:177–186, 2002.
101. JK Reddy. Nonalcoholic steatosis and steatohepatitis. III. Peroxisomal beta-oxidation, PPAR alpha, and steatohepatitis. Am J Physiol Gastrointest Liver Physiol 281:G1333–G1339, 2001.

14

Food, Fads, Foolishness, and the Future: Immune Function and Functional Foods

Miriam H. Watson, M. Eric Gershwin, and Judith S. Stern
University of California, Davis, Davis, California, U.S.A.

I. INTRODUCTION

Medical interventions, not widely taught in U.S. medical schools and not available in U.S. hospitals, are generally received as complementary alternative therapies. Interestingly, the use of complementary and alternative medicines (CAMs) in the United States has been on the rise. Eisenberg and associates (1) reported that an estimated one in three Americans used alternative medical therapies. In addition, the estimated annual total of 425 million visits to providers of CAM was greater that the 388 million visits to all U.S. primary care physicians (2). A reported 83% of people who use CAM also seek treatment for the same condition from an allopathic physician (1).

The availability of the Internet has fostered a take charge approach to health care (3). It is estimated that 37% of adult Internet users have sought health information from on-line sources (4). More than 70% of patients who acknowledge using alternative therapy never mention it to their health care providers (1). Inconclusive evidence about, and lack of professional training on, the safety and effectiveness of these therapies make advising patients who use or seek alternative therapies a challenge. Yet ignoring the issue will not make it go away. Some danger lies in CAM, and traditional allopathic medicine must not be abandoned in pursuit of one another (Table 1).

Nutrition supplement sales are a multibillion-dollar industry. In 1996 alone it was estimated that $453 million was spent on weight loss aids, $720

Table 1 Definitions of Terms Used in Complementary and Alternative Medicine

Term	Definition
Ayurveda	A wholistic system of medicine from India. Aim is to provide guidance regarding food and lifestyle so that healthy people can stay healthy and individuals with health challenges can improve their health Unique aspects include: 1) Recommendations are often different for each person. 2) Ayurveda is validated by observation, inquiry, direct examination, and knowledge derived from ancient texts. 3) It understands that energetic forces influence nature and human beings. 4) It sees and acknowledges a strong connection between mind and body.
Botanical	A plant part or extract used especially in skin and hair products.
Herb	1) A seed-producing annual, biennial, or perennial that does not develop persistent woody tissue but dies down at the end of a growing season. 2) A plant or plant part valued for its medicinal, savory, or aromatic qualities.
Homeopathy	A system of medical practice that treats a disease especially by the administration of minute doses of a remedy that would in healthy persons produce symptoms similar to those of the disease.
Naturopathy	A system of treatment of disease that avoids drugs and surgery and emphasizes the use of natural agents (as air, water, and sunshine) and physical means (as manipulation and electrical treatment).
Osteopathy	A system of medical practice based on a theory that diseases are due chiefly to loss of structural integrity, which can be restored by manipulation of the parts supplemented by therapeutic measures (as use of medicine or surgery).
Reflexology	1) The study and interpretation of behavior in terms of simple and complex reflexes. 2) Massage of the hands or feet based on the belief that pressure applied to specific points, on these extremities, benefit other parts of the body.
Vitamin	Any of various organic substances that are essential in minute quantities to the nutrition of most animals and some plants, act especially as coenzymes and precursors of coenzymes in the regulation of metabolic processes but do not provide energy or serve as building units, and are present in natural foodstuffs or sometimes produced within the body

million on minerals, \$1.0 billion on herbals, \$1.4 billion on sports nutrition aids, and \$3.6 billion on vitamins. Furthermore, it has been projected that herbal supplement sales will double to almost \$4 billion annually (5). This increase in sales reflects the increase in public awareness and use of CAM. As health care providers we have a responsibility to become informed of all food–drug interactions and toxicities of these remedies (5). Plants and herbals are reported to have fewer side effects than drugs however, the U.S. Food and Drug Administration does not have a reliable means to evaluate this ascercion (6).

Health care providers need to take a proactive approach to the discussion of CAM. It is true that talking to patients about alternative therapies requires additional training. However, health care professionals need to ask the questions that will produce reliable answers. Dietary interviews typically do not include questions prompting patients to disclose use of alternative therapies; perhaps instead of inquiring only about supplements we need to become more specific in our questioning. Many alternative therapies are associated with health promotion and disease prevention; we can use this association as a backbone for our questions. For example, an individual who has arthritis could be asked, "Many people with arthritis have been trying alternate approaches to medical therapy to find relief from their symptoms. Have you ever thought about using herbs, vitamins, teas, or any other kinds of methods for treating your these symptoms, or for any other reason?" It has become vital, in the health care field, that we become fully informed about patient practices, including the use of CAM, to ensure quality of care. If we avoid the issues of CAM, the consequences could be devastating.

For example, a practitioner of alternative medicine was charged in the death of a diabetic girl. In this case, the girl's mother had stopped administering insulin to her daughter in the belief that the diabetic condition could instead be treated through the use of herbs (7). Although this case may be extreme, it is an excellent example of the misuse of CAM and its consequences.

Complementary and alternative medicines may interfere with conventional therapy or may themselves cause detrimental effects. In 1995, a study examining the effects of a combination of 30 mg of β-carotene per day and 25,000 IU of retinol (vitamin A) on lung cancer and cardiovascular disease was halted prematurely (8). The study involved 18,314 smokers, former smokers, and workers exposed to asbestos, during follow-up of about 4 years; 388 new cases of lung cancer were diagnosed. When compared to the placebo group the active treatment group had a statistically significant relative risk of lung cancer of 1.28; the relative risks of death of any cause, death of lung disease, and death of cardiovascular disease were 1.27, 1.46, and 1.26, respectively. The study was stopped 21 months early. Hence, safety and quality control studies must be done to establish the role of alternative medicines. Research needs to be performed to examine the alternative methods. Studies must be evidence-based and include randomized clinical trials, metaanalysis, case studies, and testimonials. In addition, evaluations of dietary and nutritional therapies, which affect both biochemical and physiological processes, require examination. These studies will help to establish a level of control and understanding of the use of CAM. So the real questions remain: Is CAM efficacious and is it safe?

An ad hoc advisory panel to the National Center for Complementary and Alternative Medicine (NCCAM) developed a categorization scheme that separates CAM into seven major categories (Table 2). Within each category CAM practices are those that are not commonly used accepted, or available in conventional medicine. Those practices that fall mainly into the category of conventional medicine are identified as behavioral medicine, and practices that can be either CAM or behavioral are designated as overlapping (9).

In addition to a classification system of CAM practices there are seven fields of practice. The first is mind–body control: It is accepted that the mind and body have the power to affect each other (10). This knowledge has led researchers to explore the mind's ability to affect the body. Clinically this is called *mind–body medicine* (11). The placebo effect is an example of a mind–body interaction in which the belief that a treatment will heal enables the healing to occur. Throughout history people have been using the placebo effect as a therapeutic management tool. However, it has changed from a harmless management tool, without curative or symptomatic consequences, to a "nondrug" with superpowers that can mimic potent drugs (12). In fact, many skeptics believe that CAM is no more than simply a means of prescribing placebos (13). Although this claim may, in part, be true, should we not begin to harness the use of the placebo (14)? For some patients this may be a useful effort. For example, in a 1998 study by Nickel, a drug was examined to determine its effectiveness in reducing benign prostate hypertrophy. The subjects taking the placebo suffered an increase in prostate size, as would be expected, but they also experienced an improvement in symptoms as well as an improvement in urinary flow. These patients also experienced adverse effects, and 13% discontinued the placebo because of these effects. This result shows that a placebo may not only influence the subjective signs and symptoms, but also affect the objective (15). In another study, by Khoo and colleagues, the effects of evening primrose oil in the treatment of premenstrual syndrome (PMS) were examined (16). This trial was randomized, double-blinded, and placebo-controlled and was crossed-over after three cycles. Although the results of this study showed an improvement in PMS symptoms during the trial, there were no significant differences between the active and placebo groups. The findings of this study indicated that the improvement experiences of these women with moderate PMS were solely a placebo effect. Spirituality and religion are two other areas of therapeutic potential that are not taught or used in the practice of medicine. Specific mind–body interventions include psychotherapy, support groups, meditation, imagery, hypnosis, biofeedback, yoga, dance therapy, music therapy, art therapy, and prayer and mental healing.

The second field of practice is alternative systems of medical practice (17). Approximately 10%–30% of health care is provided by conventional, biomedically oriented practitioners (18). 70% to 90% of health care is based on practices ranging from self-care, to use of folk principles, to care in a health care system that is based on alternative tradition or practice. Health care that most people practice and receive at home, such as the use of an herbal tea or chicken soup to treat a cold, is known as *popular health care*. The nonprofessionalized but specialized health care practices of rural and urban people, reflecting the health needs, beliefs, and natural environment, are referred to as *community-based health care*. The more formalized care, in which practitioners undergo standardized training and work in established locations, is called *professionalized health care*. The alternative systems of medical practice include professionalized health care practices, such as acupuncture, ayurvedic medicine, homeopathy, anthroposophy, naturopathy, and environmental medicine, and community-based health care practices, such as Native American medicine, Latin American medicine, and programs such as Alcoholics Anonymous.

The third field includes diet, nutrition, and lifestyle changes (19). Although we may not think of CAM this way, health care professionals must acknowledge the role of nutrition and dietetics in CAM. It is understandable that there is a fear surrounding the inclusion of the word *nutrition* in alternative medicine; nutrition science has always been a part of conventional medicine and when used in alternative practices it has been, for the most part, used by undertrained individuals. Nutrition is too important to leave to alternative practitioners who have had limited or no training in the field. Dietitians have the responsibility to step into the field and "reclaim" our territory. The field of diet, nutrition, and lifestyle changes can be divided into four areas: The first is preventing and treating chronic disease. Throughout history humans have evolved to adapt to changes in diet. Although the types of foods eaten during the seasons may have changed, the mix of macronutrients has remained relatively constant in terms of protein, carbohydrate, and fat. In the past, starvation was also a constant threat because of the uncertain nature of the food supply. More recently, however, this consistency in dietary macronutrient composition has begun to change. The relative contribution

Table 2 The National Institutes of Health Office of Alternative Medicine Classification of Categories of Complementary and Alternative Medicine

Mind/body control	Involving behavioral, psychosocial, social, and spiritual approaches to health. These methods explore the mind's ability of affect the body. They are based on traditional medical systems that use the interconnection of mind and body.	Art therapy Biofeedback Counseling Dance therapy Guided imagery Humor therapy Hypnotherapy Meditation Music therapy Prayer therapies Psychotherapy Relaxation techniques Support groups Yoga
Alternate systems of practice	Involves complete systems of theory and practice developed outside the Western biomedical approach. This type of health care ranges from self-administered care following folk principles, to health care in an organized system based on alternative traditions or practices.	Acupuncture Anthroposophically extended medicine Ayurveda Community-based health care practices Environmental medicine Homeopathic medicine Latin America rural practices Native American practices Natural products Naturopathic medicine Past life therapy Shamanism Tibetan medicine Traditional Oriental medicine
Diet/nutrition/lifestyle changes	Involving the theories and practices to prevent illness and to identify and treat risk factors, or support healing and recovery. The use of the knowledge of how to prevent illness, maintain health and reverse the effects of chronic disease through dietary or nutritional intervention.	Changes in lifestyle Diet Gerson therapy Macrobiotics Megavitamins Nutritional supplements

Table 2 Continued

Biologically-based	Including natural and biologically based practices, intervention, and products, many of which overlap with conventional medicine and the use of dietary supplements. This category is divided into four subcategories: phytotherapy or herbalism; special diet therapies; orthomolecular medicine; pharmacological, biological, and instrumental interventions.	Echinacea (purple cornflower) Ginger rhizome *Ginkgo biloba* extract Ginseng root Wild chrysanthemum flower Witch hazel Yellowdock Antioxidizing agents Cell treatment Chelation therapy Metabolizing therapy Oxidizing agents (ozone, hydrogen peroxide)
Manual healing methods	Includes systems that are based on the manipulation and/or movement of the body. Using touch and manipulation with the hands as a diagnostic and therapeutic tool.	Acupressure Alexander technique Aromatherapy Biofield therapeutics Chiropractic medicine Feldenkrais method Massage therapy Osteopathy Reflexology Rolling
Biofield	Involves systems that use energy fields in and around the body for unconventional uses and medical purposes.	
Bioelectromagnetic applications	The unconventional use of electromagnetic fields for medical purposes.	Blue light treatment and artificial lighting Electroacupuncture Electromagnetic fields Electrostimulation and neuromagnetic stimulation devices Magnetoresonance spectroscopy

Source: Ref. 9.

of nutrients from carbohydrate has decreased, and that from fat has increased (20). Additionally, there is evidence to support the belief that we are eating more "sugar" and less fiber. As the capacity to produce and store large amounts of food became possible, people became freer to choose what and when they eat; as a result major dietary changes began to occur (21). In developed countries, such as the United States, the changes have been so rapid that there has been little or no time for biological adaptation to take place. Additionally, with advances in the treatment of infectious diseases and trauma and surgery we are now living longer. As a result of this increase in life span, our diet has a longer period in which to have an adverse effect, and chronic diseases have become complicated. Together the lack of adaptation and the increased life span have enabled the long-term adverse health effects of the diets predominant in the developed countries to become increasingly apparent. The rapid rise in chronic illnesses that are either directly or indirectly related to diet and lifestyle has caused the focus of nutrition research to shift from eliminating nutritional deficiencies to studying chronic diseases caused by nutritional excesses. Data supporting the relationship between diet habit and nutritional intake and these chronic illnesses have been growing (20,22–25), supporting the need for the prevention and treatment of chronic illness through diet management.

The second category of diet, nutrition, and lifestyle changes are the dietary supplements.We currently do not recommend supplementation of the typical American diet with vitamins or nutrients beyond the recommended dietary allowances (RDAs). The alternative approaches postulate that it is necessary to supplement the typical American diet beyond the RDAs to promote optimal health and that supplementation well above the RDA is necessary to reverse the effects of long-term deficiencies. These approaches also may encourage dietary modification, eliminating or adding foods or macronutrients to treat conditions such as cardiovascular disease and cancer. It is believed that some of the major contributors to the American diet, such as meat, high in saturated fats, and high-fat dairy foods, are unhealthy and should be avoided. There is evidence that suggests that a diet high in fruits, vegetables, nuts, and low-fat dairy products and that emphasizes fish and chicken rather than red meat and is low in saturated fat, cholesterol, sugar, and refined carbohydrates may actually lower blood pressure (26–31). This diet is known as the *DASH diet* and has been suggested as a means of preventing hypertension. Evidence suggesting that the RDAs for some minerals, such as calcium and magnesium, may be too low to prevent the onset of chronic disease and the fact that the RDAs for some other micronutrients and vitamins may also be inadequate to prevent chronic illness have highlighted an important fact: That is, the RDAs were initially designed to be "safe and adequate for virtually all healthy people in the population." They were not intended to prevent or treat illnesses. The acknowledgment of this fact has prompted the Food and Nutrition Board's reevaluation of the RDAs.

A. Dietary Reference Intake

The dietary reference intake (DRI) is designed to include four components: (a) the RDA, used as a goal level; (b) the tolerable upper level (UL), used to assist in understanding whether the level of intake may result in adverse effects; (c) the adequate intake (AI), a level to meet the needs for all individuals in the group; and (d) the estimated average intake (EAR), a level at which the needs of 50% of the population will not be met. The purpose of the DRIs is to provide an intake range for all nutrients, thereby preventing both deficiencies and toxicities. The DRIs have been established to define the role of nutrients and other components in food in long-term health, beyond diseases of nutritional deficiencies (32). Whether or not further supplementation over the DRIs will be required remains to be shown.

The third area considered is orthomolecular medicine or therapeutic use of high-dose vitamins in the treatment of chronic diseases. By supplementing diet beyond the optimal concentration of substances normally present in the body, this approach promotes improvement in health for the treatment of disease. However, how do we determine optimal concentration? In most cases it is defined as the amount that prevents symptoms of deficiency. This increased intake above and beyond the level associated with the prevention of deficiency may have health benefits for some individuals. Some evidence suggests that this method may be effective in the management of certain conditions, for example, acquired immunodeficiency syndrome (AIDS), bronchial asthma, and cancer. There are numerous alternative diets reported to aid in the treatment of chronic illnesses and food allergies. The Ornish diet is one example; a low-fat, plant-based (vegetarian) diet is an important part of the Ornish plan, and stress management and exercise are also key. One problem with these diets, however, is that most of the research available on their use and effectiveness is either difficult to find or anecdotal. The majority of the diets focus on increasing the consumption of fruits and vegetables, whole grains, and legumes. Areas of investigation include food allergies (33–36), food intolerance associated with rheumatoid arthritis (37–44), and food-elimination diets (33,45–48) for use by hyperactive children. Studies of alternate dietary lifestyles that are believed to increase the resistance to chronic illness include research on vegetarian (49–52) and macrobiotic diets, both of which lower the risk of heart disease, obesity, and cancer, and on Asian and Mediterranean diets, which appear to lower the risk for heart disease and certain forms of cancer (53–59). Evaluating the effectiveness of dietary and nutritional therapy is complex because both biochemical and physiological processes are involved.

Herbal medicine is a biologically based practice that includes the folk medicine traditions of all cultures including the use of plants and plant products. Ancient cultures began the collection of information on herbs and their uses. This information has been used for the derivation of much of the present day scientific medicine (60). Many prescription drugs contain at least one plant-derived active ingredient. It has been estimated by the World Health Organization that 80% of the world's population use herbal medicine for some aspect of health care. In spite of the knowledge of useful therapeutic uses of some plants, many people remain skeptical.

In the United States herbal products may only be marketed as food supplements. The U.S. Food and Drug Administration's approval is required to make specific health claims related to an herbal product. The European market is more hospitable to natural remedies. The cost and time required to approve medicine as safe and effective are less. When a substance has a long history of use and effectiveness, it can be approved under the "doctrine of reasonable certainty." In the absence of scientific evidence to the contrary, a substance's historical use is a valid way to document safety and efficacy. European phytomedicine has been part of clinical practice for more than 10 years; this form of botanical medicine closely resembles American medicine.

Manual healing, an important diagnostic and therapeutic tool, has been used since the beginning of medical practice. These methods of practice utilize the knowledge that the dysfunction of a part of the body may secondarily affect the function of another part of the body, which may not be obviously related. Realigning or correcting the secondary parts can restore the body to optimal function and health. These practices include osteopathic medicine, chiropractic science, massage therapies, and biofield therapeutics (61).

Bioelectromagnetic applications examine how living organisms interact with electromagnetic fields. Living organisms contain electrical energy, and electrical currents in the body produce magnetic fields that extend outside the body. These fields can be influenced by external magnetic and electromagnetic fields. Physical and behavioral changes caused by changes in the body's natural fields have been reported (62). External fields occur naturally or may be artificial (power lines, appliances). Low-frequency electromagnetic fields may have beneficial biological effects (63,64). Specific effects on body tissues can result from certain frequencies. Electromagnetic fields and the way in which they produce biological effects are under study.

Pharmacological and biological treatments include drugs and vaccines not yet accepted by mainstream medicine. Examples include antineoplasms (used for tumors and AIDS), cartilage products (used for cancer and arthritis), ethylene diamine tetraacetic acid (EDTA) chelation therapy (used for heart disease, circulatory problems, rheumatoid arthritis, and cancer prevention), immunoaugmentive therapy (experimental form of cancer immunotherapy), 714-X (used for cancer), Coley's toxins (used for cancer), MTH-68n (for cancer), neural therapy (for chronic pain), apitherapy (for a number of diseases), and iscador (for tumors) (65). Randomized clinical trials need to be done in order for these treatments to be accepted by allopathic medicine. The high expense of running these trials, however, is a major roadblock to research of alternative pharmacology.

II. IMMUNITY ENHANCERS

Provided the information we have responsibility as health care providers to determine whether immunity enhancement methods are actually alternative medicines or fiction. We need to examine the evidence provided by science-based research. The U.S. Food and Drug Administration (FDA) requires that randomized clinical trials be performed on new drugs, biologicals, and medical devices. They must be proved safe and effective to gain approval. The clinical trials compare the results observed in patients who receive the treatment with the results in similar patients who receive a different treatment or placebo. These studies are usually double-blind. The safety and effectiveness criteria must be met for all products, regardless of whether they are labeled alternative or are categorized as mainstream American medical practice. Additionally, not all substances undergoing FDA clinical trials, whether alternative or mainstream, are proved effective. Only a small fraction are. In addition to clinical trials we can also gain knowledge on the efficacy of a treatment by examining the case studies and testimonials presented. By these means we hope to maintain safety and quality control.

Standards of quality ("good manufacturing practices" such as standardization, expiration dates, and contaminant screening) are lacking in the manufacturing of herbal products. Therefore, there is no way of ensuring the quality of these products. Products may contain contaminants that may actually be causing the side effects that are observed. Some products may be contaminated with heavy metals, resulting in detrimental effects. It has been speculated that experimentation on the use of ginseng and other herbal products in humans has not been conducted because of lack of reliable or standardized preparations, the possibility of obtaining patent protection for discoveries, the fundamental differences between Western and Oriental medicine, and a lack of information on the proper dosage required to prevent side effects.

To ensure the highest quality it is, therefore, essential to use a product that is under regulation. For example, ginseng has been used for centuries in

Asia and the former Soviet Union as a cure-all; it is described as useful in the treatment of all afflictions. In these areas ginseng is not used to cure a particular disease; it has a supportive role to maintain health. In China, Asian ginseng is used as a tonic and aphrodisiac. Ginseng consumption has more recently entered the Western world. Here, American ginseng is believed to strengthen normal body functions and overcome disease by building up general vitality. These effects are difficult to document scientifically. However, the stress effects of temperature changes, diet, restraint, and exercise have been recorded. The active components of ginseng are believed to be saponins, which are present in the root in large numbers. There have been 28 ginsenosides, and 7 eleutherosides (both saponins) identified to date. The American and Oriental ginseng species differ in composition of saponins. In addition to varying compositions between species, the composition of a given ginseng may vary with age, location where grown, season harvested, and method of curing or drying. Authentic high-quality ginseng is not easily obtained. High-quality root is extremely expensive; this high cost and lack of quality control have resulted in commercial ginseng products that are highly variable. Products are available in teas, powders, capsules, extracts, roots, tablets, etc. The actual concentration of ginseng in these products is highly variable (66). An analysis of 54 ginseng products showed that 60% contained very little ginseng and little bioactivity, and 25% contained no ginseng at all (67). Is ginseng a safe herb to use, or are there toxic side effects? A number of studies have reported that ginseng use may result in estrogenic effects; however, these studies do not provide any experimental evidence to support the claim. Documented side effects include insomnia, diarrhea, and skin eruptions (68). The side effects of ginseng may be specific to the species and dosage used. Ginseng use does, however, appear to involve a low risk even over prolonged periods or in excessive amounts (69). The plant that is the origin of Siberian ginseng is native to eastern Siberia, Korea, and the Shansi and Hopei provinces of China. The drug is known as *eleutherococc* in the former USSR. The name *Siberian* ginseng is actually very misleading because the plant does not belong to the same genus as the true ginsengs; the name was given in order to enhance the appearance of this inexpensive drug. Knowledgeable individuals usually refer to the plant as eleuthero. This drug has been reported to have androgenic effects; on further scrutiny it was revealed that much of the eleurthero imported into the United States is misidentified and may actually be derived from six or more closely related plants and some unrelated species. Because of the lack of experimentally supported information on this herb, ginseng has no proven efficacy for humans (70).

Another popular medicinal herb is garlic, a member of the genus *Allium*. The onion, leek, and shallot also belong to this family. Used historically as both food and medicine, the bulbs and leaves of these plants are collectively known as *alliums*. The garlic bulb has been recognized as a medicinal remedy in both Chinese and Western cultures. Its therapeutic effects are released when it is eaten raw or cooked. There are more than 100 different medical uses reported. The active ingredient in garlic is an odorless, sulfur-containing amino acid derivative known as *alliin*. Alliin itself has no antibacterial properties; however, when it is ground, the enzyme allinase converts alliin into allicin. Allicin is responsible for the herb's pungent smell and is an effective antibacterial agent (71). In China garlic is used not only for colds and coughs but also for intestinal and digestive disorders. Some Chinese herbalists also use garlic as an external antibiotic to relieve skin infections; Western herbalists also prescribe garlic for these ailments (71,72). It is also thought, however, to strengthen the cardiovascular system. Studies in experimental animals have demonstrated the effectiveness of garlic in treatment of digestive disturbances and cardiovascular disorders (72–74). Garlic has been shown to prevent hypercholesterolemia and reduce high blood pressure in both humans and experimental animals (75–77). Additionally, small doses of garlic have been shown to increase intestinal peristalsis, whereas large doses inhibit movement. There is some evidence that garlic can inhibit the development of cancer by stimulating the immune system (78–81). The ability of garlic to inhibit the aggregation of blood platelets is believed to offer protection against atherosclerosis, coronary thrombosis, and (82–84). Ajoene, a self-condensation product of allicin, is the compound responsible for this property.

Ajoene is antithrombotic and believed to be as potent as aspirin (85–87). Numerous studies have evaluated the effects of garlic and onions on a number of cardiovascular risk factors in humans (88). The claims in these studies are valid for the beneficial effects on blood cholesterol levels, platelet aggregation, and fibrinolytic activities, but only at high dosage levels: 5 to 20 average-size cloves of fresh garlic daily. Studies examining the effectiveness of garlic from different commercial preparations produced contradictory findings. Examination of the stability of the active components in garlic has shown that although dried garlic preparations contain alliin, they contain neither allicin nor ajoene (71). Although the enzyme alliinase may be present in garlic dried at a low temperature, it is not stable in the acidic environment of the stomach and is destroyed before conversion to the active compounds can take place. Garlic preparations, protected by an enteric coating, pass through the stomach into the small intestine. In the small intestine the enzyme can function, producing the active compounds. Fresh garlic releases its active properties in the mouth during chewing (70). It is currently believed that daily consumption of garlic should not pose any health risk. However, consumption of large quantities may result in undesirable side effects such as heartburn, flatulence, and gastrointestinal upset (70,71). Whereas the most effective means of assuring effectiveness is to eat large quantities of raw garlic, this practice may not enhance one's social activities. Finally, because of the anticoagulation properties of the herb, individuals who take anticoagulation drugs should avoid eating large quantities of garlic (70).

The use of glucosamine sulfate and chondroitin sulfate for knee osteoarthritis is currently under study by the National Institute of Health (NIH) (89). Glucosamine, a substance synthesized in the body, is important in the repair and maintenance of joint cartilage. Glucosamine sulfate and chondroitin sulfate are synthetic versions of these substances, both of which are available as supplements and are not regulated by the FDA. Studies in Europe on the use of glucosamine sulfate and chondroitin sulfate in the treatment of osteoarthritis reveal that patients derived some short-term benefits in symptomatic relief from pain (44,90). There are no long-term, double-blind controlled studies assessing the long-term benefits or safety of this drug. Osteoarthritis results from the breakdown of the cartilage cushioning the ends of bone joints. This deterioration causes joint pain, stiffness, and deformity. A limited number of experimental animal studies have shown that glucosamine sulfate can potentially retard the degradation of cartilage. In a Portuguese study the efficacy of oral glucosamine was greater than that of anti-inflammatory agents, cartilage extracts, and vitamins. Glucosamine was easy to use and was effective and well tolerated (91). One study examined the effectiveness of glucosamine sulfate in the treatment of osteoarthritis. For this study, three double-blind, controlled parallel groups were randomized into 4- to 6- week trials of glucosamine sulfate versus placebo or the nonsteroidal anti-inflammatory drug (NSAID) ibuprofen. The subjects were 606 gonarthrosic outpatients. Efficacy goals were strictly predetermined and movement limitation and pain were scored. When given NSAIDs, glucosamine sulfate, or a placebo for the treatment of osteoarthritis, 37% of the patients suffered adverse reactions to the NSAID medication, as compared to 7% of both the glucosamine and placebo groups This same study showed that NSAIDS and glucosamine provided equal relief of arthritis pain. However, 3–4 weeks of continuous glucosamine use was necessary to provide the relief received from NSAIDs in 1 week (92).

Although studies are clearly showing the potential effectiveness in the use of the alternative medicines for the treatment of chronic conditions, we should not forget the need for patient instruction in their use and potential drug interactions. Additionally, we need to be mindful that not all alternative medications may be effective.

The U.S. population is increasing in weight. Over the past 20 years the percentage of the adult population that is classified as overweight has increased from about 25% to more than 55%. As a result of this increase, weight-loss supplement sales have also increased. Chromium picolinate is one of these hot-selling supplements; this product is marketed to both body builders, to aid in the loss of body fat while increasing muscle size, and individuals who are trying to lose weight. Several weight-loss products use chromium as their fat-burning ingredient. Does the product actually work? Chromium plays an important role in the function of the hormone insulin and therefore assists in the metabolism of carbohydrates, proteins, and fats. Research has shown that a low dietary intake of chromium may impair the body's ability to use glucose. Indeed, in some people chromium supplements can improve glucose homeostasis (93,94). Some researchers are concerned that we may not be getting enough chromium in our diet and, therefore, risk the development of prediabetic conditions. The RDA of the mineral chromium has not been defined; further research needs to be performed before it is. The question remains, however, whether we are ingesting enough chromium in our diet? Many studies have shown that a supplement containing 200 μg of chromium improves the ability of the body to handle sugar (93). A highly refined diet, high in pasta, bread, candy, and soda, is common among the U.S. population and may be low in the mineral chromium. The reported beneficial effect of chromium, therefore, may be a response to the normally low intake of chromium dietary habits. Additionally, athletes may require more chromium because strenuous exercise can result in more urinary loss of chromium (95). When a low dietary intake is combined with an increased urinary loss, is a supplement necessary, and would a supplement help to improve athletic performance? Many controlled studies examining the effects of chromium supplementation on athletic performance have shown that this type of supplementation does not improve performance; nor does it help to lower body fat or increase muscle size (95,96). So why did chromium picolinate become a hot-selling item if research has shown it to be ineffective? Basically through successful marketing strategies.

A similar approach was taken in the marketing of chitosan, an n-decetylated derivative of the polysaccharide chitin, found in shells of invertebrates. Between February 1999 and February 2000, U.S. sales of the top three brands of chitosan exceeded $6 million despite lack of evidence of efficacy (S. Squires. The risk of fat busters. NY Times 2000:March 28: HE 14). The weight loss claims for chitosan included "Lose weight without restructuring food intake on a high fat diet." The mechanism of action proposed was that chitosan bound with dietary fat in the stomach and prevented the absorption of up to 120 g of fat per day.

III. HEALTH CLAIMS

Dietitians and other health care professionals need to become familiar with identifying claims and interpreting the meaning of advertisements. The supplement industry has been rapidly expanding. In an attempt to ensure that consumers get accurate information about dietary supplements and about the use of these products, the Federal Trade Commission (FTC) and the Food and Drug Administration work together. The FDA is responsible for claims on product labeling, including packaging, inserts, and other promotional materials distributed at the point of sale. The FTC has primary responsibility for the claims in advertising, including print and broadcast ads, infomercials, catalogues, and similar direct marketing materials. Marketing on the Internet is subject to the same regulation as in other media. According to the Federal Trade Commission advertising for any dietary supplement must be truthful, not misleading, and substantiated. In the application of FTC law the marketer is required to identify all expressed and implied claims that an ad communicates to the consumer. The ad must be accurate and cannot suggest claims that cannot be made directly. The advertisement may also not be deceptive in what it fails to say. In the case that the disclosure of qualifying information is necessary to prevent the ad from being deceptive, the information must be clearly and prominently presented so that it is noticed and understood by the consumer. In addition to conveying the product claims clearly and accurately, marketers also need to substantiate these claims by verifying that there is adequate support for them. Advertisers should not make claims based on consumer experiences or expert endorsements. To evaluate a claim about the safety and efficacy of foods, dietary supplements, and drugs, the FTC has applied a substantiation standard of competent and reliable scientific evidence (97,98).

The only way to provide necessary evidence to allow the approval of herbal medicine for medical conditions is to perform long-term clinical trials. Although expensive, these trials provide information necessary for FDA and FTC approval. The NIH is currently funding a multicenter randomized clinical trial examining the efficacy and safety of St. John's wort (*Hypericum perforate*) in the treatment of depression. Past studies examined whether extracts of (a) were more effective than placebo in the treatment of depression, (b) were as effective as standard antidepressive treatments, and (c) resulted in fewer side effects than the standard antidepressant drugs. These studies revealed that the extracts were significantly more effective than the placebo and as effective as the standard antidepressants. Side effects occurred in 50 patients on extracts and 84 patients on standard antidepressants. The overall conclusion was that although the extracts were effective in the treatment of mild to moderately severe depressive disorders, further studies were needed (77,99).

The process of providing evidence before approval is very long and may perhaps be preventing individuals from obtaining valuable help from herbal medicines. As suggested by Dr. David Kritchefsky, "If you wait for the evidence the patients will be dead!" So we need to make judgment calls; keeping efficacy in mind, we need to weigh the risks of use against the risk of the disease. Take, for example, childhood leukemia; this acute illness must be treated in a conventional manner. Successful treatment requires very rapid treatment response, and the treatment must be reliable. When treated in this manner childhood leukemia is almost 100% curable. In a case such as this, there should be no question as to whether a CAM approach would be appropriate; the answer is No. The use of CAM may be considered bad practice, in that it may delay treatment and potentially lead to death. On the other hand, the use of CAM for the treatment of a disorder such as Alzeihmer's disease, which is a chronic disease that is progressively degenerative and has to date no known cure, may be beneficial by providing some amelioration of the disease process. The use of ginkgo is said to relieve some of the symptoms typical of cerebral insufficiency, including difficulties of concentration, difficulties of memory, absentmindedness, confusion, lack of energy, tiredness, decreased physical performance, depressive mood, anxiety, dizziness, tinnitus, and headache (100). This is a case in which medicinal use of CAM is not harmful and may in fact be beneficial.

In the treatment of medical and nonmedical conditions we turn to the use of drugs for a cure. This is true whether we use a conventional or alternative approach to care. As health care practitioners we are, for the most part, aware of the many drug–drug and drug–nutrient interactions that may occur when using conventional medications that have been highly studied and approved by the FDA. This, however, is not the case with many of the CAM treatments. The simultaneous use of more than one drug known as *polypharmacy*. There may be drug–drug or drug–nutrient interactions that result from combining conventional and CAM medical treatments; some of these combinations may be lethal. For example, ginkgo has been shown to promote vasodilatation and to improve blood flow in both arteries and capillaries (100). Ginkgo also reduces blood clotting time and therefore may be contraindicated for individuals who are already taking anticoagulants, such as aspirin and vitamin E. Whereas alone each of these may not be harmful, when combined they may result in a stroke. This is a dramatic example of a fact that cannot be ignored. As health care practitioners we must be aware of interactions and their hazards. We should, therefore, avoid the polypharmacy of CAM and the combination of conventional and CAM medications.

REFERENCES

1. Eisenberg, D. M., R. C. Kessler, C. Foster, F. E. Norlock, D. R. Calkins, and T. L. Delbanco. 1993. Unconventional medicine in the United States: prevalence, costs, and patterns of use. *N Engl J Med 328:246*.

2. Eisenberg, D. M. 1997. Advising patients who seek alternative medical therapies. *Ann Intern Med 127:61.*
3. 1996. *USA Today*. In the wired world, a trove of health data.
4. 1997. *USA Today*. Snapshot.
5. Cowley, G., and M. Hager. 1996. Herbal warning: Health food stores have built a new natural-drug culture. How safe are their wares? *Newsweek*, p. 60.
6. Stehlin, I. B. 1995. An FDA guide to choosing medical treatments. *FDA Consumer.*
7. Credit, A. 1999. Health adviser charged in death of diabetic girl. *The New York Times*, New York, p. 37.
8. Omenn, G. S., G. E. Goodman, and M. D. Thornquist. 1996. Effects of a combination of beta carotene and vitamin A on lung cancer and cardiovascular disease. *New Engl J Med 334:1150.*
9. NIH/NCCAM. 1992. What is CAM? Classification of Alternative Medicine Practices. National Institutes of Health.
10. NIH/NCCAM. Fields of Practice/Mind-Body Control. National Institutes of Health.
11. Cerrato, P. L. 1998. Understanding the mind/body link. *RN 61:28.*
12. Kaptchuk, T. J. 1998. Powerful placebo: the dark side of the randomized controlled trial. *Lancet 351:1722.*
13. Joyce, C. R. B. 1994. Placebo and complementary medicine. *Lancet 344:1279.*
14. Tausk, F. A. 1998. Alternative medicine: Is it all in your mind? *Arch Dermatol 134:1422.*
15. Nickel, J. C. 1998. Placebo therapy in benign prostatic hyperplasia: a 25-month study: Canadian PROSPECT study group. *Brit J Urol 81:383.*
16. Khoo, S. K., C. Munro, and D. Battistutta. 1990. Evening primrose oil and the treatment of premenstrual syndrome. *Med J Austr 153:189.*
17. NIH/NCCAM. Fields of Practice/Alternative Systems of Medical Practice. NIH.
18. Eisenberg, D. M., R. B. Davis, S. L. Ettner, S. Appel, S. Wilkey, M. Van Rompay, and R. C. Kessler. 1998. Trends in alternative medicine use in the United States, 1990–1997: results of a follow-up national survey. *JAMA 280:1569.*
19. NIH/NCCAM. Fields of Practice/Diet, Nutrition, Lifestyle Changes. NIH.
20. Willett, W. C. 1994. Diet and health: What should we eat? *Science 264:532.*
21. Rodriguez, A. F., J. R. Banegas, M. A. Graciani, V. R. Hernandez, and C. J. Rey. 1996. Food nutrient composition in Spain in the period 1940-1988: analysis of its consistency with the Mediterranean diet. *Med Clin 106:161.*
22. Toniolo, P., E. Riboli, F. Protta, M. Charrel, and A. P. Cappa. 1989. Calorie-providing nutrients and risk of breast cancer. *J Natl Cancer Instit 81.*
23. La Vecchia, C. A. D., and R. Pagano. 1998. Vegetable consumption and risk of chronic disease. *Epidemiology 9:208.*
24. Green, S. M., and J. E. Blundell. 1996. Effect of fat- and sucrose-containing foods on the size of eating episodes and energy intake in lean dietary restrained and unrestrained females: potential for causing overconsumption. *Eur J Clin Nutr 50:625.*

25. Mera, S. L. 1994. Diet and disease. *Br J Biomed Sci 51:189.*
26. Blacher, J., J. M. Tartier, G. Amah, P. Campa, J. Raison, and M. Safar. 1998. Arterial hypertension, non-drug treatment and cardiovascular disease. *Ann Cardiol Angeiol 47:81.*
27. Svetkey, L. P., F. M. Sacks, E. Obarzanek, W. M. Vollmer, L. J. Appel, P. H. Lin, N. M. Karanja, D. W. Harsha, G. A. Bray, M. Aickin, M. A. Proschan, M. M. Windhauser, J. F. Swain, P. B. McCarron, D. G. Rhodes, and R. L. Laws. 1999. The DASH Diet, Sodium Intake and Blood Pressure Trial (DASH-sodium): rationale and design. DASH-Sodium Collaborative Research Group. *J Am Diet Assoc 99:S96.*
28. Windhauser, M. M., D. B. Ernst, N. M. Karanja, S. W. Crawford, S. E. Redican, J. F. Swain, J. M. Karimbakas, C. M. Champagne, K. P. Hoben, and M. A. Evans. 1999. Translating the Dietary Approaches to Stop Hypertension diet from research to practice: dietary and behavior change techniques. DASH Collaborative Research Group. *J Am Diet Assoc 99:S90.*
29. Sacks, F. M., L. J. Appel, T. J. Moore, E. Obarzanek, W. M. Vollmer, L. P. Svetkey, G. A. Bray, T. M. Vogt, J. A. Cutler, M. M. Windhauser, P. H. Lin, and N. Karanja. 1999. A dietary approach to prevent hypertension: a review of the Dietary Approaches to Stop Hypertension (DASH) Study. *Clin Cardiol 22:III6.*
30. Kolasa, K. M. 1999. Dietary Approaches to Stop Hypertension (DASH) in clinical practice: a primary care experience. *Clin Cardiol 22:III16.*
31. Fraser, G. E. 1999. Nut consumption, lipids, and risk of a coronary event. *Clin Cardiol 22:III11.*
32. Yates, A. A., S. A. Schlicker, and C. W. Suitor. 1998. Dietary Reference Intakes: the new basis for recommendations for calcium and related nutrients, B vitamins, and choline. *J Am Diet Assoc 98:699.*
33. Arvola, T., and D. Holmberg-Marttila. 1999. Benefits and risks of elimination diets. *Ann Med 31:293.*
34. Kirjavainen, P. V., E. Apostolou, S. J. Salminen, and E. Isolauri. 1999. New aspects of probiotics—a novel approach in the management of food allergy. *Allergy 54:909.*
35. Cantani, A., and D. Gagliesi. 1998. Prediction and prevention of allergic disease in at risk children. *Eur Rev Med Pharmacol Sci 2:115.*
36. Fox, D., and M. Gaughan. 1999. Food allergy in children. *Paediatr Nurs 11:28.*
37. Kjeldsen-Kragh, J. 1999. Rheumatoid arthritis treated with vegetarian diets. *Am J Clin Nutr 70:594S.*
38. Mangge, H., J. Hermann, and K. Schauenstein. 1999. Diet and rheumatoid arthritis—a review. *Scand J Rheumatol 28:201.*
39. Martin, R. H. 1998. The role of nutrition and diet in rheumatoid arthritis. *Proc Nutr Soc 57:231.*
40. Ballard, A. E. 1996. Traditional and complementary therapies used together in the treatment, relief and control of Crohn's disease and polyarthritis. *Complement Ther Nurs Midwifery 2:52.*
41. Shapiro, J. A., T. D. Koepsell, L. F. Voigt, C. E. Dugowson, M. Kestin, and J.

L. Nelson. 1996. Diet and rheumatoid arthritis in women: a possible protective effect of fish consumption. *Epidemiology 7:256.*

42. Hansen, G. V., L. Nielsen, E. Kluger, M. Thysen, H. Emmertsen, K. Stengaard-Pedersen, E. L. Hansen, B. Unger, and P. W. Andersen. 1996. Nutritional status of Danish rheumatoid arthritis patients and effects of a diet adjusted in energy intake, fish-meal, and antioxidants. *Scand J Rheumatol 25:325.*
43. Kjeldsen-Kragh, J., M. Haugen, C. F. Borchgrevink, and O. Forre. 1994. Vegetarian diet for patients with rheumatoid arthritis—status: two years after introduction of the diet. *Clin Rheumatol 13:475.*
44. Horstman, J. 1998. Glucosamine and chondroitin. *Arthritis Today.*
45. Breakey, J. 1997. The role of diet and behaviour in childhood. *J Paediatr Child Health 33:190.*
46. Krummel, D. A., F. H. Seligson, and H. A. Guthrie. 1996. Hyperactivity: is candy causal? *Crit Rev Food Sci Nutr 36:31.*
47. Harley, J. P., C. G. Matthews, and P. Eichman. 1978. Synthetic food colors and hyperactivity in children: a double-blind challenge experiment. *Pediatrics 62: 975.*
48. Stine, J. J. 1976. Symptom alleviation in the hyperactive child by dietary modification: a report of two cases. *Am J Orthopsychiatry 46:637.*
49. Ritter, M. M., and W. O. Richter. 1995. Effects of a vegetarian life style on health. *Fortschr Med 113:239.*
50. Weitzman, S. 1998. Alternative nutritional cancer therapies. *Int J Cancer Suppl 11:69.*
51. Hillman, H. 1981. Vegetarianism—an alternative diet or a crazy fad? *Nurs Times 77:444.*
52. Phillips, R. L. 1975. Role of life-style and dietary habits in risk of cancer among seventh-day adventists. *Cancer Res 35:3513.*
53. Berrino, F., and P. Muti. 1989. Mediterranean diet and cancer. *Eur J Clin Nutr 43:49.*
54. Strom, S. S., Y. Yamamura, C. M. Duphorne, M. R. Spitz, R. J. Babaian, P. C. Pillow, and S. D. Hursting. 1999. Phytoestrogen intake and prostate cancer: a case–control study using a new database. *Nutr Cancer 33:20.*
55. Wu, A. H., R. G. Ziegler, P. L. Horn-Ross, A. M. Nomura, D. W. West, L. N. Kolonel, J. F. Rosenthal, R. N. Hoover, and M. C. Pike. 1996. Tofu and risk of breast cancer in Asian-Americans. *Cancer Epidemiol Biomarkers Prev 5:901.*
56. de Lorgeril, M., P. Salen, J. L. Martin, I. Monjaud, P. Boucher, and N. Mamelle. 1998. Mediterranean dietary pattern in a randomized trial: prolonged survival and possible reduced cancer rate. *Arch Intern Med 158:1181.*
57. Ghiselli, A., A. D'Amicis, and A. Giacosa. 1997. The antioxidant potential of the Mediterranean diet. *Eur J Cancer Prev 6 (Suppl 1):S15.*
58. Khlat, M. 1995. Cancer in Mediterranean migrants—based on studies in France and Australia. *Cancer Causes Control 6:525.*
59. Tavani, A., and C. La Vecchia. 1995. Fruit and vegetable consumption and cancer risk in a Mediterranean population. *Am J Clin Nutr 61:1374S.*

60. NIH/NCCAM. Fields of Practice/Herbal Medicine. National Institutes of Health.
61. NIH/NCCAM. Fields of Practice/Manual Healing. National Institutes of Health.
62. NIH/NCCAM. Fields of Practice/Bioelectromagnetic applications. National Institutes of Health.
63. Papatheofanis, F. J. 1984. A review on the interaction of biological systems with magnetic fields. *Physiol Chem Phys Med NMR 16:251.*
64. Bassett, C. A. 1987. Low energy pulsing electromagnetic fields modify biomedical processes. *Bioessays 6:36.*
65. NIH/NCCAM. Fields of Practice/Pharmacological and Biological Treatments. National Institutes of Health.
66. Harkey, M. R., G. L. Henderson, M. E. Gershwin, J. S. Stern, and R. M. Hackman. 2001. Variability in commercial ginseng products: an analysis of 25 preparations. *Am J Clin Nutr 73:1101.*
67. Liberti, L. E., and A. Der Marderosian. 1978. Evaluation of commercial ginseng products. *J Pharm Sci 67:1487.*
68. Baldwin, C. A., L. A. Anderson, and J. D. Phillipson. 1986. *Pharm J 237:583.*
69. Chandler, R. F. 1988. *Can Pharm J 121:36.*
70. Foster, S., and V. E. Tyler. 1999. Tyler's Honest Herbal: *A sensible guide to the use of herbs and related remedies.* Pharmaceutical Products Press, New York.
71. *The review of natural products: Facts and comparisons.* 1997. St. Louis.
72. Reuter, H. D. 1995. *Allium sativum* and *Allium ursinum*, Part 2. Pharmacology and medicinal application. Phytomedicine 2:73.
73. Warshafsky, S., R. S. Kamer, and S. L. Sivak. 1993. Effect of garlic on total serum cholesterol: a meta-analysis. *Ann Intern Med 119:599.*
74. Lau, B. H., F. Lam, and R. Wang-Chen. 1987. Effect of an odor modified garlic preparation on blood. *Nutr Res 7:139.*
75. Kandziora, J. 1988. Antihypertensive effectiveness and tolerance of a garlic medication. *Artztl Forsch 35:1.*
76. Silagy, C. A., and H. A. Neil. 1994. A meta-analysis of the effect of garlic on blood pressure. *J Hypertens 12:463.*
77. Klepser, T. B., and M. E. Klepser. 1999. Unsafe and potentially safe herbal therapies. *Am J Health Syst Pharm 56:125.*
78. Sumiyoshi, H. 1997. New pharmacological activities of garlic and its constituents. *Nippon Yakurigaku Zasshi 110 (Suppl 1):93P.*
79. Riggs, D. R., J. I. DeHaven, and D. L. Lamm. 1997. *Allium sativum* (garlic) treatment for murine transitional cell carcinoma. *Cancer 79:1987.*
80. Romano, E. L., R. F. Montano, B. Brito, R. Apitz, J. Alonso, M. Romano, S. Gebran, and A. Soyano. 1997. Effects of Ajoene on lymphocyte and macrophage membrane-dependent functions. *Immunopharmacol Immunotoxicol 19: 15.*
81. Salman, H., M. Bergman, H. Bessler, I. Punsky, and M. Djaldetti. 1999. Effect of a garlic derivative (alliin) on peripheral blood cell immune responses. *Int J Immunopharmacol 21:589.*

82. Kiesewetter, H., F. Jung, G. Pindur, E. M. Jung, C. Mrowietz, and E. Wenzel. 1991. Effect of garlic on thrombocyte aggregation, microcirculation, and other risk factors. *Int J Clin Pharmacol Ther Toxicol 29:151.*
83. Ali, M., T. Bordia, and T. Mustafa. 1999. Effect of raw versus boiled aqueous extract of garlic and onion on platelet aggregation. *Prostaglandins Leukot Essent Fatty Acids 60:43.*
84. Tyler, V. E., and S. Foster. 1996. Herbs and phytomedicinal products. In *Handbook of non-prescription drugs.* T. R. Covington, R. R. Berardi, and L. L. Young, eds. American Pharmaceutical Association, Washington, D.C., p. 695.
85. Milner, J. A. 1996. Garlic: its anticarcinogenic and antitumorigenic properties. *Nutr Rev 54:S82.*
86. Srivastava, K. C., and O. D. Tyagi. 1993. Effects of a garlic-derived principle (ajoene) on aggregation and arachidonic acid metabolism in human blood platelets. *Prostaglandins Leukot Essent Fatty Acids 49:587.*
87. Block, E. 1985. The chemistry of garlic and onions. *Sci Am 252:114.*
88. Kleijnen, J., P. Knipschild, and G. ter Riet. 1989. Garlic, onions and cardiovascular risk factors: a review of the evidence from human experiments with emphasis on commercially available preparations. *Br J Clin Pharmacol 28:535.*
89. NIH/NCCAM. 1999. Questions and Answers: NIH Study on Glucosamine/Chondroitin Sulfate for Knee Osteoarthritis. National Institutes of Health.
90. Rye, D. 1999. Glucosamine. Alternate ways with Debbie Rye.
91. Foundation, A. 1999. Glucosamine sulfate and chondroitin sulfate.
92. Rovati, L. C. 1992. Clinical research in osteoarthritis: design and results of short-term and long-term trials with disease-modifying drugs. *Int J Tissue React 14:243.*
93. Wilson, B. E., and A. Gondy. 1995. Effects of chromium supplementation on fasting insulin levels and lipid parameters in healthy, non-obese young subjects. *Diabetes Res Clin Pract 28:179.*
94. Anderson, R. A., N. Cheng, N. A. Bryden, M. M. Polansky, J. Chi, and J. Feng. 1997. Elevated intakes of supplemental chromium improve glucose and insulin variables in individuals with type 2 diabetes. *Diabetes 46:1786.*
95. Lefavi, R. G., R. A. Anderson, R. E. Keith, G. D. Wilson, J. L. McMillan, and M. H. Stone. 1992. Efficacy of chromium supplementation in athletes: emphasis on anabolism. *Int J Sport Nutr 2:111.*
96. Trent, L. K., and D. Thieding-Cancel. 1995. Effects of chromium picolinate on body composition. *J Sports Med Phys Fitness 35:273.*
97. Commission, F. T. 1999. Fraudulent health claims: don't be fooled. Federal Trade Commission.
98. Commission, F. T. 1999. Dietary supplements: an advertising guide for industry. Federal Trade Commission.
99. Linde, K., G. Ramirez, C. D. Mulrow, A. Pauls, W. Weidenhammer, and D. Melchart. 1996. St John's wort for depression—an overview and meta-analysis of randomised clinical trials. *BMJ 313:253.*
100. Kleijnen, J., and P. Knipschild. 1992. *Ginkgo biloba. Lancet 340:1136.*

15
Immunology and Inflammation

Eduardo J. Schiffrin and Stephanie Blum
Nestlé Research Center, Lausanne, Switzerland

I. INTRODUCTION

The understanding of both mucosal immune defenses and intestinal inflammation requires some consideration of the interactions between the host and the intestinal commensal microbiota. Humans live in a bacterial world as microbes from the environment are in constant contact with the skin and have access to the mucosal compartments of the body. The latter can remain sterile or, in contrast, become colonized. Whereas the distal respiratory tract is sterile under physiological conditions, the distal gastrointestinal (GI) tract is heavily colonized. It has been calculated that there are more bacterial cells in the indigenous microbiota of the gut than own eucaryotic cells in the body (1).

The intestinal microbiota is a very complex mixture of different bacterial genera and species. Although in permanent interaction with environmental microorganisms, partially food-derived, the indigenous microbiota has a remarkably stable composition throughout most of its life span. However, there are a variety of complex societies of different microorganisms along the GI tract determined by different habitats and ecological conditions. The mechanisms of control that maintain the equilibrium of the GI microbiota are mainly unknown and may differ at different levels of the intestine. The composition of the gut microbiota may transiently vary as a consequence of an important bacterial inoculum in the diet (2,3). Also, the administration of certain ingredients, mainly oligosaccharaides, nondigestible by human enzymes but fermented by bacteria in the colon, may result in enhanced colonic growth of bifidobacteria (4).

A drastic change in microbiota composition certainly occurs immediately after birth, when the currently sterile fetus becomes suddenly exposed to the unfriendly external environment. In addition to the maternal bacteria encountered during delivery and skin contact during breast-feeding, environmental microbes contribute to buildup of the neonatal microflora (5). It has been observed that before attainment of a the steady state around weaning, there is a defined sequence of dominant bacterial genera. Nevertheless, it appears that the type of neonatal feeding may influence the composition of the microflora as clear differences have been found in the intestinal microflora of breast-fed and formula-fed babies (6). During adulthood perturbations of GI microbiota equilibrium are rare and mainly associated with pathological conditions such as enteral infections, antibiotherapy, or immune suppression. Another significant change in the composition of the microbiota is observed during the aging process. A reduction in the putative protective bacteria, such as bifidobacteria, has been reported (7) in the elderly.

Bacterial colonization of the gut by commensal bacteria was shown to modulate the physiological processes of the host by modification of gene expression, which affects both nutritional and protective functions of the intestine (8). Interestingly, different commensal bacteria appear to induce a different profile of gene activation in the host. This recent field of research will help to broaden our understanding of host–microbe interactions in health and disease. As a consequence, an appropriate manipulation of the host microflora might become a new tool in the prevention or management of GI pathophysiological disorders.

In the last decades a lower exposure of the neonate and infant to intestinal microbial challenge has been indicated by epidemiological data. This observation is associated with a higher incidence of allergic diseases, which has generated the hygiene hypothesis. It appears that a certain level of bacterial challenge or bacterial inoculum is an important prerequisite to achieve normal homeostasis in host reactivity against external antigens and pathogens (9–11).

Mucosal surfaces are exposed to a plethora of microbes other than commensals, including a proportion of pathogens. Although the real magnitude of the host exposure to pathogens at the mucosal sites is difficult to assess, it appears that these encounters rarely result in infectious diseases. In any case, when mechanisms of defense are activated at mucosal surfaces, the control of the inflammatory reaction is an important condition to preserve the integrity of the mucosal interface with the external world.

II. THE INTESTINAL COMMENSAL MICROBIOTA, AN "ORGAN" WHOSE CELLULAR COMPOSITION IS INDEPENDENT OF HOST CONTROL?

The host and microbial factors that direct the establishment and composition of the intestinal microbiota are largely unknown (12). It is well established that colonization of the intestinal mucosa by a pathogen often results in host tissue damage. In response to this, different mechanisms of host defense are switched on to eliminate the noxious newcomer. In contrast, the subtle interactions between the host and its commensal intestinal microbiota are less characterized. To date, little is known about the factors that maintain the steady-state situation of the microbiota and, moreover, about whether the host plays an active role in its control. Some authors have tried to incorporate, in theoretical models, all the relevant interbacterial interactions that could contribute to the ecological equilibrium of the colonic microbiota (13).

However, it is difficult to assume that the host remains passive in the control of the microbiota composition. There is increasing evidence that an immunological surveillance of intestinal bacteria seems to take place. But nevertheless, the overall picture of the host immune response to commensals and the way it alters intestinal colonization and functionality remains to be elucidated.

According to current knowledge two types of control mechanisms relevant to the preservation of the intestinal microbiota stability can be distinguished:

1. Intermicrobial interactions (host-independent)
2. Host–bacterial interactions

Examples of the first type of control are bacterial metabolic activities that modify the intestinal ecological characteristics through production of short-chain fatty acids and acidification, O_2 consumption, and modifications of the redox potential, which create an ecosystem optimal for the growth of some bacterial genera and hostile for others. The production of antimicrobial compounds by some microbes and competition for mucus binding or receptors on intestinal epithelial cells are additional factors contributing to the establishment of ecological niches within the intestine (14). Increasing importance has been attributed to the communication among bacteria in biofilms or definite microenvironments, called *quorum sensing*, that might help to explain basic aspects of colonization (12).

In addition to pure intermicrobial interactions, there are host–microbe interactions of extreme complexity that are starting to be characterized. Host–bacteria interactions appear to be relevant to the control of the intestinal microbiota composition. In turn, the indigenous microflora participates in the development and maturation of the gut (8) and in the regulation of intestinal functions.

III. HOST-MICROBIOTA CROSS-TALK

It has become evident that the intestinal microbiota plays a role in the development of innate and adaptive immune responses (15). In this review particular emphasis is put on the influence of nonpathogenic bacteria on preservation of intestinal tissue homeostasis.

Some authors have proposed that the immune system does not react to but is tolerant of components of the intestinal microflora (16,17). Moreover, it is thought that the breakdown of tolerance could lead to or perpetuate inflammatory bowel disease (18). In contrast to this, it has also been reported that commensal bacteria can be recognized by the host immune system and elicit systemic antibody responses (19,20). A wide proportion of the resident microflora is covered by antibodies in the lumen of the gut, mainly of the immunoglobulin A (IgA) isotype (21). The same study indicates that a variable proportion of the intestinal microflora is devoid of antibody coating. The secretory or systemic immune responses to components of the intestinal microbiota, found in physiological conditions, does not appear to have any pathogenic consequences for the host. Interestingly the antibody response also does not seem to correlate with the elimination of bacteria from the intestine (22). Thus, it is still unknown whether the immune response may play a role in the constitution of the commensal microbiota. It is possible that nonspecific binding of antibodies to bacteria occurs in the lumen of the gut and may represent a mechanism of clearance of an excess of antibodies generated by a permanently challenged intestinal mucosal by microbes, food antigens, and proinflammatory molecules.

To date, the mechanisms implicated in the prevention of a deleterious inflammatory-immune response to nonpathogenic luminal antigens is not clear. Antigen-specific $CD4^+$ regulatory T cells were shown to be an important lymphocyte subset for the balance of protective and pathogenic immune responses. These cells predominantly produce interleukin 10 (IL-10) or transforming growth factor-β (TGFβ) upon stimulation and promote the downregulation of the inflammatory response (23–25).

Regulation of intestinal homeostasis by innate defense mechanisms has received increasing interest and might be involved in the blunting of the inflammatory reaction at the lamina propria. The generation of arachidonic acid metabolites by cyclooxygenase 2 (COX-2) was shown in 1999 to be an essential modulator of the intestinal response to dietary antigens (26). The

COX-2 is expressed in intestinal epithelial cells (IECs) (27) and in macrophages (28). Induction of the enzyme activity results in the production of different metabolites, including prostaglandin E2 (PGE2) (29), which has important immunomodulatory activities such as downregulation of major histocompatibility (MHC) class II antigens (30) and induction of IL-10 production (31). It has been proposed that induction of COX-2 in lamina propria macrophages depends on luminal bacterial stimuli such as lipopolysaccharide (LPS) (26). In addition, the small intestinal mucosa is different from other mucosal sites as non-bone-marrow-derived lamina propria cells were shown to have the capacity to express basal levels of COX-2 in the absence of exogenous stimuli (32). This finding suggests that commensal bacteria may exert a dual function. In addition to the stimulation of mucosal mechanisms of defense (15), they may play a role in the maintenance of the homeostasis of the immune response (26).

Although several mechanisms seem to work in concert to prevent the inflammatory reaction of the intestinal barrier to commensal microbiota and external antigens, intestinal inflammatory disease is a common pathological condition of humans. The development of intestinal inflammation in knockout and transgenic rodents has confirmed the way genetic and environmental factors are responsible for mucosal inflammation. Spontaneous inflammation of the GI tract has been demonstrated in transgenic rats expressing the human human leukocyte antigen B27 (HLA-B27) and β-microglobulin, in mice deficient for IL-2, IL-2Rα, IL-10, T cell receptor α (TCRα), TGF-β1, and others. In most of the murine models no inflammation occurs if the animals are maintained under germ-free conditions (33). The diversity of genetic murine alterations that lead to models of inflammatory bowel disease (IBD) seems to suggest an impairment of the physiological regulation of tissue homeostasis by a complex network of cytokines and cells.

IV. MUCOSAL RECOGNITION OF BACTERIA: HOMEOSTASIS, DEFENSE, AND IMMUNE-INFLAMMATORY CONFLICT

The capacity of a microorganism to cause infectious or inflammatory disease in the host depends not only on the microbe virulence traits but also on the host's health since immunodeficiency or malnutrition may account for increased microbial pathogenesis (34).

The diversity in the genetic makeup of bacterial genera and species (35) to which humans are exposed at the mucosal surfaces is extremely wide. Gastric acid pH, the mucus layer, production of bactericidal peptides such as α-defensins, and the integrity of the epithelial layer are nonspecific factors that can confer wide protection to the host. However, the concept of the mucosal barrier has progressively moved from that of a stable barrier against microbes, to a reactive compartment with the capacity to sense environmental changes and respond to them.

Host protective functions against microbes at mucosal sites require a molecular machinery for microbial recognition. This function may rely on either a wide repertoire of specific molecules or on fewer less specific molecules that recognize microbial traits common to different genera and species, called *pattern recognition receptors* (PRRs) (36,37).

The specific molecules comprise immunoglobulins (Igs) and T cell receptors (TCRs), which are part of the adaptive immune response and need molecular adaptation to bacterial molecules in order to accomplish recognition successfully.

In contrast, PRRs are germ-line-encoded glycoproteins and recognize bacterial molecules shared by a variety of microorganisms. In the gut mucosa PPRs are found on macrophages that are strategically distributed beneath the epithelial basal membrane, where they play a key role, as they guard potential sites of epithelial layer breaching by luminal bacteria. The PRRs are also expressed on IECs (38). CD14 and the family of molecules called *Toll-like receptors* (TLRs) are examples of PRRs (39). CD14 was shown to react with TLR4 to deliver signals to eucaryotic cells (40) Different members of the TLR family react with different types of bacteria or bacterial compounds. Whereas TLR4 participates in the recognition of gram-negative bacteria or their products, e.g., lipopolysaccharide (LPS), TLR2 is involved in signaling immune cells upon reaction with gram-positive bacteria or their products, e.g., lipotheichoic acid (LTA) and peptidoglycan (PG) (41,42). Ten TLR molecules have been described so far in mammalians, and it is assumed that each of the TLRs recognizes a discrete subset of molecules widely shared by microbial pathogens. Thus, together, TLRs afford protection against an important number of pathogenic bacteria (43). Reaction of bacterial products with a given TLR results in host cellular signaling, leading to activation of κB (NF-κB) or AP-1 (40).

A major feature of the intestine's defensive function is the capacity to differentiate between commensal bacteria, which may be beneficial for the host, and the colonization by pathogens. It is possible that highly specific molecules, Igs or TCRs, could account for the discrimination between pathogens and commensals, although the latter can also evoke a humoral immune response (19–21). Discrimination between commensals and pathogens seems a difficult task for PRR as they recognize common traits across bacterial genera and species. It is likely that in addition to sensing of the presence of bacteria via PRR, a second danger signal has to be delivered to initiate an appropriate response to pathogenic bacteria (44,45). Endogenous ligands of PRR, and more specifically TLRs, may initiate an immune activation upon detection of cell damage or cellular stress. These endogenous "danger signals" may be heat shock proteins, nucleotides, or extracellular matrix breakdown products. This possibility has given origin to the so-called danger model of the immune response (46,47), in which activation occurs if host tissue damage is detected. One feature of bacterial pathogenicity is the capacity to invade host epithelial cells (43). The defensive systems can also recognized noninvading pathogens when they induce cytopathic or enterotoxic effects or when they target membrane cell compartments not exposed to the luminal environment (39). The recognition of pathogens before induction of tissue damage, in the lumen of the gut, remains for the moment speculative, but some bacterial stimulation of airway and intestinal epithelia may result in the production of bactericidal proteins such as defensins. Interestingly epithelial expression of defensins seems to depend on the recognition bacterial molecules by PRRs such as CD14 (48,49).

A. The Epithelium, a Major Player for Sensing the Bacterial Content of the Gut and Danger of Infection

The epithelium of the intestinal tract is a constitutive component of the innate response of the host to the luminal microbiota.

Some authors have demonstrated an epithelial innate response to invasive pathogens (50,51). In fact, the intracellular detection of LPS by a protein called caspase-activating and -recruitment domain 4 (CARD4; also known as nucleotide binding oligomerization domain (*NOD1*)) was shown to be implicated in NF-κB-activation-dependent proinflammatory pathway in IECs triggered by invasive bacteria (52–54).

In addition, the innate epithelial host response also seems to be able to discriminate between commensals and pathogens in the absence of invasion. The molecular bases of discrimination between commensals and pathogens may depend on different molecular mechanisms. It has been postulated that conserved bacterial molecules, present in pathogens as well as nonpathogens, promote an epithelial response when detected in combination with a second danger signal (39). An example reported in 2001 is the recognition of flagellin in the basolateral membrane of the intestinal epithelium, which may initiate a defensive response subsequent to TLR5 binding (55,56).

Furthermore, the IEC also recognizes and responds to nonpathogenic commensal bacteria, an issue that was neglected until 2000 (57). It has been demonstrated in vitro that the epithelial cell response to nonpathogens was enhanced upon coculturing IEC with professional immune cells. Epithelial cell cultures were performed in a two compartment transwell system, with enterocyte-like human epithelial cells on the apical compartment, and peripheral blood mononuclear cells in the basolateral compartment. Bacterial stimulation was performed adding bacteria to the apical-epithelial compartment. Notably, a differential response to nonpathogenic bacteria could be observed, distinguishing two major cytokine–chemokine profiles of IEC (Fig. 1):

1. Gram-negative *Escherichia coli* and specific strains of gram-positive lactobacilli triggered a NF-κB-mediated inflammatory response. This initial proinflammatory event was transient and self-limiting (57). It was shown that immune-modulatory cells in the lamina propria, particularly the macrophage,

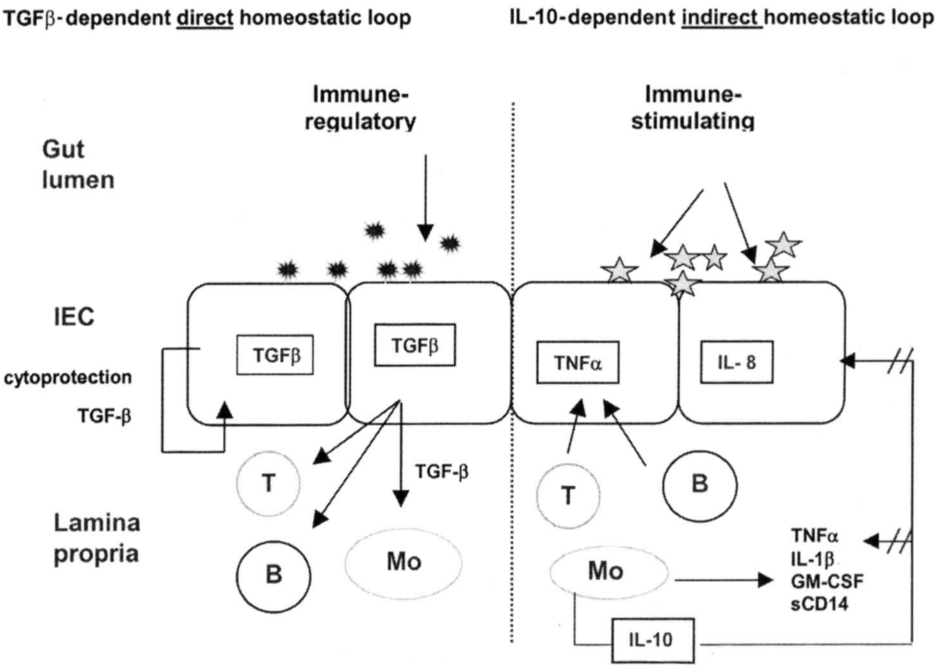

Figure 1 Proposed model for an autoregulatory loop to ensure homeostasis after activation of IEC by nonpathogenic bacteria. Left side, Direct TGFβ-dependent epithelial homeostatic loop induced by immune-regulatory nonpathogenic bacteria. TGFβ exerts cytoprotective effects on the epithelium and strengthens barrier integrity. In addition, TGFβ promotes regulatory T cells (Th_3) and prevents T cell death. Right side, Indirect IL-10-dependent homeostatic loop induced by immune-stimulatory bacteria. The transient proinflammatory response is blunted by lamina propria (LP) cells, particularly immunosuppressive LP macrophages, which predominantly secrete IL-10. IEC, intraepithelial cell; IL-10, interleukin 10; TNFα, tumor necrosis factor α; GM-CSF, granulocyte macrophage stimulating factor; IE, intraepithelial; T, T lymphocyte; B, B-lymphocyte; Mo, macrophage.

seem to play a crucial role in switching off the initial inflammatory response of the epithelium. This integrated epithelium–lamina propria response requires the participation of an intricate network of cells and cellular mediators that assure tissue homeostasis.

2. A second type of response was evoked by gram-positive bacteria that failed to induce stimulation of proinflammatory mediators, but instead triggered gene activation of the regulatory cytokine TGF-β (57). This type of response could constitute one of the mechanisms underlying induction of immunological tolerance to components of the microflora.

Both types of epithelial responses against nonpathogenic bacteria seem to indicate the importance of either a self-limiting (type 1) or a noninflammatory (type 2) cellular immune response in the context of the antigen-rich intestinal environment (58).

In contrast, if bacterial invasion occurs, or other cytopathic effects are induced by pathogens (59), the inflammatory reaction persists in order to recruit leukocytes for the clearance of the pathogen. Thus, epithelial recognition and response to pathogens may require multistep stimulation, comprising recognition of bacteria and cellular damage. Once the pathogen has been cleared from the body, regulatory mechanisms intervene to stop an idle inflammatory response and restore the intestinal barrier integrity. Thus, integrated epithelium–lamina propria cell interaction may also be crucial for tissue restitution.

The NF-κB activation is a pivotal event in defensive inflammatory reactions, inducing gene activation such as IL-2, IL-6, IL-8, and tumor necrosis factor α (TNF-α) (60). Pathogens and stress are common activators of this pathway. In addition, NF-κB activation may play a pathogenic role in chronic inflammatory diseases including arthritis (61) and inflammatory bowel disease (62), in particular Crohn's disease. The NF-κB proteins are a family of heterodimers and homodimers with different activatory capacities. The p50 and p65 heterodimers are involved in IL-1 and TNF-αgene activation and are blocked by IL-10 (63,64). The NF-κB exists in the cytoplasm in an inactive form associated with regulatory proteins called *inhibitors of κB* (IκBs). Phosphorylation of IκB is an important step in NF-κB activation. Phosphorylated IκB is then ubiquitinated and degraded in the 26S proteosome (63). Then NF-κB is released from the complex with IκB and translocated to the nucleus, where proinflammatory gene activation is initiated.

In 2001 an association between Crohn's disease and a mutation in NOD-2 gene was documented. The NOD-2 encodes a cytosolic protein whose expression is restricted to monocytes (65). The NOD-2 protein is a member of the CED4/APAF1 superfamily of apoptosis regulators. It is composed of two caspase recruitment domains (CARDs), a nucleotide binding domain, and a leucine-rich repeat (LRR) in its carboxy terminus side (66,67). The LRR domain binds lipopolysaccharide and activates the NF-κB pathway; however, this induction is deficient in the mutant NOD-2. This genetic defect may be related to perdurable pathogenic inflammatory-immune response. However, the mechanisms that link the carriage of the mutated NOD-2 variant and the persistent tissue inflammation in Crohn's disease have not been unraveled yet.

Modulation of NF-κB has become an obvious target for anti-inflammatory therapies, and specific compounds may act at different levels of its complex regulatory pathway. It will be of interest to investigate whether luminal components, such as bacterial products, can stimulate a cellular response that blunts the pathogenic activation of NF-κ(B. It has been reported that some nonpathogenic bacteria can prevent NF-κB activation through inhibition of IκB-α ubiquitination. This type of bacteria would have direct anti-inflammatory activity at the level of epithelial cells. This relation shows that luminal microflora can send positive and negative signals to mucosal epithelial cells as part of the symbiotic interactions with the host (68).

V. CONCLUSION

Bacterial strains, which induce immunoregulatory cytokines at the IEC level, may modulate inflammation without the participation of lamina propria cells, such as macrophages or lymphocytes. This may represent a strategy to control inflammation in the gut when the integrated epithelium–lamina propria cell loop is impaired by genetic defects.

The homeostatic response to a limited epithelial inflammation controlled by immunomodulatory cytokines such as TGF-β, IL-10 (64), PGE2, or an appropriate NF-κB regulation may be impaired in IBD, as many studies seem to indicate.

Therefore, use of commensal bacterial strains, including probiotic bacteria, selected for their properties to activate an epithelial inhibitory response may be a rational nutritional approach for patients suffering from IBD.

REFERENCES

1. MJ Blaser, JM Musser. Bacterial polymorphisms and disease in humans. J Clin Invest 107:391–392, 2001.
2. P Pochart, P Marteau, Y Bouhnik, I Goderel, P Bourlioux, JC Rambaud. Survival of bifidobacteria ingested via fermented milk during their passage through the human small intestine: an in vivo study using intestinal perfusion Am J Clin Nutr 55:78–80, 1992.

3. Y Bouhnik, P Pochart, P Marteau, G Arlet, I Goderel, JC Rambaud. Fecal recovery of viable *Bifidobacterium* sp ingested in fermented milk. Gastroenterology 102:875–878, 1992.
4. Y Bouhnik, K Vahedi, L Achour, A Attae, J Salfati, P Pochart, P Marteau, B Flourié, F Bornet, JC Rambaud. Short-chain fructo-oligosaccharide administration dose-dependently increases fecal bifidobacteria in healthy humans. J Nutr 129:113–116, 1999.
5. MC Moreau, M Thomasson, R Ducluzeau, P Raibaud. Cinétique d'établissement de la microflore digestive chez le nouveau-né humain en fonction de la nature du lait. Reprod Nutr Dev 26:745–753, 1986.
6. Y Benno, K Sawada, T Mitsuoka. The intestinal microflora of infants: composition of fecal flora in breast-fed and bottle-fed infants. Microbiol Immunol 28:975–986, 1984.
7. MJ Hopkins, R Sharp, GT Macfarlane. Age and disease related changes in intestinal bacterial populations assessed by cell culture, 16S rRNA abundance, and community cellular fatty acids profile. Gut 48:198–205, 2001.
8. LV Hooper, MH Wong, A Thelin, L Hansson, PG Falk, JI Gordon. Molecular analysis of commensal host–microbial relationships in the intestine. Science 291:881–884, 2001.
9. WJ Anderson, L Watson. Asthma and the hygiene hypothesis. N Engl J Med 344:1643–1644, 2001.
10. PM Matricardi, S Bonini. High microbial turnover rate preventing atopy: a solution to inconsistencies impinging on the hygiene hypothesis? Clin Exp Allergy 30:1506–1510, 2000.
11. JH Droste, MH Wieringa, JJ Weyler, VJ Nelen, PA Vermeire, HP Van Bever. Does the use of antibiotics in early childhood increase the risk of asthma and allergic disease? Clin Exp Allergy 30:1547–1553, 2000.
12. LV Hooper, JI Gordon. Commensal host–bacterial relationship in the gut. Science 292:1115–1118, 2001.
13. R Freter, E Stauffer, D Cleven, LV Holdeman, WE Moore. Continuous-flow cultures as in vitro models of the ecology of large intestinal flora. Infect Immun 39:666–675, 1983.
14. D Brassart, EJ Schiffrin. The use of probiotics to reinforce mucosal defence mechanisms.Trends Food Sci Technol 8:321–326, 1997.
15. JJ Cebra. Influences of microbiota on intestinal immune system development. Am J Clin Nutr 69:1046S–1051S, 1999.
16. MC Foo, A Lee. Antigenic cross-reaction between mouse intestine and a member of autochthonous microflora. Infect Immun 9:1066–1069, 1974.
17. MC Foo, A Lee. Immunologic response of mice to members of the autochthonous intestinal microflora. Infect Immun 6:525–532, 1972.
18. R Duchmann, I Kaiser, E Hermann, W Mayet, K Ewe, KH Meyer-zum-Buschenfelde. Tolerance exists towards resident intestinal flora but is broken in active inflammatory bowel disease. Clin Exp Immunol 102:448–455, 1995.
19. K Kimura, AL McCartney, MA McConnell, GW Tannock. Analysis of fecal populations of bifidobacteria and lactobacilli and investigation of the immuno-

logical responses of their human hosts to the predominant strains. Appl Environ Microbiol 63:3394–3398, 1997.
20. HZ Apperloo-Renkema, TG Jagt, RH Tonk, D van der Waaij. Healthy individuals possess circulating antibodies against indigenous faecal microflora as well as against allogenous microflora: an immunomorphometrical study. Epidemiol Infect 111:273–285, 1993.
21. LA van der Waaij, PC Limburg, G Mesander, D van der Waaij. In vivo IgA coating of anaerobic bacteria in human faeces. Gut 38:348–354, 1996.
22. HZ Apperloo-Renkema, D van der Waaij. Study on the colonization resistance for Enterobacteriaceae in man by antibodies in the clearance of these bacteria from the intestine. Epidemiol Infect 107:619–626, 1991.
23. J Braunstein, L Qiao, F Autschbach, G Schurmann, S Meuer. T cells of the human lamina propria are high producers of interleukin-10. Gut 41:215–220, 1997.
24. P Gonnella, Y Chen, J Inobe, Y Komagata, M Quartulli, HL Weiner. In situ immune response in gut-associated lymphoid tissue (GALT) following antigen in TCR-transgenic mice. J Immunol 160:4708–4718, 1998.
25. H Groux, A O'Garra, M Bigler, M Rouleau, S Antonenko, JE de Vries, MG Roncarolo. A CD4+ T-cell subset inhibits antigen-specific T-cell responses and prevents colitis. Nature 389:737–742, 1997.
26. RD Newberyy, WF Stenson, RG Lorenz. Cyclooxygenase-2-dependent arachidonic acid metabolites are essential modulators of the intestinal immune response to dietary antigens. Nat Med 5: 900–906, 1999.
27. RN DuBois, MTsujii, P Bishop, JA Awad, K Makita, A Lanahan. Cloning and characterization of a growth factor–inducible cyclooxygenase gene from rat intestinal epithelial cells. Am J Physiol 266:G822–G827, 1994.
28. M Giroux, A Descoteaux. Cyclooxygenase-2 expression in macrophages: modulation by protein kinase C-alpha. J Immunol 165:3985–3991, 2000.
29. WL Smith, RM Garavito, DL DeWitt. Prostaglandin endoperoxide H synthases (cyclooxygenases)-1 and -2. J Biol Chem 271:33157–33160, 1996.
30. DS Snyder, DI Beller, ER Unanue. Protaglandins modulate macrophage Ia expression. Nature 299:163–165, 1982.
31. CE Demeure, LP Yang, C Desjardins, P Raynaud, G Delespesse. Prostaglandin E2 primes naive T cells for the production of anti-inflammatory cytokines. Eur J Immunol 27:3526–3530, 1997.
32. RD Newberry, JS McDonough, WF Stenson, RG Lorenz. Spontaneous and continuous cyclooxygenase-2-dependent prostaglandin E2 production by stromal cells in the murine small intestine lamina propria: directing the tone of the intestinal immune response. J Immunol 166:4465–4472, 2001.
33. AK Bhan, E Mizoguchi, RN Smith, A Mizoguchi. Colitis in transgenic and knockout animals as models of human inflammatory bowel disease. Immunol Rev 169:195–207, 1999.
34. A Casadevall, LA Pirofski. Host–pathogen interactions: redefining the basic concepts of virulence and pathogenicity. Infect Immun 67:3703–3713, 1999.
35. BG Spratt, MCJ Maiden. Bacterial population genetics, evolution and epidemiology. Philos Trans R Soc Lond B Biol Sci 354:701–710, 1999.

36. R Medzhitov, EB Kopp. The toll-receptor family and control of innate immunity. Curr Opin Immunol 11:13–18, 1999.
37. O Takeuchi, K Hoshino, T Kawai, H Sanjo, H Takada, T Ogawa, K Takeda, S Akira. Differential roles of TLR2 and TLR4 in recognition of gram-negative and gram-positive bacterial cell wall components. Immunity 11:443–451, 1999.
38. E Cario, D Podolsky. Differential alteration in intestinal epithelial cell expression of Toll-like receptor 3 (TLR3) and TLR4 in inflammatory bowel disease. Infect Immun 68:7010–7017, 2000.
39. DJ Philpott, SE Girardin, PJ Sansonetti. Innate immune responses of epithelial cells following infection with bacterial pathogens. Curr Opin Immunol 13:410–416, 2001.
40. G Zhang, S Ghosh. Toll-like receptor-mediated NF-κB activation: a phylogenetically conserved paradigm in innate immunity. J Clin Invest 107:13–19, 2001.
41. S Akira, K Takeda, T Kaisho. Toll-like receptors: critical proteins linking innate and acquired immunity. Nat Immunol 2:675–680, 2001.
42. DM Underhill, A Ozinsky, AM Hajjar, A Stevens, CB Wilson, M Basetti, A Aderem. The Toll-like receptor 2 is recruited to macrophage phagosomes and discriminates between pathogens. Nature 401:811–815, 1999.
43. DA Kimbrell, B Beutler. The evolution and genetics of innate immunity. Nat Rev Gen 2:256–267, 2001.
44. CA Janeway, CC Goodnow, R Medzhitov. Immunological tolerance: danger—pathogen on the premises! Curr Biol 6:519–522, 1996.
45. P Matzinger. Tolerance, danger, and the extended family. Annu Rev Immunol 12:991–1045, 1994.
46. P Matzinger. An innate sense of danger. Semin Immunol 10:399–415, 1998.
47. S. Gallucci, P Matzinger. Danger signals: SOS to the immune system. Curr Opin Immunol 13:114–119, 2001.
48. MN Becker, G Diamond, MW Verghese, SH Randell. CD-14 dependent lipopolysaccharide-induced beta-defensin-2 expression in human tracheobronchial epithelium. J Biol Chem 275:29731–29736, 2000.
49. DA O'Neil, EM Porter, D Elewaut, GM Anderson, L Echmann, T Ganz, MF Kagnoff. Expression and regulation of the human β-defensins hBD-1 and hBD-2 in intestinal epithelium. J Immunol 163:6718–6724, 1999.
50. HC Jung, L Eckmann, SK Yang, A Panja, J Fierer, E Morzycka-Wroblewska, MF Kagnoff. A distinct array of proinflammatory cytokines is expressed in human colon epithelial cells in response to bacterial invasion. J Clin Invest 95:55–65, 1995.
51. L Eckmann, MF Kagnoff, J Fierer. Epithelial cells secrete the chemokine interleukin-8 in response to bacterial entry. Infect Immun 61:4569–4574, 1995.
52. J Bertin, WJ Nir, CM Fischer, OV Tayber, PR Errada, JR Grant, JJ Keilty, ML Gosselin, KE Robinson, GH Wong, MA Glucksmann, PS DiStefano. Human CARD4 protein is a novel CED-4IApaf-1 cell death family member that activates NF-kappaB. J Biol Chem 274:12955–12958, 1999.
53. N Inohara, T Koseki, L del Peso, Y Hu, C Yee, S Chen, R Carrio, J Merino, D Liu, J Ni, G Nuñez. Nod1, and Apaf-1 like activator of caspase-9 and nuclear factor-κ. B J Biol Chem 274:14560–14567, 1999.

54. N Inohara, Y Ogura, FF Cen, A Muto, G Nuñez. Human Nod1 confers responsiveness to bacterial lipopolysaccharide. J Biol Chem 276:2551-2554, 2001.
55. AT Gerwitz, PO Simon, CK Schmidtt, LJ Taylor, CH Hagedorm, AD O'Brien, AS Neish, JL Madara. *Salmonella typhimurium* translocates flagellin across the intestinal epithelia, inducing a proinflammatory response. J Clin Invest 107:99–109, 2001.
56. F Hayashi, KD Smith, A Ozinsky, TR Hawn, EC Yi, DR Goodlett, JK Eng, S Akira, DM Underhill, A Aderem. The innate immune response to bacterial flagellin is mediated by Toll-like receptor 5. Nature 410:1099–1103, 2001.
57. D Haller, C Bode, WP Hammes, AMA Pfeifer, EJ Schiifrin, S Blum. Non-pathogenic bacteria elicit a differential cytokine response by intestinal epithelial cell/leucocyte co-cultures. Gut 47:79–87, 2000.
58. Madara J L. Warner-Lambert/Parke-Davis Award Lecture. Pathobiology of the intestinal epithelial barrier. Am J Pathol 137:1273–1281, 1990.
59. BB Finlay, P Cossart. Exploitation of mammalian host cell functions by bacterial pathogens. Science 276:718–726, 1997.
60. AS Baldwin Jr. Series introduction: the transcription factor NF-κB and human disease. J Clin Invest 107:3–6, 2001.
61. ZN Han, DL Boyle, AM Manning, GS Firestein. AP-1 and NF-kappa B regulation in rheumatoid arthritis and murine collagen-induced arthritis. Autoimmunity 28:197–208, 1998.
62. M Neurath, S Petterson, KH Meyer-zum-Buschenfelde, W Strober. Local administration of antisense phophorothioate oligonucleotides to the p65 subunit of NF-kappaB abrogates established experimental colitis. Nat Med 2:998–10041, 1996.
63. PP Tak, GS Firestein. NF-κB: a key role in inflammatory diseases. J Clin Invest 107:7–11, 2001.
64. AJ Schottelius, MW Mayo, RB Sartor, AS Blaldwin Jr. Interleukin-10 signaling blocks inhibitors of kappaB kinase activity and nuclear factor kappaB DNA binding. J Biol Chem 274:31868–31874, 1999.
65. Y Ogura, N Inohara, A Benito, FF Chen, G Nuñez. Nod1/Apaf-1 family member that is restricted to monocytes and activates NF-κB. J Biol Chem 276:4812–4818, 2001.
66. JP Hugot, M Chamaillard, H Zouall, S Lesage, JP Cézard, J Belaiche, S Almer, C Tysk, CA O'Morain, M Gasull, V Binder, Y Kinkel, A Cortot, R Modigliani, P Laurent-Puig, C Gower-Rousseau, J Macry, JF Colombel, M Sahbatou, G Thomas. Association of NOD2 leucine-rich repeat variants with susceptibility to Crohn's disease. Nature 411:599–603, 2001.
67. Y Ogura, DK Bonen, N Inohara, DL Nicolae, FF Chen, R Ramos, H Britton, T Moran, R Karaliuskas, RH Duerr, JP Achkar, SR Brant, TM Bayless, BS Kirschner, SB Hanauer, G Nuñez, JH Cho. A frameshift mutation in NOD2 associated with susceptibility to Crohn's disease. Nature 411:603–606, 2001.
68. AS Neish, AT Gewirtz, H Zeng, AN Young, ME Hobert, V Karmali, AS Rao, JL Madara. Prokaryotic regulation of epithelial responses by inhibition of IκB- α ubiquitination. Science 289:1560–1563, 2000.

16

Influence of Diet on Aging and Longevity

Katherine Mace
Nestlé Research Center, Lausanne, Switzerland

Barry Halliwell
National University of Singapore, Singapore

I. INTRODUCTION

Aging can be defined as a time-dependent decrease in the ability of an organism to maintain homeostasis or react to external and internal changes (1,2). Therefore, aging can be highlighted by a loss of functionality and resistance or adaptability to stress that leads to increased susceptibility of individuals to age-associated diseases. In developed countries, the large majority of the aging population die of degenerative diseases such as cancers and cardiovascular diseases, which account for almost 50% of all deaths in persons 85 years and older (3).

There are many theories that have been put forward to explain the process of aging, and none of them gives a complete satisfactory explanation (for review see Ref. 1). In general, theories of aging can be classified into two categories: (a) Aging is caused by random detrimental events that cause accumulation of errors in the cell over time (i.e., stochastic theory) and (b) aging is directly controlled by a genetic program.

A. Stochastic Theory

In the stochastic theory, one major mechanism involving free radical damage seems to emerge in a quite consensual manner. Reactive oxygen species

(ROS), produced by a variety of processes, including mitochondrial electron transport, react with macromolecules and cause extensive oxidative damage, including deoxyribonucleic acid (DNA) base modifications (e.g., hydroxylation or ring opening), peroxidation of membrane lipids, carbonylation, and loss of sulfydryl groups in proteins. This oxidative damage, if unrepaired, would be responsible for age-associated loss of functional capacity.

As reviewed by Lindsay (4), an age-associated elevation of oxidative stress has been demonstrated by an increase in the exhalation of alkanes (products of ROS-induced peroxidation of lipids) and in the concentration of oxidatively damaged protein as well as an increase in the concentration of 8-hydroxy 2′-deoxyguanosine (8-OHdG), indicator of oxidative DNA damage. Oxidative damage in mitochondrial proteins is increased with age, and in damaged mitochondria, free radical generation may increase (5). The increase in oxidative damage may result from an increase in the rate of ROS generation or a decline in antioxidant defenses. The decreased age-related function of the mitochondrial respiratory chain has been proposed to be responsible for increased electron leakage associated with enhanced ROS production (6,7). Nevertheless, studies investigating the relationship between age and respiratory chain function have generated conflicting data: some demonstrating decreases of different respiratory complex enzymes (8–11) and others showing no correlation (12–14). There is little evidence that aging is associated with changes in the activity of enzymes involved in antioxidant defense, such as superoxide dismutase (SOD) and catalase (CAT) (15,16). Interestingly, transgenic *Drosophila* species overexpressing both SOD and CAT have been reported to live up to 30% longer than wild-type flies (17). Specific overexpression of SOD alone in adult motoneurons is able to extend *Drosophila* sp. life span by 40% (18) but not when expressed all over (19). Therefore, cumulative ROS toxicity in the nervous system may be an important factor in determining longevity, at least in flies. In rodents, overexpression of SOD alone in various organs or tissues, including brain, is not sufficient to extend the life spans of animals (20,21).

Other sources of damage are generated by metabolic reactions. For example, glucose reacts with primary amine groups on proteins, DNA, and lipids throughout the body. This complex reaction, called the *Maillard reaction*, occurs at high rates in diabetics and at a lower rate in healthy humans (22). Extensive accumulation of a variety of Maillard reaction products was shown to occur in cartilage collagen with age (23), consistently with a role for glycation in aging. Nevertheless, definitive proof that active ROS generation and glycation product formation are causative of rather than a result of aging remains to be obtained.

B. Genetic Control

Since different organs are controlled by different genes and each organ follows its own pattern of aging, it initially appeared uncertain that one single gene can be responsible for overall aging. However, defects in a single gene involved in genomic stability, in DNA repair pathways, in control of transcription, or in antioxidant defense could theoretically contribute to the alteration of other genes and accelerate the aging process. The theory of genetic control of aging has gained interest since the finding that single genes can have large effects on the longevity of yeast, worms, flies, and even mammals (for review see Ref. 24). Interesting findings have raised the possibility that insulin or insulin-like growth factor (IGF) signaling pathways somehow limit life span. *C. elegans*, mutated in the daf-2 gene, which encodes an insulin/IGF receptor homologue, are able to live more than twice as long as wild-type animals (25). Similarly, mutation of the *Drosophila melanogaster* InR gene, which is homologous to the daf-2 gene, leads to a 85% extension of longevity (26). A 50% extension of the median life span is also observed in *D. melanogaster* mutated in the CHICO gene encoding an insulin receptor sub-strate (27).In 1996, the first life extension mutant in mice was reported (28). Mice, called *Ames dwarf mice*, exhibited a life span extended by 50% (28). In these mice, mutation in the df gene was demonstrated and associated with a defective function of the anterior pituitary gland leading to a deficiency in growth hormone (GH) and insulin-like growth factor I (IGF-I) (29). Subsequently other mutations extending murine life span were found, such as a mutation in the gene encoding p66shc, involved in a signaling pathway controlling stress apoptotic responses (30).

Since 1975, nine senescence-accelerated mouse (SAM) strains have been generated by selective inbreeding at Kyoto University (for review see Ref. 31). The SAM strains are an interesting murine model of senescence acceleration and age-associated disorders that should contribute to a better understanding of the role of genes in longevity. The nature of the genetic defect(s) associated with the different SAM phenotypes is unknown. Nevertheless, loci analysis revealed that chromosomes 4, 14, 16 and 17 could contain the genes responsible for accelerated senescence in the SAM strains (32). The development of a mouse model deficient in the catalytic subunit of the telomerase gene (TERT) (33) is increasing our understanding in the longevity field, especially for the aging of highly proliferative organs. The TERT gene is known to block cellular senescence by maintaining the length of telomeres (34). In the absence of telomerase activity, each chromosome loses a few telomere repeats, at each cell division, making the chromosome shorter, unstable, and subject to fusion as well as rearrangements. The mice lacking TERT exhibit shortened life span, defective spermatogenesis, compromised proliferative capacity of hematopoietic progenitors, small intestine atrophy, impaired wound healing, alopecia, and hair graying but no greater incidence of thinning bones, blocked arteries, cataracts, or diabetes than normal mice (35–37). Therefore, telomere shortening does not cause generalized premature aging but rather a reduced capacity to respond to stress, as in wound healing and hematopoietic ablation (36). The elevated genomic instability, in the absence of telomerase activity, also leads to an increase in the incidence of spontaneous cancers originating from highly proliferative cell types [i.e., germ cells (teratocarcinomas), leukocytes (lymphomas), and keratinocytes (squamous cell carcinoma)] (36). Nevertheless, the role of the telomerase gene as a cause of cancers appears to be complex and not always beneficial since telomerase is activated in most human cancers (38,39). The genome integrity is preserved in normal cells by the p53 tumor suppressor protein, known to induce apoptosis, cell cycle arrest, and cellular senescence in stress conditions (40). The p53 is activated in telomerase-deficient mice (41). In 2002, Tyner and associates demonstrated that mutant mice showing increased p53 activity display enhanced resistance to spontaneous tumors, as expected but also display early aging-associated phenotypes and reduced longevity (42). As mentioned by the authors, this finding supports the concept that senescence is a mechanism of tumor suppression (42).

Werner's syndrome (WS) is a rare autosomal recessive disease of rapid aging (for review see Ref. 43). In WS the loss of functional capacity is restricted mainly to tissues that undergo continuous regeneration such as the skin, gut, connective tissues, and bone marrow. The WS gene (WRN) encodes a member of the RecQ family of DNA helicases, which when mutated lead to genomic unstability and a distinctive mutator phenotype (44). Whereas normal human fibroblasts achieve about 60 population doublings during their life span in culture, WS fibroblast cultures achieve no more than about 20 population doublings (45). Ectopic expression of TERT in WS fibroblasts is able to prevent the premature replicative senescence of cultured WS fibroblasts (46). Therefore, one consequence of the WS defect seems to be the acceleration of normal telomere-driven replicative senescence.

Aging appears as a complex process involving both intrinsic (genetic) and extrinsic (environmental) factors. These factors could have a different influence, depending on the tissue and its renewal capacity. Gut and skin tissues show an elevated turnover rate that requires a continuous recruitment of stem cells. Loss of stem cell pools or decreased capacity of these cells to proliferate, via, for example, telomere shortening controlled replicative senescence, may be the major aging process for these tissues. Muscle and brain tissues, which display both slow turnover of regeneration and high-energy metabolism, may be the main targets for oxidative damage. In fact, the effects of genetic and environmental factors could combine in both types of tissues.

With aging and senescence of stem cells, gut and skin regeneration declines and the nonrenewed terminally differentiated cells (i.e., postmitotic cells) may could be subjected to stochastic events such as oxidative stress. For muscle tissue, the increased numbers of damaged myotubes over time, in response to stochastic events, may not be compensated for by regeneration processes from muscle stem cells (satellite cells), presenting a senescence phenotype with age.

II. NUTRITION AND AGING

Throughout history, claims have been made for dietary substances that influence aging. Unfortunately, the supporting evidence for many of these claims is less than compelling. Today only three dietary interventions, discussed in the following sections, show potential benefits for aging.

A. Calorie Restriction

1. Effects on Aging

Calorie restriction (CR), which is defined as undernutrition without malnutrition, is the only experimental intervention known clearly to retard aging. It extends both mean (survival) and maximal (longevity) life span and delays the onset of a number of life-shortening diseases such as diabetes, hypertension, and cancer (for review see Ref. 47). Its effects on aging have been demonstrated in several animal species such as *D. melanogaster*, *C. elegans*, and rodents. Middle-aged monkeys, fed for 10 years with a calorie-restricted diet, show low blood pressure, reduced intraabdominal fat, and decreased levels of circulating insulin, glucose, cholesterol, and triacylglycerols (48–50). Therefore, CR puts these monkeys at a lower risk for diabetes and cardiovascular diseases than the ad libitum fed animals (49,51). The survival (median life span) of the restricted monkeys appears to be exceeding that of ad libitum–fed monkeys (51), but effects on longevity (maximal life span) in nonhuman primates will be only demonstrated conclusively, by this ongoing study, around 2020.

2. Mechanisms of Action

The mechanism by which CR produces these beneficial changes is not understood, but pathological and molecular observations have provided clues on the ways CR that could decelerate aging. Today three mechanisms are being explored: (a) attenuation of oxidative damage, (b) insulin and glucose exposure, and (c) genomic stability.

Attenuation of Oxidative Damage. In mice fed ad libitum, levels of oxidized DNA (8-OHdG) and proteins (carbonyls) increase with age in brain as well as in cardiac and skeletal muscle (52). The CR (60% of the caloric intake of the ad libitum regimen) prevents the increase of oxidative damage in these tissues (52–54). Age-dependent increase of oxidative damage in skeletal muscle of nonhuman primates is attenuated by CR (55). A CR-induced decrease in oxidative stress appears to be most profound in postmitotic tissues and may derive from lower mitochondrial production of free radicals. Whether CR decreases the generation of oxidants by increasing antioxidant defense enzymes (e.g., SOD and catalase) or/and by causing less free radical production via improved electron transport is still a matter of debate.

Insulin and Glucose Exposure. Rodents as well as nonhuman primates, on a low-calorie diet, have increased insulin sensitivity and both lowered fasting glucose and lowered insulin levels (50, 51, 56–58).

Different observations suggest that insulin levels throughout life influence aging. Hyperinsulinemia is associated with an increased prevalence of coronary heart diseases (59), one of the primary causes of death in the developed world. Therefore, preserved insulin sensitivity and normoinsulinemia by reducing pathological condition onset positively influence life span. Interestingly, healthy centenarians have a preserved insulin effect on glucose and fat metabolism (60,61). As previously mentioned, impairment of genes encoding components of the insulin–IGF-I pathways increases the life span of animal models. In these models, it is speculated that insulin signaling pathways, by modulating neuroendocrine functions involved in growth, energy metabolism, and reproductive state, may influence aging. A theory of aging based on an imbalance of hormonal and growth factor exposure has also been proposed in mammals (62–64). The CR modulates not only insulin levels but several other endocrine factors. Among them, CR induces glucocorticoid (65) and growth hormone (GH) (66) concentrations, and IGF-I (67) and triiodothyronine (T3) (68) levels are reduced, at least in rodents. By potentiating cellular and organismic homeostasis throughout life, this endocrine compensatory mechanism certainly plays a role in the antiaging action of CR.

Protein glycation and accumulation of advanced glycation end products, caused by elevated blood glucose levels, are suspected to play an important role in aging. Although it is clear that CR reduces fasting glucose concentrations, conflicting data exist regarding its effects on the levels of glycosylated hemoglobin in rodents (69, 70). In middle-aged non human primates submitted to CR, the levels of glycosylated hemoglobin do not differ from those of the nonrestricted animals (48,58).

3. Increase of Genome Stability

Yeast silent information regulator 2 (Sir-2) is on oxidized nicotinamide-adenine dinucleotide– (NAD^+)-dependent histone deacetylase involved in transcriptional silencing and control of genomic stability (71,72). Yeast cells lacking Sir-2 have a reduced replicative life span, whereas cells with an extra copy of Sir-2 display a longer life span than wild-type (72,73). When yeast is grown in the presence of low concentrations of glucose, to mimic calorie restriction, extention of life span is observed (72). The increased longevity induced by CR in yeast is associated with the activation of Sir-2 gene expression and subsequent chromosome stability (72). In Sir-2 knockout yeast or in cells showing a disrupted NAD pathway, the effect of lowered glucose level on life span is abolished (74). One mechanism proposed is that Sir-2 competes with metabolic enzymes for NAD^+ utilization. By reducing metabolic activity, CR would reduce the utilization of NAD^+, which could then be used by Sir-2 to silence genes and stabilize the genome (74). In mammals, the evidence that genomic stability plays a role in the effects of CR on aging is only correlative: That is the antiaging action of calorie-restricted rodents is correlated with decreased DNA damage and mutation and increased DNA repair capacity (75). Therefore, experiments that employ more accurate assays on the regulatory pathways involved in chromosome stability are needed in order to understand the antiaging action of CR.

B. Antioxidants

As previously discussed, sustained oxidative damage of cell constituents is suspected to be a major factor in the decline of tissues and organs associated with aging and age-related degenerative diseases. Several epidemiological and clinical studies have revealed potential roles for certain dietary antioxidants in the reduction of risk of morbidity and mortality from cancer and heart diseases (76). Nevertheless, supplementation studies with antioxidants and their effects on life span in different cellular, animal, and human studies have produced inconsistent results. Ad libitum feeding of middle-aged C57BL/6NIA male mice with diets supplemented with different antioxidants (vitamin E, glutathione, melatonin, and strawberry extract) had no effect on the pathological outcome or on mean (survival) nor maximal life span (longevity) of the mice, despite the reduced level of lipid peroxidation products (77). Beta-carotene, of which the in vivo fre-radical scavenger activity is a matter of debate, was not shown to be an effective means for prolonging life span in C57BL/6J male mice, even if the supplementation began at 29 days of age (78). The addition of vitamin E to the diet of male mice, starting shortly after weaning, was shown to increase the average life span by 7% (79), whereas antioxidant supplementation initiated during middle age did not appear to affect age-associated pathological markers and longevity of mice (80). The few animal studies performed in the past and showing beneficial effects of antioxidants on longevity have to be considered cautiously as, at that time, rodent diets were often suboptimal. Corrections of deficiency rather than supplementation effects were then studied. Interestingly, brussels sprouts decrease DNA damage in rats and humans under conditions in which supplementation with vitamin E, vitamin C, or beta-carotene does not (81). Therefore, other known (e.g., polyphenols, micronutrients) or unknown compounds in fruits and vegetables may be responsible for the beneficial effects of diet on survival and should be further investigated.

C. Polyunsaturated Fatty Acids

Diet rich in n-3 polyunsaturated fatty acids (PUFAs) seems to modulate many age-associated degenerative diseases (i.e., atherosclerosis, coronary heart disease, and inflammatory disease) favorably (82). Long-term feeding of normal rats with diet rich in alpha-linolenate (18:3n-3) or linoleate (18:2n-6) induced a significant difference in the n-3/n-6 ratios of highly unsaturated fatty acids in brain. Rats fed the alpha-linolenate-rich diet had a longer mean survival time and an increased learning ability in advanced age (83). Similarly, diets with fish oil containing n-3 PUFAs dramatically extend the life span and delay the onset and progression of disease in autoimmune-prone female mice as compared to those fed corn oil rich in n-6 lipids (84,85). A study in 2000 was undertaken to investigate the effects of dietary PUFAs on longevity in the senescence-accelerated SAMP8 mouse model (86). Male mice were fed either an n-3 PUFA-rich (9 g/100 g perilla oil) or an n-6 PUFA-rich (9 g/100 g safflower oil) diet beginning at 6 wk of age. The results showed that the animals fed perilla oil had significantly lower scores of aging markers, represented by changes in behavior (e.g., reactivity or passivity), appearance (e.g., glossiness and coarseness of coat, hair loss), and pathological condition onset (e.g., ulcers, ophthalmic lesions, cancers) than those fed safflower oil (86). Surprisingly, their mean life span was significantly reduced compared to that of those fed with safflower oil. Therefore, even if supplementation of diet with n-3 PUFAs shows beneficial effects in certain animal models and in some target tissues such as immune system and brain, further investigations are required to demonstrate and to understand the potential benefits of PUFAs in aging.

III. CONCLUSIONS

Although the mechanism of aging is not elucidated, there is growing evidence that genetic controls as well as increased tissue damage are involved. Single genes appear to regulate aging, meaning that intervention may be possible. Nevertheless, because of the interplay between cellular senescence and cancer formation, development of pharmacological interventions, aimed to delay aging, and targeting proteins involved in genome stability (e.g., telomerase, p53, WRN) appears rather challenging.

Among all the nutritional claims vaunting beneficial influences on aging, only caloric restriction in laboratory animals has been clearly shown to slow senescence. Nevertheless, dietary interventions with antioxidants or PUFAs seem to demonstrate some potential benefits on the average life expectancy, probably by reducing the onset of age-related diseases. As the majority of the aging population die of degenerative diseases, research in dietary ingredients able to delay the appearance of these disorders appears to be of particular interest. Further work will be needed to determine whether specific compounds in the diet may influence aging per se.

REFERENCES

1. D Harman. Aging: phenomena and theories. Ann N Y Acad Sci 854: 1–7, 1998.
2. SM Jazwinski. Longevity, genes, and aging: a view provided by a genetic model system. Exp Gerontol 34: 1–6, 1999.
3. JE Sutherland, VW Persky, JA Brody. Proportionate mortality trends: 1950 through 1986. JAMA 264: 3178–3184, 1990.
4. DG Lindsay. Diet and ageing: the possible relation to reactive oxygen species. J Nutr Health Aging 3: 84–91, 1999.
5. ER Stadtman. Protein oxidation and aging. Science 257: 1220–1224, 1992.
6. MK Shigenaga, TM Hagen, BN Ames. Oxidative damage and mitochondrial decay in aging. Proc Natl Acad Sci U S A 91: 10771–10778, 1994.
7. BN Ames, MK Shigenaga, TM Hagen. Mitochondrial decay in aging. Biochim Biophys Acta 1271: 165–170, 1995.
8. I Trounce, E Byrne, S Marzuki. Decline in skeletal muscle mitochondrial respiratory chain function: possible factor in ageing. Lancet 1: 637–639, 1989.
9. JM Cooper, VM Mann, AH Schapira. Analyses of mitochondrial respiratory chain function and mitochondrial DNA deletion in human skeletal muscle: effect of ageing. J Neurol Sci 113: 91–98, 1992.
10. D Boffoli, SC Scacco, R Vergari, G Solarino, G Santacroce, S Papa. Decline with age of the respiratory chain activity in human skeletal muscle. Biochim Biophys Acta 1226: 73–82, 1994.
11. RH Hsieh, JH Hou, HS Hsu, YH Wei. Age-dependent respiratory function decline and DNA deletions in human muscle mitochondria. Biochem Mol Biol Int 32: 1009–1022, 1994.
12. A Barrientos, J Casademont, A Rotig, O Miro, A Urbano-Marquez, P Rustin, F Cardellach. Absence of relationship between the level of electron transport chain activities and aging in human skeletal muscle. Biochem Biophys Res Commun 229: 536–539, 1996.

13. D Chretien, J Gallego, A Barrientos, J Casademont, F Cardellach, A Munnich, A Rotig, P Rustin. Biochemical parameters for the diagnosis of mitochondrial respiratory chain deficiency in humans, and their lack of age-related changes. Biochem J 329: 249–254, 1998.
14. JA Kent-Braun, AV Ng. Skeletal muscle oxidative capacity in young and older women and men. J Appl Physiol 89: 1072–1078, 2000.
15. JB de Haan, F Cristiano, RC Iannello, I Kola. Cu/Zn-superoxide dismutase and glutathione peroxidase during aging. Biochem Mol Biol Int 35: 1281–1297, 1995.
16. S Oh-Ishi, T Kizaki, H Yamashita, N Nagata, K Suzuki, N Taniguchi, H Ohno. Alterations of superoxide dismutase iso-enzyme activity, content, and mRNA expression with aging in rat skeletal muscle. Mech Ageing Dev 84: 65–76, 1995.
17. WC Orr, RS Sohal. Extension of life-span by overexpression of superoxide dismutase and catalase in Drosophila melanogaster. Science 263: 1128–1130, 1994.
18. TL Parkes, AJ Elia, D Dickinson, AJ Hilliker, JP Phillips, GL Boulianne. Extension of Drosophila lifespan by overexpression of human SOD1 in motorneurons (comment) (see comments). Nat Genet 19: 171–174, 1998.
19. WC Orr, RS Sohal. Effects of Cu-Zn superoxide dismutase overexpression of life span and resistance to oxidative stress in transgenic Drosophila melanogaster. Arch Biochem Biophys 301: 34–40, 1993.
20. TT Huang, EJ Carlson, AM Gillespie, Y Shi, CJ Epstein. Ubiquitous overexpression of CuZn superoxide dismutase does not extend life span in mice. J Gerontol A Biol Sci Med Sci 55: B5–9, 2000.
21. IM Gallagher, P Jenner, V Glover, A Clow. CuZn-superoxide dismutase transgenic mice: no effect on longevity, locomotor activity and 3H-mazindol and 3H-spiperone binding over 19 months. Neurosci Lett 289: 221–223, 2000.
22. T Niwa, N Takeda, H Yoshizumi, A Tatematsu, M Ohara, S Tomiyama, K Niimura. Presence of 3-deoxyglucosone, a potent protein crosslinking intermediate of Maillard reaction, in diabetic serum. Biochem Biophys Res Commun 196: 837–843, 1993.
23. N Verzijl, J Degroot, E Oldehinkel, RA Bank, SR Thorpe, JW Baynes, MT Bayliss, JW Bijlsma, FP Lafeber, JM Tekoppele. Age-related accumulation of Maillard reaction products in human articular cartilage collagen. Biochem J 350: 381–387, 2000.
24. L Guarente, C Kenyon. Genetic pathways that regulate ageing in model organisms. Nature 408: 255–262, 2000.
25. C Kenyon, J Chang, E Gensch, A Rudner, R Tabtiang. A *C. elegans* mutant that lives twice as long as wild type. Nature 366: 461–464, 1993.
26. M Tatar, A Kopelman, D Epstein, MP Tu, CM Yin, RS Garofalo. A mutant Drosophila insulin receptor homolog that extends life-span and impairs neuroendocrine function. Science 292: 107–110, 2001.
27. DJ Clancy, D Gems, LG Harshman, S Oldham, H Stocker, E Hafen, SJ Leevers, L Partridge. Extension of life-span by loss of CHICO, a Drosophila insulin receptor substrate protein. Science 292: 104–106, 2001.

28. HM Brown-Borg, KE Borg, CJ Meliska, A Bartke. Dwarf mice and the ageing process (letter). Nature 384: 33, 1996.
29. SA Camper, TL Saunders, RW Katz, RH Reeves. The Pit-1 transcription factor gene is a candidate for the murine Snell dwarf mutation. Genomics 8: 586–590, 1990.
30. E Migliaccio, M Giorgio, S Mele, G Pelicci, P Reboldi, PP Pandolfi, L Lanfrancone, PG Pelicci. The p66shc adaptor protein controls oxidative stress response and life span in mammals. Nature 402: 309–313, 1999.
31. T Takeda. Senescence-accelerated mouse (SAM): a biogerontological resource in aging research. Neurobiol Aging 20: 105–110, 1999.
32. C Xia, K Higuchi, M Shimizu, T Matsushita, K Kogishi, J Wang, T Chiba, MF Festing, M Hosokawa. Genetic typing of the senescence-accelerated mouse (SAM) strains with microsatellite markers. Mamm Genome 10: 235–238, 1999.
33. MA Blasco, HW Lee, MP Hande, E Samper, PM Lansdorp, RA DePinho, CW Greider. Telomere shortening and tumor formation by mouse cells lacking telomerase RNA. Cell 91: 25–34, 1997.
34. JR Smith, OM Pereira-Smith. Replicative senescence: implications for in vivo aging and tumor suppression. Science 273: 63–67, 1996.
35. HW Lee, MA Blasco, GJ Gottlieb, JW Horner, 2nd, CW Greider, RA DePinho. Essential role of mouse telomerase in highly proliferative organs. Nature 392: 569–574, 1998.
36. KL Rudolph, S Chang, HW Lee, M Blasco, GJ Gottlieb, C Greider, RA DePinho. Longevity, stress response, and cancer in aging telomerase-deficient mice. Cell 96: 701–712, 1999.
37. E Herrera, E Samper, J Martin-Caballero, JM Flores, HW Lee, MA Blasco. Disease states associated with telomerase deficiency appear earlier in mice with short telomeres. EMBO J 18: 2950–2960, 1999.
38. SE Artandi, S Chang, SL Lee, S Alson, GJ Gottlieb, L Chin, RA DePinho. Telomere dysfunction promotes non-reciprocal translocations and epithelial cancers in mice. Nature 406: 641–645, 2000.
39. D Hanahan. Benefits of bad telomeres. Nature 406: 573–574, 2000.
40. AJ Levine. p53, the cellular gatekeeper for growth and division. Cell 88: 323–331, 1997.
41. L Chin, SE Artandi, Q Shen, A Tam, SL Lee, GJ Gottlieb, CW Greider, RA DePinho. p53 deficiency rescues the adverse effects of telomere loss and cooperates with telomere dysfunction to accelerate carcinogenesis. Cell 97: 527–538, 1999.
42. SD Tyner, S Venkatachalam, J Choi, S Jones, N Ghebranious, H Igelmann, X Lu, G Soron, B Cooper, C Brayton, S Hee Park, T Thompson, G Karsenty, A Bradley, LA Donehower. p53 mutant mice that display early ageing-associated phenotypes. Nature 415: 45–53, 2002.
43. JO Nehlin, GL Skovgaard, VA Bohr. The Werner syndrome: a model for the study of human aging. Ann N Y Acad Sci 908: 167–179, 2000.
44. MD Gray, JC Shen, AS Kamath-Loeb, A Blank, BL Sopher, GM Martin, J

Oshima, LA Loeb. The Werner syndrome protein is a DNA helicase. Nat Genet 17: 100–103, 1997.
45. RG Faragher, IR Kill, JA Hunter, FM Pope, C Tannock, S Shall. The gene responsible for Werner syndrome may be a cell division "counting" gene. Proc Natl Acad Sci U S A 90: 12030–12034, 1993.
46. FS Wyllie, CJ Jones, JW Skinner, MF Haughton, C Wallis, D Wynford-Thomas, RG Faragher, D Kipling. Telomerase prevents the accelerated cell ageing of Werner syndrome fibroblasts. Nat Genet 24: 16–17, 2000.
47. EJ Masoro. Caloric restriction and aging: an update. Exp Gerontol 35: 299–305, 2000.
48. WT Cefalu, JD Wagner, ZQ Wang, AD Bell-Farrow, J Collins, D Haskell, R Bechtold, T Morgan. A study of caloric restriction and cardiovascular aging in cynomolgus monkeys (Macaca fascicularis): a potential model for aging research. J Gerontol A Biol Sci Med Sci 52: B10–19, 1997.
49. WT Cefalu, JD Wagner, AD Bell-Farrow, IJ Edwards, JG Terry, R Weindruch, JW Kemnitz. Influence of caloric restriction on the development of atherosclerosis in nonhuman primates: progress to date. Toxicol Sci 52: 49–55, 1999.
50. JJ Ramsey, RJ Colman, NC Binkley, JD Christensen, TA Gresl, JW Kemnitz, R Weindruch. Dietary restriction and aging in rhesus monkeys: the University of Wisconsin study. Exp Gerontol 35: 1131–1149, 2000.
51. MA Lane, DK Ingram, GS Roth. Nutritional modulation of aging in nonhuman primates. J Nutr Health Aging 3: 69–76, 1999.
52. RS Sohal, S Agarwal, M Candas, MJ Forster, H Lal. Effect of age and caloric restriction on DNA oxidative damage in different tissues of C57BL/6 mice. Mech Ageing Dev 76: 215–224, 1994.
53. A Dubey, MJ Forster, H Lal, RS Sohal. Effect of age and caloric intake on protein oxidation in different brain regions and on behavioral functions of the mouse. Arch Biochem Biophys 333: 189–197, 1996.
54. C Leeuwenburgh, P Wagner, JO Holloszy, RS Sohal, JW Heinecke. Caloric restriction attenuates dityrosine cross-linking of cardiac and skeletal muscle proteins in aging mice. Arch Biochem Biophys 346: 74–80, 1997.
55. TA Zainal, TD Oberley, DB Allison, LI Szweda, R Weindruch. Caloric restriction of rhesus monkeys lowers oxidative damage in skeletal muscle. FASEB J 14: 1825–1836, 2000.
56. N Kalant, J Stewart, R Kaplan. Effect of diet restriction on glucose metabolism and insulin responsiveness in aging rats. Mech Ageing Dev 46: 89–104, 1988.
57. NL Bodkin, HK Ortmeyer, BC Hansen. Long-term dietary restriction in older-aged rhesus monkeys: effects on insulin resistance. J Gerontol A Biol Sci Med Sci 50: B142–147, 1995.
58. MA Lane, SS Ball, DK Ingram, RG Cutler, J Engel, V Read, GS Roth. Diet restriction in rhesus monkeys lowers fasting and glucose- stimulated glucoregulatory end points. Am J Physiol 268: E941–948, 1995.
59. M Pyorala, H Miettinen, M Laakso, K Pyorala. Hyperinsulinemia predicts

coronary heart disease risk in healthy middle- aged men: the 22-year follow-up results of the Helsinki Policemen Study. Circulation 98: 398–404, 1998.
60. G Paolisso, A Gambardella, S Ammendola, A D'Amore, V Balbi, M Varricchio, F D'Onofrio. Glucose tolerance and insulin action in healty centenarians. Am J Physiol 270: E890–894, 1996.
61. G Paolisso, A Gambardella, S Ammendola, MR Tagliamonte, MR Rizzo, A Capurso, M Varricchio. Preserved antilipolytic insulin action is associated with a less atherogenic plasma lipid profile in healthy centenarians. J Am Geriatr Soc 45: 1504–1509, 1997.
62. WE Sonntag, C Lynch, P Thornton, A Khan, S Bennett, R Ingram. The effects of growth hormone and IGF-1 deficiency on cerebrovascular and brain ageing. J Anat 197: 575–585, 2000.
63. T Parr. Insulin exposure controls the rate of mammalian aging. Mech Ageing Dev 88: 75–82, 1996.
64. T Parr. Insulin exposure and aging theory. Gerontology 43: 182–200, 1997.
65. JF Nelson, K Karelus, MD Bergman, LS Felicio. Neuroendocrine involvement in aging: evidence from studies of reproductive aging and caloric restriction. Neurobiol Aging 16: 837–843; discussion 855–836, 1995.
66. WE Sonntag, X Xu, RL Ingram, A D'Costa. Moderate caloric restriction alters the subcellular distribution of somatostatin mRNA and increases growth hormone pulse amplitude in aged animals. Neuroendocrinology 61: 601–608, 1995.
67. CR Breese, RL Ingram, WE Sonntag. Influence of age and long-term dietary restriction on plasma insulin- like growth factor-1 (IGF-1), IGF-1 gene expression, and IGF-1 binding proteins. J Gerontol 46: B180–187, 1991.
68. JT Herlihy, C Stacy, HA Bertrand. Long-term food restriction depresses serum thyroid hormone concentrations in the rat. Mech Ageing Dev 53: 9–16, 1990.
69. MM Scrofano, J Jahngen-Hodge, TR Nowell, Jr, X Gong, DE Smith, G Perrone, G Asmundsson, G Dallal, B Gindlesky, CV Mura, A Taylor. The effects of aging and calorie restriction on plasma nutrient levels in male and female Emory mice. Mech Ageing Dev 105: 31–44, 1998.
70. M Novelli, P Masiello, M Bombara, E Bergamini. Protein glycation in the aging male Sprague-Dawley rat: effects of antiaging diet restrictions. J Gerontol A Biol Sci Med Sci 53: B94–101, 1998.
71. CE Fritze, K Verschueren, R Strich, R Easton Esposito. Direct evidence for SIR2 modulation of chromatin structure in yeast rDNA. EMBO J 16: 6495–6509, 1997.
72. S Imai, CM Armstrong, M Kaeberlein, L Guarente. Transcriptional silencing and longevity protein Sir2 is an NAD- dependent histone deacetylase. Nature 403: 795–800, 2000.
73. M Kaeberlein, M McVey, L Guarente. The SIR2/3/4 complex and SIR2 alone promote longevity in Saccharomyces cerevisiae by two different mechanisms. Genes Dev 13: 2570–2580, 1999.
74. SJ Lin, PA Defossez, L Guarente. Requirement of NAD and SIR2 for life-span extension by calorie restriction in Saccharomyces cerevisiae. Science 289: 2126–2128, 2000.

75. JJ Raffoul, Z Guo, A Soofi, AR Heydari. Caloric restriction and genomic stability. J Nutr Health Aging 3: 102–110, 1999.
76. A Ghiselli, A D'Amicis, A Giacosa. The antioxidant potential of the Mediterranean diet. Eur J Cancer Prev 6: S15–19, 1997.
77. M Meydani. Dietary antioxidants modulation of aging and immune-endothelial cell interaction. Mech Ageing Dev 111: 123–132, 1999.
78. HR Massie, JR Ferreira, Jr, LK DeWolfe. Effect of dietary beta-carotene on the survival of young and old mice. Gerontology 32: 189–195, 1986.
79. EA Porta, NS Joun, RT Nitta. Effects of the type of dietary fat at two levels of vitamin E in Wistar male rats during development and aging. I. Life span, serum biochemical parameters and pathological changes. Mech Ageing Dev 13: 1–39, 1980.
80. RD Lipman, RT Bronson, D Wu, DE Smith, R Prior, G Cao, SN Han, KR Martin, SN Meydani, M Meydani. Disease incidence and longevity are unaltered by dietary antioxidant supplementation initiated during middle age in C57BL/6 mice. Mech Ageing Dev 103: 269–284, 1998.
81. B Halliwell. Establishing the significance and optimal intake of dietary antioxidants: the biomarker concept. Nutr Rev 57: 104–113, 1999.
82. WE Connor. Importance of n-3 fatty acids in health and disease. Am J Clin Nutr 71: 171S–175S, 2000.
83. N Yamamoto, Y Okaniwa, S Mori, M Nomura, H Okuyama. Effects of a high-linoleate and a high-alpha-linolenate diet on the learning ability of aged rats: evidence against an autoxidation-related lipid peroxide theory of aging. J Gerontol 46: B17–22, 1991.
84. JT Venkatraman, B Chandrasekar, JD Kim, G Fernandes. Effects of n-3 and n-6 fatty acids on the activities and expression of hepatic antioxidant enzymes in autoimmune-prone NZBxNZW F1 mice. Lipids 29: 561–568, 1994.
85. CA Jolly, A Muthukumar, CP Avula, D Troyer, G Fernandes. Life span is prolonged in food-restricted autoimmune-prone (NZB × NZW)F(1) mice fed a diet enriched with (n-3) fatty acids. J Nutr 131: 2753–2760, 2001.
86. M Umezawa, T Takeda, K Kogishi, K Higuchi, T Matushita, J Wang, T Chiba, M Hosokawa. Serum lipid concentrations and mean life span are modulated by dietary polyunsaturated fatty acids in the senescence-accelerated mouse. J Nutr 130: 221–227, 2000.

17

Foods and Food Components in the Prevention of Cancer

Gary D. Stoner and Mark A. Morse
The Ohio State University, Columbus, Ohio, U.S.A.

Gerald N. Wogan
Massachusetts Institute of Technology, Cambridge, Massachusetts, U.S.A.

I. INTRODUCTION

Epidemiological studies clearly indicate a protective effect of fruit and vegetable consumption against the occurrence of several types of cancer in humans (1). This has led to a recommendation by the U.S. National Cancer Institute that all citizens consume at least four to six servings of fruit and vegetables per day. If this were to occur, it is estimated that the overall reduction in cancer incidence in the United States would be approximately 20%–30% (2), with similar (or possibly greater) reductions worldwide. The protective effects of fruits and vegetables against cancer, broadly referred to as *chemoprevention*, is attributed to their content of fiber and to multiple nutritive and nonnutritive compounds.

The goal of this chapter is to define chemoprevention, identify stages in carcinogenesis amenable to prevention strategies, summarize investigations on the chemopreventive effects of various food components and foods, and, where known, discuss the mechanisms of action of these foodstuffs in cancer prevention. The chapter also summarizes current human clinical trials to assess the preventative effects of dietary factors. It is not possible to include all known dietary inhibitors of cancer in this report; more information can be found in comprehensive reviews that have been published in recent years (3–6).

A. Chemoprevention

Chemoprevention is defined as prevention of cancer by the administration of one or more chemical entities, either as individual drugs or as dietary supplements (7). To date, more than 800 individual compounds, many of which have been isolated from dietary fruits and vegetables, have been shown to inhibit the development of cancer in animals. Several of these have been or are being evaluated for efficacy in humans (6), among whom the most likely target populations are individuals from high-risk groups. These include those who (a) engage in risk-taking behavior or lifestyle (e.g., smokers and heavy alcohol users), (b) have had occupational exposure to known carcinogens (e.g., asbestos workers, uranium miners), (c) are genetically predisposed to the development of cancer (e.g., individuals with familial colonic polyposis (FAP), xeroderma pigmentosum), (d) have premalignant lesions (e.g., colonic polyps, Barrett's esophagus, oral leukoplakia), (e) are survivors of primary cancers with a high degree of recurrence or a high tendency toward formation of second primary tumors (head and neck cancer, breast cancer, etc.), or (f) received significant chemotherapy and/or radiation therapy when treated for cancer (4–6). We do not foresee the administration of chemopreventive agents on a routine basis to the general population, since there is always the potential for toxicity after long-term administration. Further, the general population is probably not motivated to take chemopreventives on a routine basis for long periods. However, changes in dietary habits, such as those recommended by the National Cancer Institute mentioned, have the potential to reduce cancer incidence through broad application in general populations.

B. Multistep Nature of Carcinogenesis

The chemopreventive effects of food-derived and synthetic agents may act at several points in the progression of cancer, since carcinogenesis studies in animal model systems have shown that the development of cancer in several organs is preceded by multiple genetic changes. Early investigations, using the mouse skin model of tumorigenesis, defined three distinct stages of tumor development, i.e., initiation, promotion, and progression (8). Initiation is a process whereby chemical and physical carcinogenic agents produce a permanent structural change in the deoxyribonucleic acid (DNA) of a target cell. The target cell is then designated an *initiated cell.* At least three processes are important in initiation, namely, carcinogen metabolism, DNA repair, and cell proliferation. Metabolism may activate or inactivate the carcinogen; DNA repair may either correct the damage or introduce an altered base into the genome; and cell proliferation is essential to stabilize heritable changes in the genome. Initiation is an irreversible process, but not all initiated cells go on to establish a tumor, since many die through the process of programmed cell death (apoptosis) or undergo normal cellular differentiation. Promotion is an experimentally defined part of the carcinogenic process that is thought to involve epigenetic events, selectively influencing the clonal expansion of initiated cells. One group of chemicals, known as *tumor promoters*, may contribute to carcinogenesis by nongenotoxic mechanisms. For example, tumor promoters stimulate the proliferation of initiated cells, resulting in their clonal expansion and the formation of benign focal lesions such as adenomas and papillomas. Many of these lesions regress, but a few cells may acquire additional genetic changes and progress to cancer. Promotion is considered to be a reversible process, since cells return to their previous state if the promoting agent is withdrawn. Although there is no simple unifying mechanism to explain the functions of promoting agents, they appear to cause a transient increase in cell division and to decrease apoptosis, thus giving a selective growth advantage to initiated cells.

The third stage of carcinogenesis, progression, defines the stepwise development of a cancer cell as it acquires additional heritable changes leading to malignancy and metastatic potential. During progression, tumors acquire the ability to grow uncontrollably, invade locally, and establish distant metastases. A principal hallmark of progession is karyotypic instability. Chromosomes in tumors are often disrupted; afterward segments are translocated to other chromosomes, are present in multiple copies, or are partially or totally deleted. In parallel with these structural changes, there are changes in critical cellular genes, such as the activation of oncogenes through mutation and the inactivation of tumor suppressor genes through mutation, deletion, or hypermethylation. These changes lead to the additional selection of cells suited to neoplastic growth and to a more pronounced karyotypic instability. Analysis of tumor development at the molecular level clearly shows that multiple changes in the genome are associated with tumorigenesis. For example, studies on the stepwise development of colon cancer have shown that multiple genes on different chromosomes must be altered to allow a normal colon cell to acquire neoplastic and metastatic potential (9).

II. FOOD COMPONENTS AND FOODS IN CHEMOPREVENTION

As indicated, findings from epidemiological studies seeking to relate dietary patterns to disease risk have provided strong support for the conclusion that consumption of diets rich in fruits and vegetables is protective against cancers at many sites (1,2,10). Components of these foods include many substances that have been shown to have promise as cancer chemopreventive agents

(11,12). These agents exhibit inhibitory effects in vitro and in animal model tumor systems, and many have been evaluated for efficacy in phase II and III human clinical trials. Some of the major categories of dietary inhibitors of cancer are discussed in the following sections.

A. Carotenoids

Carotenoids comprise a large class of lipid-soluble compounds that include xanthophylls, lycopene, and carotenes (Fig. 1). *β*-Carotene, abundantly available in the diet as a component of green and yellow–red and yellow–orange vegetables and fruits, acts as an antioxidant in many in vitro systems and is also a precursor of vitamin A (13). An extensive literature shows that β-carotene is not genotoxic, teratogenic, or carcinogenic and does not induce organ toxicity when administered to animals in subacute, subchronic, or chronic oral toxicity studies at doses up to 1 g/animal/day (13). Consequently,

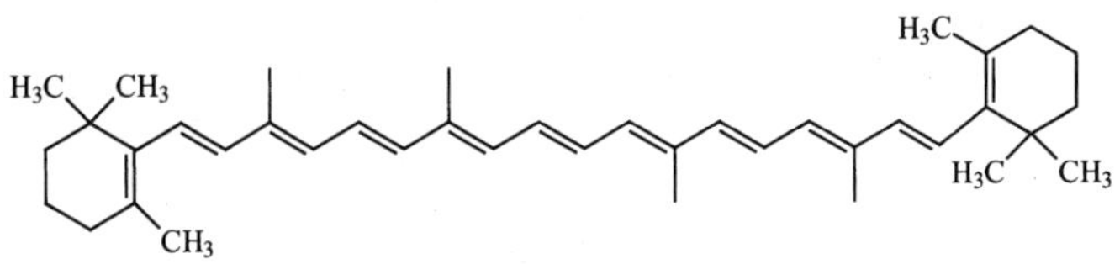

β-Carotene

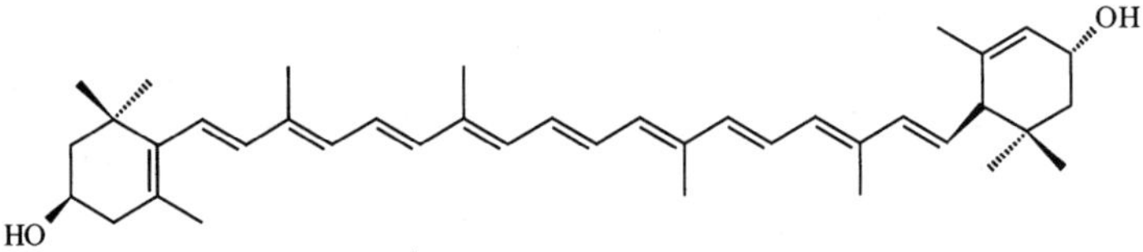

Xanthophyll

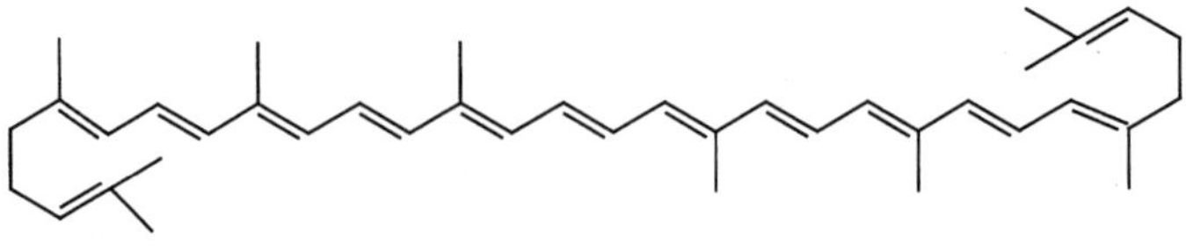

Lycopene

Figure 1 Representative carotenoids with chemopreventive activity.

β-carotene has been designated as a *generally recognized as safe* (GRAS) substance by the U.S. Food and Drug Administration.

Observational epidemiological studies that have correlated either high intakes of β-carotene-rich vegetables and fruits or high blood concentrations of β-carotene with cancer risk have consistently found evidence of a significant inverse association with lung cancer risk (14,15). Although it is unclear whether observed effects were attributable to replenishment of vitamin A in deficient diets, to its inherent antioxidant activity, or to confounding by other compounds in the diet, these data and animal model data indicated both chemopreventive promise and safety in the case of β-carotene. In particular, it was suggested that β-carotene might be an effective inhibitor of lung cancer (16). Large-scale interventions using β-carotene are presented in Sec. IV. Unexpectedly, results from these interventions indicated that β-carotene supplementation may increase lung cancer risk in high-risk individuals (17–19).

Importantly, accumulating evidence indicates that other carotenoids, including retinoids, as summarized later, and lycopene show promise as possible cancer chemopreventive agents. Several case-control and large prospective studies focusing on dietary assessment suggest inverse associations between intake of tomato products and risk of prostate, lung, and stomach cancers (20). Lycopene, a carotenoid found at relatively high concentration in tomatoes and tomato products, may contribute to the lower cancer risk, although the possible contributions of other carotenoids and phytochemicals present in tomato products have not been assessed. Lycopene has been shown to inhibit the action of insulin-like growth factor (IGF-1) in cultured breast cancer cells (21). Frequent consumption of tomato products as part of an overall healthy dietary pattern may therefore reduce risks of prostate and other cancers as well as other chronic diseases.

B. Vitamin A and Retinoids

Retinoids comprise vitamin A (retinol and its fatty acid esters, 3,4 didehydroretinol and retinal) together with a series of synthetic analogs (Fig. 2). Vitamin A is not synthesized by humans but is supplied through the diet; particularly rich sources are liver, milk and other dairy products, egg yolk, and fish liver oils. Vitamin A is required for growth and bone development, vision, reproduction, epithelial tissue differentiation, immune system integrity, and a variety of biochemical processes such as cholesterol biosynthesis and steroid metabolism. With respect to chemoprevention, the natural analogues of vitamin A can elicit significant toxicity when administered at high doses and/or for long periods. During the past few decades, therefore, numerous retinoid analogues of vitamin A have been synthesized and evaluated for toxicity and for efficacy as chemopreventive agents.

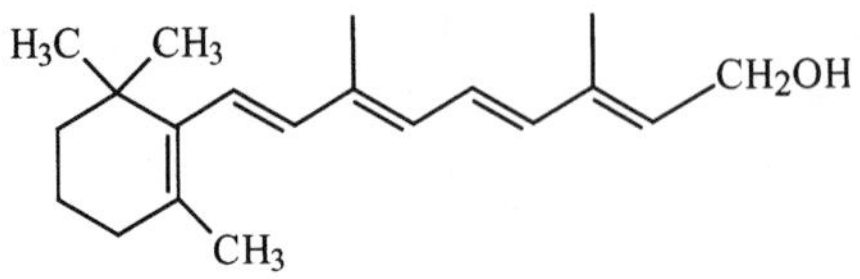

Retinol (Vitamin A)

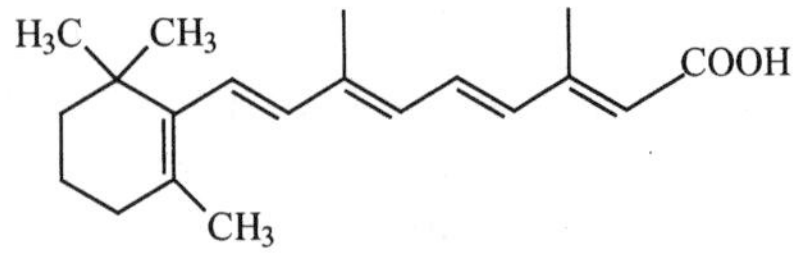

Retinoic acid

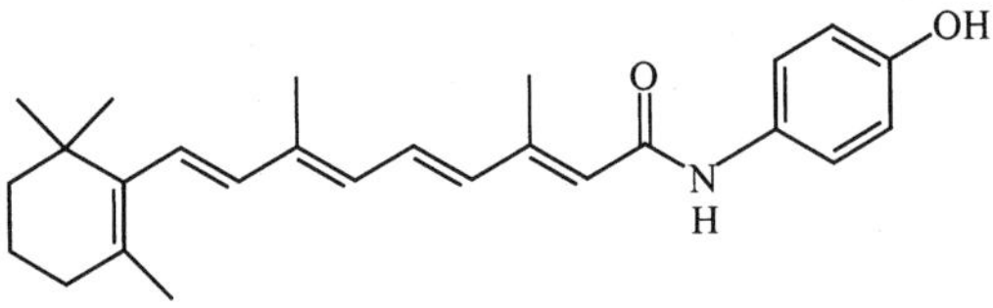

4-Hydroxyphenylretinamide

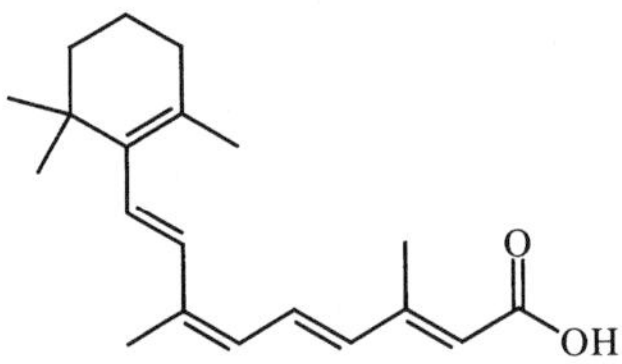

9-*cis* Retinoic acid

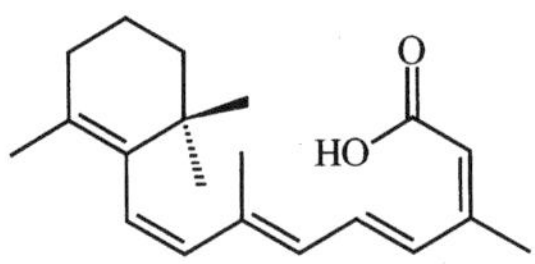

13-*cis* Retinoic acid

Figure 2 Vitamin A (retinol), its metabolite (retinoic acid), and representative retinoids (4-hydroxyphenylretinamide, 9-*cis*-retinoic acid, 13-*cis*-retinoic acid) with chemopreventive activity.

Multiple in vitro and in vivo studies have shown that retinoids are effective in suppressing tumor development in models of skin, breast, oral cavity, lung, prostate, bladder, liver, and pancreatic cancers (22–24). Retinoids are particularly effective chemopreventive agents for tumors of epithelial origin, such as those that arise in the aerodigestive tract (25). For example, 13-*cis*-retinoic acid inhibits chemically induced carcinogenesis in the buccal pouch and tongue of hamsters and oral squamous cell carcinoma in mice (26,27). Similarly, *N*-(4-hydroxyphenyl)retinamide (4-HPR) can reverse premalignant lesions and can inhibit the development of adenomas and adenocarcinomas in animal models of lung cancer (28,29). On the basis of these preclinical studies, several retinoids, including 13-*cis*-retinoic acid and 4-HPR, are being tested in clinical trials of epithelial cancers in humans (25,30).

Interestingly, in a 2001 report from our laboratory, the dietary administration of 4-HPR resulted in a marked enhancement of tumorigenesis in the rat esophagus induced by the carcinogen *N*-nitrosomethylbenzylamine (NMBA) (31). The 4-HPR was found to stimulate the metabolic activation of NMBA in explant cultures of rat esophagus by upregulating cytochrome P4502A3 and to increase formation of O^6-methylguanine adducts from NMBA in esophageal DNA in vivo. In addition, 4-HPR increased esophageal tumor size, presumably by enhancing cell proliferation. Thus, we suggest caution in the use of this agent in chemoprevention trials.

Vitamin A and its retinoid analogues have shown activity against many carcinogenesis-associated molecular targets, primarily in signal transduction pathways (6,32). At the cellular level, they act primarily by inhibiting proliferation and by inducing differentiation and apoptosis. Many of their effects are likely mediated by their interaction with retinoid receptors belonging to the superfamily of steroid and hormone receptors. All of these receptors function as transcription factors that regulate the expression of specific genes. Transcriptional changes that have been associated with chemopreventive functions of retinoids include genes involved in intercellular and intracellular signaling; growth factors, receptors, and oncogenes; modulation of hormones and immune response; altered intercellular matrix and proteolytic enzymes; and inhibition of viral replication. Differences in activity among the retinoids are thought to result from differential binding affinity with retinoid receptors (RARs and RXR's), through which they activate the receptors to regulate expression of genes associated with carcinogenesis (23,29). Unfortunately, retinoid toxicity has also been associated with binding to RARs, limiting the usefulness of certain of these analogues as cancer chemopreventive agents.

Overall, evidence that retinoids inhibit human carcinogenesis is compelling, although not definitive, since it is largely from uncontrolled trials (6). Low dietary intake of retinol has been associated with increased risk for breast and colon cancer in prospective studies and with lung and prostate cancers in retrospective studies. Serum retinol levels have been inversely associated with risk for gastrointestinal cancers in prospective studies and with sarcoma and lung, prostate, bladder, and ovarian cancers in retrospective studies. Work is in progress to develop retinoids as agents that take advantage of the preventive activities associated with the RXR-mediated pathway while avoiding the toxicity associated with RAR-mediated pathways.

C. Other Vitamins and Selenium

Vitamin C (Fig. 3), a water-soluble vitamin with antioxidant properties, is an important free-radical scavenger widely present in fruits, vegetables, and tubers. Epidemiological as well as biochemical evidence indicate that vita-

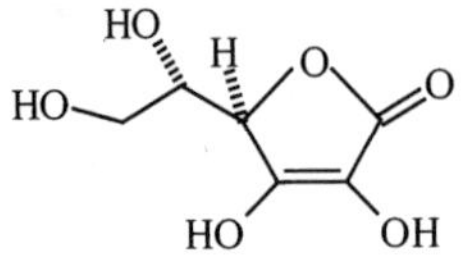

Vitamin C (Ascorbic acid)

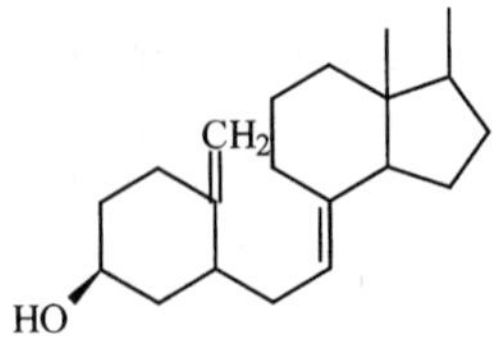

Vitamin D (Calciferol)

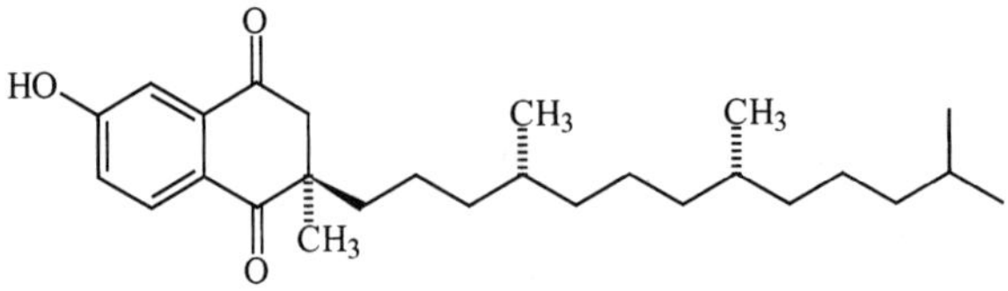

Vitamin E (α-tocopherol)

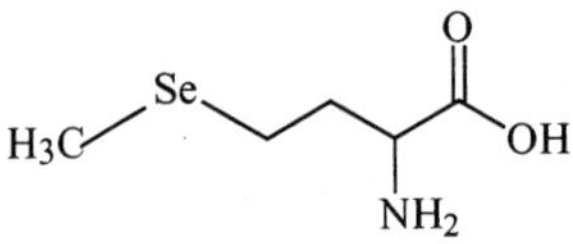

Selenomethionine

Figure 3 Vitamins C, D, and E and a representative selenium compound (selenomethionine) with chemopreventive activity.

min C probably decreases risk for stomach cancer and possibly decreases risk for cancers of the mouth, pharynx, esophagus, lung, pancreas, and cervix (33). In experimental animals, it blocks in vivo formation of carcinogenic *N*-nitroso compounds and inhibits carcinogenesis in several animal model systems (34).

Vitamin E (Fig. 3), a fat-soluble vitamin present at substantial levels in vegetable oils, whole-wheat products, and nuts, acts as a free-radical scavenger and prevents lipid oxidation in cell membranes. α-Tocopherol is the most active member of a series of structurally related compounds with vitamin E activity. One review of the epidemiological data suggested that vitamin E decreases risk for lung cancer (33); however, this decrease was not confirmed in another study (35). Results from a large-scale clinical intervention suggested a protective effect of vitamin E against cancers of prostate and colon (17).

Moreover, data from the Health Professionals Follow-up Study indicated an inverse association between supplemental vitamin E and prostate cancer risk among smokers (36). Vitamin E succinate, a derivative of vitamin E, has been shown to stimulate apoptosis of human prostate cancer cells in vitro (37).

Vitamin D (calciferol) (Fig. 3) is a fat-soluble vitamin obtained from dietary sources such as liver, oily saltwater fish, and eggs. In the context of cancer chemoprevention, it inhibits cell proliferation and DNA synthesis, alters expression of several oncogenes, reduces lipid peroxidation and angiogenesis, and induces differentiation (38). Epidemiological studies support that there is an inverse association between vitamin D intake and colorectal cancer risk and have also shown that vitamin D deficiency and low plasma levels increase the risk of prostate cancer (39,40). The overall evidence, however, is insufficient to provide definitive evidence of cancer chemopreventive activity for vitamin D. The situation with respect to folic acid is similar. It is abundantly available in many of the fruits and vegetables that are rich in carotenoids, and, together with vitamin B_{12}, methionine, and choline, it is involved in methyl group metabolism. Since DNA hypomethylation has been associated with neoplastic development, attention has been focused on a possible role for folic acid in cancer chemoprevention. An inverse correlation between folic acid intake and adenomatous polyps or colorectal cancer has been reported, but definitive evidence for a causal role has yet to be established.

Selenium, a trace element abundant in seafood, liver, and meat, is required as a cofactor for glutathione peroxidase, the major antioxidant enzyme catalyzing the oxidation of hydroperoxides (41,42). It is also involved in cell signaling in immune responses, which may contribute to its cancer chemopreventive potential. Experiments in animal models have demonstrated that selenium, in many forms, can inhibit carcinogenesis (43). For example, the selenium in high-selenium broccoli significantly reduced the incidence of chemically induced aberrant crypt foci in the colon of rats (44). Aberrant crypt foci are preneoplastic lesions that are thought to progress to adenomas and adenocarcinomas. Data from epidemiological studies suggest a protective effect of selenium on the incidence of lung cancer (33,45), but data have not been convincing overall for cancer at other sites, including prostate, skin, breast, pancreas, ovary, colon, and cervix (40).

D. Polyphenols

Polyphenolic compounds found in tea leaves, vegetables, and fruits are anticarcinogenic in several animal model systems. Those that have been investigated most extensively as chemopreventive agents include the tea polyphenols, curcumin, resveratrol, ellagic acid, and apigenin.

1. Tea Polyphenols

Teas made from dried leaves of *Camellia sinensis* are widely consumed globally, and the possible health benefits of tea have been extensively investigated. Of the approximately 2.5 million metric tons of dried tea manufactured annually, about 20% is green tea, 2% is oolong tea, and the remainder is black tea (46). Green tea is manufactured by drying fresh leaves, whereas black tea is made from crushed leaves. Black tea is commonly consumed in Western countries and green tea principally in Asia.

Green tea contains 35%–52% (measured in weight percentage of solid extract) catechins and flavonols combined. The four major catechins in green tea (Fig. 4) are (-)-epicatechin (EC), (-)-epicatechin-3-gallate (ECG), (-)-epigallocatechin (EGC), and (-)-epigallocatechin-3-gallate (EGCG). The major component is EGCG, which accounts for greater than 40% of the total polyphenolic mixture (47). During the process of black tea preparation, the polyphenols in green tea undergo oxidative polymerization by a process known as *fermentation*, which results in the conversion of catechins to the

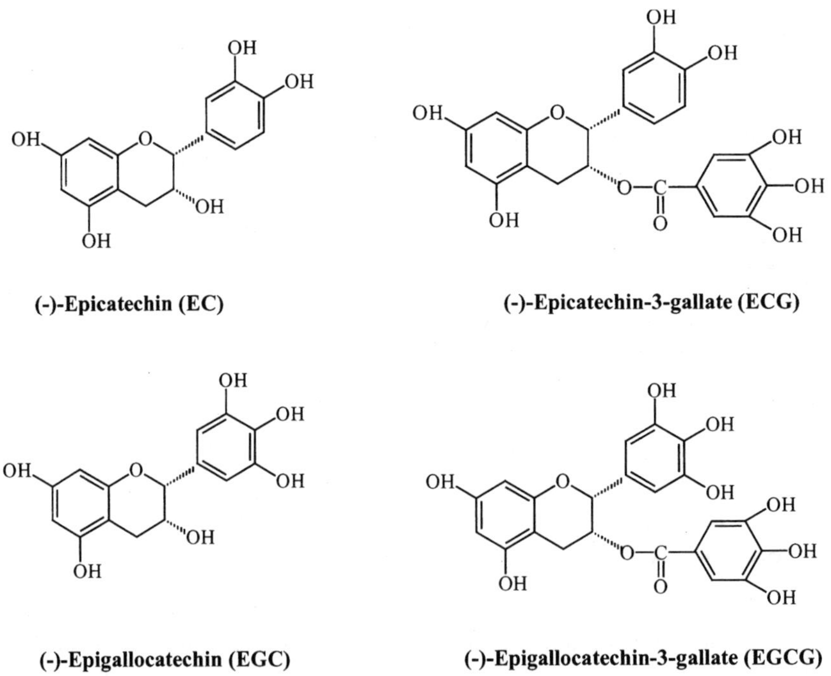

Figure 4 Catechins in green tea with chemopreventive activity.

theoflavins and thearubigins (Fig. 5). A typical black tea beverage contains 3%–10% catechins, 3%–6% theaflavins, 12%–18% thearubigins, and other components.

Various green and black tea preparations possess strong anticarcinogenic effects against cancer at different organ sites in animals, including skin, lung, oral cavity, esophagus, stomach, small intestine, colon, pancreas, and mammary gland (47). The constitutive polyphenols in these preparations, particularly EGCG, have been shown to influence carcinogen metabolism by inhibiting phase I P450 enzyme activities and stimulating phase II enzyme activities (48,49). In addition, tea components influence tumor promotion

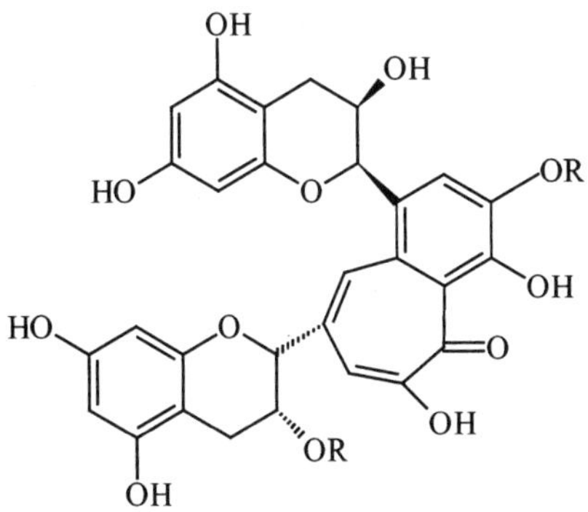

R = H or Galloyl

Theoflavins (TFS)

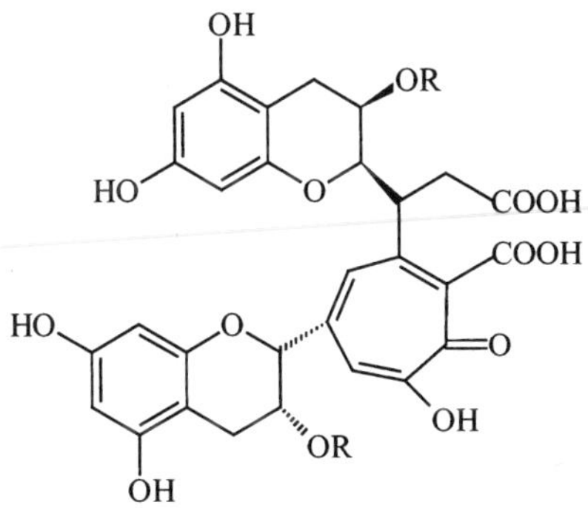

R = H or Galloyl

Thearubigins (TRS)

Figure 5 Theaflavins and thearubigins in black tea with chemopreventive activity.

and progress events. For example, EGCG and theaflavins block the signal transduction pathway leading to the activation of nuclear transcription factor κB (NF-κB) (50).

Epidemiological studies, on the other hand, have produced only inconclusive evidence concerning possible protective effects of tea consumption against cancer formation in humans (51,52). For example, several studies suggested a protective effect of black tea against oral, pharyngeal, laryngeal, digestive tract, and urinary bladder cancers, and of green tea against cancers of the esophagus and stomach (52,53). On the other hand, others found no evidence of protective effects against stomach, colorectal, lung, and breast cancers (54). Most reports showing positive cancer preventive effects were from studies of Asians who drink predominantly green tea, whereas studies of black tea drinkers infrequently observed protective effects (55,56).

Many factors may plausibly be responsible for the observed discrepancies between animal and human data. For example, dose levels that are effective in animal models, in which tea is provided as the sole form of liquid, are typically much higher than levels of human tea consumption. Moreover, the underlying mechanisms of tumorigenesis in animal models, generally induced by chemical carcinogens, may be unrelated to those responsible for human cancers of the same tissues. Additionally, bioavailability of tea constituents may be an important factor in determining its chemopreventive activity in animals and may confound extrapolation of animal data to humans. Diverse mechanisms have been suggested for inhibition of tumorigenesis by tea, including modulation of signaling pathways responsible for cell proliferation and transformation, induction of apoptosis in preneoplastic and neoplastic cells, and inhibition of both angiogenesis and tumor invasiveness (55,56).

2. Curcumin

Curcumin (diferuloylmethane) (Fig. 6) is a plant phenol widely used as a spice (curry) and food coloring agent. It is the major phenolic antioxidant and anti-inflammatory agent in turmeric and can be extracted from turmeric with ethanol and other organic solvents. Studies in in vitro and in vivo experimental models have shown that curcumin may prevent DNA damage by inhibiting phase I and stimulating phase II enzyme activities (57). It also inhibits tumor promotion and progression events in mouse epidermis, e.g., arachidonic acid metabolism via lipoxygenase and cyclooxygenase pathways (58) and TPA-induced inflammation (59). Animal studies have shown that curcumin is effective in inhibiting carcinogenesis in skin, colon, oral cavity, stomach, and mammary gland (47). Human trials with curcumin are under way.

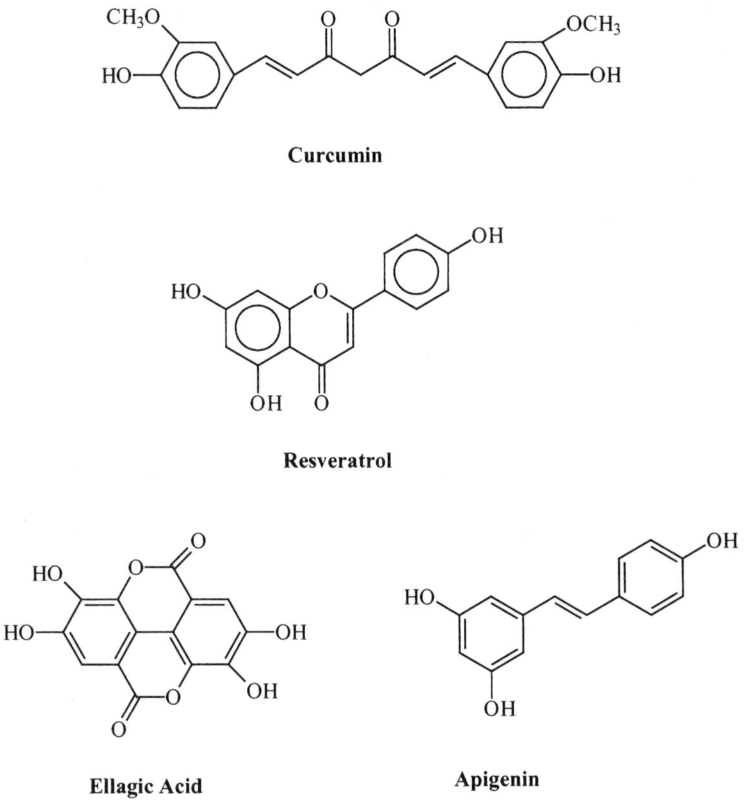

Figure 6 Naturally occurring polyphenols with chemopreventive activity.

3. Resveratrol

Resveratrol (3, 4′,5-trihydroxystilbene) (Fig. 6) is a polyphenol phytoalexin found in grapes and wines (60). It has been reported to exhibit a wide range of pharmacological properties and is believed to play a role in the prevention of human cardiovascular disease (61). It is synthesized by a wide variety of plants in response to injuries. Red wines are particularly rich sources of resveratrol, which is present in grape skin, not the flesh of the berry, and is therefore much more concentrated in red than in white wines. Fresh grape skins contain 50–100 mg/g resveratrol (62). Resveratrol is a promising cancer chemopreventive agent and also has potential as a chemotherapeutic agent (63). Numerous experiments have demonstrated its ability to block each step in the carcinogenesis process. These are summarized extensively by Gusman and associates

(64). Through inhibition of the activity of cyclooxygenases and the cytochrome P450 complex, resveratrol can simultaneously inhibit promutagen activation, stimulate carcinogen detoxification, and protect against diverse environmental toxicants. Through inhibition of prostaglandins and nitric oxide formation, it can prevent their stimulating effects on tumor development. Since formation of reactive oxygen species is also thought to be involved in tumor development, the antioxidant activity of resveratrol may also play a protective role. Suitably designed epidemiological studies will be required to assess the potential importance of resveratrol as a modulator of cancer risk as they have in showing an inverse correlation between red wine consumption and the incidence of cardiovascular disease.

4. Ellagic Acid

Ellagic acid (Fig. 6) is a naturally occurring phenolic constituent of many species from a diversity of flowering plant families. It is present in plants in the form of hydrolyzable tannins called *ellagitannins*. We determined the content of ellagic acid in a series of fruits and nuts; the highest amounts were found in blackberries, raspberries, strawberries, cranberries, walnuts, and pecans (65). It is pharmacologically active and has been found to control hemorrhage in animals and in humans, presumably as a result of its ability to activate Hageman factor.

In vivo and in vitro experiments in our laboratory and others have demonstrated the ability of ellagic acid to influence events involved in tumor initiation (47). For example, it inhibits the metabolic activation of carcinogens into forms that induce DNA damage. It can also promote carcinogen detoxification by stimulating the activity of various isoforms of the enzyme glutathione-S-transferase. A third mechanism by which ellagic acid may inhibit tumor initiation is through its potential role as a scavanger of the reactive metabolites of carcinogens, e.g., benzo(a)pyrene diolepoxide, discussed later. Finally, it has been shown to occupy sites in DNA that might otherwise react with carcinogens or their metabolites. Ellagic acid has also been shown to influence promotion and progression events in mouse skin through its inhibition of TPA-induced ornithine decarboxylase (ODC) activity, hydroperoxide production, and DNA synthesis (59), and of free-radical-induced lipid peroxidation (66). In vitro studies have demonstrated its ability to inhibit prostate cancer cell proliferation by activating *Waf1/Cip1* and by downregulating IGF-II in a p53-independent manner (67,68). Studies in 2002 using DNA microarray showed that ellagic acid influences the expression of clusters of genes that control cell proliferation, differentiation, and apoptosis in cultured prostate cancer cells (69).

Animal studies have demonstrated the preventative effect of ellagic acid on chemically induced tumors of the lung, skin, esophagus, and liver (47). It appears that the compound must be present in the target tissue at significant concentrations in order to exhibit a protective effect. Thus, it was effective in inhibiting chemically induced lung tumorigenesis in mice when it was injected along with the carcinogen benzo(a)pyrene, but not when fed in the diet (70,71). The most likely explanation for this is the relatively poor bioavailability of ellagic acid when administered in the gastrointestinal tract (71). Human studies of ellagic acid as a preventative agent have not been undertaken.

5. Apigenin

Another natural polyphenol with promising chemopreventive properties is apigenin (Fig. 6). Apigenin is found abundantly in celery root (72) and in *Digitaria elixis,* a variety of millet commonly found in semiarid regions (73). In vitro studies have demonstrated the ability of apigenin to influence several factors associated with tumor promotion and progression. It inhibits the growth of v-Ha-*ras*-transformed NIH3T3 cells, in part, by downregulating various mitogen-activated protein kinases and their associated oncogenes (74). It is a strong inhibitor of ornithine decarboxylase, a marker of tumor promotion in mouse skin (75). Apigenin has been shown to increase gap junction formation in rat epithelial cells, thus enhancing cell–cell communication (76). Furthermore, it inhibits the growth of human colon, breast, prostate, and thyroid cancer cells by inducing cell cycle arrest and stimulating apoptosis (77,78). Apigenin also suppresses the growth of endothelial cells, thus inhibiting angiogenesis (79), and it inhibits matrix metalloproteinases involved in tumor cell metastasis (80). Finally, a 1999 study has shown that apigenin is a potent inhibitor of the activation of nuclear transcription factor B (NF-κB), which plays a key role in the regulation of cell growth and apoptosis (81).

In vivo studies have demonstrated the ability of apigenin to inhibit tumor necrosis factor–induced upregulation of intercellular adhesion molecule-1 in mouse skin (82). Topical application of apigenin was found to reduce tumor development in SENCAR mouse skin as well as ultraviolet B– (UVB)-radiation-induced squamous cell carcinomas in hairless SKH mice (83,84). More recently, apigenin was shown to induce a reversible G2/M arrest in both epidermal and fibroblast cells in mouse skin by inhibition of $p34^{cdc2}$ kinase activity (85). Apigenin was also shown to increase the stability and transactivational activity of wild-type p53 in cultured mouse skin Reratinocytes (86). To date, there have been no human clinical trials to evaluate the putative chemopreventive effects of apigenin.

E. Isothiocyanates

Epidemiological studies suggest that consumption of cruciferous vegetables may be particularly effective, compared with total fruit and vegetable consumption, in reducing cancer risk at several organ sites (87). Crucifers are the family of plants that include broccoli, cauliflower, cabbage, kale, and brussel sprouts, as well as others such as daikon and radishes widely consumed in various areas of the world. These plant foods are especially rich in glucosinolates, which are converted by plant myrosinase and the gut microflora to isothiocyanates (88,89).

Some isothiocyanates are potent inducers of phase II detoxification enzymes such as the glutathione-S-transferases, and substantial evidence indicates that induction of phase II enzymes is highly effective in reducing susceptibility to carcinogens (90). This conclusion was supported in 2001 by molecular evidence from experiments on mice in which nrf2, the transcription factor essential for induction of phase II proteins, was knocked out (90). These mice exhibit greatly enhanced sensitivity to induction of forestomach cancer by benzo(a)pyrene, demonstrating the importance of phase II enzymes as a determinant of carcinogen metabolism. Sulforaphane (1-isothiocyanato-4-(methyl-sulfinyl)butane) (Fig. 7) is an isothiocyanate of particular interest in the context of cancer chemoprevention. It was isolated from one variety of broccoli as the major and potent inducer of phase II detoxication enzymes in murine hepatoma cells in culture (91). Subsequently, it was shown that sulforaphane as well as synthetic analogues, designed as phase II enzyme inducers, blocked the chemically induced formation of mammary cancer in rats, demonstrating the potential for phase II enzyme inducers as effective chemopreventive agents (92). These initial observations have been confirmed and extended in many subsequent investigations, including the 2002 discovery that sulforaphane is a potent bacteriostatic agent against *Helicobacter pylori*, the bacterium responsible for gastric ulcers and a strong risk factor for stomach cancer, irrespective of its resistance to conventional antibiotics (93). Further, brief exposure to sulforaphane was bactericidal and eliminated intracellular *H. pylori* from a human epithelial cell line. Gastric infection by *H. pylori* is common globally but is particularly high in developing regions, where there is also a high prevalence of gastric cancer. Consequently, the dual actions of sulforaphane in inhibiting *H. pylori* infections and blocking gastric tumor formation offer hope that these mechanisms might function synergistically to provide diet-based protection against gastric cancer in humans (93).

Another series of isothiocyanates prevent chemically induced cancer in animals principally through their ability to inhibit P450 enzymes involved in the metabolic activation of carcinogens (94,95). These include benzyl

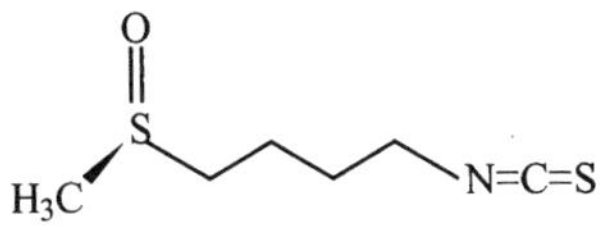

Sulforaphane

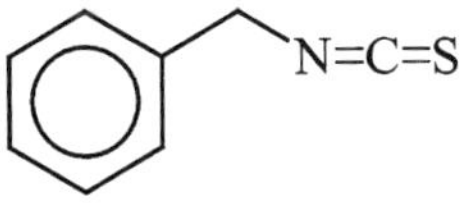

Benzyl isothiocyanate (BITC)

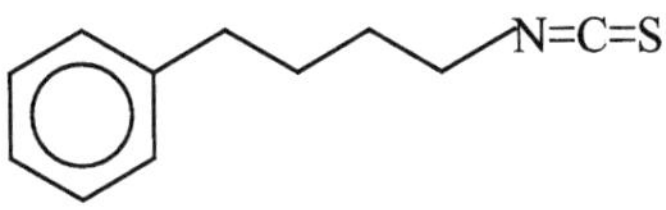

4-Phenylbutyl isothiocyanate (PBITC)

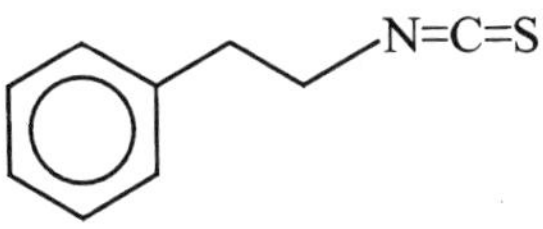

Phenylethyl isothiocyanate (PEITC)

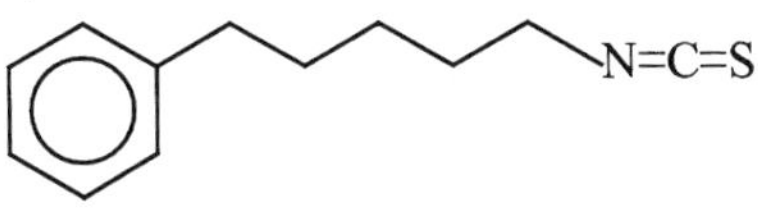

5-Phenylpentyl isothiocyanate (PPeITC)

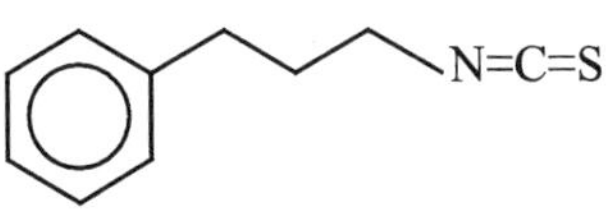

3-Phenylpropyl isothiocyanate (PPITC)

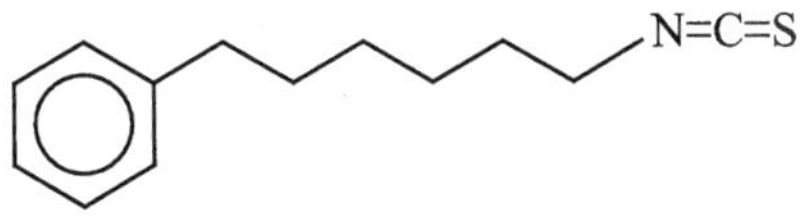

6-Phenylhexyl isothiocyanate (PHITC)

Figure 7 Naturally occurring (sulforaphane, benzyl-, phenylethyl-) and synthetic (phenylpropyl-, phenylbutyl-, phenylpentyl-, phenylhexyl-) isothiocyanates with chemopreventive activity.

isothiocyanate (BITC) and phenethyl isothiocyanate (PEITC) (Fig. 7), which are found in a series of glucosinolates in cruciferous vegetables such as Chinese watercress, cabbage, brussels sprouts, turnips, and cauliflower (88). The PEITC is rapidly absorbed and distributed throughout the body when administered by gavage to mice. In addition, PEITC and its glucuronide conjugates can be detected in the urine of humans after the consumption of watercress (96). Along with related isothiocyanates (Fig. 7) BITC and PEITC are highly effective inhibitors of tumorigenesis in animal models induced by polycyclic aromatic hydrocarbons and nitrosamines (89). Further, BITC is a potent inhibitor of benzo(a)pyrene-induced forestomach and lung tumors in mice (7). Both PEITC and synthetic isothiocyanates of longer chain length, i.e., phenylpropyl isothiocyanate (PPITC), phenylbutyl isothiocyanate (PBITC), phenylpentyl isothiocyanate (PPeITC), and phenylhexyl isothiocyanate (PHITC), produced up to a 100% inhibition of lung tumors in strain A/J mice induced by the tobacco-specific nitrosamine, 4-(methylnitrosamino)-1-(3-pyridyl)-1-butanone (NNK) (97,98). The efficacy of the isothiocyanates to inhibit NNK-induced lung tumors increased with increasing chain length, presumably as a result of better absorption and distribution of the longer-chain compounds and their higher binding affinity for P450 enzymes (99). On the basis of these and other studies in animals, PEITC is being evaluated in a phase I trial for its ability to inhibit nitrosamine metabolism in tobacco smokers. In the rat esophagus model in which tumors are induced by the carcinogen *N*-nitrosomethylbenzylamine (NMBA), PEITC is a very effective inhibitor of esophageal tumorigenesis (100). Whereas PPITC was found to be even more potent than PEITC, as expected, the longer-chain isothiocyanates (PBITC and PHITC) failed to yield a significant inhibitory effect (101). In fact, PHITC actually enhanced tumorigenesis, presumably by causing a low-grade toxicity in the basal epithelium leading to an increased proliferation of premalignant cells (102,103). In addition, PHITC enhanced azoxymethane carcinogenesis in the rat colon (104), suggesting that when administered in the diet, PHITC may be absorbed locally where it produces toxicity and enhances tumorigenesis.

Studies in 1998, 1999, and 2000 in cell culture and in animals have shown that in addition to their action on phase I and II enzymes, isothiocyanates, in particular, BITC, PEITC, sulforaphane, and their conjugates, can induce apoptosis (105–107). They activate mitogen-activated protein (MAP) kinases and activator protein 1 (AP-1) in cultured human tumor cells (106,107), and PEITC has been shown to induce p53 activation in mouse epidermal cells (108). Together, these studies suggest that isothiocyanates may have potential for inhibiting the tumor promotion and progression stages of carcinogenesis.

F. Chlorophyllins

Chlorophylls and chlorophyllins, their water-soluble salts, are components of many common foods and have been found to be effective anticarcinogens in several animal models (109). Chlorophyllin, a mixture of nontoxic sodium–copper salts of chlorophyll commonly used as a food colorant, has potent antimutagenic activity in a range of short-term in vitro and in vivo genotoxicity assays. Mechanistic studies suggest that chlorophyllin forms tight molecular complexes with carcinogens such as aflatoxin B_1 and may thereby diminish the bioavailability of dietary carcinogens by preventing their absorption and promoting fecal excretion. It also acts as an antioxidant to inhibit lipid peroxidation. On the basis of its efficacy as an anticarcinogen, antimutagen, and antioxidant; its lack of known toxicity; and its ready availability, chlorophyllin has been subjected to a randomized, double-blind, placebo-controlled chemoprevention trial among residents of Qidong county, People's Republic of China, who are exposed to high dietary levels of aflatoxins and who are at high risk for subsequent development of liver cancer (110). The primary endpoint of the trial was modulation of levels of aflatoxin B_1-guanine adducts in urine, a biomarker of the biologically effective dose of aflatoxin B_1, elevation of which is associated with increased risk of liver cancer. Chlorophyllin consumption at each meal led to an overall 55% reduction in median urinary adduct levels compared to levels excreted by placebo controls. Comparable reduction of this biomarker in experimental animals dosed with aflatoxin is associated with very significant reductions in liver cancer incidence. These results suggest that this type of intervention can have a favorable impact on the toxicokinetics of unpreventable exposures to dietary aflatoxin and other carcinogens. Natural chlorophylls also exert protective effects against carcinogen exposure in animals, suggesting that supplementation of diets with foods rich in chlorophylls may be an effective approach to chemoprevention readily implementable in many regions of the world (110).

G. Soy Products

Soy products contain large quantities of isoflavones (Fig. 8), structural isomers of flavonoids, with which they share similar biological properties. Genistein and other members of this group have weak estrogenic activity and in some test systems also show antiestrogenic properties (111–113). These findings have led to the suggestion that they may have potential as chemopreventive agents against hormone-dependent cancers. Although health claims for soy products have been approved by the U.S. Food and Drug Administration (FDA) on the basis of lowering of serum cholesterol levels, the

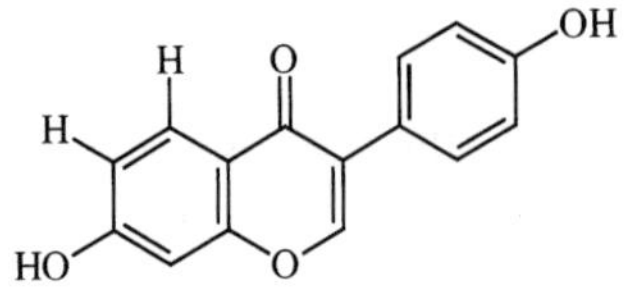

Daidzein

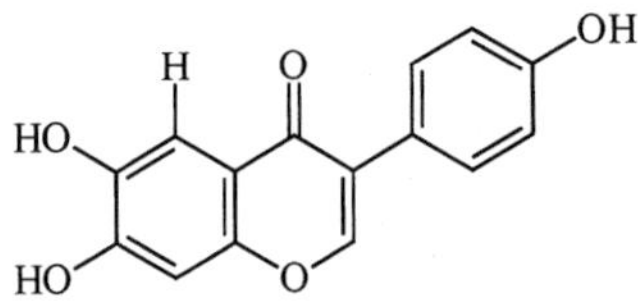

Genistein

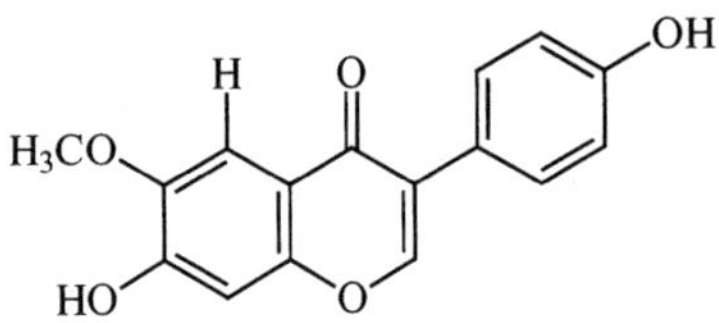

Glycitein

Figure 8 Soy compounds with chemopreventive activitiy.

extent to which isoflavones contribute to this effect has not been determined. Epidemiological studies suggest that in Asian countries where high amounts of soy products are consumed the population may have a reduced risk of breast, bladder, and prostate cancers (114–116). Isoflavones isolated from soy products also have been shown to prevent certain cancers in experimental animals (112,113,117–119). For example, preweanling rats given genistein, a major isoflavone, are resistant to later chemically induced mammary cancer (120) and soy-isoflavone extracts suppressed promotion of chemically induced liver cancer (111). Mechanisms underlying cancer chemopreventive effects of isoflavones may involve not only estrogenic or antiestrogenic properties, but also the ability of isoflavones to inhibit some tyrosine kinases and to inhibit angiogenesis required for tumor development and growth (121,122). In vitro studies suggest the ability of nutritionally relevant amounts of isoflavones to increase human natural killer cell activity, an innate tumor surveillance mechanism (123). The literature on health effects of purified isoflavones is limited, and no studies in humans appear to have been done. Other relevant soy constituents are saponins, which are coextracted with isoflavones and are present at similar levels in most soy foods. They have been shown to inhibit chemically induced colon cancer in rats (123). However, it remains to be determined whether isoflavones themselves can produce human health benefits and to what extent they contribute to the potential health benefits of soybean foods or soy-derived food ingredients or dietary supplements.

H. Berries

In the mid-1980s, our laboratory and others were involved in studies of the chemopreventive potential of the natural occurring polyphenol ellagic acid. As indicated, as part of these studies, we evaluated several fruits for their content of ellagic acid and found that the compound was particularly abundant in black and red raspberries, strawberries, and, to a lesser extent, cranberries (65). On a weight basis, the ellagic acid content of berry seeds is higher than in the pulp; berry juice contains only small amounts of ellagic acid. On the basis of these observations, we decided to freeze-dry berries and pulverize the freeze-dried material (seed and pulp) into a powder. Since berries are approximately 90% water, this would concentrate any preventative components, including ellagic acid, approximately 10-fold. Administration of freeze-dried strawberries and black raspberries at 5% and 10% of the diet before, during, and after treatment of Fischer 344 rats with the esophageal carcinogen NMBA resulted in a 40%–60% reduction in the number of NMBA-induced esophageal tumors (124–126). Both berry types were equally effective in preventing tumor development, and the inhibitory effect of the berries could not be attributed to ellagic acid alone. Berries were found to reduce the formation of O^6-methylguanine adducts from NMBA in esophageal DNA in a dose-dependent manner. They also reduced the excretion of 8-hydroxy-deoxyguanosine adducts in the urine of chemically treated animals, indicating protection against oxidative DNA damage (127). Thus, they appear to inhibit tumor initiation by influencing carcinogen metabolism and reducing oxidative stress. Strawberries and black raspberries were also shown to inhibit esophageal tumorigenesis when administered in the diet after treatment of rats with NMBA. One mechanism of their protective effect on tumor progression is inhibition of the growth rate of preneoplastic cells. Other mechanisms are currently under investigation. When fed at 2.5%, 5%, and 10% of the diet to rats that had been pretreated with azoxymethane, black raspberries produced up to an 80% inhibition of adenocarcinoma development in the rat colon (127). However, when administered to A/J mice at 10% of the diet, strawberries failed to inhibit NNK- and benzo(a)pyrene-induced lung tumorigenesis (128). These results suggest that the protective effect of berries in the diet could be limited to the gastrointestinal tract, where absorption of berry components is likely to be optimal. Chemical analysis indicates that berries contain multiple components with chemopreventive activity, including vitamins A, C, E, and folic acid; calcium and selenium; numerous polyphenolic compounds, including coumaric acid, ferulic acid, quercitin, and several anthocyanins; and phytosterols, including β-sitosterol (127).

In 2001 in vitro studies with organic extracts of freeze-dried berries demonstrated the ability of certain berry extracts to inhibit benzo(a)pyrene-induced transformation of Syrian hamster embryo cells (129). These same extracts were effective in downregulating benzo(a)pyrene diol epoxide induction of the transcription activators, nuclear factor-kappa B (NF-kB), and activator protein 1 (AP-1) in JB-6 mouse epidermal cells (130). Unpublished results from a phase I clinical trial in 2002 indicate that freeze-dried black raspberries are well tolerated by humans when administered orally at 90 grams per day. Phase II prevention trials of berries have not, as yet, been conducted.

III. MECHANISMS BY WHICH DIETARY COMPONENTS EXERT CHEMOPREVENTIVE EFFECTS

In this section we describe, in more detail, known mechanisms by which dietary components have been shown to inhibit both the initiation and promotion and progression stages of cancer. Largely on the basis of studies in animal models of carcinogenesis, Wattenberg (7) classified chemopreventive agents, including compounds in the diet, into three functional categories: (a) inhibitors of carcinogen exposure, (b) *anti-initiating* or "blocking" agents, and (c) *antipromotion/antiprogression* or "suppressing" agents (Table 1). All three categories are subject to further stratification on the basis of the specific mechanistic principles by which each agent acts.

A. Inhibitors of the Effects of Carcinogen Exposure

Wattenberg originally defined a class of chemopreventive agents that inhibit carcinogen formation (7). These include the dietary components vitamin C, caffeic acid, and ferulic acid, which inhibit the in vivo nitrosation of primary

Table 1 Representative Dietary Chemopreventive Agents and Their Mechanisms of Action[a]

Mechanism of action	Examples
Inhibitors of carcinogen exposure	
Inhibitors of carcinogen formation (e.g., inhibitors of nitrosation)	Ascorbic acid γ-Tocopherol Dietary phenols
Inhibitors of carcinogen absorption and uptake	Calcium Sodium thiosulfate
Anti-initiating agents	
Cytochrome P450 inhibitors	Phenylisothiocyanate Phenylpropylisothiocyanate Diallyl sulfide Monoterpenes
Cytochrome P450 inducers	Indole-3-carbinol
Phase II enzyme inducers	Isothiocyanates Polyphenols
Scavengers of electrophiles	Ellagic acid *N*-acetylcysteine
Inducers of DNA repair	Vanillin *N*-acetylcysteine
Antipromotion/antiprogression agents	
Inducers of apoptosis	Phenylisothiocyanate EGCG
Inducers of differentiation	Retinoids Carotenoids Calcium Deltanoids
Inhibitors of angiogenesis	Resveratrol EGCG
Inhibitors of protease activity	Bowman-Birk inhibitor
Inhibitors of reactive oxygen species	EGCG *N*-acetylcysteine Other polyphenols

[a] DNA, deoxyribonucleic acid; EGCG, (-)-epigallo-catechin-3-gallate.

and secondary amines to form nitrosamines in the stomach (34,132). Calcium inhibits the uptake of bile acids by colon epithelium, thus reducing exposure to these tumor promoting agents (133). Sodium thiosulfate can prevent the absorption of reactive electrophilic metabolites of carcinogens in the diet by reacting with such compounds, detoxifying them before uptake by the gastrointestinal tract (134).

B. Anti-Initiating or "Blocking" Agents

Blocking agents intervene in the stage of tumor initiation. We have already described some of the known blocking agents in the diet. Most blocking agents can be assigned to one or more of five major subclasses, namely, (a) inhibitors of cytochrome P450 enzymes, (b) inducers of cytochrome P450 enzymes, (c) inducers of phase II detoxification enzymes, (d) scavengers of electrophiles, and (e) inducers of DNA repair (Table 1).

1. Inhibitors of Cytochrome P450 Enzymes

Most chemical carcinogens are procarcinogens: That is, they must be converted into metabolites that cause mutational events in DNA to produce cancer. The major enzymes involved in the metabolic activation of carcinogens are the cytochrome P450s. Several dietary components, including the isothiocyanates and ellagic acid, mentioned previously, have been shown to prevent chemically induced cancer in animals by inhibiting the activities of cytochrome P450s. In addition, diallyl sulfide, a naturally occurring constituent of *Allium* sp. vegetables, inhibits carcinogen activation (135) and tumorigenesis in a number of animal model systems (136–138).

2. Inducers of Cytochrome P450 Enzymes

Inducers of cytochrome P450 can act as blocking agents by increasing the production of activated carcinogen metabolites in resistant nontarget tissues or by enhancing oxidative detoxification at any tissue site. The most thoroughly investigated dietary factor with P450 inductive activity is indole-3-carbinol (I3C), which is present in numerous vegetables, is a potent inducer of cytochrome P450 enzymes, and exhibits chemopreventive activity in several animal models (139–141). The induction of P450 activity by I3C, however, can lead to a shift in the target organ through enhanced activation of carcinogen elsewhere (142); that effect may account for the known ability of I3C to enhance tumorigenesis in some organs (143,144).

3. Inducers of Phase II Detoxification Enzymes

Among the most attractive anti-initiating agents are the inducers of phase II detoxification enzymes. Phase II enzyme inducers have historically received preference over cytochrome P450 inducers as chemopreventive agents, since they appear to inhibit a wider range of chemical carcinogens and are regarded as less likely to enhance tumorigenesis themselves. Significant interest in phase II enzyme inducers has focused on inducers of those enzymes that detoxify electrophiles, e.g., glutathione S-transferases (GSTs) and reduced nicotinamide-adenine dinucleotide phosphate NADPH quinone oxidoreductases (NQOs). However, such compounds usually induce UDP glucuronosyltransferases as well. One of the first dietary factors whose mechanism was determined to be partially due to phase II enzyme induction, namely, induction of glutathione S-transferase(s) (145), is BITC. As mentioned, other dietary isothiocyanates, such as PEITC (91), and sulforaphane (92) are known inducers of glutathione S-transferase, although most of the inhibitory activity of PEITC can be attributed to its inhibitory effects on cytochrome P450s.

4. Scavengers of Electrophiles

Some dietary components chemically react with the activated (electrophilic) forms of carcinogens and, in so doing, protect DNA from electrophilic attack. In general, such components contain nucleophilic moieties such as hydroxyl groups and sulfhydryl functions. Examples include ellagic acid and *N*-acetylcysteine. Ellagic acid has been shown to react directly with the diol epoxide of benzo(a)pyrene (i.e., BPDE) to form a conjugate (146); such activity may account for its inhibition of BPDE-induced carcinogenicity in mouse skin (147,148). The sulfhydryl moiety of *N*-acetylcysteine (NAc) can react with a number of electrophilic species; the reaction may account for some of the antimutagenic and anticarcinogenic effects of NAc (149–151). Scavenging agents must be maintained at relatively substantial concentrations in target tissues at all times during which carcinogens are present to elicit protective effects.

5. Inducers of DNA Repair

Enhancement in the rate and/or fidelity of DNA repair in an individual exposed to genotoxic carcinogens should result in a decreased incidence of initiating events. In a prior review, we cited vanillin as an example of a dietary component that affects DNA repair (3). Although vanillin shows significant inhibition of mammalian cell mutagenicity (152,153), the utility of vanillin appears to be largely limited to inhibition of UVB and X-ray-induced mutagenesis. Kelloff has listed *N*-acetylcysteine and dietary protease inhibitors among enhancers of poly [adenosine diphosphate (ADP)-ribose] polymerase (PARP) (6). Activity of PARP normally increases in the presence of DNA damage (i.e., DNA strand breaks), and PARP is believed to be important in DNA repair (154).

C. Antipromotion and Antiprogression Agents

Classification of dietary factors that inhibit tumor promotion and progression is more difficult since the precise sequence of events in the stages of promotion and progression is not known with certainty. However, as indicated in our previous reviews (3,5) and those of Kelloff (6) and De Flora and Ramel (155), most antipromotion and antiprogression agents can be classified into one or more of the following categories: (a) inducers of apoptosis, (b) inducers of differentiation, (c) inhibitors of angiogenesis, (d) inhibitors of protease activity, and (e) inhibitors of reactive oxygen species. Representative dietary agents possessing these types of activity are listed in Table 1. This list of mechanistic categories should not be considered to be exhaustive, and as for the anti-initiating agents, it is clear that nearly all of these agents possess multiple mechanisms of action.

1. Inducers of Apoptosis

Apoptosis, or programmed cell death, is a means by which damaged cells can be removed when genomic DNA damage is excessive. Since tumor cells tend to lose their apoptotic responses, restoration of apoptotic response mechanisms in tumor cells represents a potential means of inhibiting cancer. The principal apoptotic pathways are the death receptor pathway and the mitochondrial pathway, both of which lead to the activation of caspases, i.e., cysteine proteases that cleave other proteins involved in the maintenance of nuclear integrity and cell cycle progression (156). Because both pathways can be influenced by a number of stimuli, a substantial number of chemopreventive agents likely induce apoptosis. In addition to its anti-initiating activities, PEITC at high doses can induce apoptosis in vitro and in vivo (106,107). Another dietary factor that stimulates apoptosis is the green tea polyphenol EGCG, which appears to induce apoptosis by the mitochondrial pathway; it specifically reduces expression of apoptotic inhibitors such as *bcl*-2 while increasing expression of the apoptotic enhancer *bax*, resulting in caspase-9 activation (157).

2. Inducers of Differentiation

Tumor cells are phenotypically dedifferentiated in nature (158). The induction of terminal differentiation in a proliferating cell results in inhibition of proliferation and produces a cell that is more susceptible to apoptosis. Inducers of differentiation, therefore, offer the possibility of arresting the growth of a neoplasm at the premalignant stage. There are a number of mechanisms that can induce terminal differentiation in cells, including upregulation of cellular gap junctions (159), as well as inhibition of histone deacetylase (160). As stated, retinoids induce the differentiation of epithelial cells in vivo and in vitro, presumably by binding to nuclear receptors and regulating gene expression. The effects of retinoids on cell growth and on in vitro transformation are well correlated with their abilities to affect gap junctional communication; in this regard, carotenoids have similar effects in vitro (161–163). Calcium and vitamin D agonists (deltanoids), as do retinoids, stimulate cell differentiation (164) and inhibit tumor promotion in animal models, particularly when high-fat diets are employed (165–167). It has been known for some time that calcium is an important regulator of the growth and differentiation of squamous epithelial cells (164).

3. Inhibitors of Angiogenesis

Tumor cells require access to adequate levels of nutrients and removal of cellular wastes in order to support their growth and proliferation. When solid tumors approach a size of approximately 2 to 3 mm in diameter, further growth tends to be inhibited unless the tumor is able to develop its own blood supply. The process by which new blood vessels develop from existing vessels is termed *angiogenesis*, and tumor cells produce factors that stimulate angiogenesis (168). Thus, an important approach to the prevention of tumor development is inhibition of angiogenesis. Several dietary factors have been demonstrated to exhibit antiangiogenic activity. The tea polyphenol EGCG inhibits endothelial cell proliferation as well as tumor vascularization (169,170). Such antiangiogenic activities may make a significant contribution to the chemopreventive activity of EGCG. Similarly, resveratrol exhibits antiangiogenic activity both in vitro and in vivo (64). A number of other factors that are normally placed in other mechanistic categories also display antiangiogenic activities, at least in vitro. Such agents include the retinoids such as all-*trans*-retinoic acid, 9-*cis*-retinoic acid, and 13-*cis*-retinoic acid and the Bowman-Birk inhibitor (171).

4. Inhibitors of Protease Activity

Protease inhibitors are active inhibitors of the early stages of promotion in mouse skin (172) and are also effective in the suppression of breast, bladder, colon, liver, and lung tumorigenesis (173–176). They have been shown to prevent transformation of NIH-3T3 cells by the H-*ras* oncogene (177) and inhibit the formation of promoter-induced oxygen radicals and hydrogen peroxide (178). Protease inhibitors that can inhibit chymotrypsin are more potent in the inhibition of oxygen radicals and appear to be better inhibitors of tumor promotion (179).

The protease inhibitor that has been examined to the greatest extent is the Bowman-Birk inhibitor. The Bowman-Birk inhibitor (BBI) is a soybean-derived peptide protease inhibitor with a molecular weight of approximately 8000 daltons (180–182). It is part of a larger family of polypeptides isolated from soybeans and other plants that are referred to collectively as *Bowman-Birk inhibitors* (180–183). The BBI is a potent inhibitor of the proteases trypsin and chymotrypsin (180). In addition to its protease inhibitory activity, BBI possesses anti-inflammatory effects (181). It is believed that BBI may also reverse the initiating event(s) in carcinogenesis, as well as affecting the promotional stage (182). Indeed, protease inhibitors are normally classified as suppressing agents. Both BBI and Bowman-Birk inhibitory concentrate (BBIC) decrease the expression of oncogenes such as c-*myc* and c-*fos* and inhibit certain types of proteolytic activities that are found to be elevated in carcinogen-treated tissues or transformed cells (181–183). Although BBI is known to inhibit apoptosis, this activity appears to be primarily due to the phospholipids that copurify with BBI (184,185). In experimental animals (186–188) BBI inhibits tumors of the intestine, cheek pouch, and esophagus. The U.S. Food and Drug Administration granted BBIC Investigational New Drug status in April 1992 (180). In patients with oral leukoplakia BBIC has been employed in phase I and IIa trials (189,190). In the phase IIa study, BBIC elicited a complete or partial response in approximately one-third of the individuals (190); BBIC is still under evaluation in clinical trials.

5. Inhibitors of Reactive Oxygen Species

The generation of reactive oxygen species (ROS) and free radicals is facilitated by a number of promoters and complete carcinogens (191). At low concentrations, ROS serve as regulatory molecules in various cellular signaling processes; at high concentrations, these moieties promote cellular damage and inflammatory processes (192). The ROS are involved in initiation, the early stages of promotion, as well as tumor progression. The production of superoxide, hydroxyl radicals, and hydrogen peroxide can result in the formation of DNA adducts, including 8-oxodeoxyguanosine, thymine glycol, and 5-hydroxymethyl uracil (191). Although we have chosen to classify ROS inhibitors as antipromotion and antiprogression agents, it is clear that antioxidants that scavenge ROS can provide protection at all stages of carcinogenesis.

Dietary antioxidants are particularly attractive as potential chemopreventive agents since they are potent inhibitors of ROS. Antioxidants provide protection of cells against damage caused by free radicals produced at low levels in the course of aerobic metabolism, and in large excess by inflammatory cells. They act by either preventing free radical formation or scavenging radicals after they are formed. Various dietary antioxidants have been shown to inhibit carcinogenesis in experimental animals, acting at the stages of tumor initiation, promotion, and progression.

It is apparent that many of the effects of green tea or its semipurified polyphenol fraction can be ascribed to EGCG. Despite the fact that EGCG exhibits apoptotic and antiangiogenic effects, many of the preventative effects of EGCG are attributed to its antioxidant activities. In most in vitro test systems, EGCG exhibits considerably greater antioxidant potential than the other catechins (193). The antioxidant effect is due to the metal ion chelating potential of the catechol moieties, the ability to scavenge superoxide and hydroxyl radicals, and the ability to quench lipid peroxidation chain reactions (193). EGCG inhibits TPA-induced increases in hydrogen peroxide as well as the formation of oxidized DNA bases (191). Under conditions in which EGCG does not affect NNK-induced DNA alkylation, EGCG reduces both 8-oxodeoxyguanosine formation and tumorigenesis in lung induced by NNK (194). In addition to tea, many other foods are rich in polyphenolic content, which is undoubtedly important for their antioxidant effects.

IV. HUMAN CLINICAL TRIALS

Clinical trials designed to determine the efficacy of chemopreventive agents, including dietary components, have been conducted in several different countries. In the United States, the principal sponsor of such trials is the Division of Cancer Prevention of the National Cancer Institute (NCI), National Institute of Health (NIH). A comprehensive listing of studies currently in progress is available on the NCI Clinical Trials web page (http://www.nci.nih.gov/search/clinical_trials). Those specifically designed to evaluate foods or food components are summarized in Table 2. We shall not attempt to address these comprehensively, but rather shall discuss several illustrative examples, including one well-known past failure, and provide a brief discussion of trials currently under way. (For more comprehensive reviews of human clinical trials of chemopreventive agents against organ-specific cancers, the reader is referred for example, to Ref. 195 and Ref. 196.)

A. β-Carotene: A Tragic Failure

Perhaps among the most interesting and most disappointing clinical trials were those trials conducted with β-carotene (13,197). A large number of epidemiological studies, both retrospective and prospective, strongly suggested that β-carotene might be an effective inhibitor of lung cancer (16). Such studies demonstrated that individuals who ate more fruits and vegetables, which are rich in carotenoids, and people who had higher serum β-carotene levels had a lower risk of lung cancer. Several trials employing β-carotene were designed to evaluate this possibility, including the α-Tocopherol, β-Carotene Cancer Prevention Study (ATBC Study), the Carotene and Retinol Efficacy Trial (CARET Trial), and the Physicians' Health Study. The ATBC

Table 2 Active U.S. Clinical Trials on Dietary Chemopreventive Agents

Agent	Type of study	Cancer targeted	Cohort	Endpoint	Site
Dietary soy	Phase II	Prostate cancer	Patients with elevated PSA levels	PSA levels and other biomarkers	Cancer and Leukemia Group B
Doxercalciferol	Phase II	Prostate cancer	Patients with localized prostate cancer	Intermediate endpoint biomarkers	University of Wisconsin Cancer Center, Madison, WI
Isoflavones	Phase III	Prostate cancer	Men with grade 1 or 2 prostate cancer	Tumor progression, surrogate endpoint biomarkers	Moffitt Cancer Center, Tampa, FL
Isotretinoin and vitamin E	Phase II	Lung cancer	Current smokers or former smokers at high risk for lung cancer	Surrogate endpoint biomarkers	University of Colorado Cancer Center, Denver, CO
Selenium	Phase III	Lung cancer	Previously resected stage I non–small cell lung cancer	Incidence of secondary primary tumors	American College of Surgeons Oncology Group, Cancer and Leukemia Group B, Eastern Cooperative Oncology Group, NCIC-Clinical Trials Group, North Central Cancer Treatment Group, Southwest Oncology Group
Selenium	Phase III	Prostate cancer	Men with high-grade prostatic intraepithelial neoplasia	Prostate cancer incidence and surrogate endpoint biomarkers	Cancer and Leukemia Group B, Eastern Cooperative Oncology Group, Southwest Oncology Group
Selenium and vitamin E	Phase III	Prostate cancer	Males above the age of 50–55 years	Incidence of prostate cancer and other cancers	Cancer and Leukemia Group B, Eastern Cooperative Oncology Group, NCIC-Clinical Trials Group, North Central Cancer Treatment Group, Radiation Therapy Oncology Group, Southwest Oncology Group

Source: NCI Clinical Trials web page (http://www.nci.nih.gov/search/clinical_trials/).

Study, conducted in Finland, was a randomized trial of more than 29,000 male smokers (198) who received α-tocopherol (50 mg), β-carotene (20 mg), both, or neither daily for periods of 5–8 years. Although a 19% reduction in lung cancer incidence was observed in subjects with highest versus lowest quintile of serum α-tocopherol, β-carotene administration was associated with increased lung cancer incidence. The CARET Trial was a randomized trial involving about 18,000 smokers in the Pacific Northwest, performed at the Fred Hutchinson Cancer Center in Seattle (199). Subjects were randomized to treatment arms of 30 mg β-carotene or 25,000 IU retinyl palmitate daily. As in the ATBC study, smokers receiving β-carotene also suffered a 46% excess in lung cancer mortality rate compared to controls. The Physicians' Health Study included β-carotene supplementation (50 mg every other day) in a group of more than 22,000 U.S. male physicians, which included very few current smokers (200). In this trial, no decrease in overall cancer incidence or in lung cancer incidence was observed after β-carotene administration. Thus, in three trials, an increased risk of lung cancer incidence or mortality rate occurred in two studies, with no evidence of a protective effect in the third.

A number of factors may account for the disappointing failure of these trials. One relates to the current paradigm for drug development in the United States, which, then and now, normally includes a requirement for demonstration of efficacy in preclinical studies in experimental animals. This requirement was waived in the case of β-carotene, for which there is no evidence of chemopreventive activity against lung cancer in an experimental animal model. Additionally, limitations inherent in the observational epidemiological data suggesting chemopreventive activity were not fully appreciated. Despite their value in identifying potential beneficial or deleterious effects of environmental exposures, statistical associations in such studies cannot prove causality. In this context, it is important to recognize that fruits and vegetables are complex mixtures of hundreds of chemicals, and that β-carotene may simply have served as a surrogate biomarker for high fruit and/or vegetable consumption in the epidemiological studies suggesting chemoprotective efficacy, leaving the active chemopreventive constituents unidentified. It is also known that many antioxidants can act as potent prooxidants under some conditions. For example, at high oxygen partial pressures, conditions that plausibly may have occurred in active smokers in the ATBC and CARET trials, β-carotene actually displays a dose-dependent prooxidant effect (201). Although the results of these β-carotene trials are disappointing, they do emphasize a need for the following minimal benchmarks in the future development of cancer chemopreventive agents: (a) clear demonstration of preclinical efficacy, (b) thorough understanding of the mechanisms of action of the prospective agent, and (c) initiation of phase III trials only after successful completion of phase I trials and short-term phase II trials utilizing reliable surrogate endpoint biomarkers, even in the case of a substance that is considered generally recognized as safe (GRAS).

B. Current Human Trials

Current human clinical chemoprevention trials that involve foods or food components are listed in Table 2. All studies are phase II or phase III clinical trials and are supported by the NCI. The targets include prostate cancer (five trials) and lung cancer (two trials). Two trials involve the use of soy or soy isoflavones for the prevention of prostate cancer, and there are three trials of selenium for prevention of lung and prostate cancer. None of the agents under evaluation is a purely anti-initiating agent, and the vast majority of agents are more properly classified as antipromotion and/or antiprogression agents.

V. CONCLUSIONS

Chemoprevention continues to merit consideration as an effective means of controlling cancer incidence. As we have already pointed out, a large number of identifiable groups at high risk for the development of cancer stand to benefit from chemopreventive approaches. A large number of chemopreventive agents, including many dietary components, have been and are currently being examined in preclinical experiments, with a correspondingly fewer number of agents under consideration in current human clinical trials. As our understanding of the molecular mechanisms of action of these agents increases, we should expect to develop more rational combinations of agents, for it is apparent that the most fruitful approach will be to employ such combinations, rather than individual agents. In this respect, many foods, such as tomatoes, berries, and soybeans, contain multiple chemopreventive agents that exhibit activity in preclinical models. The concept of administering these foodstuffs in concentrated form (e.g., freeze-dried preparations) to humans may well be a viable approach to cancer prevention. In addition, the long-term administration of foodstuffs to humans is less likely to produce toxic effects than the high-dose administration of single chemopreventive agents or combinations of chemopreventive agents. Finally, since a disproportionate share of the burden of funding chemoprevention efforts falls upon the public sector and since chemoprevention is relatively underfunded in comparison to other means of cancer control and treatment, it is incumbent upon us to step up our efforts to increase government funding for chemopreventive efforts and to attempt to enlist greater support from the private sector as well.

REFERENCES

1. Cancer Facts and Figures—2002. American Cancer Society, New York, pp 1–48, 2002.
2. Cancer Facts and Figures—1992. American Cancer Society, New York, pp 1–3, 1992.
3. MA Morse, GD Stoner. Cancer chemoprevention: principles and prospects. Carcinogenesis 14:1737–1746, 1993.
4. S De Flora, A Izzotti, F D'Agostini, RM Balansky, D Noonan, A Albini. Multiple points of intervention in the prevention of cancer and other mutation-related diseases. Mut Res 9:480–481, 2001.
5. GD Stoner, MA Morse, G Kelloff. Perspectives in cancer chemoprevention. Environ Health Perspect 105:945–954, 1997.
6. GJ Kelloff. Perspectives on cancer chemoprevention research and drug development. Adv Cancer Res 78:199–334, 2000.
7. LW Wattenberg. Chemoprevention of cancer. Cancer Res 45:1–8, 1985.
8. L Foulds. The experimental study of tumor progression: review. Cancer Res 14:327–339, 1954.
9. ER Fearon, B Vogelstein. A genetic model for colorectal tumorigenesis. Cell 61:759–767, 1990.
10. JD Potter, K Steinmetz. Vegetables, fruit and phytoestrogens as preventive agents. In: BW Steward, D MacGregor, P Kleihues, eds. Principles of Chemoprevention. Lyon: IARC Scientific Publications, No. 139, pp 61–90, 1996.
11. P Greenwald, CK Clifford, JA Milner. Diet and cancer prevention. Eur J Cancer 37:948–965, 2001.
12. A Murakami, H Ohigashi, K Koshimizu. Chemoprevention: insights into biological mechanisms and promising food factors. Food Rev Int 15:335–395, 1999.
13. WA Pryor, W Stahl, CL Rock. Beta carotene: from biochemistry to clinical trials. Nutr Rev 58:39–53, 2000.
14. G van Poppel, RA Goldbohm. Epidemiologic evidence for β-carotene and cancer prevention. Am J Clin Nutr 62:1393S–1402S, 1995.
15. DA Cooper, AL Eldridge, JC Peters. Dietary carotenoids and lung cancer: a review of recent research. Nutr Rev 57:133–145, 1999.
16. G van Poppel. Carotenoids and cancer: an update with emphasis on human intervention studies. Eur J Cancer 29A:1335, 1993.
17. Alpha-Tocopherol Beta-Carotene Cancer Prevention Study Group, OP Heinonen, JK Huttunen, D Albanes. The effect of vitamin E and beta carotene on the incidence of lung cancer and other cancers in male smokers. N Engl J Med 330:1029–1035, 1994.
18. GS Omenn, GE Goodman, MD Thornquist, J Balmes, MR Cullen, A Glass, JP Keogh, FL Meyskens, B Valanis, JH Williams, S Barnhart, S Hammar. Effects of a combination of beta carotene and vitamin A on lung cancer and cardiovascular disease. N Engl J Med 334:1150–1155, 1996.
19. CH Hennekens, JE Buring, JE Manson, M Stampfer, B Rosner, NR Cook, C Belanger, F LaMotte, JM Gaziano, PM Ridker, W Willett, R Peto. Lack of

effect of long-term supplementation with beta carotene on the incidence of malignant neoplasms and cardiovascular disease. N Engl J Med 334:1145–1149, 1996.

20. EC Miller, E Giovannucci, JW Erdman, Jr, R Bahnson, SJ Schwartz, SK Clinton. Tomato products, lycopene, and prostate cancer risk. Urol Clin North Am 29:83–93, 2002.
21. M Karas, H Amir, D Fishman, M Danilenko, S Segal, A Nahum, A Koifmann, Y Gait, J Levy, Y Sharoni. Lycopene interferes with cell cycle progression and insulin-like growth factor I signaling in mammary cancer cells. Nutr Cancer 36:101–111, 2000.
22. RC Moon, RG Mehta, KJ Rao. Retinoids and cancer in experimental animals. In: Sporn MB, Roberts AB, Goodman DS, eds. The retinoids: biology, chemistry and medicine. New York: Raven Press, pp 573–595, 1994.
23. R Lotan. Retinoids in cancer chemoprevention. FASEB J 10:1031–1039, 1996.
24. GJ Kelloff, CW Boone, JA Crowell, VE Steele, RA Lubet, LA Doody, WF Malone, ET Hawk, CC Sigman. New agents for cancer chemoprevention. J Cell Biochem Supply 26:1–28, 1996.
25. SC Lippman, R Lotan. Advances in the development of retinoids as chemopreventive agents. J Nutr 130(2S Suppl):479S–482S, 2000.
26. G Shklar, J Schwartz, D Grau, DP Trickler, KD Wallace. Inhibition of hamster buccal pouch carcinogenesis by 13-*cis*-retinoic acid. Oral Surg Oral Pathol 50:45–52, 1980.
27. I Inoue, Y Yamamoto, T Ito, H Takahashi. Chemoprevention of tongue carcinogenesis in rats. Oral Surg Oral Med Oral Pathol 76:608–615, 1993.
28. I Lasnitzki, W Bollag. Prevention and reversal by non-polar arotinoid (Ro 15-0778) of 3,4-benzopyrene- and cigarette smoke condensate-induced hyperplasia and metaplasia of rodent respiratory epithelia grown in vitro. Eur J Cancer Clin Oncol 23:861–865, 1987.
29. R Lotan. Retinoids and chemoprevention of aerodigestive tract cancers. Cancer Metastasis Rev 16:349–356, 1997.
30. WK Hong, SC Lippman, WN Hittelman, R Lotan. Retinoid chemoprevention of aerodigestive cancer: from basic research to the clinic. Clin Cancer Res 1:677–686, 1995.
31. A Gupta, R Nines, KA Rodrigo, RM Aziz, PS Carlton, DL Gray, VE Steele, MA Morse, GD Stoner. Effects of dietary *N*-(4-Hydroxyphenyl)retinamide on *N*-nitrosomethylbenzylamine metabolism and esophageal tumorigenesis in the Fischer 344 rat. J Natl Cancer Institute 93:990–998, 2001.
32. R Sankaranarayanan, B Mathew. Retinoids as cancer-preventive agents. In: BW Steward, D MacGregor, P Kleihues, eds. Principles of Chemoprevention. Lyon: IARC Scientific Publications, No. 139, pp 47–59, 1996.
33. World Cancer Research Fund. Food, Nutrition and the Prevention of Cancer: A Global Perspective. Washington, DC, American Institute for Cancer Research, 1997.
34. SS Mirvish. Ascorbic acid inhibition of N-nitrosocompound formation in chemical, food and biological systems. In Zedeck MS, Lipkin M, eds. Inhibition

of Tumor Induction and Development. New York: Plenum Publishing, p 101, 1981.
35. LE Voorrips, RA Goldbohm, HA Brants, GA van Poppel, F Sturmans, RJ Hermus, PA van den Brandt. A prospective cohort study on antioxidant and folate intake and male lung cancer risk. Cancer Epidemiol Biomarkers Prev 9:357–365, 2000.
36. JM Chan, MJ Stampfer, J Ma, EB Rimm, WC Willett, EL Giovanucci. Supplemental vitamin E intake and prostate cancer risk in a large cohort of men in the United States. Cancer Epidemiol Biomarkers Prev 8:893–899, 1999.
37. K Israel, W Yu, BG Sanders, K Kline. Vitamin E succinate induces apoptosis in human prostate cancer cells: role for Fas in vitamin E succinate-triggered apoptosis. Nutr Cancer 36:90–100, 2000.
38. GS Omenn. Micronutrients (vitamins and minerals) as cancer-preventive agents. In: BW Steward, D MacGregor, P Kleihues, eds. Principles of Chemoprevention. Lyon: IARC Scientific Publications, No. 139, pp 33–45, 1996.
39. EA Platz, E Giovannucci. Vitamin D and calcium in colorectal and prostate cancers. In Heber D, Blackburn GL, Go VLW, eds. Nutr Oncology. San Diego, Academic Press, pp 223–252, 1999.
40. RM Ramimi, P Lagiou, H-O Adami, D Trichopolous. Prospects for chemoprevention of cancer. J Intern Med 251:286–300, 2002.
41. DH Holben, AM Smith. The diverse role of selenium within selenoproteins: a review. J Am Diet Assoc 99:836–843, 1999.
42. A Azzi, I Breyer, M Feher, M Pastori, R Ricciarelli, S Spycher, M Staffieri, A Stocker, S Zimmer, JM Zingg. Specific cellular responses to α-tocopherol. J Nutr 130:1649–1652, 2000.
43. HE Ganther. Selenium metabolism, selenoproteins and mechanisms of cancer prevention: complexities with thioredoxin reductase. Carcinogenesis 20:1657–1666, 1999.
44. JW Finley, CD Davis, Y Feng. Selenium from high selenium broccoli protects rats from colon cancer. J Nutr 130:2384–2389, 2000.
45. RG Ziegler, ST Mayne, CA Swanson. Nutrition and lung cancer. Cancer Causes Control 7:157–177, 1996.
46. H Mukhtar, SK Katiyar, R Agarwal. Green tea and skin—anticarcinogenic effects. J Invest Dermatol 102:3–7, 1994.
47. GD Stoner, H Mukhtar. Polyphenols as cancer chemopreventive agents. J Cell Biochem Suppl 22:169–180, 1995.
48. ZY Wang, M Das, DR Bickers, H Mukhtar. Interaction of epicatechins derived from green tea with rat hepatic cytochrome P-450. Drug Metab Dispos 16:98–103, 1988.
49. SG Khan, SK Katiyar, R Agarwal, H Mukhtar. Enhancement of antioxidant and phase II enzymes by oral feeding of green tea polyphenols in drinking water to SKH-1 hairless mice: possible role in cancer chemoprevention. Cancer Res 52:4050–4052, 1992.
50. M Nomura, W Ma, N Chen, AM Bode, Z Dong. Inhibition of 12-*O*-tetra-

decanoylphorbol-13-acetate-induced NF-κB activation by tea polyphenols,(–)-epigallocatechin gallate and theaflavins. Carcinogenesis 21:1885–1890, 2000.

51. Y Kuroda, Y Hara. Antimutagenic and anticarcinogenic activity of tea polyphenols. Mutat Res 436:69–97, 1999.
52. SI Trevisanato, Y-I Kim. Tea and health. Nutr Rev 58:1–10, 2000.
53. M Inoue, K Tajima, K Hirose, N Hamajima, T Takezaki, T Kuroishi, S Tominaga. Tea and coffee consumption and the risk of digestive tract cancers: data from a comparative case-referent study in Japan. Cancer Causes Control 9:209–216, 1998.
54. W Zheng, TJ Doyle, LH Kushi, TA Sellers, C-P Hong, AR Folsom. Tea consumption and cancer incidence in a prospective cohort study of postmenopausal women. Am J Epidemiol 144:175–182, 1996.
55. CS Yang, JM Landau. Effects of tea consumption on nutrition and health. J Nutr 130:2409–2412, 2000.
56. CS Yang, P Maliakal, XF Meng. Inhibition of carcinogenesis by tea. Annu Rev Pharmacol Toxicol 42:25–54, 2002.
57. MA Azuine, SV Bhide. Chemopreventive effect of turmeric against stomach and skin tumors induced by chemical carcinogens in Swiss mice. Nutr Cancer 17:77–83, 1992.
58. M-T Huang, T Lysz, T Ferraro, TF Abidi, JD Laskin, AH Conney. Inhibitory effects of curcumin on in vitro lipoxygenase and cyclooxygenase activities in mouse epidermis. Cancer Res 51:813–819, 1991.
59. HU Gali, EM Perchellat, J-P Perchellet. Inhibition of tumor promoter-induced ornithine decarboxylase activity by tannic acid and other polyphenols in mouse epidermis in vivo. Cancer Res 51:2820–2825, 1991.
60. P Langcake, RJ Pryce. The production of resveratrol by *Vitis vinifera* and other members of the *Vitaceae* as a response to infection or injury. Physiol Plant Pathol 9:77–86, 1976.
61. S Nonomura, H Kanagawa, A Makimoto. Chemical constituents of polygonaceous plants. I. Studies on the components of Ko-jo-kon (Polygonum cuspidatum SIEB et ZUCC). Yakugaku Zasshi 83:983–988, 1963.
62. M Foroozesh, G Primrose, Z Guo, LC Bell, WL Alworth, FP Guengerich. Aryl acetylenes as mechanism-based inhibitors of cytochrome P450-dependent monooxygenase enzymes. Chem Res Toxicol 10:91–102, 1997.
63. M Jang, L Cai, GO Udeani, KV Slowing, CF Thomas, CW Beecher, HH Fong, NR Farnsworth, AD Kinghorn, RG Mehta, RC Moon, JM Pezzuto. Cancer chemopreventive activity of resveratrol, a natural product derived from grapes. Science 275:218–220, 1997.
64. J Gusman, H Malonne, G Atassi. A reappraisal of the potential chemopreventive and chemotherapeutic properties of resveratrol. Carcinogenesis 22:1111–1117, 2001.
65. EM Daniel, AS Krupnick, Y-H Heur, JA Blinzler, RW Nims, GD Stoner. Extraction, stability, and quantitation of ellagic acid in various fruits and nuts. J Food Comp Anal 2:338–349, 1989.
66. S Majid, KL Khanduja, RK Gandhi, S Kapur, RR Sharma. Influence of

ellagic acid on antioxidant defense system and lipid peroxidation in mice. Biochem Pharmacol 42:1141–1145, 1991.

67. BA Narayanan, O Geoffroy, CM Willingham, GG Re, DW Nixon. P53/p21$^{WAF1/CIP1}$ expression and its possible role in G1 arrest and apoptosis in ellagic acid treated cancer cells. Cancer Lett 136:215–221, 1999.
68. BA Narayanan, GG Re. IGF II down regulation associated cell cycle arrest in colon cancer cells exposed to phenolic antioxidant ellagic acid. Anticancer Res 21:359–364, 2001.
69. BA Narayanan, NK Narayanan, GD Stoner, BP Bullock. Interactive gene expression pattern in prostate cancer cells exposed to phenolic antioxidants. Life Sci 70:1821–1839, 2002.
70. P Lesca. Protective effects of ellagic acid and other plant phenols on benzo-(a)pyrene-induced neoplasia in mice. Carcinogenesis 4:1651–1653, 1983.
71. G Stoner. unpublished data.
72. O Daniel, MS Meier, J Schlatter, P. Frischknecht. Selected phenolic compounds in cultivated plants: ecologic functions, health implications, and modulation by pesticides. Environ Health Perspect 107:109–114, 1999.
73. A King, G Young. Characteristics and occurrence of phenolic phytochemicals. J Am Diet Assoc 99:213–218, 1999.
74. ML Kuo, JK Lin, TS Huang, NC Yang. Reversion of the transformed phenotypes of v-H-*ras* NIH3T3 cells by flavonoids through attenuating the content of phosphotyrosine. Cancer Lett 87:91–97, 1994.
75. SK Lee, JM Pezzuto. Evaluation of the potential of cancer chemopreventive activity mediated by inhibition of 12-O-tetradecanoyl phorbol 13-acetate-induced ornithine decarboxylase activity. Arch Pharm Res 22:559–564, 1999.
76. C Chaumontet, V Bex, I Gaillard-Sanchez, C Seillan-Heberden, M Suschetet, P Martel. Apigenin and tangeretin enhance gap junctional intercellular communication in rat liver epithelial cells. Carcinogenesis 15:2325–2330, 1994.
77. W Wang, L Heideman, CS Chung, JC Pelling, KJ Koehler, DF Birt. Cell-cycle arrest at G2/M and growth inhibition by apigenin in human colon carcinoma cell lines. Mol Carcinog 28:102–110, 2000.
78. F Yin, AE Giuliano, AJ Van Herle. Signal pathways involved in apigenin inhibition of growth and induction of apoptosis of human anaplastic thyroid cancer cells (ARO). Anticancer Res 19:4297–4303, 1999.
79. V Trochon, E Blot, F Cymbalista, C Engelmann, RP Tang, A Thomaidis, M Vasse, J Soria H Lu, C Soria. Apigenin inhibits endothelial-cell proliferation in G(2)/M phase whereas it stimulates smooth-muscle cells by inhibiting P21 and P27 expression. Int J Cancer 85:691–696, 2000.
80. KB Reddy, JS Krueger, SB Kondapaka, CA Diglio. Mitogen-activated protein kinase (MAPK) regulates the expression of progelatinase B (MMP-9) in breast epithelial cells. Int J Cancer 82:268–273, 1999.
81. YC Liang, YT Huang, SH Tsai, SY Lin-Shiau, CF Chen, JK Lin. Suppression of inducible cyclooxygenase and inducible nitric oxide synthase by apigenin and related flavonoids in mouse macrophages. Carcinogenesis 20:1945–1952, 1999.

82. J Panes, ME Gerritsen, DC Anderson, M Miyasaka, DN Granger. Apigenin inhibits tumor necrosis factor-induced intercellular adhesion molecule-1 upregulation in vivo. Microcirculation 3:279–286, 1996.
83. H Wei, L Tye, E Bresnick, DF Birt. Inhibitory effect of apigenin, a plant flavonoid, on epidermal ornithine decarboxylase and skin tumor promotion in mice. Cancer Res 50:499–502, 1990.
84. DF Birt, D Mitchell, B Gold, P Pour, HC Pinch. Inhibition of ultraviolet light induced skin carcinogenesis in SKH-1 mice by apigenin, a plant flavonoid. Anticancer Res 17:85–91, 1997.
85. DM Lepley, B Li, DF Birt, JC Pelling. The chemopreventive flavonoid apigenin induces G2/M arrest in keratinocytes. Carcinogenesis 17:2367–2375, 1996.
86. M McVean, H Xiao, K Isobe, JC Pelling. Increase in wild-type p53 stability and transactivational activity by the chemopreventive agent apigenin in keratinocytes. Carcinogenesis 21:633–699, 2000.
87. LE Voorrips, RA Goldbohm, DT Verhoeven, GA van Poppel, F Sturmans, RJ Hermus, PA van den Brandt. Vegetable and fruit consumption and lung cancer risk in the Netherlands Cohort Study on diet and cancer. Cancer Causes Control 11:101–115, 2000.
88. G van Poppel, DTH Verhoeven, H Verhagen, RA Goldbohm. Brassica vegetables and cancer prevention: epidemiology and mechanisms. Adv Exp Med Biol 472:159–168, 1999.
89. SS Hecht. Inhibition of carcinogenesis by isothiocyanates. Drug Metab Rev 32:395–411, 2000.
90. P Talalay, JW Fahey. Phytochemicals from cruciferous plants protect against cancer by modulating carcinogen metabolism. J Nutr 131:3027s–3033s, 2001.
91. Y Zhang, P Talalay, CG Cho, G Posner. A major inducer of anticarcinogenic protective enzymes from broccoli: Isolation and elucidation of structure. Proc Natl Acad Sci USA 89:2399–2403, 1992.
92. Y Zhang, TW Kensler, CG Cho, GH Posner, P Talalay. Anticarcinogenic activities of sulforaphane and structurally related synthetic norbornyl isothiocyanates. Proc Natl Acad Sci USA 91:3147–3150, 1994.
93. JW Fahey, X Haristoy, PM Dolan, TW Kensler, I Scholtus, KK Stephenson, P Talalay, A Lozniewski. Sulforaphane inhibits extracellular, intracellular, and antibiotic-resistant strains of *Helicobacter pylori* and prevents benzo(a)pyrene-induced stomach tumors. Proc Natl Acad Sci USA 99:7610–7615, 2002.
94. TC Goosen, UM Kent, L Brand, PF Hollenberg. Inactivation of cytochrome P450 2B1 by benzyl isothiocyanate, a chemopreventive agent from cruciferous vegetables. Chem Res Toxicol 13:1349–1359, 2000.
95. RL Moreno, T Goosen, UM Kent, F-L Chung, PH Hollenberg. Differential effects of naturally occurring isothiocyanates on the activity of cytochrome P450 2E1 and the mutant P450 2E1 T303A. Chem Res Toxicol 391:99–110, 2001.
96. SM Getahun, F-L Chung. Conversion of glucosinolates to isothiocyanates in humans after ingesting cooked watercress. Cancer Epidemiol Biomarkers Prev 8:447–451, 1999.

97. SS Hecht. Chemoprevention of cancer by isothiocyanates, modifiers of carcinogen metabolism. J Nutr 129:768S–774S, 1999.
98. MA Morse, SG Amin, SS Hecht, F-L Chung. Effects of aromatic isothiocyanates on tumorigenicity, O^6-methylguanine formation, and metabolism of the tobacco-specific nitrosamine 4-(methylnitrosamino)-1-(3-pyridyl)-1-butanone in A/J mouse lung. Cancer Res 49:2894–2897, 1989.
99. MA Morse, KI Eklind, SG Amin, SS Hecht, F-L Chung. Effects of alkyl chain length on the inhibition of NNK-induced lung neoplasia in A/J mice by arylalkyl isothiocyanates. Carcinogenesis 10:1757–1759, 1989.
100. GD Stoner, DT Morrissey, Y-H Heur, EM Daniel, AJ Galati, SA Wagner. Inhibitory effects of phenethyl isothiocyanate on N-nitrosobenzylmethylamine carcinogenesis in the rat esophagus. Cancer Res 51:2063–2068, 1991.
101. JT Wilkinson, MA Morse, LA Kresty, GD Stoner. Effect of alkyl chain length on inhibition of N-nitrosomethylbenzylamine-induced esophageal tumorigenesis and DNA methylation by isothiocyanates. Carcinogenesis 16:1011–1015, 1995.
102. GD Stoner, JC Siglin, MA Morse, DH Desai, SG Amin, LA Kresty, AL Toburen, EM Heffner, DJ Francis. Enhancement of esophageal carcinogenesis in male F344 rats by dietary phenylhexyl isothiocyanate. Carcinogenesis 16:2473–2476, 1995.
103. TS Hudson, PS Carlton, A Gupta, GD Stoner, MA Morse. Investigation of the enhancement of NMBA-induced esophageal tumorigenesis by 6-phenylhexyl isothiocyanate. Cancer Lett 162:19–26, 2001.
104. CV Rao, A Rivenson, B Simi, E Zang, R Hamid, GJ Kelloff, V Steele, BS Reddy. Enhancement of experimental colon carcinogenesis by dietary 6-phenylhexyl isothiocyanate. Cancer Res 55:4311–4318, 1995.
105. Y-R Chen, W Wang, A-NT Kong, T-H Tan. Molecular mechanisms of c-Jun N-terminal kinase-mediated apoptosis induced by anticarcinogenic isothiocyanates. J Biol Chem 273:1769–1775, 1998.
106. L Gamet-Payrastre, P Li, S Lumeau, G Cassar, M-A Dupont, S Chevolleau, N Gasc, J Tulliez, F Terce. Sulforaphane, a naturally occurring isothiocyanate, induces cell cycle arrest and apoptosis in HT29 human colon cancer cells. Cancer Res 60:1426–1433, 2000.
107. R Yu, S Mandlekar, JJ Haney, DS Ucker, A-N Kong. Chemopreventive isothiocyanates induce apoptosis and caspase-3-like protease activity. Cancer Res 58:402–408, 1999.
108. C Huang, W-Y Ma, J Li, SS Hecht, Z Dong. Essential role of p53 in phenethyl isothiocyanate-induced apoptosis. Cancer Res 58:4102–4106, 1998.
109. R Dashwood, T Negishi, H Hyatsu, V Breinholt, J Hendricks, G Bailey. Chemopreventive properties of chlorophylls towards aflatoxin B_1. Mutat Res 399:245–253, 1998.
110. PA Egner, J-Bin Wang, Y-R Zhu, B-C Zhang, Y Wu, Q-N Zhang, G-S Qian, S-Y Kuang, SJ Gange, LP Jacobson, KJ Helzlsouer, GS Bailey, JD Groopman, TW Kensler. Chlorophyllin intervention reduced aflatoxin-DNA adducts in individuals at high risk for liver cancer. Proc Natl Acad Sci USA 98:14601–14606, 2001.

111. S Barnes, J Sfakianos, L Coward, M Kirk. Soy isoflavonoids and cancer prevention. Underlying biochemical and pharmacological issues. Adv Exp Med Biol 401:87–100, 1996.
112. WA Fritz, L Coward, J Wang, CA Lamartiniere. Dietary genistein: perinatal mammary cancer prevention, bioavailability and toxicity testing in the rat. Carcinogenesis 19:2151–2158, 1998.
113. CA Lamartiniere, JB Moore, NM Brown, R Thompson, MJ Hardin, S Barnes. Genistein suppresses mammary cancer in rats. Carcinogenesis 16:2833–2840, 1995.
114. AH Wu, RG Ziegler, AM Nomura, DW West, LN Kolonel, PL Horn-Ross, RN Hoover, MC Pike. Soy intake and risk of breast cancer in Asians and Asian Americans. Am J Clin Nutr 68:1437S–1443S, 1998.
115. MT Goodman, LR Wilkens, JH Hankin, L-C Lyu, AH Wu, LN Kolonel. Association of soy and fiber consumption with the risk of endometrial cancer. Am J Epidemiol 146:294–306, 1997.
116. SS Strom, Y Yamamura, CM Duphorne, MR Spitz, RJ Babaian, PC Pillow, SD Hursting. Phytoestrogen intake and prostate cancer: a case-control study using a new database. Nutr Cancer 33:20–25, 1999.
117. T Gotoh, K Yamada, H Yin, A Ito, T Kataoka, K Dohi. Chemoprevention of N-nitroso-N-methylurea-induced rat mammary carcinogenesis by soy foods or biochanin A. Jpn J Cancer Res 89:137–142, 1998.
118. AI Constantinou, RG Mehta, A Vaughan. Inhibition of N-methyl-N-nitrosourea-induced mammary tumors in rats by the soybean isoflavones. Anticancer Res 16:3293–3298, 1996.
119. M Pollard, PH Luckert. Influence of isoflavones in soy protein isolates on development of induced prostate-related cancers in L-W rats. Nutr Cancer 28:41–45, 1997.
120. WB Murrill, NM Brown, JX Zhang, PA Manzolillo, S Barnes, CA Lamartiniere. Prepubertal genistein exposure suppresses mammary cancer and enhances gland differentiation in rats. Carcinogenesis 17:1451–1457, 1996.
121. J-R Zhou, ET Gugger, T Tanaka, Y Guo, GL Blackburn, SK Clinton. Soybean phytochemicals inhibit the growth of transplantable human prostate carcinoma and tumor angiogenesis in mice. J Nutr 129:1628–1635, 1999.
122. JN Davis, B Singh, M Bhuiyan, FH Sarkar. Genistein-induced upregulation of $p21^{WAFI}$, down regulation of cyclin B, and induction of apoptosis in prostate cancer cells. Nutr Cancer 32:123–131, 1998.
123. DB Fournier, JW Erdman Jr, GB Gordon. Soy, its components, and cancer prevention: a review of the in vitro, animal, and human data. Cancer Epidemiol Biomark Prev 7:1055–1065, 1998.
124. GD Stoner, LA Kresty, PS Carlton, JC Siglin, MA Morse. Isothiocyanates and freeze-dried strawberries as inhibitors of esophageal cancer. Toxicological Sciences. 52 (Suppl):95–100, 1999.
125. PS Carlton, LA Kresty, JC Siglin, C. Morgan, J Lu, GD Stoner. Inhibition of *N*-nitrosomethylbenzylamine-induced tumorigenesis in the rat esophagus by dietary freeze-dried strawberries. Carcinogenesis 22:441–446, 2001.

126. LA Kresty, MA Morse, C Morgan, PS Carlton, J Lu, A Gupta, M Blackwood, GD Stoner. Chemoprevention of esophageal tumorigenesis by dietary administration of lyophilized black raspberries. Cancer Res 61:6112–6119, 2001.
127. GK Harris, A Gupta, RG Nines, LA Kresty, SG Habib, WL Frankel, K LaPerle, DD Gallaher, SJ Schwartz, GD Stoner. The effects of lyophilized black raspberries on azoxymethane-induced colon cancer and 8-hydroxy-2′-deoxyguanosine levels in the Fischer 344 rat. Nutr Cancer 40:125–133, 2001.
128. PS Carlton, LA Kresty, GD Stoner. Failure of dietary strawberries to inhibit 4-(methylnitrosamino)-1-(3-pyridyl)-1-butanone-and benzo(a)pyrene-induced lung tumorgenesis in strain A/J mice. Cancer Lett 159:113–117, 2000.
129. H Xue, RM Aziz, N Sun, JM Cassady, LM Kamendulis, Y Xu, GD Stoner, JE Klaunig. Inhibition of cellular transformation by berry extracts. Carcinogenesis 22:351–361, 2001.
130. C Huang, Y Huang, J Li, W Hu, R Aziz, M-S Tang, N Sun, J Cassady, GD Stoner. Inhibition of BPDE-induced transactivation of AP-1 and NFkB by black raspberry extracts. Cancer Res 62:6857–6863, 2002.
131. Reference deleted.
132. W Kuenzig, J Chau, E Norkus, H Holowaschenko, H Newmark, W Mergens, AH Conney. Caffeic acid and ferulic acid as blockers of nitrosamine formation. Carcinogenesis 5:309–513, 1984.
133. CN van der Veere, B Schoemaker, R van der Meer, AK Groen, PL Jansen, RP Oude Elferink. Rapid association of unconjugated bilirubin with amorphous calcium phosphate. J Lipid Res 36:1697–1707, 1995.
134. LW Wattenberg, JB Hochalter, AR Galbraith. Inhibition of β-propiolactone-induced mutagenesis and neoplasia by sodium thiosulfate. Cancer Res 47: 4351–4354, 1987.
135. J-Y Hong, T Smith, M-J Lee, W Li, B-L Ma, SM Ning, JF Brady, PE Thomas, CS Yang. Metabolism of carcinogenic nitrosamines by rat nasal mucosa and the effect of diallyl sulfide. Cancer Res 51:1509–1514, 1991.
136. MJ Wargovich. Diallyl sulfide, a flavor component of garlic (*Allium sativum*), inhibits dimethylhydrazine-induced colon cancer. Carcinogenesis 8:487–489, 1987.
137. VL Sparnins, G Barany, LW Wattenberg. Effects of organosulfur compounds from garlic and onions on benzo(a)pyrene-induced neoplasia and glutathione S-transferase activity in the mouse. Carcinogenesis 9:131–134, 1988.
138. MN Wargovich, C Woods, VWS Eng, LC Stephens, K Gray. Chemoprevention of N-nitrosomethylbenzylamine-induced esophageal cancer in rats by the naturally-occurring thioether, diallyl sulfide. Cancer Res 48:6872–6875, 1988.
139. MA Morse, SD LaGreca, SG Amin, FL Chung. Effects of indole-3-carbinol on lung tumorigenesis and DNA methylation induced by 4-(methylnitrosamino)-1-(3-pyridyl)-1-butanone (NNK) and on the metabolism and disposition of NNK in A/J mice. Cancer Res 50:2613–2617, 1990.
140. N Takahashi, RH Dashwood, LF Bjeldanes, DE Williams, GS Bailey. Mechanisms of indole-3-carbinol (I3C) anticarcinogenesis: inhibition of aflatoxin B1-

DNA adduction and mutagenesis by I3C acid condensation products. Food Chem Toxicol 33:851–857, 1995.

141. MA Morse, CX Wang, SG Amin, SS Hecht, FL Chung. Effects of dietary sinigrin or indole-3-carbinol on O^6-methylguanine-DNA-transmethylase activity and 4-(methylnitrosamino)-1-(3-pyridyl)-1-butanone-induced DNA methylation and tumorigenicity in F344 rats. Carcinogenesis 9:1891–1895, 1988.

142. RH Dashwood, DN Arbogast, AT Fong, C Pereira, JD Hendricks, GS Bailey. Quantitative interrelationships between aflatoxin B1 carcinogen dose, indole-3-carbinol anti-carcinogen dose, target organ DNA adduction and final tumor response. Carcinogenesis 10:175–181, 1989.

143. BC Pence, F Buddingh, SP Yang. Multiple dietary factors in the enhancement of dimethylhydrazine carcinogenesis: Main effect of indole-3-carbinol. J Natl Cancer Inst 77:269–276, 1986.

144. GS Bailey, JD Hendricks, DW Shelton, JE Nixon, NE Pawlowski. Enhancement of carcinogenesis by the natural anticarcinogen indole-3-carbinol. J Natl Cancer Inst 78:931–934, 1987.

145. VL Sparnins, LW Wattenberg. Enhancement of glutathione S-transferase activity of the mouse forestomach by inhibitors of benzo[a]pyrene-induced neoplasia of the forestomach. J Natl Cancer Inst 66:769–771, 1981.

146. JM Sayer, H Yagi, AW Wood, AH Cooney, DM Jerina. Extremely facile reaction between the ultimate carcinogen benzo[a]pyrene-7,8-diol 9,10-epoxide and ellagic acid. J Am Chem Soc 104:5562, 1982.

147. AW Wood, M-T Huang, RL Chang, HL Newmark, RE Lehr, H Yagi, JM Sayer, DM Jerina, AH Conney. Inhibition of the mutagenicity of bay-region diol epoxides of polycyclic aromatic hydrocarbons by naturally occurring plant phenols: exceptional activity of ellagic acid. Proc Natl Acad Sci USA 79:5513–5517, 1982.

148. RL Chang, M-T Huang, AW Wood, C-Q Wong, HL Newmark, H Yagi, JM Sayer, DM Jerina, AH Conney. Effect of ellagic acid and hydroxylated flavonoids on the tumorigenicity of benzo[a]pyrene and (+/-)-7 beta, 8 alpha-dihydroxy-9 alpha, 10 alpha-epoxy-7,8,9,10-tetrahydrobenzo[a]pyrene on mouse skin and in the newborn mouse. Carcinogenesis 6:1127–1133, 1985.

149. S DeFlora, C Bennicelli, P Zannachi, A Camoirano, A Morelli, A DeFlora. In vitro effects of *N*-acetylcysteine on the mutagenicity of direct-acting compounds and procarcinogens. Carcinogenesis 5:505–510, 1984.

150. S DeFlora, C Bennicelli, A Camoirano, D Serra, M Romano, GA Rossi, A Morelli, A DeFlora. In vivo effects of *N*-acetylcysteine on glutathione metabolism and on the biotransformation of carcinogenic and/or mutagenic compounds. Carcinogenesis 6:1735–1745, 1985.

151. S DeFlora, M Astengo, D Serra, C Bennicelli. Inhibition of urethan-induced lung tumors in mice by dietary *N*-acetylcysteine. Cancer Lett 32:235–241, 1986.

152. YF Sasaki, H Imanishi, T Ohta, Y Shirasu. Effects of vanillin on sister-chromatid exchanges and chromosome aberrations induced by mitomycin C in cultured Chinese hamster ovary cells. Mutat Res 191:193–200, 1987.

153. H Imanishi, YF Sasaki, K Matsumoto, M Watanabe, T Ohta, Y Shirasu, K

Tutikawa. Suppression of 6-TG-resistant mutations in V79 cells and recessive spot formations in mice by vanillin. Mutat Res 243:151–158, 1990.
154. WM Tong, U Cortes, ZQ Wang. Poly(ADP-ribose) polymerase: a guardian angel protecting the genome and suppressing tumorigenesis. Biochim Biophys Acta 1552:27–37, 2001.
155. S De Flora, C Ramel. Classification of mechanisms of inhibitors of mutagenesis and carcinogenesis. Basic Life Sci 52:461–462, 1990.
156. S Gupta. Molecular steps of death receptor and mitochondrial pathways of apoptosis. Life Sci 69:2957–2964, 2001.
157. M Masuda, M Suzui, IB Weinstein. Effects of epigallocatechin-3-gallate on growth, epidermal growth factor receptor signaling pathways, gene expression, and chemosensitivity in human head and neck squamous cell carcinoma cell lines. Clin Cancer Res 7:4220–4229, 2001.
158. M Leszczyniecka, T Roberts, P Dent, S Grant, PB Fisher. Differentiation therapy of human cancer: basic science and clinical applications. Pharmacol Ther 90:105–156, 2001.
159. JE Trosko, CC Chang. Modulation of cell-cell communication in the cause and chemoprevention/chemotherapy of cancer. Biofactors 12:259–263, 2000.
160. M Jung. Inhibitors of histone deacetylase as new anticancer agents. Curr Med Chem 8:1505–1511, 2001.
161. PP Mehta, JS Bertram, WT Lowenstein. The actions of retinoids on cellular growth correlate with their actions on gap-junctional communication. J Cell Biol 108:1053–1065, 1989.
162. MZ Hossain, LR Wilkens, PP Mehta, W Loewenstein, JS Bertram. Enhancement of gap junctional communication by retinoids correlates with their ability to inhibit neoplastic transformation. Carcinogenesis 10:1743–1848, 1989.
163. JS Bertram. Role of gap junctional cell/cell communication in the control of proliferation and neoplastic transformation. Radiat Res 123:252–256, 1990.
164. JF Whitfield. Calcium switches, cell cycles, differentiation, and death. In: Lipkin M, Newmark HL, Kelloff GJ, eds. Calcium, Vitamin D, and Prevention of Colon Cancer. Boca Raton, FL: CRC Press, p 31, 1991.
165. BC Pence. Calcium and vitamin D effects of tumor promotion in the colon and mouse skin. In: Lipkin M, Newmark HL, Kelloff GJ, eds. Calcium, Vitamin D, and Prevention of Colon Cancer. Boca Raton, FL: CRC Press, p. 191, 1991.
166. KK Carroll, LA Eckel, LJ Fraher, JV Frei, HL Newmark. Dietary calcium, phosphate, and vitamin D in relation to mammary carcinogenesis. In: Lipkin M, Newmark HL, Kelloff GJ, eds. Calcium, Vitamin D, and Prevention of Colon Cancer. Boca Raton, FL: CRC Press, p. 229, 1991.
167. DJ Skrypec. Effect of dietary calcium on azoxymethane-induced intestinal carcinogenesis in male F344 rats fed high-fat diets. In: Lipkin M, Newmark HL, Kelloff GJ, eds. Calcium, Vitamin D, and Prevention of Colon Cancer. Boca Raton, FL: CRC Press, p. 241, 1991.
168. R Hasina, MW Lingen. Angiogenesis in oral cancer. J Dent Educ 65:1282–1290, 2001.
169. Y Cao, R Cao. Angiogenesis inhibited by drinking tea. Nature 398:381, 1999.
170. YD Jung, LM Ellis. Inhibition of tumor invasion and angiogenesis by epi-

gallocatechin gallate (EGCG), a major component of green tea. Int J Exp Pathol 82:309–316, 2001.
171. S Sharma, M Ghoddoussi, P Gao, GJ Kelloff, VE Steele, L Kopelovicyh. A quantitative angiogenesis model for efficacy testing of chemopreventive agents. Anticancer Res 21:3829–3837, 2001.
172. W Troll, A Klassen, A Janoff. Tumorigenesis in mouse skin: inhibition by specific inhibitors of proteases. Science 169:1211–1213, 1970.
173. W Troll, S Garte, K Frenkel. Suppression of tumor promotion by inhibitors of poly(ADP)ribose formation. Basic Life Sci 52:225–232, 1990.
174. M Hozumi, M Ogawa, T Sugimura, T Takeuchi, H Umezama. Inhibition of tumorigenesis in mouse skin by leupeptin, a protease inhibitor from *Actinomycetes*. Cancer Res 32:1725–1728, 1972.
175. T Nomura, S Hata, T Enomoto, H Tanaka, K Shibata. Inhibiting effects of antipain on urethane-induced lung neoplasia in mice. Br J Cancer, 42:624–626, 1980.
176. J Yavelow, TH Finlay, AR Kennedy, W Troll. Bowman-Birk soybean protease inhibitor as an anticarcinogen. Cancer Res (Suppl) 43:2454s–2459s, 1983.
177. W Troll, JS Lim, S Belman. Protease inhibitors suppress carcinogenesis in vivo and in vitro. In: Wattenberg L, Lipkin M, Boone CW, Kelloff GJ, eds. Cancer Chemoprevention. Boca Raton, FL: CRC Press, p 503, 1992.
178. K Frenkel, K Chrzan, CA Ryan, R Wiesner, W Troll. Chymotrypsin-specific protease inhibitors decrease H_2O_2 formation by activated human polymorphonuclear leukocytes. Carcinogenesis 8:1207–1212, 1987.
179. W Troll. Protease inhibitors interfere with the necessary factors of carcinogenesis. Environ Health Perspect 81:59–62, 1989.
180. AR Kennedy. The Bowman-Birk inhibitor from soybeans as an anticarcinogenic agent. Am J Clin Nutr 68(Suppl):1406s–1412s, 1998.
181. AR Kennedy. Chemopreventive agents: protease inhibitors. Pharmacol Ther 78:167–209, 1998.
182. AR Kennedy. Prevention of carcinogenesis by protease inhibitors. Cancer Res 54 (Suppl):1999s–2005s, 1994.
183. AR Kennedy, BF Szuhaj, PM Newberne, PC Billings. Preparation and production of a cancer chemopreventive agent, Bowman-Birk inhibitor concentrate. Nutr Cancer 19:281–302, 1993.
184. MW Foehr, LD Tomei, JG Goddard, PA Pemberton, IC Bathurst. Antiapoptotic activity of the Bowman-Birk inhibitor can be attributed to copurified phospholipids. Nutr Cancer 34:199–205, 1999.
185. IP Gladysheva, NG Balabushevich, NA Moroz, NI Larionova. Isolation and characterization of soybean Bowman-Birk inhibitor from different sources. Biochemistry (Mosc) 65:198–203, 2000.
186. HG Weed, RB McGandy, AR Kennedy. Protection against dimethylhydrazine-induced adenomatous tumors of the mouse colon by the dietary addition of an extract of soybeans containing the Bowman-Birk protease inhibitor. Carcinogenesis 6:1239–1241, 1985.
187. PV Messadi, P Billings, G Shklar, AR Kennedy. Inhibition of oral carcinogenesis by a protease inhibitor. J Natl Cancer Inst 76:447–452, 1986.

188. E von Hofe, PM Newberne, AR Kennedy. Inhibition of N-nitrosomethylbenzylamine induced esophageal neoplasms by the Bowman-Birk protease inhibitor. Carcinogenesis 12:2147–2150, 1991.
189. FL Meyskens Jr. Development of difluoromethylornithine and Bowman-Birk inhibitor as chemopreventive agents by assessment of relevant biomarker modulation: some lessons learned. IARC Sci Publ 154:49–55, 2001.
190. WB Armstrong, AR Kennedy, XS Wan, TH Taylor, QA Nguyen, J Jensen, W Thompson, W Lagerberg, FL Meyskens, Jr. Clinical modulation of oral leukoplakia and protease activity by Bowman-Birk inhibitor concentrate in a phase IIa chemoprevention trial. Clin Cancer Res 6:4684–4691, 2000.
191. K Frenkel. Carcinogen-mediated oxidant formation and oxidative DNA damage. Pharmacol Ther 53:127–166, 1992.
192. W Droge. Free radicals in the physiological control of cell function. Physiol Rev 82:47–95, 2002.
193. CS Yang, ZY Wang. Tea and cancer. J Natl Cancer Inst 85:1038–1049, 1993.
194. Y Xu, C-T Ho, SG Amin, C Han, F-L Chung. Inhibition of tobacco-specific nitrosamine-induced lung tumorigenesis in A/J mice by green tea and its major polyphenol as antioxidants. Cancer Res 52:3875–3879, 1992.
195. K Gwyn, FA Sinicrope. Chemoprevention of colorectal cancer. Am J Gastroenterol 97:13–21, 2002.
196. SM Lippman, MR Spitz. Lung cancer chemoprevention: an integrated approach. J Clin Oncol 19(18 Suppl):74s–82s, 2001.
197. H Vainio. Chemoprevention of cancer: a controversial and instructive story. Br Med Bull 55:593–599, 1999.
198. D Albanes, OP Heinonen, JK Huttunen, PR Taylor, J Virtamo, BK Edwards, J Haapakoski, M Rautalahti, AM Hartman, J Palmgren, et al. Effects of alpha-tocopherol and beta-carotene supplements on cancer incidence in the Alpha-Tocopherol Beta-Carotene Cancer Prevention Study. Am J Clin Nutr 62(6 Suppl):1427S–1430S, 1995.
199. GE Goodman. Prevention of lung cancer. Crit Rev Oncol Hematol 33:187–197, 2000.
200. NR Cook, IM Le, JE Manson, JE Buring, CH Hennekens. Effects of beta-carotene supplementation on cancer incidence by baseline characteristics in the Physicians' Health Study (United States). Cancer Causes Control 11:617–626, 2000.
201. S Yeh, M Hu. Antioxidant and pro-oxidant effects of lycopene in comparison with beta-carotene on oxidant-induced damage in Hs68 cells. J Nutr Biochem 11:548–554, 2000.

18

Food Biotechnology and U.S. Products Liability Law: The Search for Balance Between New Technologies and Consumer Protection

Steven H. Yoshida
Consultant, Davis, California, U.S.A.

I. INTRODUCTION

Someone once said to me that whenever people find new ways of looking at things, these new ways inevitably meet resistance. In this context, I recall that during the 1980s, the growing popularity of desktop computers engendered fears that these machines would take over our lives and destroy our humanity. Now in the twenty-first century, we use these machines to buy goods, do business, communicate with others, and entertain ourselves. We even use them to write manuscripts for books. Fears that personal computers would destroy our lives have given way to an appreciation of their usefulness.

Genetically modified organisms (GMOs) are now encountering that resistance. As with past worries over computers, GMOs represent unfamiliar technologies with unknown risks. However, there are some differences between computers and GMOs. Admittedly, unlike computers, which are inert materials when not in use, GMOs are living organisms that literally do have lives of their own. GMOs or their components may reproduce or move in physical space regardless of our intentions. Also, GMOs were sold to consumers without their knowledge whereas desktop computers were highly publicized products.

Many issues concerning GMOs are widely discussed in print. Perhaps foremost, critics argue that the development and use of GMOs were less than ethical because the means by which GMOs were produced, released into the environment, and sold to consumers were unfair (1,2). For example, anti-GMO groups feel that consumers should have the right to know what they are buying and have the choice to avoid GMOs. These issues are central to the issue of the labeling of foods and food products. Thus, fairness is achieved by establishing correct processes for the manufacturing and distribution of products to their consumers. In contrast, regulatory agencies are said to base their decisions (e.g., food safety) on the end product rather than the process (3). In essence, if the food product is safe, it is approved for use regardless of the methods by which it was produced. Despite these generalizations, one might argue that formal procedures are applied to GMOs. The institutions that produce GMOs (e.g., Monsanto) must use some form of standardized research and development methods, and regulatory agencies [e.g., the U.S. Environmental Protection Agency (EPA)] examine these novel products before their release into the environment or to the markets. The argument actually is one of sufficiency, transparency, and goals of the process.

Another ethical problem is the supposed unnaturalness of organisms created by recombinant DNA technology (4). In the most objective sense, this is a nonissue. At least in terms of known scientific concepts, we are part of the natural world. Traveling at velocities greater than the speed of light or living forever is beyond any current human being. Thus, everything we do (including making GMOs) is, by definition, natural. However, many people disagree and this disagreement creates some interesting paradoxes. Herbal remedies are often sold without regulation or testing (5). They are in part exempt from the regulatory controls applied to GMOs because of their aura of naturalness (6,7). But herbal medicines are essentially not biologically different from GM foods as they both are derived from living plant materials. In addition, the fact that something is deemed more natural than another does not necessarily make it a better product. If it is natural to die of smallpox and unnatural to be vaccinated for smallpox, which would you choose? The real question should not be the naturalness of GMOs, but the wisdom of their creation and use.

Another source of contention is substantial equivalence. Comparisons between the key constituents of a GMO and those of its nongenetically modified (non-GM) counterpart are used as one factor in assessing the safety of the GMO. The more similar the GMO is to its traditional counterpart, the safer it is considered to be (8). But the definition of "substantial equivalence" and the sufficiency of process have become issues (9,10).

One issue related to substantial equivalence is that evidence of differences between a GMO and its non-GMO counterpart may depend on the extent of testing and study. Indeed, studies have shown that secondary metabolites may result from genetic modification (11,12). However, modification by selective breeding or mutagenesis is also likely to cause secondary or unknown changes in agricultural organisms. For example, a study compared the breeding efficiency of a GM form of *Arabidopsis thaliana* and counterparts created by a mutagen with their wild-type cousin (10). Results from the study showed that the GMO was relatively more efficient at breeding with the wild type than the mutated plant was. Researchers concluded that this enhanced outcrossing by a GMO could increase the probability of transgene escape. There is another interpretation of this finding. One might expect that the method of mutagenesis would affect many physiological systems within the plant, including reproduction, because mutagenesis is a relatively nonselective means of modifying a plant's genetic makeup. However, a specific advantage of recombinant DNA methods over traditional breeding methods is the ability to select more reliably for useful changes to be made to an organism (13). Thus, a GMO might be expected to reproduce more faithfully with its wild-type counterpart if reproduction were not a target of modification. This supports the view that in the long run, recombinant DNA techniques will probably create more predictable and less risky products than other methods of product development (14).

A further objection to GMOs is the claim that GMOs are less safe to eat than non-GMOs. Proponents of this view use the frequently referenced brazilnut protein and potato lectin studies to demonstrate the potential harm from transgenics (15,16). But causing harm is a possibility for many foods, whether developed by recombinant DNA techniques or not (11,17,18). This is the reason for studying all new food products before release to consumers. Again, the relative precision of transgenic techniques will in the long run probably lead to safer and more predictable products.

There is also a dispute over the need for agricultural biotechnology and GMOs. The GMO critics argue that there is enough food in the world to feed everyone and no one presently alive need go hungry. The argument is that the real problem is politics and not the lack of food (19,20). However, proponents of GMOs believe that biotechnology will be needed to meet the needs of the larger world population expected in the future (20,21). In addition, many products of GMOs (e.g., edible vaccines) are targeted to developing countries where current agricultural or medical supplies are unsuitable for those regions or are not available. Since simply moving donated food from one part of the world to another might lead to a system of dependency and unreliable and temporary fixes rather than a more permanent form of self-sufficiency, it is better to develop a crop that will grow well in sub-Saharan Africa than constantly to ship these items from a more developed nation (21). The current fear is that since there is already relatively less economic incentive for companies to perform research and development (R&D) for poor countries, the effects of anti-GMO activities might easily stop these research efforts (22,23). One might argue that since politics is a reality, the fact that there is theoretically enough food in the world for everyone cannot be relied on to fight world hunger. Many anti-GMO organizations should realize this, since their strategies and tactics are often political in nature (24,25). Thus, perhaps science and technology could be used to sidestep the counterproductiveness of political conflicts and suboptimal distribution of the world's food for the sake of humankind. However, there are indications that even absent the resistance posed by anti-GMO groups, economics and politics may still be major obstacles to the implementation of biotechnology for the world's poor (20,26,27).

On the basis of these and many other points of dispute, one major issue emerges, that of risk. This includes the risk that the food a consumer eats is not safe, the risk that the environment may be harmed through the release of GM crops, and the risk that an ethical belief or religious practice may be violated. The ways we define, predict, manage, and live with these risks are central to our level of comfort with GMOs or any other product or activity. Two widely discussed forms of risk management are found in the GMO literature. First, there are government regulations. Generally, governmental agencies assess and manage risk through regulation by determining risks and risk factors, deciding on acceptable levels of risk, and regulating human behavior. There are many reviews on this subject, particularly as they apply to GMOs (14,28–30).

A second is the Precautionary Principle (PP). This principle is thought to have evolved in Germany and culminated in the Principle of Precautionary Action of German federal law. This law states that actions must be taken to prevent environmental harm even when evidence of causality is limited (31). In the context of GMOs, the PP is used in a case-by-case basis to determine the acceptability of uncertainties that a newly modified organism may create an environmental risk. The PP allows decisions to be made in the absence of evidence of harm, or harmful characteristics of the GMO, and shifts the burden onto the promoter of the GMO to prove that it is safe (32). The basis for burden shifting has been analogized to the principle of criminal law that a defendant need not prove his or her innocence, but that the prosecutor must prove guilt (33).

> Just as society does not require a defendant to prove his innocence, so it should not require objectors to prove that a technology is harmful. It is up to those who want to introduce something new to prove, not with certainty but beyond reasonable doubt that it is safe. Society balances the trial in favour of the defendant because we believe that convicting an innocent person is far worse than failing to convict someone who is actually guilty. In the same way, we should balance the decision on risks and hazards in favour of safety, especially in those cases where the damage, should it occur, is serious and irredeemable (33).

As with other decision-making tools, the PP has its share of problems and detractors. First, there are multiple interpretations and definitions of the PP (34,35). Second, since decisions may be based on the avoidance of uncertainties and not on measurable risks or known risk factors, the PP may lead to overregulation, the elimination of highly beneficial products, and the stifling of innovation. Thus, decisions based on the PP may be unfair.* Perhaps to prevent some of the problems that may result from a rigid and narrow application of the PP, guidelines and factors that include non-discrimination, examinations of costs and benefits, and the use of scientific evidence have been proposed (33,37).

A third form of risk management that is less widely discussed as regulation and the PP is litigation in court (38). Litigation in the context of GMOs could involve many issues, including antitrust (21), the insufficiency of process, and even the interpretation of the PP (35). Litigation is often seen as a means of reparation after harm is already done. Perhaps less appreciated is that the opportunity or threat of litigation can deter wrongful behavior and encourage correct behavior. Section II discusses how an awareness of the potential of litigation may lead to the lowering of risks associated with GMO development and use. The balance of this chapter primarily focuses on a brief review of products liability laws and a hypothetical legal challenge to a GMO. Specifically, the author's prior examination of the safety of the Roundup Ready GM soybeans is used to demonstrate how these laws might be applied to hypothetical problems of safety.

---

* Although interesting, the comparison of burden shifting by the PP to that in criminal law [as stated in Ref. 33] is not completely satisfying. In a criminal case, there must first be evidence that a wrong or harm was actually committed. The prosecutor must then prove beyond a reasonable doubt that the defendant is guilty of the crime. This may include proof that the defendant intended to harm the victim and that the defendant's actions did indeed cause the harm suffered by the victim. The PP does not require a showing that any harm had occurred before barring of the use of the GMO. Thus, the PP can effectively put an innocent (or perhaps an immensely productive and respectable) GMO behind bars without any proof that the defendant had committed a wrong or that any wrongdoing had occurred. Similarly, U.S. constitutional laws on prior restraint may be applicable. In the United States, the freedom of speech is of such high value that in general, one cannot be restrained from speaking, although a person may be punished after the act. Also, the government or opponent of the proposed speech usually has the burden to prove that such speech must be prevented before it occurs. In the United States, people may be criminally liable for their acts (including speech), but not solely for their thoughts, their appearance of foreignness or strangeness, or their existence. Therefore, most criminal liability is imposed after the act has already occurred. However, certain acts may be prohibited (e.g., injunctions) before their occurrence, for example, if there is proof of irreparable harm (actual or potential) to land.

II. LITIGATION AS A FORM OF PRECAUTIONARY PRINCIPLE

There is a rationale for giving consumers and other interest groups the opportunity to litigate over GMOs. First, politics, business, trade, employment, environmentalism, and many other examples of human interactions are increasingly being conducted on a global scale. Fairness suggests that this inexorable process of globalization include litigation and rules of liability. Second, human activities may be accompanied by uncertainties, hazards, and sometimes unfair attempts to gain at the expense of others. Therefore, avenues for just compensation of injuries ought to be available. Third, statutes and regulations are usually general rules that become more specifically defined through a case-by-case series of lawsuits (39). In this sense, the PP is similar to regulations or statutes. Case law (e.g., rulings from lawsuits) can be used to refine the PP's meanings through the application of its provisions in specific situations (37). From a scientific point of view, court trials seem analogous to scientific experiments that test hypotheses or theories. Fourth, lawsuits vindicate personal autonomy by providing private citizens a means to participate in the making of laws and rules that affect their personal lives. Perhaps consumers will more readily accept novel products if there is a formal and legal forum in which they may voice their complaints after the appearance of product defects (40).

The last assertion suggests a conceptual link between the availability of litigation after harm has occurred and the need for caution before actual harm occurs. In theory at least, the opportunity to litigate puts organizations on notice of the availability of legal repercussions for deliberate or inadvertent wrongdoing. This notice could encourage care while minimizing overregulation and the stifling of innovation (41). The ability of litigation to balance the risks and benefits of any particular technology or product is a result of the fact-finding process and its use on a case-by-case basis, the refinement of laws through a series of related cases, and the exacting of penalties on the losers of trials. In this way, risks may be minimized, benefits maximized, and, ideally, fairness achieved.

III. U.S. PRODUCTS LIABILITY LAWS

Lawsuits involving GMOs can be based on a number of legal theories (e.g., nuisance); this chapter focuses on products liability. Since most GMOs are developed as consumer products and these products result from modern technology, products liability laws are among the most readily available means to challenge GMOs legally. Thus, a review of U.S. products liability laws is presented. In addition, the author of this chapter has previously written about the evidence of safety of GM soybeans (42). This provides an opportunity to examine the application of such laws in a hypothetical situation within the context of the food safety of GM soybeans. Since the theme of this chapter is the use of litigation as a form of deterrence, the analysis of GM soybeans and products liability laws is intended to provide some insight on the extent of liability that could be imposed on GMO makers and the types of premarket scientific studies that may lead to successful legal defenses as well as the production of safe products.

A. Evolution of Products Liability Laws

1. Contract Privity

Today products liability laws are considered to be part of the law of torts (43,44). However, at least one historical review of products liability in the United States begins with the law of contracts (45). The seminal case for the contractual origins of products liability law is the 1842 English decision of *Winterbottom v. Wright*. A mail coach driver was injured by a defective coach during the course of his employment. The manufacturer of the coach had a contractual relationship (i.e., contract privity) with the postmaster-general to maintain the coaches in good repair. Although the driver had no direct contractual agreement with the manufacturer of the coach, he sued the manufacturer because his employer (the postmaster, with whom he had legal ties) was immune from lawsuits. The court decided that liability for the injuries of the driver was limited to his employer (who could not be sued) and not the manufacturer of the coach because the driver had no direct contractual agreement with the manufacturer (46).

> There is no privity of contract between these parties; and if the plaintiff can sue, every passenger, or even any person passing along the road, who was injured by the upsetting of the coach, might bring a similar action. Unless we confine the operation of such contracts as this to the parties who entered into them, the most absurd and outrageous consequences, to which I can see no limit, would ensue (46).

This limitation on liability applied to situations in which a party to a contract simply did not do what he or she contracted to do (as opposed to instances of improper performance of obligations).

2. Negligence

The 1916 decision of *MacPherson v. Buick Motor Co.* removed contractual privity as a requirement for liability for products that become dangerous when defective. In that case, a purchaser who bought an automobile from a dealer was injured by the collapse of a defective wheel. There was no contract between the consumer and the manufacturer. Before this case, only consumers of "inherently dangerous" products (e.g., poison and dynamite) could sue manufacturers for injuries derived from their products. But the court in *MacPherson* argued that many things become inherently dangerous if they are made defectively and that manufacturers should be potentially liable for injuries caused by the defects in such products. To balance the risks to manufacturers, the court added that the inherent dangers in potentially defective products must be reasonably foreseeable (47). In addition, since manufacturers and not retail sellers are often most responsible for the condition of the sold goods, fairness to the consumer should guarantee recourse against the manufacturer for redress (48). Although *MacPherson* opened a path through which manufacturers could be held liable for consumer injuries, plaintiffs still had to prove both that products were defective and that manufacturers were negligent (49).

3. Warranty

Breach of express and implied warranties also became sources of manufacturers' liability. Warranty-based litigation has elements of both tort and contract law. A breach of a warranty may be interpreted as a form of misrepresentation as well as a violation of the terms of a contract of sale (50). The form of liability for breach of warranty is strict liability. This no-fault form of liability does not require proof of negligence or deliberate wrongdoing of a manufacturer. Instead, liability is based on the product's inability to meet the standards guaranteed by the manufacturer or seller.

Since the evolution of products liability laws generally led away from the need for direct contractual agreements between manufacturers and consumers, warranty litigation became increasingly reliant on implied rather than express warranties. Initially, the Uniform Sales Act of 1906 provided that a warranty of "merchantability" was implied in the sale of goods (51). Thus, if a good was not at least of average quality and fit for the general purpose for which it was manufactured and sold, the seller breached this warranty and was liable for personal injuries incurred through the defect in the product. Again, strict liability was based on the performance of the product and not the manufacturer or seller. Since this implied warranty existed between only the seller and the buyer and not a third-party user, contract privity was still required for a seller to be legally responsible to a purchaser. Also, a written disclaimer within a contract could eliminate an implied warranty.

The protection for consumers initiated by the Uniform Sales Act was extended by the 1960 decision of *Henningsen v. Bloomfield Motors, Inc.* Mr.

Henningsen bought a car as a gift for his wife. The contract for the purchase of the car included a disclaimer that the car dealer gave no warranties on the car, express or implied, other than replacement of defective parts. A steering failure then led to the injury of Mrs. Henningsen. First, the court removed the requirement for a contractual agreement by holding manufacturers liable to virtually all users and not only to people who purchased the product. This was based on the idea that many products, including cars, are used or consumed by people other than the purchasers (52). Also, proof of negligence was not required because the product defect constituted a breach of warranty, and liability for this breach did not require proof of fault (i.e., strict liability). Second, the court ruled that the disclaimer of implied warranties of merchantability was unconscionable and illegal (53).

> The gross inequality of bargaining position occupied by the consumer in the automobile industry is thus apparent.... Because there is no competition among the motor vehicle manufacturers with respect to the scope of protection guaranteed to the buyer, there is no incentive on their part to stimulate good will in that field of public relations. Thus, there is lacking a factor existing in more competitive fields, one which tends to guarantee the safe construction of the article sold. Since all competitors operate in the same way, the urge to be careful is not so pressing (53).

4. Strict Liability in Tort

In the 1963 case of *Greenman v. Yuba Power Products, Inc.*, the court specifically ruled that there could be liability without negligence. The user of a power tool was injured when a piece of wood flew out of the machine and hit him on his forehead. Evidence showed that the injuries were caused by the defective design and construction of the power tool. The court used this case to impose strict liability on the makers of products for injuries incurred by their users (54). Essentially, there is a duty to supply safe products. Strict liability is imposed if the court finds that this duty was breached, the user sustained actual damages or injuries, and the breach of duty led to these injuries.

The results of *Greenman* agreed with those of *Henninsen* but considered the implied warranty rationale a camouflage for the creation of a new tort: strict product liability in tort. That is, strict liability through breach of warranty and strict liability in tort are the same forms of liability but under different names. As with breach of warranty, strict liability in tort does not require fault by the manufacturer or seller, and the same evidence may be used for either theory of liability. Thus, there is no legal difference between an implied warranty from the manufacturer that a product is safe for its intended use and a duty imposed on the manufacturer to make the product safe for its intended use.

B. Specific Considerations in Products Liability

The evolution of U.S. products liability laws shows a tendency toward increased protection for purchasers, users, and bystanders. This trend was achieved by imposing liability without the need for a contractual agreement between the injured plaintiff and defendant manufacturer, and the creation of implied warranties or duties to guarantee the safety and utility of the products sold.

Before we can apply products liability laws to a hypothetical problem with GM soybeans, other specific details must be discussed. One of the most highly debated and contentious areas in products liability litigation, particularly when science and technology are involved, is causation. If a user of a product sues the manufacturer on the basis of alleged injuries from the product, the plaintiff must show proof that the particular product did in fact cause the harm. In fairness to manufacturers, defenses are available to limit the extent to which liability may be imposed. In particular, the foreseeability of harm becomes important. Both causation and foreseeability are discussed as part of the review of negligence.

Another area to be clarified is the definition of a defective product. Previous discussions of negligence and strict liability in torts included the concept of defective products. Under product liability laws, there are three potential types of defects in a product: manufacturing, design, and failure to warn of potential problems.

1. Negligence

The requirements for a case of negligence are usually analyzed in four interrelated steps. The first step is that of duty of care. In a civilized society, we are supposed to interact in ways that minimize harm to each other. But to whom does one owe a duty to be cautious and careful? In general, there is no duty to act affirmatively. For example, a bystander is not required to help a drowning person if the bystander did not create the dangerous situation or add a new risk of harm to the victim. However, if an actor acts in a way that may create an unreasonable risk of harm to someone, a duty of care is established. Courts have also imposed duties of care to other situations, including (a) a certain special relationship (e.g., parent and child), (b) the need to control a person from injuring others, and (c) a situation in which a person who is not legally required to act voluntarily goes to the aid of another (55).

In order to establish a duty, one must ask, What was the defendant supposed to do, or how was he supposed to act? One approach to answering this question is the use of the *reasonable prudent person* standard. This standard is based on a hypothetical average person with the same general physical characteristics as the actual person or defendant who had the duty of care. The standard of care then becomes what this hypothetical reasonably prudent person would have done in the same situation. There are legally accepted variations on this reasonable person theme. For example, children may be compared to a hypothetical child of the same age, and the standard of care of professionals or those with special occupational skills may be established through the use of people with equivalent levels of training and experience. Standards may also be created through customs, statutes, or case law (56).

Second, we look for a breach of duty or standard of care. This is a failure to conform to the required standard (57).

Third, proof of causation must show a reasonably close causal link between the breach of duty and the plaintiff's injuries. An analysis of causation usually begins with actual cause or causation in fact. A person's act is considered a cause in fact if the injury would not have occurred "but for" the act. Thus, there must be some actual link between the defendant's act and the plaintiff's injury. Other rules were developed for cases in which more than one person caused the injuries of a victim or the contributions of multiple defendants to the plaintiff's injuries are unknown (58,59).

After the establishment of causation in fact, an analysis of proximate or legal cause is done to set boundaries on the extent of liability. The rationale for this limitation is the idea that although a defendant's act may be a cause in fact of the plaintiff's injuries, there may be circumstances that make the imposition of liability unfair to the defendant. There are two major questions that guide the proximate cause analysis. The first is, Was the harm to the plaintiff foreseeable? The second is, Was the injury of the type that the duty of care was designed to avoid? Glannon (1995) uses a hypothetical situation to illustrate the use of these two questions. Suppose that a car owner leaves his car unlocked and the keys in the ignition switch. Such a driver may negligently breach a duty of care if a thief steals the car and then uses it to injure someone. Harm to an innocent victim is a foreseeable result of leaving a car unattended. This satisfies the foreseeability factor. But the type of harm is also a factor in imposing liability on the car owner. If the thief hits a pedestrian, the car owner is liable for this injury, since this injury is of the type of risk of harm that the duty of care is designed to minimize. But if the thief uses the car as part of a bomb to blow up a building, the car owner is probably not liable for that result (60). There is liability even if the injuries to the victim become more extensive than anticipated (e.g., there is liability if small scratches lead to a lethal infection) or if the foreseeable injury comes about in an unusual or unforeseeable manner (61).

> In [*United Novelty Co. v. Daniels*, 42 So. 2d 395 (Miss. 1949)] the defendant allowed its employee to clean some machines with gasoline in a small room heated by a heater with an open flame. A rat, drenched with gas, ran from under one of the machines over to the heater, caught fire, and ran for refuge back to the machine, causing an explosion that killed the employee. The court concluded that, while the manner in which the accident took place was unusual, an explosion was exactly the type of accident to be anticipated from using a volatile, flammable liquid in a small room with an open flame. The defendant was held liable (61).

Thus, foreseeability of the type of harm, rather than the extent of harm, seems to be more important in limiting liability.

Fourth, the plaintiff must prove damages and prove that she sustained some kind of injury to her person or property.

2. Product Defects

The *Greenman* case mentioned earlier used the defectiveness of a product to justify the imposition of strict or absolute liability on a manufacturer. Subsequent cases and legal commentary have refined the concept of defectiveness to encompass three types of product defects: manufacturing defects, design defects, and failure to warn of potential dangers adequately.

A manufacturing defect in a product occurs when "the product departs from its intended design even though all possible care was exercised in the preparation and marketing of the product" (62). Thus, the intended design serves as an objective standard. Generally, strict liability is imposed on the manufacturer if the product is proved to be defective at the time of sale and it proximately caused the plaintiff's injuries. Under strict liability, there is no requirement to prove that the defendant acted negligently or that less than reasonable care was used in the manufacturing process. The rationale for strict liability is based on several policy considerations, including that (a) the product defect was probably due to a negligent act that would be very difficult to prove, (b) the manufacturer is in a better position relative to the plaintiff to develop methods to reduce the frequency of manufacturing defects, and (c) the manufacturer is better able than the plaintiff to absorb the costs of injuries that cannot be eliminated in a cost-effective manner (63). The primary defense to an alleged manufacturing defect is to show that the product did conform to the intended design when it was sold or manufactured but that it was altered after it left the control of the manufacturer.

The second type of defect, design, occurs "when the foreseeable risks of harm posed by the product could have been reduced or avoided by the adoption of a reasonable alternative design by the seller or the distributor, or a predecessor in the commercial chain of distribution, and the omission of the alternative design renders the product not reasonably safe" (64). A determination that a design is defective suggests that all members of that product line are inherently defective. For example, a product may be dangerously defective because a bolt was not placed in a particular location in the product. A design defect occurs when such bolts were intentionally not used in the entire product line. A manufacturing defect occurs when a bolt that was supposed to be included was inadvertently left out. These situations require different standards or tests for defectiveness.

Two tests have been used to determine whether a product design is defective: the consumer expectations test and the risk–benefit test. The consumer expectations test states that the product is defective if it is not as safe as expected. However, problems are associated with this test, including (a) whether the consumer or the jury is to decide what was expected and (b) how safe a product must be before it is not defective, particularly if the product is known by the consumer to be dangerous or its usefulness is related to its dangerousness (e.g., a knife, car, or cigarette) (65,66). As a result, this test is used primarily for very simple or obvious expectations. For example, a car should not overturn when driven at 2 miles per hour.

In contrast, the risk–benefit or risk–utility test inquires whether the risk that accompanies the use of a product outweighs its benefits or utility. Since this test requires the balancing of risks and benefits, it creates a partial return to a negligence type of analysis but without the requirement of proving that a consumer's injury was foreseeable (67). The two major problems associated with this balancing test are the need for expert knowledge and the availability of alternative designs. The first problem forces jurors to make decisions about products, concepts, or information about which they may have little knowledge or experience (68). The second problem requires jurors to decide whether a safer product design was available and not used, and, on occasion, whether it was cost-effective or feasible. If jurors are required to determine the reasonableness or feasibility of an alternative design, this calls for a return to a negligence standard since such an analysis may include an examination of foreseeable harm by the product (69). The analysis of design defects would also include other considerations. For example, courts generally argue that the manufacturer rather than the consumer (a) has more knowledge of the risks that accompany a product, (b) had the power to decide on its design and (c) chose the manner of advertising its products (70). The *polycentric* nature of a design alludes to the interrelatedness of the components and functions within a finished product (71). A design that accentuates a particular desired characteristic of a product may unavoidably alter or compromise other characteristics. Therefore, a balancing of pros and cons is recognized as an integral part of the development of a reasonable and feasible product.

The third type of defect, warning, is based on the idea that manufacturers rather than consumers have superior knowledge of the characteristics of their products. A warning defect occurs "because of inadequate instructions or warnings when the foreseeable risks of harm posed by the product could have been reduced or avoided by the provision of reasonable instructions or warnings by the seller or other distributor, or a predecessor in the commercial chain of distribution, and the omission of the instructions or warnings renders the product not reasonably safe" (72). By providing consumers with adequate product information, warnings enable users to perform their own risk–benefit analyses. A failure by a manufacturer to warn of potential hazards deprives a consumer of the information to choose to avoid or accept the risk associated with the use of the product. In addition, the manufacturer has a duty to warn of any foreseeable misuse of a dangerous product (73). As with a design defect, liability due to a failure to warn is based on negligence.

Two issues stand out in a discussion of a failure to warn. The first is the adequacy of the warning. A jury determines the adequacy of a warning, and Abraham (1997) has suggested that this could be no easier a task for a jury than deciding whether a design was defective (74). Second, in an attempt to provide complete and adequate warnings, an excessive amount of information may distract the consumer from the precise warning that she needs to avoid injury.

Abraham (1997) also identified interactions between the adequacy of warnings and design defects (75). An adequate warning may eliminate liability by sufficiently reducing the risks stemming from what would otherwise be a defective design. Less predictable results arise from cases in which useful and desired products have no reasonable alternative design, and adequate warnings would not be sufficient to decrease the risks relative to benefits. Then there are products with obvious associated hazards, and the inclusion of warnings make the hazards even more obvious, and yet courts may still impose liability on manufacturers (76). In addition, the obviousness of a hazard may depend on the user's sophistication, professional expertise, and level of education (77).

Finally, the materiality of the lack of warning is a measure of the importance of the warning in the consumer's decision to use or avoid a product. Essentially, the question asks whether a warning would have affected a plaintiff's choice. This is also a causation question, since a material lack of warning would have influenced a person's actions (78).

3. Defenses

A manufacturer can make several defenses against an allegation of product liability, including some in common with general negligence or strict liability claims. Contributory and comparative negligence are potential defenses in situations in which the plaintiff is partly responsible for the injuries that are associated with the use of a product. Contributory negligence may preclude any award to the injured plaintiff. Comparative negligence gives to the plaintiff a percentage of a damage award equivalent to the fault of the manufacturer. Therefore, if a court decides that a consumer is personally responsible for 40% of his own injuries, he will be awarded 60% of the claimed damages.

Product misuse is a defense unique to products liability claims, and here foreseeablility becomes an issue (79). Generally, a manufacturer is not liable for injuries caused by unforeseeable misuses of a product even if it contains a manufacturing or design flaw. Avoidance of liability from foreseeable misuses by the consumer requires additional and adequate warnings from the manufacturer.

As mentioned earlier, the primary defense to an alleged manufacturing defect is a showing that the product conformed to the intended design when it was sold or manufactured but that it was altered after it left the control of the manufacturer.

IV. THE APPLICATION OF PRODUCTS LIABILITY LAWS TO A HYPOTHETICAL FOOD SAFETY SITUATION

A major concern of consumers is the safety of GM products when consumed. As mentioned earlier in this chapter, I have previously written on the evidence of safety of GM soybeans. On the basis of published scientific data, a review of the United States' Federal Rules of Evidence as they apply to scientific evidence, and pertinent regulations of the U.S. Food and Drug Administration, the GM soybean appeared to be as safe as non-GM soybeans. But the preparation of that paper did result in a few caveats concerning GMOs. Although the soybean is only one GM product among many, it was chosen for the paper because more information was readily available on the soybean than on other GMOs. In general, information on GMOs may not be as accessible to the public as one might hope. In the previous paper, I suggested that safety studies include immunological assays. Thus, the range of methods for studying products of transgenic technology might be broadened, if not to increase the actual safety of the foods then perhaps to ensure the peace of mind of consumers.

In this present work, my intention is to review some of the routes of liability that might reach a product such as the GM soybean. In this regard, the accessibility and sufficiency of the GM soybean literature provide bases for assessing the reasonable foreseeability of hazards that might accompany the consumption of this food. Thus, my previous analysis of the safety of GM soybeans can be used as a tool to use to decide whether liability should be imposed in a hypothetical food safety issue. Since potential defendants are as capable of generating hypothetical problems as potential plaintiffs, such exercises would also be useful in deciding what types of GMOs and GMO-derived products to develop and in determining the types of studies to conduct to ensure that these products are reasonably safe. And perhaps one form of defense in litigation is the use of good faith efforts to develop reasonably safe products. Thus, the choice of studies to perform and data to retrieve during product development is central to the creation of safe and useful products, the level of confidence that consumers have in those products, and the vuner-ability of these products to lawsuits.

The rest of this chapter includes a brief review of food allergies and discussions on (a) the foreseeability of allergies, (b) product defects, (c) implied warranties, and (d) strict liability as applied to GM soybeans.

A. Food Allergies

As one trained in immunology, I am more familiar with immunological problems that might result from the consumption of foods than other types of reactions (e.g., acute toxicity or carcinogenicity). These immune system responses would most likely include allergic, hypersensitivity, or autoimmune types of reactions to foods. Therefore, I limit the rest of this chapter to a hypothetical problem related to an immune response to GM soybeans.

We begin with an allegation that consumption of soybeans or a soy product made from GM soybeans caused some type of allergic response. There might be questions about increased levels of known soy allergens or the creation of new allergens. A hypothetical new allergen could be the direct protein product of the introduced genes or a secondary metabolite that is indirectly derived from the transgene's activity. The overt manifestation of the allergy might include rashes, gastrointestinal upsets, anaphylactic shock, pain, swollen mucous membranes, and other signs of inflammation. But proof that the soy caused the allergy must include evidence that the reaction is allergic in nature and was directly caused by the soy product.

To comprehend the nature of true allergic reactions caused by food, one must understand the underlying structure and function of the immune system (80). Briefly, the immune system's main function is to protect the individual's body from damage caused by invading microorganisms. This protective response results from complex interactions between various white blood cell types, antibodies that are released by B lymphocytes, the numerous means by which cells signal or communicate with each other, the particular genetic composition of the individual, and contacts with environmental components that might be physical, chemical, or biological in nature. Foods definitely fall within the category of environmental contacts. Allergies result from the activation of immune defenses to what would otherwise be benign contacts with the environment.

True allergies (as opposed to other types of hypersensitivities or adverse reactions to foods) involve a particular class of antibody, the immunoglobulin E (IgE). This type of antibody literally acts as a physical bridge between allergens and immune cells (i.e., eosinophils) that release inflammatory chemicals. About one-third of the human population may have allergies of the IgE type (81).

The ability to predict the allergenicity of any particular antigen to a particular person is confounded by the complex interactions alluded to previously. For example, scientists are attempting to identify and characterize "atopy genes" that may predispose people to allergies (81). These genes may influence the nature of intercellular communication, the interactions between IgE and immune cells, and the activity of cells (e.g., transcription regulation). Also, the *hygiene hypothesis* suggests that vaccinations and other public health measures to reduce the prevalence of communicable diseases may increase the incidence of allergic diseases (82). Cross-reactivity among allergens is also implicated in food allergies. Latex, such as that used in gloves, contains more than 30 proteins similar to food proteins. In cross-reactions, the IgE generated by an allergic response to latex may recognize certain food components as well and thereby cause reactions to foods (83). A study of the structures of protein allergens also suggests that there is no particular amino acid sequence or protein conformation that is more likely to be allergenic than another (84). Allergenicity is probably more closely associated with other factors such as amount and route of exposure, digestibility, and avoidance of mechanisms that suppresses helper T-cell 2. Therefore, the ability to predict the allergenicity of a particular protein on the basis of its conformation or amino acid sequence may be limited. The effects of nutrition on the immune system may also include one's susceptibility to allergies (85). The phenomenon of *oral tolerance* adds more complexity to the understanding of food allergies (86). The oral intake of antigens may lead to a suppression of immune responses to those specific antigens, a situation that appears to conflict with the notion of food-initiated allergic reactions. Finally, other factors such as age, exercise, and drugs also affect the allergic response (87).

The diagnosis of a food allergy includes distinguishing an IgE-mediated response from the many other types of adverse reactions to foods. Food intolerances not associated with IgE include enzyme deficiencies (e.g., lactose intolerance), pharmacological substances in foods (e.g., caffeine in coffee and histamine in some fish), toxins naturally occurring in foods and those generated through storage, and psychological aversions (88). There are several standard tests used to differentiate true IgE-mediated food allergies from these other reactions to foods (88,89). Some involve controlled contacts between patients and allergens. These examine the reactions of patients when challenged (or fed) suspect foods or when such foods are eliminated from their diets. Other tests examine patients' blood for IgE that is specific for the allergens in question. Purified allergens are used in skin tests to examine patients' dermal reactions. Medical histories also examine patients' recollections of their experiences with certain foods and adverse reactions. However, these tests are not without their limitations. At least one author has stated that "reports of food allergy or intolerance from individuals or parents of children are notoriously unreliable" and that blood and skin tests are "unreliable because of a large number of false positive and false negative reactions" (87). The reliability of a diagnosis increases when these tests are integrated and a broader spectrum of information about the patient is available.

B. Foreseeability of Allergies from Genetically Modified Soybeans

Presuming that allegations of GM soybean allergies have been made and confirmed by medical tests, any determination of negligence requires a discussion of the reasonable foreseeability that genetic modification will increase the allergenicity of the soybean. Genetic modification might increase the allergenicity of a food by increasing the food's content of a known and preexisting allergen. In fact, non-GM soybeans are already among the most allergenic of foods (90). Trypsin inhibitor, a known soybean allergen, is a protein that has been found at similar levels in GM and non-GM soybeans, as have other constituents such as amino acids, fats, and carbohydrates (84,91). Thus, on the basis of a comparison of the components of GM and non-GM soybeans, GM soybeans could reasonably be foreseen to be as allergenic as their non-GM counterparts. Notably, for regulatory purposes, these comparisons also helped to establish that GM soybeans are substantially equivalent to the already existing non-GM soybeans.

Genetic modification may also result in the appearance of allergens absent in the unmodified soybean. In the first case, genetic modification may increase the allergenicity of the soybean through the transfer of the gene of an allergen. This potential hazard was illustrated by the previously mentioned transfer of a brazil nut allergen to soy plants (15). This study is often used as an example of the dangers of transgenic foods. However, this paper also demonstrates how safety studies successfully prevent hazards from entering our food supply. It may also serve as an example of how cutting edge science works. The published safety studies on the equivalency of GM and non-GM soybeans, and the brazil-nut soy transgenics, were published in the same year, 1996. The common date suggests that these experiments were being done at about the same time and that one scientific group was not necessarily aware of what the other group was doing or had found. Without further information on this issue, there is no reason to believe that Padgett et al. (91) should have been on notice of the actual transfer of an allergen and its allergenic properties until the brazil-nut paper was published. But generally, the transfer of the allergenic potential of a known allergen along with its gene is reasonably foreseeable. The objective of most genetic modification is to transfer the function of a gene in order to give the GMO some desired characteristic. With the Roundup Ready soy plant, glyphosate resistance was an expected result of the insertion of the glyphosate-resistant gene. Thus, the transfer of other characteristics, including allergenicity, should also be expected with gene transfer.

In the case of the glyphosate-resistant GM soybean, the question is whether a known allergen was introduced into the plant. Known food allergens share a number of characteristics (92). Allergenic proteins are usually within a range of physical size that is optimal for recognition by the immune system. They are often glycosylated and demonstrate physical stability when heated or in contact with digestive enzymes. Certain amino acid sequences are associated with allergenicity. Finally, food allergens tend to be abundant in food. For example, the egg allergen ovalbumin constitutes 54% of egg white protein. To confirm that the glyphosate-resistance protein would probably not be allergenic to consumers, studies were done to show that it (a) was not glycosylated, (b) rapidly degraded with digestive enzymes and heat, (c) had no amino acid similarities with known food allergens, and (d) constituted only about 0.08% of the soybeans' total protein (93,94). Thus, the gene and protein that conferred glyphosate resistance to the soy plant could not reasonably be foreseen as an allergen in the edible soybean. Indeed, study in 2000 by Aalberse suggests that amino acid sequence data may be less reliable indicators of allergenicity than previously thought (84). In effect, this indefinite reference retrospectively reduced the number of reliable tests of allergenicity that were done on the glyphosate-resistance protein and arguably lowered the study's actual ability to predict its potentially allergenic character.

Unexpected secondary metabolites may also generate novel allergens. The internal complexity of any individual organism, not only of its immune system but of all its interconnected biochemical, cellular, and physiological systems, precludes the ability to predict all results of a change such as the introduction of one or more genes. Thus, genetic modification is likely to lead to unintended and unpredictable alterations in the soybean and other GMOs. But as discussed in Sec. I, the products of other forms of plant production such as selective breeding and mutagenesis may be equally predictable if not more unpredictable than those derived from recombinant DNA techniques.

But GMO critics counter that recombinant DNA technology allows for another form of complexity and uncertainty through the introduction of genes from any one species into potentially any other species. Reproductive barriers generally prevent a similar free mixing of genes. For the purpose of a legal discussion on proximate cause, perhaps the question is whether on the basis of existing knowledge and reasonably obtainable additional information, the generation of allergenic secondary metabolites is reasonably foreseeable. Since allergic reactions are derived from the interactions of consumers and their foods, reasonable forseeability of allergies would also require knowledge of the biological characteristics of the consumers. As with many other difficult issues, the determination of proximate causation in a hypothetical trial would probably be done on a case-by-case basis and rely on the judgment of a jury.

C. Product Defects

As discussed earlier, a case for a manufacturing defect must show that a particular individual product deviated from the intended design as a result of some problem or negligence in production. This may be of little use in litigation against a GMO such as the soybean. Genetic mechanisms within eucaryotic organisms such as plants and animals minimize the occurrence of mutations during DNA replication. Therefore, soy plants derived from a single and stable transgenic event are unlikely to deviate from the intended design because of variability in their genetic makeup. However, the components of soybeans, both GM and non-GM, may vary because of differences in growing conditions such as light, water, insect pests, and soil nutrients. Thus, variability in food quality and quantity would more likely be due to everyday growing conditions than to some inherent genetic instability with the soybean. If the introduction of foreign genes to create a GMO somehow perturbs the mechanisms that ensure that fidelity of DNA replication, and if some individual soy plants do deviate from the intended design, perhaps these factors could open a case for a manufacturing defect, or a design defect.

Liability based on design defects is determined by the consumer expectation test and the risk–benefit test. As discussed earlier, problems that juries face with these tests include an understanding of what was expected, the need for expert knowledge, and the availability of alternate designs. In terms of consumer expectations within the context of allergies, the GM soybean should be no more allergenic than the non-GM soybean. Although human subjects were not used in the published safety studies of GM soybeans, the published reports did show substantial equivalence between these two types of soybeans. Also, experiments on the character of the glyphosate-resistance protein showed that it would probably be nonallergenic. Thus, consumer expectations for a GM soy product that was no more allergenic than non-GM soybeans were met as well as possible. As for alternate designs, they are limited by the state of scientific and technical knowledge and the genes that are available in nature for use.

The balancing of risks and benefits of the GM soybean is complicated by the fact that there may be more than one class of consumers. The two relevant groups of consumers are the people who eat the soy products and the farmers who grow the crops. Among GMO critics, unfairness arises because farmers gain all the benefit from the glyphosate resistance of the GM soybeans and people who directly eat the soy products bear all the risk. In other words, the risks and benefits are unfairly distributed and thus more difficult to balance. As an additional consideration, the analysis of risks and benefits of GM soy should also include the gains to be realized in future GM products. As with all other consumer goods, knowledge from the experience with GM soybeans will be applied to improve and refine the designs of subsequent products. Thus, risks and benefits must also include the contributions from the use of current products to the development of better ones.

Liability based on a failure to warn would probably be associated with the use of food labels. The conflict between the failure to warn and the proper use of labels is that warnings presume that there is some risk to the consumer from the use of the product (95). However, the safety studies conducted on the GM soybean show no evidence of any risk to human health above that of non-GM soybeans. Thus, no warning is required. However, one might argue that labels provide information of all sorts (e.g., nutritional content) that help consumers decide whether or not to buy a product. The addition of labels designating the contents of a product as including or being free of GM soy components might help consumers to make their purchase decisions. These labels could materially affect a person's choice, and the purchase of a GM food might be interpreted as consent to, or assumption of, any risk associated with the product. But if decisions to avoid the product are based on a perceived but unfounded risk associated with GM products in general, they would be unfair to the manufacturers of such products. Perhaps the best policy would be to indicate on the label of a GM food that it contains GM components approved by the FDA. Although this might help some consumers while confusing others, one might argue that such information is balanced and impartial.

D. Implied Warranties

Successful litigation based on the theory of a breach of implied warranty of merchantability must show that the product used was below average quality and unfit for the general purpose for which it was sold. In the hypothetical case of an allergic reaction, the allergenicity of the GM soy product in question could be compared to the allergenicities of "average" GM or non-GM soy products. Comparisons with GM soy products might be complemented by an argument that a manufacturing defect could have caused this particular product to deviate from the average. Comparisons with non-GM soy could be interpreted as an attempt to define all GM soy products as below the quality of the average non-GM soy product. This might then be supplemented with a design defect case. However, the published studies on the composition and allergenicity of GM soybeans showed that they are very similar to non-GM soybeans in these respects. Perhaps one could extrapolate from these findings and say that the average allergenicities of both types of beans would also likely be very similar. And again, more variability may be caused by growing conditions than by genetics. Incidentally, implied warranties may also be based on the concept of "fitness for a particular purpose." In this situation, if a seller knows or has reason to know of the particular purpose for which a product is bought, and the buyer relies on the seller's skill and judgment in selecting the goods, an implied warranty is created. A breach of this implied warranty would involve issues of negligence and foreseeability of the particular purpose. Since GM foods are generally bought for consumption, the particular purpose is fairly clear.

E. Strict Liability

A case for strict liability in tort might be based on a duty of manufacturers to produce and distribute safe GM soy foods, the breach of this duty, evidence of causation, and injuries to consumers. But in light of the known allergenicities of non-GM soybeans, the unknowns that remain inherent to foods in general, and the contributions of environmental conditions and human biological processes to allergies, what would be considered a safe GM food? As with implied warranties, perhaps comparisons of the allergenicity of the GM product to that of other GM and non-GM products could provide criteria for a determination that the product is unsafe or less safe. A safe GM soy food might then be one that is no more allergenic than a similar non-GM product. Or perhaps as with the risk–benefit analysis for a design defect, allergenicity might be balanced with the utility of the product.

Some cases have imposed strict liability on injuries through foods on the basis of a *foreign-natural* test (96). Under this theory, strict liability is applied if the injurious substance is considered foreign to the food (e.g., glass in a chicken dinner), whereas legal recovery from injuries from a natural material (e.g., chicken bone in a chicken dinner) requires a showing of negligence. Genetically modified soybeans were developed through the integration of genes and gene products from a variety of organisms. In a hypothetical allergy case, the use of this theory of liability may generate a lively discussion on what is foreign or natural in a domesticated organism whose genetic makeup is derived from the deliberate insertion of genes from other species.

Incidentally, the earlier discussion of the characteristics of implied warranty and strict liability in tort suggested that they are very similar if not the same theory of liability. The concurrent use of both theories in a suit against a manufacturer is not advisable since a jury may acquit a defendant on one theory and convict on the other. The contradictory nature of this result might require that another trial be held.

V. CONCLUSIONS

The thesis of this chapter, that the existence of products liability laws may help to foster the exercise of caution as does the Precautionary Principle, is based largely on the concept of reasonableness. Admittedly, finding a workable definition of reasonableness may itself be a source of contention. Courts seem generally to define *reasonableness* as that which characterizes the ordinary, average, law-abiding person or entity. This move toward reasonable decisions is noted by the importance placed on reasonable foreseeability, the balancing of risks and benefits, and the notion of proximate causation to limit liabilities.

With reasonableness, we may try to find the best compromises for the development of valuable, practical, and safe products. Good faith efforts at compromise will be required to balance the many concerns that confront any attempt at creating something new. With GMOs, these will include the complexity and unknowns inherent to biological systems, the press for innovation to help solve our many problems and fulfill our wants and needs, the safety of products derived from new technologies, the financial profits spurring the development of new knowledge and products, and the opportunities to litigate that provide necessary protections for consumers but that can also be abused.

Courts are forums for litigation, symbols of fairness, and a source of laws. Therefore, they serve as mediators of compromise and reasonable decision making. Litigation, or perhaps the preference to prevent litigation, could encourage the developers of GMOs to proceed with caution in the production and use of their products (97). The threat of litigation would be balanced by legal protections and freedoms needed to promote innovation and risk taking, and to allow businesses to function. Courts and laws are based on reasonableness and fairness. And as with all other matters in life, reasonableness and fairness are important considerations in facing the controversies over GMOs.

ACKNOWLEDGMENT

The author thanks Professors James Hogan and Edward Imwinkelried (School of Law, University of California at Davis) for their very helpful comments.

REFERENCES

1. PB Thompson. Agricultural biotechnology, ethics, food safety, risk, and individual choice. In: TH Murray, MJ Mehlman, eds. Encyclopedia of Ethical, Legal, and Policy Issues in Biotechnology. New York: John Wiley & Sons, 2000, pp 17–26.
2. D Jackson. Labeling products of biotechnology: Towards communication and consent. Journal of Agricultural and Environmental Ethics. 12:319–330, 2000.
3. DR MacKenzie. Agricultural biotechnology law, APHIS regulation. In: TH Murray and MJ Mehlman, eds. Encyclopedia of Ethical, Legal, and Policy Issues in Biotechnology. New York: John Wiley & Sons, 2000, pp 56–66.
4. JC Polkinghorne. Ethical issues in biotechnology. Trends in Biotechnology. 18:8–10, 2000.
5. J Carroll, B McKay. New FDA letters warn manufacturers of tougher stance on herbal additives. Wall Street Journal, p. B2, June 8, 2001.
6. M Higgins. Hard to swallow. American Bar Association Journal. 85:60–63, 1999.
7. SH Zeisel. Policy Forum: Health. Regulation of "nutraceuticals." Science. 285:1853–1855, 1999.
8. D. Hoover, BM Chassy, RL Hall, HJ Klee, JB Luchansky, HI Miller, I Munro, R Weiss, SL Hefle, CO Qualset. IFT expert report on biotechnology and foods: Human food safety evaluation of rDNA biotechnology-derived foods. Food Technology. 54:53–61, 2000.
9. E Millstone, E Brunner, S Mayer. (Commentary). Beyond "substantial equivalence" Nature. 401:525–526, 2000.
10. S Pouteau. Beyond substantial equivalence: ethical equivalence. Journal of Agricultural and Environmental Ethics. 13:273–291, 2000.
11. AJ Conner, JME Jacobs. Food risks from transgenic crops in perspective. Nutrition. 16:709–711, 2000.
12. HP Noteborn, A Lommen, RC van der Jagt, JM Weseman. Chemical fingerprinting for the evaluation of unintended secondary metabolic changes in transgenic food crops. Journal of Biotechnology. 77:103–114, 2000.
13. J Bergelson, CB Purrington, G Wichmann. Scientific correspondence. Promiscuity in transgenic plants. Nature. 395:25, 1998.
14. IFT expert report on biotechnology and foods: Introduction. Food Technology. 54:124–136, 2000.
15. JA Nordlee, SL Taylor, JA Townsend, LA Thomas, RK Bush. Identification of

a brazil-nut allergen in transgenic soybeans. New England Journal of Medicine. 334:688–692, 1996.
16. SW Ewen, A Pusztai. Effect of diets containing genetically modified potatoes expressing galanthus nivalis lectin on rat small intestine. Lancet. 354:1353–1354, 1999.
17. P Firth. Leaving a bad taste. Scientific American. 280:34–35, 1999.
18. HI Miller. Labeling of gene-spliced foods: a label we don't need. Nutrition. 16:706–709, 2000.
19. B Halweil. Testimony from Senate Subcommittee on International Economic Policy, Export and Trade Promotion at the Hearing on "The Role of Biotechnology in Combating Poverty and Hunger in Developing Nations." 07/12/00. http://www.worldwatch.org/biotech/bhtest.htm (Visited on January 13, 2001).
20. NJ Taylor, CM Fauquet. Biotechnology's greatest challenge. Forum for Applied Research and Public Policy. 1:51, 2000 (Electronic version from the University of California at Davis).
21. IFT expert report on biotechnology and foods: Benefits and concerns associated with recombinant DNA biotechnology-derived foods. Food Technology. 54:61–80, 2000.
22. HI Miller. UN-based biotechnology regulation: scientific and economic havoc for the 21st century. Trends in Biotechnology. 17:185–190, 1999.
23. The voice of reason in the global food fight. Fortune. 141:164–172, 2000.
24. R Bailey. Dr. Strangelunch. Or: why we should learn to stop worrying and love genetically modified food. Reason. January 2001. Reason Online. http://reasona.com/0101/fe.rb.dr.html. (Visited on December 19, 2000).
25. JH. Business and Regulatory News. GMO Roundup. Nature Biotechnology. 18:911, 2000.
26. T Smith. Biotechnology and global justice. Journal of Agricultural and Environmental Ethics. 11:219–242, 1999.
27. CJ Peters. Genetic engineering in agriculture: who stands to benefit? Journal of Agricultural and Environmental Ethics. 13:313–317, 2000.
28. S Kasanmoentalib. Science and values in risk assessment: The case of deliberate release of genetically engineered organisms. Journal of Agricultural and Environmental Ethics. 9:42–60, 1996.
29. L Levidow, S Carr, D Wield, R von Schomberg. European biotechnology regulation: framing the risk assessment of a herbicide-tolerant crop. Science, Technology & Human Values. 22:472–505, 1997.
30. HI Miller. Agricultural biotechnology, law, and food biotechnology regulation. In: TH Murray, MJ Mehlman, eds. Encyclopedia of Ethical, Legal, and Policy Issues in Biotechnology. New York: John Wiley & Sons, 2000, 2000, pp 37–46.
31. B Gremmen, H van den Belt. The precautionary principle and pesticides. Journal of Agricultural and Environmental Ethics. 12:197–205, 2000.
32. M Matthee, D Vermersch. Are the precautionary principle and the international trade of genetically modified organisms reconcilable? Journal of Agricultural and Environmental Ethics. 12:59–70, 2000.

33. PT Saunders. Use and abuse of the precautionary principle. http://www.twnside.org.sg/title/service2.ht (2001).
34. JA Nettleton. Amber light: the precautionary principle. Food Technology. 53: 30, 1999.
35. KR Foster, P Vecchia, MH Repacholi. Science and the precautionary principle. Science. 288:979–981, 2000.
36. Reference Deleted.
37. AG Shalit, B Jank, J Rath, H Miller, G Conko. Correspondence: the precautionary principle. Nature Biotechnology. 18:697, 2000.
38. S Morris, C Adley. Evolving European GM regulation: an example of biopolitics at work. Trends in Biotechnology. 18:325–326, 2000.
39. E Bodenheimer, JB Oakley, JC Love. Fundamentals of statutory interpretation. In: An Introduction to the Anglo-American Legal System, 2nd ed. St. Paul, MN: West Publishing Company, 1988, p 132.
40. S Jasanoff. Science at the Bar. Cambridge, MA: Harvard University Press, 1995, pp 11–12.
41. KS Abraham. The Forms and Functions of Tort Law. New York: Foundation Press, 1997, pp 15–16.
42. SH Yoshida. The safety of genetically modified soybeans: evidence and regulation. Food and Drug Law Journal. 55:193–208, 2000.
43. JW Wade, VE Schwartz, K Kelly, DF Partlett. Prosser, Wade and Schwartz's Torts. Cases and Materials, 9th ed. Westbury, NY: Foundation Press, 1996, p 1.
44. JW Wade, VE Schwartz, K Kelly, DF Partlett. Prosser, Wade and Schwartz's Torts. Cases and Materials, 9th ed. Westbury, NY: Foundation Press, 1996, p 694.
45. KS Abraham. The Forms and Functions of Tort Law. New York: Foundation Press, 1997, p 186.
46. Lord Abinger. *Winterbottom v. Wright*. In: JW Wade, VE Schwartz, K Kelly, DF Partlett. Prosser, Wade and Schwartz's Torts: Cases and Materials, 9th ed. Westbury, NY: Foundation Press, 1996, p 440.
47. J Cardozo. *MacPherson v. Buick Motor Co*. In: JW Wade, VE Schwartz, K Kelly, DF Partlett. Prosser, Wade and Schwartz's Torts, 9th ed. Westbury, NY: Foundation Press, 1996, pp 697–698.
48. KS Abraham. The Forms and Functions of Tort Law. New York: Foundation Press, 1997, p 190.
49. KS Abraham. The Forms and Functions of Tort Law. New York: Foundation Press, 1997, p 190.
50. JW Wade, VE Schwartz, K Kelly, DF Partlett. Prosser, Wade and Schwartz's Torts, 9th ed. Westbury, NY: Foundation Press, 1996, p 700.
51. KS Abraham. The Forms and Functions of Tort Law. New York: Foundation Press, 1997, pp 191–102.
52. J Francis. *Henningsen v. Bloomfield Motors, Inc*. In: JW Wade, VE Schwartz, K Kelly, DF Partlett. Prosser, Wade and Schwartz's Torts: Cases and Materials, 9th ed. Westbury, NY: Foundation Press, 1996, pp 707–708.
53. J Francis. *Henningsen v. Bloomfield Motors, Inc*. In: JW Wade, VE Schwartz, K

Kelly, DF Partlett. Prosser, Wade and Schwartz's Torts: Cases and Materials, 9th ed. Westbury, NY: Foundation Press, 1996, p 710.

54. Justice Traynor. *Greenman v. Yuba Power Products, Inc.* In: JW Wade, VE Schwartz, K Kelly, DF Partlett. Prosser, Wade and Schwartz's Torts: Cases and Materials, 9th ed. Westbury, NY: Foundation Press, 1996, pp. 715–716.
55. JW Glannon. The Law of Torts. Boston: Little, Brown and Co., 1995, pp 171–180.
56. JW Glannon. The Law of Torts. Boston: Little, Brown and Co., 1995, pp 61–70.
57. JW Wade, VE Schwartz, K Kelly, DF Partlett. Prosser, Wade and Schwartz's Torts: Cases and Materials, 9th ed. Westbury, NY: Foundation Press, 1996, p 131.
58. JW Glannon. The Law of Torts. Boston: Little, Brown and Co., 1995, pp 121–127.
59. KS Abraham. The Forms and Functions of Tort Law. New York: Foundation Press, 1997, pp 108–114.
60. JW Glannon. The Law of Torts. Boston: Little, Brown and Co., 1995, pp 150–151.
61. JW Glannon. The Law of Torts. Boston: Little, Brown and Co., 1995, p 154.
62. American Law Institute. Section 2(a). Categories of Product Defect. In: Restatement of the Law Third, Torts, Products Liability. St. Paul, MN: American Law Institute Publishers, 1998, p 14.
63. KS Abraham. The Forms and Functions of Tort Law. New York: Foundation Press, 1997, pp 195–196.
64. American Law Institute. Section 2(b). Categories of Product Defect. In: Restatement of the Law Third, Torts, Products Liability. St. Paul, MN: American Law Institute Publishers, 1998, p 14.
65. KS Abraham. The Forms and Functions of Tort Law. New York: Foundation Press, 1997, pp 196–197.
66. MS Shapo. Products Liability and the Search for Justice. Durham, NC: Carolina Academic Press, 1993, pp 120–124.
67. KS Abraham. The Forms and Functions of Tort Law. New York: Foundation Press, 1997, p 197.
68. KS Abraham. The Forms and Functions of Tort Law. New York: Foundation Press, 1997, pp 197–198.
69. MS Shapo. Products Liability and the Search for Justice. Durham, NC: Carolina Academic Press, 1993, p 125.
70. MS Shapo. Products Liability and the Search for Justice. Durham, NC: Carolina Academic Press, 1993, p 137.
71. MS Shapo. Products Liability and the Search for Justice. Durham, NC: Carolina Academic Press, 1993, pp 128–130.
72. American Law Institute. Section 2(c). Categories of Product Defect. In: Restatement of the Law Third, Torts, Products Liability. St. Paul, MN: American Law Institute Publishers, 1998, p 14.
73. MS Shapo. Products Liability and the Search for Justice. Durham, NC: Carolina Academic Press, 1993, p 142.

74. KS Abraham. The Forms and Functions of Tort Law. New York: Foundation Press, 1997, p 201.
75. KS Abraham. The Forms and Functions of Tort Law. New York: Foundation Press, 1997, p 200.
76. KS Abraham. The Forms and Functions of Tort Law. New York: Foundation Press, 1997, p 202.
77. MS Shapo. Products Liability and the Search for Justice. Durham, NC: Carolina Academic Press, 1993, pp 142–143.
78. MS Shapo. Products Liability and the Search for Justice. Durham, NC: Carolina Academic Press, 1993, p 148.
79. KS Abraham. The Forms and Functions of Tort Law. New York: Foundation Press, 1997, p 203.
80. There are numerous books and review articles on the immune system in general and allergic responses in particular. Among them are SH Yoshida, CL Keen, AA Ansari, ME Gershwin. Nutrition and the Immune System. In: ME Shils, JA Olson, M Shike, AC Ross, eds. Modern Nutrition in Health and Disease. Baltimore: Williams & Wilkins, 1999, pp 725–750.
81. SJ Ono. Molecular genetics of allergic diseases. Annual Review of Immunology. 18:347–366, 2000.
82. RM Helm, AW Burks. Mechanisms of food allergy. Current Opinion in Immunology. 12:647–653, 2000.
83. JE Perkins. The latex and food allergy connection. Journal of the American Dietetic Association. 100:1381–1384, 2000.
84. RC Aalberse. Structural biology of allergens. Journal of Allergy and Clinical Immunololgy. 106:228–238, 2000.
85. SH Yoshida, RV Perez, JB German, ME Gershwin. Nutritional modification of T lymphocyte subsets: clinical implications and applications. Seminars in Clinical Immunology. 12:5–24, 1996.
86. KM Smith, AD Easton, LM Finlayson, P Garside. Oral tolerance. American Journal of Respiratory and Critical Care Medicine 162:5175–5178, 2000.
87. TJ David. Adverse reactions and intolerance to foods. British Medical Bulletin. 56:34–50, 2000.
88. JC Guarderas. Is it food allergy? Differentiating the causes of adverse reactions to foods. Postgraduate Medicine. 109:125–134, 2001.
89. W Burks. Diagnosis of allergic reactions to foods. Pediatrics Annals. 29:744–752, 2000.
90. SL Taylor, SL Hefle. Will genetically modified foods be allergenic? Journal of Allergy and Clinical Immunology. 107:767, 2001.
91. SR Padgette, NB Taylor, DL Nida, MR Bailey, J MacDonald, LR Holden, RL Fuchs. The composition of glyphosate-tolerant soybean seeds is equivalent to that of conventional soybeans. Journal of Nutrition. 126:702–716, 1996.
92. SB Lehrer, WE Horner, G Reese. Why are some proteins allergenic? Implications for biotechnology. Critical Reviews in Food Science and Nutrition. 36:553–564, 1996.
93. LA Harrison, MR Bailey, MW Naylor, JE Ream, BG Hammond, DL Nida,

BL Burnett, TE Nickson, TA Mitsky, ML Taylor, RL Fuchs, SR Padgette. The expressed protein in glyphosate-tolerant soybean, 5-enolypyruvylshikimate-3-phosphate synthase from *Agrobacterium* sp. Strain CP4, is rapidly digested in vitro and is not toxic to acutely gavaged mice. Journal of Nutrition. 126:728–740, 1996.

94. SR Padgette, DB Re, GF Barry, DE Eichholtz, X Delannay, RL Fuchs, GM Kishore, RT Fraley. New weed control opportunities: development of soybeans with a Roundup Ready TM gene. In: SO Duke, ed. Herbicide Resistant Crops. Boca Raton, FL: Lewis Publishers, 1996, pp 53–84.
95. JW Wade, VE Schwartz, K Kelly, DF Partlett. Prosser, Wade and Schwartz's Torts: Cases and Materials, 9th ed. Westbury, NY: Foundation Press, 1996, p 754.
96. *Mexicali Rose v. Superior Court*, 1 Cal. 4th, pp 617–653, 1992.
97. WE Parmet, RA Daynard. The new public health litigation. Annual Review of Public Health. 21:437–454, 2000.

19

Scientific Concepts of Functional Foods in the Western World

Steven H. Yoshida
Consultant, Davis, California, U.S.A.

I. INTRODUCTION

For the average consumer, the rationale for the development of functional foods is relatively obvious. There may be a need to create crops that contain higher protein levels for human populations who face protein deficiency. Or the manipulation of a plant's ability to synthesize a particular fatty acid may lead to enhanced immune system or cardiovascular health for the consumer.

What is not so obvious to the average person are the scientific studies and information that underlie the known efficacy, safety, and problems associated with these products. One result of the relative difficulty in understanding the nature (as opposed to the need) of bioengineered products is the manipulation of public perceptions over their usefulness and safety by people or organizations with particular policy or political views.

A large proportion of the general population, including those with little scientific training or understanding, were reported to have a high regard for scientific research and information (1). Perhaps if consumers had a greater understanding of the types of studies and tests that are conducted during the research and development of products derived from new technologies, and the information derived from such studies, they would be more prepared to decide personally on the usefulness and safety of these functional foods.

Many of the premarket and postmarket scientific studies on functional foods are done to comply with various governmental regulations that exist to control the quality, safety, manufacture, dosage, and other characteristics of consumer products. Therefore, a review of the regulatory requirements that guide the research and development of a functional food would include a review of the scientific concepts and issues that are relevant to that particular product. Thus, this chapter first defines edible vaccines as functional foods. Second, some of the issues specific to bioengineered foods are briefly reviewed. Third, the requirements imposed by the U.S. Food and Drug Administration (U.S. FDA) concerning the premarket and postmarket studies of one type of vaccine [deoxyribonucleic acid (DNA) vaccines] are used to illustrate the types of scientific information that are needed for regulatory compliance. Finally, existing scientific information on edible vaccines is reviewed to describe the ongoing assessment of these functional foods.

II. EDIBLE VACCINES AS FUNCTIONAL FOODS

A 1996 review emphasized the basic tenets that characterize the nature and development of functional foods (2). In essence, foods that are similar in appearance to conventional foods and that are consumed as part of a normal diet are developed to enhance human health or reduce the risk of disease by enhancing specific physiological functions. The functionality of these foods must be based on sound interdisciplinary scientific studies and evidence, and such evidence must be communicated to the general public.

Edible vaccines are functional foods as they appear similar to conventional foods, they function to stimulate the immune system to provide protection against specific infectious diseases, and a wide range of studies and disciplines are required for their development and application. For example, an edible vaccine may be derived from the tomato. Such a fruit appears similar to, and is consumed in the same way as, a conventional tomato. However, the edible vaccine is engineered to deliver an immunogen (i.e., antigen that generates an immune response) to give the consumer resistance to a particular disease. The development of a single edible vaccine can conceivably encompass the inputs of many disciplines, including molecular biology, agriculture, medicine, ecology, nutrition, food science, and toxicology. Thus, edible vaccines fulfill the definition of functional foods.

III. SCIENTIFIC ISSUES SPECIFIC TO BIOENGINEERED FOODS

Perhaps the principal consumer and scientific issue is the safety of the bioengineered food. Essentially, there must be evidence that the food is as safe as its conventional counterpart (3). One example of the analysis of safety of a bioengineered food is the present author's published review of studies performed on Monsanto's glyphosate-resistant soybeans (4). These studies showed that genetically modified soybeans were substantially equivalent or similar to nonmodified soybeans in nutritional content, toxicity and allergenic potential, and acceptability to domestic animals (5–7).

Stein and Webber described other regulatory issues that concern bioengineered plants (8). The authors wrote that environmental regulations control the cultivation, harvesting, and transportation of plants to ensure that their genes and genetically influenced propensities (e.g, production of immunogens) do not contaminate other crops or inadvertently enter the human food supply.

Stein and Webber also reviewed chemistry and manufacturing controls (8). They stated that the production of bioengineered plants, including transfection methods and genetic constructs used to modify plants, should be thoroughly described. Seed banks should also be established to ensure consistency between batches of plants that are cultivated. In the United States, the manufacture of biologics (e.g., vaccine) derived from bioengineered plants is overseen by the U.S. Food and Drug Administration, and compliance with current good manufacturing practices must be enforced. Also, analytical methods must be used to maintain consistency in the potency and dose of biologics derived from fruits and vegetables that are eaten without further manufacturing or processing. The stage of ripeness of the fruit or vegetable, the maturity of the plant, and the shelf life or stability of the fruit or vegetable may influence the biological characteristics of the product. Finally, the structure of the engineered protein product within the plant must be analyzed to determine its immunogenicity (i.e., ability to generate a useful immune response as a vaccine) and allergenicity.

IV. SCIENTIFIC CONCEPTS UNDERLYING THE PREMARKET AND POSTMARKET LIFE CYCLE OF A VACCINE

As this chapter was written, there was no edible vaccine yet on the market. Therefore, a complete review of the scientific studies of a particular edible vaccine was not possible. However, another type of vaccine, the DNA vaccine, is further along in its research and development. Since several useful reviews on the regulation of this vaccine were published, studies on the efficacy and safety of the DNA vaccine can provide a model for analogous work that will probably be required for the edible vaccine.

A. Quality Control of Manufacture

The first issue is that there should be a product that both is of high quality and can be obtained in consistent form. The information required for regulatory compliance includes descriptions of (a) the intended use of the vaccine, (b) the development of the product (e.g., origins of genes and cloning methods), (c) the starting materials used to produce the unrefined product (e.g., cells used to generate the genetic material to be used as immunogens), (d) the manufacturing process used to purify the product, and (e) the product in its final and usable form (9).

B. Preclinical Study for Safety

The second issue concerns the safety of DNA vaccines. Some of the safety concerns are unique to DNA vaccines since they involve the direct administration of genetic material into the host to be immunized. Apprehension stems from the possibility that the incorporation and expression of foreign genetic material by the host may lead to adverse immunological reactions such as inflammation, autoimmunity, or immunosuppression (10). However, in vivo animal tests are also done to examine other potential toxicities as well as to determine dosages and routes of administration for clinical studies (11). These studies are of relevance to edible vaccines. Preclinical studies for safety may require at least 5 years for completion, and in the United States, these data are presented in an Investigational New Drug application to the U.S. FDA before approval for human clinical trials (12).

C. Prelicensure Clinical Trials

Clinical trials are divided into four phases, the first three of which are conducted before marketing of the product. The final phase occurs after release and marketing to the general public.

Phase I consists of an initial evaluation of the safety and immunogenicity of the vaccine for small groups of healthy adults. Adverse reactions, tolerable dosages, and the generation of immune responses are among the types of information sought (13).

In phase II, the vaccine is given to a larger number (50–500) of members of the designated at-risk population to obtain information on safety and immunogenicity. The purpose of this second phase is to develop the final experimental design that will be implemented in the larger phase III trial. Thus, information from phase II is used to determine such factors as the appropriate dosages, adjuvants to be used, vaccination schedules, and the number and personal characteristics of people to be included in the next phase (13).

Phase III examines the efficacy of the vaccine in the target population. Here, thousands of people at risk of contracting the particular disease for which the vaccine was developed are inoculated and their progress followed.

A typical clinical trial may require 7 years for completion and in one estimate could cost about $75 million (12). If the candidate biologic is shown to confer significant protective immunity against the disease in question, then a New Drug Application may be filed with the U.S. FDA for licensing and marketing (12). The application should contain a full description of the efficacy, safety, quality, and consistency of the vaccine as well as the final manufacturing procedure. An inspection of the manufacturing facility is also conducted before final approval of the product for use (9). The U.S. FDA approves about 20% of Investigational New Drugs for marketing (12).

D. Postlicensure Evaluation

Phase IV of a clinical trial consists of an epidemiological study of the entire vaccinated population. Rare adverse reactions not observed in previous phases most likely become evident in this final phase with the largest number of inoculated people (14). Also, only through this ongoing phase can the long-term effectiveness of the vaccine be understood. Changes in the vaccine, its manufacture, or its use may be warranted by information obtained in phase IV (9).

V. STATUS OF THE DEVELOPMENT OF EDIBLE VACCINES

Reviews of the development of edible vaccines were published in 2001 (15–17). These reviews show that efforts are being made to generate a variety of edible vaccines for both human and animal diseases. Of particular interest as functional foods are those that are directed to human illnesses such as those caused by enterotoxigenic *Escherichia coli* and *Vibrio cholerae*, hepatitis B, rabies, Norwalk virus, human cytomegalovirus, malaria, influenza, and autoimmune diabetes. At the time of the writing of this chapter, most of these vaccines were still at the early stages of development. For example, antigen-specific immune responses were generated in mice by edible vaccines to hepatitis B (18), respiratory syncytial virus (19), *E. coli* enterotoxin (20), and Norwalk virus (21). In addition, diabetic mice fed transgenic potatoes that produced human insulin–cholera toxin–subunit B fusion proteins showed reductions in pancreatic inflammation and progression of clinical diabetes (22). Interestingly, this last example demonstrated the beneficial effects of a form of antigen-specific immunosuppression or tolerance induction.

Published human clinical trials are limited to one study by Tacket and associate (23). In this phase I trial, humans fed a transgenic potato that expressed the Norwalk virus capsid protein had antigen-specific antibody responses to the viral protein. Oral vaccination resulted in increased peripheral blood immunoglobulin A–(IgA)-positive lymphocytes and intestinal virus-neutralizing IgA. Since Norwalk virus causes acute gastroenteritis and food poisoning (24), elevated IgA level in the intestines of vaccinated volunteers would be among the preferred type of immune response desired. Aside from this one published clinical trial, Walmsley and Arntzen (25) reported that trials were under way for potato-derived hepatitis B and Norwalk virus vaccines. On the basis of the length of time that is required for the completion of these trials, edible vaccines for humans will probably not be permitted for use by the general public for at least several more years.

VI. SCIENTIFIC PROBLEMS STILL TO BE RESOLVED IN EDIBLE VACCINE DEVELOPMENT

There are still numerous problems to be solved and refinements to be made in the production, application, and efficacy of edible vaccines before their widespread use is realized. Transgenic plants must produce sufficient quantities of immunogens in a consistent manner that predictable doses of vaccines are delivered to the consumer (12,15,26). Further genetic manipulation and selective breeding of plants will accomplish this goal. Other hurdles are the instability and degradation of vaccines that follow harvesting of fruits or vegetables. Types of processing of products (e.g., drying), choices of plants to be bioengineered, and locations within the plants where vaccine components are generated will all influence the stability of immunogens and their deliverability to designated populations (12,15,26). Immunogens may also succumb to digestion and degradation in the human digestive system and thereby be rendered useless as vaccines. The efficient absorption of intact immunogens may be facilitated through their encapsulation in liposomes or transgenic plant tissues, or by physical linking with other molecules (12,15). Of particular concern is the generation of oral tolerance. Since foods are potential allergens to consumers, the immune system tends to suppress or become incapable of generating immune responses to food components. Without a mechanism for the generation of oral tolerance, the consumption of any food would involve the risk of an allergic reaction. As described, one strategy for the treatment of autoimmune diseases is the feeding of autoantigens to generate tolerance to the molecular targets of an autoimmune reaction (12,22). But edible vaccines must be designed to maintain tolerance to the food vehicle and yet not generate oral tolerance to the bioactive vaccine component (27). Further studies on the design and delivery of edible vaccines, and on the basic mechanisms of oral tolerance, will ensure the beneficial functions of these bioengineered products.

VII. CONCLUSIONS

This brief review of the scientific concepts and practices that underlie one type of bioengineered product provides a basis for understanding the complex challenges that are involved in converting an idea to a useful and marketable product. In turn, an understanding of these complex challenges helps to explain the time, costs, and ingenuity that are required for the development of products such as edible vaccines. However, the potential benefits to be derived from edible vaccines are incalculable when one considers the price in human lives paid for microbial infections and other diseases.

REFERENCES

1. D Normile. Global Interest High, Knowledge Low. Science 274:1074. 1996.
2. B German, EJ Schiffrin, R Reniero, B Mollet, A Pfeifer, J-R Neeser. The Development of Functional Foods: Lessons from the Gut. TIBTECH 17:492–499, 1999.
3. R Formanek Jr. Proposed Rules Issued for Bioengineered Foods. FDA Consumer 35:9–11, 2001.
4. S Yoshida. The Safety of Genetically Modified Soybeans: Evidence and Regulation. Food Drug Law J 55:193–208, 2000.
5. SR Padgette, NB Taylor, DL Nida, MR Bailey, J MacDonald, LR Holden, RL Fuchs. The Composition of Glyphosate-Tolerant Soybean Seeds is Equivalent to That of Conventional Soybeans. J Nutr 126:702–716, 1996.
6. BG Hammond, JL Vicini, GF Hartnell, MW Naylor, CD Knight, EH Robinson, RL Fuchs, SR Padgette. The Feeding Value of Soybeans Fed to Rats, Chickens, Catfish and Dairy Cattle Is Not Altered by Genetic Incorporation of Glyphosate Tolerance. J Nutr 126:717–727, 1996.
7. LA Harrison, MR Bailey, MW Naylor, JE Ream, BG Hammond, DL Nida, BL Burnette, TE Nickson, TA Mitsky, ML Taylor, RL Fuchs, SR Padgette. The Expressed Protein in Glyphosate-Tolerant Soybean, 5-Enolypyruvylshikimate-3-Phosphate Synthase from Agrobacterium sp. Strain CP4, is Rapidly Digested in Vitro and Not Toxic to Acutely Gavaged Mice. J Nutr 126:728-740, 1996.
8. KE Stein, KO Webber. The Regulation of Biologic Products Derived from Bioengineered Plants. Curr Opin Biotechnol 12:308–311, 2001.
9. JS Robertson, E Griffiths. Review. Assuring the Quality, Safety, and Efficacy of DNA Vaccines. Mol Biotechnol 17:143–149, 2001.
10. HA Smith, DM Klinman. The Regulation of DNA Vaccines. Curr Opin Biotechnol 12:299–303, 2001.
11. LA Falk, LK Ball. Current Status and Future Trends in Vaccine Regulation—USA. Vaccine 19:1567–1572, 2001.
12. JA Zivin. Understanding Clinical Trials. Sci Am 282:69–75, 2000.

13. CP Farrington, E Miller. Review: Vaccine Trials. Mol Biotechnol 17:43–58, 2001.
14. RM Jacobson, A Adegbenro, VS Pankratz, GA Poland. Adverse Events and Vaccination—the Lack of Power and Predictability of Infrequent Events in Pre-Licensure Study. Vaccine 19:2428–2433, 2001.
15. H Daniell, SJ Streatfield, K Wycoff. Medical Molecular Farming: Production of Antibodies, Biopharmaceuticals and Edible Vaccines in Plants. Trends Plant Sci 6:219–226, 2001.
16. G Giddings, G Allison, D Brooks, A Carter. Transgenic Plants as Factories for Biopharmaceuticals. Nat Biotechnol 16:1151–1155, 2001.
17. JW Larrick, DW Thomas. Producing Proteins in Transgenic Plants and Animals. Curr Opin Biotechnol 12:411–418, 2001.
18. Q Kong, L Richter, YF Yang, CJ Arntzen, HS Mason, Y Thanavala. Oral Immunization with Hepatitis B Surface Antigen Expressed in Transgenic Plants. Proc Natl Acad Sci USA 98:11539–11544, 2001.
19. JS Sandhu, SF Krasnyanski, LL Domier, SS Korban, MD Osadjan, DE Buetow. Oral Immunization of Mice with Transgenic Tomato Fruit Expressing Respiratory Syncytial Virus-F Protein Induces a Systemic Immune Response. Transgenic Res 9:127–135, 2000.
20. HS Mason, TA Haq, JD Clements, CJ Arntzen. Edible Vaccine Protects Mice Against Escherichia coli Heat-Labile Enterotoxin (LT): Potatoes Expressing a Synthetic LT-B Gene. Vaccine 16:1336–1343, 1998.
21. HS Mason, JM Ball, JJ Shi, X Jiang, MK Estes, CJ Arntzen. Expression of Norwalk Virus Capsid Protein in Transgenic Tobacco and Potato and its Oral Immunogenicity in Mice. Proc Natl Acad Sci USA 28:5335–5340, 1996.
22. T Arakawa, J Yu, DK Chong, J Hough, PC Engen, WH Langridge. A Plant-Based Cholera Toxin B Subunit-Insulin Fusion Protein Protects Against the Development of Autoimmune Diabetes. Nat Biotechnol 16:934–938, 1998.
23. CO Tacket, HS Mason, G Losonsky, MK Estes, MM Levine, CJ Arntzen. Human Immune Responses to a Novel Norwalk Virus Vaccine Delivered in Transgenic Potatoes. J Infect Dis 182:302–305, 2000.
24. US FDA/CFSAN. Foodborne Pathogenic Microorganisms and Natural Toxins Handbook. The Norwalk Virus Family. http://vm.cfsan.fda.gov/~mow/chap34.html Visited on January 27, 2002.
25. AM Walmsley, CJ Arntzen. Plants for Delivery of Edible Vaccines. Curr Opin Biotechnol 11:126–129, 2000.
26. J K-C Ma. Genes, Greens, and Vaccines. Nat Biotechnol 18:1141–1142.
27. CO Tacker, HS Mason. A Review of Oral Vaccination with Transgenic Vegetables. Microbes Infect 1:777–783, 1999.

20

Paradigm Shift: Harmonization of Eastern and Western Food Systems

Cherl-Ho Lee
Korea University, Seoul, Korea

Chang Y. Lee
Cornell University, Geneva, New York, U.S.A.

I. INTRODUCTION

Joseph Needham (1), after lifelong comparative study, wrote in his book, *Science and Civilisation in China* as follows:

> It is my conviction that the Chinese proved themselves able to speculate about Nature at least as well as the Greeks in their earlier period. If China produced no Aristotle, it was, I would suggest, because the inhibitory factors which prevented the rise of modern science and technology there began to operate already before the time at which an Aristotle could have been produced. But apart from the vision of the Taoists, there runs throughout the Chinese history a current of rational naturalism and of enlightened scepticism, often much stronger than what was found at corresponding times in that Europe where modern science and technology in fact grew up.

From the viewpoint of analytical Westerners' knowledge, the Chinese or Eastern interpretation of scientific observations probably appeared to be based on primitive theories, such as *yin* and *yang* and the Five Phases. However, in the view of Eastern people, who emphasized the harmony of human beings and nature, simplifying the natural phenomena in an equation to manipulate nature appeared to be too dangerous. For this reason, the growth of science and technology in China was often reduced or abandoned when it experienced friction with ethical considerations. The rational naturalism and enlightened skepticism of the Chinese, as described by Needham (1) in the 1950s, would not be merely inhibitory factors preventing the growth of science and technology, but could be considered as the safeguard against the adverse effects of science and technology that we are confronting today.

John Robbins (2) warned of the adverse effects of the so-called Great American Food Machine, created by scientific research and development during the last century. His book, *Diet for a New America*, describes the effects as follows:

> Increasingly in the last few decades, the animals raised for meat, dairy products, and eggs in the United States have been subjected to ever more deplorable conditions. Merely to keep the poor creatures alive under these circumstances, even more chemicals have had to be used, and increasingly hormones, pesticides, antibiotics, and countless other chemicals and drugs end up in foods derived from animals. The more unnaturally today's livestock are raised, the more chemical residues end up in our food.

This is an example of the consequences of scientific development achieved without ethical considerations. The conventional paradigm of scientific development is facing a new challenge—recovery of harmony between humans and nature and the world is now seeking the proper model for a new paradigm.

II. THE QUAGMIRE OF THE WESTERN FOOD SYSTEM

Modern nutritional science is based on the analytical knowledge of food components. When people began to learn about chemical composition, they tried to determine the physiological effect of each component and to segregate the useful components from harmful or useless ones. It is human nature to value useful components more and to tend to concentrate them. The foodstuffs made in this way include butter, cheese, white bread, and sugar. Vitamin pills are another example. Western food technology directed by modern nutritional science for the last century achieved significant developments in mass production of high-energy and high-protein nutritious foods that are easily digested and absorbed in the body. However, it has gone too far and aggravated mistakes. Foods made with refined ingredients contain high energy density and little to excrete from the digestive channel. An average of 30% to 40% of the energy in the Western diet is derived from fat, whereas only 10% is in the traditional Korean standard meal. Overweight and obesity are the natural consequence of consumption of this high-calorie food, and it is considered the major cause of all the chronic diseases that most affluent people are suffering today. After removal of most of the undigestible matter from food, there is little left to travel to the large intestine, where material remains for a long period to lose water. Such a diet leads to constipation and colon cancer.

Statistics shows that about 70% of Americans consider themselves overweight and control their diet. They suffer from resisting their desire to eat in the midst of affluent tasty foods. According to the Report of the Selected Committee on Nutrition and Human Needs of the U.S. Senate in 1977 (3), one in six deaths in the United States is caused by incorrect eating habits, and the additional medical expenses due to incorrect eating habits of people were estimated to cost U.S.$70.9 billion in 1994 (4). Although it is well recognized in dietary guidelines that Americans must eat less fat and sugar and more fiber from cereals and vegetables, it is difficult to influence eating habits in a short period. Americans have instead chosen to solve the problem by using purified fiber and other isolated food components in pills as a dietary supplement.

Table 1 compares the causes of death in the United States, France, Japan, and Korea. It shows clearly the differences in causes between East and West. It is interesting to note that the people who eat more refined and nutritionally concentrated food have less stomach cancer and liver cancer and diseases but higher incidences of breast cancer and heart disease. Some speculate that the use of soybeans as food is related to the low incidence of breast cancer in Asian women. Aside from the genetic and environmental influences, the style of diet appeared to be the most important factor determining the cause of death. The calorie sources of both Koreans and

Table 1 Statistics of Cause of Death by Countries in 1994[a]

	United States	France	Japan	Korea
Total cancer	205.6	244.6	197.3	110.1
Stomach cancer	(5.2)	(10.4)	(38.7)	(28.8)
Liver cancer	(1.8)	(7.7)	(16.7)	(23.0)
Breast cancer	(32.7)	(36.3)	(11.3)	(3.8)
Diabetes	21.8	11.0	8.8	17.0
Hypertension	14.6	10.2	6.8	25.8
Heart disease	185.1	80.4	46.7	12.6
Cerebral blood vessel disease	58.6	74.7	96.8	84.4
Liver disease	9.9	16.0	13.4	29.2

[a] Number of deaths per 100,000 people.

Source: WHO, World Health Statistics 1977–1999.

Japanese are 58% to 66% from carbohydrates, 15% to 16% from protein, and 19% to 26% from fat, whereas those for Americans are 52% from carbohydrates, 15% from protein, and 33% from fat (5). Koreans eat 15% more grains, 70% more vegetables, 17% more fruits than Americans; Americans eat 46% more meat and meat products and 3.1 times more dairy products than Koreans (5).

The U.S. nutritional education and diet style have been the world model for the last half-century, and most of the developing countries have tried to follow the U.S. model. One of the first to pursue the model was Japan, and Korea followed thereafter. Despite the long tradition of a vegetarian food habit in Korea and Japan, the food pattern has changed substantially to Western style. Fig. 1 shows that the meat and dairy food consumption increased 2.2 times and 18 times, respectively, during three decades in Korea, and grain consumption decreased by 33% in the same period. The increase in animal food consumption resulted in a drastic reduction in food self-sufficiency from 80% in 1970 to 28% in 1995 (6).

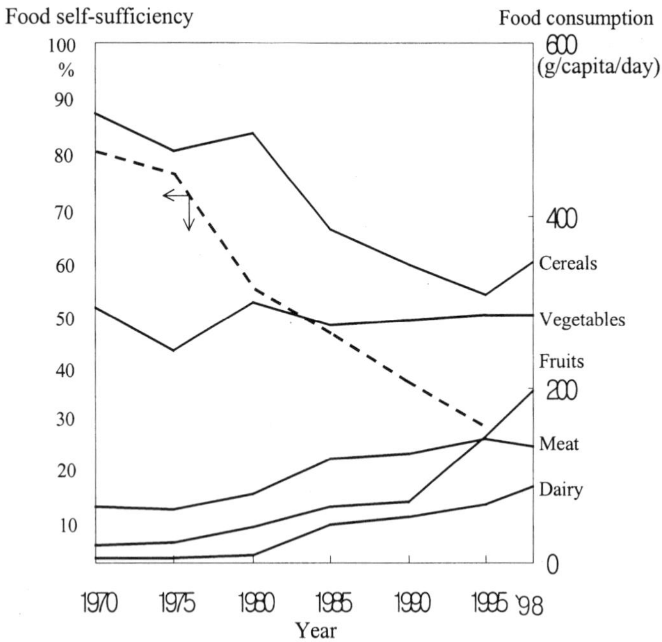

Figure 1 Changes in Korean food consumption pattern and food self-sufficiency during the past three decades. (From Refs. 6 and 7.)

Table 2 Changes in the Causes of Death in Korea per 100,000

	1989	1998	Change, percentage
Circulatory disease	161.5	123.7	−23.4
Cancer	105.0	110.8	5.5
Stomach	31.7	23.9	−24.6
Liver	23.6	20.0	−16.0
Breast	1.6	2.1	31.3
Colon	3.9	7.0	79.5
Diabetes	9.4	21.1	124.5

Source: Ref. 18.

The Westernization of the Korean diet style has a strong correlation with the prevalence of chronic disease among the population (8). Table 2 shows that the death rate caused by stomach cancer and liver cancer decreased by 26.4% and 16%; respectively, but the rates of breast cancer and colon cancer increased substantially, by 31.3% and 79.5% in Korea during the decade from 1989 to 1998. The death rate caused by circulatory disease decreased slightly (23.4%), but that caused by diabetes increased 2.2 times for the same period.

III. HEALTH CONCEPT IN EASTERN DIETARY CULTURE

Northeastern Asian thought about life and health is based on the shamanistic folk religion Taoism, which sets as the ultimate goal a healthy eternal life. Established Taoism, as developed by early Chinese philosophers, teaches that this goal can be achieved by discipline, mainly by the control of breath, sex, and food. The principle of control is the harmony of *yin* and *yang*, the negative and positive natures of the universe (9).

According to *yin–yang* and Five Phases theory, all food materials are classified by their properties and their different tastes. The properties are cool, as *yin*, neutral, and warm, as *yang*. For example, fruits on the tree are considered to have the *yang* property, whereas root crops in the soil have the *yin* property. The *yin* property also represents material entities such as nutrients; the *yang* property represents functions, such as energy. Taste is divided into five groups, representing the Five Phases: sour–wood, bitter–fire, sweet–earth, pungent–metal, and salty–water (10). Taste can be related to the human body and its organs, senses, and feelings, and even to color, the weather, and the seasons, through classification into the Five Phases. Antagonistic or affinitive relations between tastes and organs or senses are also judged or predicted by the principles of the Five Phases (10).

The basic idea of traditional Korean nutrition is to harmonize properties and tastes in the diet on the basis of *yin* and *yang* and the Five Phases. A diet that inclines toward one property or extreme taste is considered to be unhealthy. Korean meals are prepared to harmonize the properties and tastes through selecting the proper ingredients and process.

Table 3 shows a Korean meal analyzed by the *yin–yang* and Five Phases theories. The property of the main dish, cooked rice, is neutral, and the materials used for side dishes are evenly distributed among *yin* and *yang* and the Five Phases: the pattern of a balanced diet (11). The purpose of a therapeutic food in Eastern medicine is to emphasize one of the properties or tastes according to the patient's symptom of illness: cool, cold, warm, and hot. When a patient suffers from a disease caused by cool nature, a diet that emphasizes the warm property is served, and the converse.

On the basis of the philosophical ideas and medical knowledge developed in China and Korea, the Korean people have developed a standardized ideal meal, within a systematic menu program, that is called *chop bansang*.

Table 4 shows the nutritional value of a traditional Korean standard meal system in the menu of Kim Ho-Jik as calculated by the current Food Composition Table of Korean Food (12). The basic meal, which consists of a bowl of cooked rice, a bowl of soup, and a dish of kimchi, can supply 40% of the energy and 48.7% of the protein of the Recommended Dietary Allowance (RDA). When three dishes were added to the basic meal, the three-dish meal (*samchop bansang*) contained 47.2% of the energy and 94.3% of the protein of

Table 3 Analysis of a Korean Meal in Terms of *Yin–Yang* and the Five Phases

	Wood (sour)	Fire (bitter)	Earth (sweet)	Metal (pungent)	Water (salty)
Yang (warm)	Leek	Mugwort	Shepherd's purse, wheat flour	Green onion, garlic, ginger, black pepper, sesame	Salt
Neutral			Water, rice, soybeans, yellow corvina		
Yin (cool)	Vinegar	Plant root, fernbrake	Cabbage	Onion	Soy sauce, soybean paste

Table 4 Evaluation of the Nutritional Value of the Traditional Korean Standard Meal in the Menu of Kim Ho-Jik (1944)[a]

Type of menu	Basic meal	Three-dish meal	Five-dish meal	Seven-dish meal
Composition of menu	Cooked rice, soup, kimchi	Basic meal plus spinach, roasted beef, dried fish	Three-dish meal with stew plus meat jelly, fermented fish roe	Five-dish meal plus panned oysters, radish, kimchi
Total energy (kcal)	995 (40.0)	1181 (47.2)	1320 (52.8)	1672 (66.8)
Carbohydrate (%)	77.0	64.4	60.1	53.4
Protein (%)	14.7	24.0	28.0	27.7
Lipid (%)	8.3	11.6	11.9	18.9
Total protein (g)	36.5 (48.7)	70.7 (94.3)	92.5 (123.3)	115.5 (154.0)
Animal Protein (g)	28.7	59.5	69.0	72.3
Ca (mg)	161.1 (26.9)	216.3 (36.1)	255 (42.5)	596 (99.3)
Fe (mg)	12.1 (121.9)	23 (230)	26.8 (268)	40.3 (403)
Vitamin A (IU)	426.2 (17.1)	8,7616 (350.5)	9,129 (365.2)	9,965 (398.6)
Vitamin B_1 (mg)	0.62 (47.6)	0.86 (66.2)	1,08 (8.1)	2.16 (166.2)
Vitamin B_2 (mg)	1.92 (127.9)	3.03 (202.2)	3.44 (229.3)	4.35 (290.4)
Niacin (mg)	11.6 (68.3)	28.9 (169.9)	37.1 (218.2)	45.8 (269.4)
Vitamin C (mg)	19.7 (35.9)	83.7 (152.2)	86.4 (157.2)	99.6 (181.2)

[a] ( ), Percentage of recommended dietary allowance.

the RDA. Sufficient amounts of minerals and vitamins were supplied by the three-dish meal. Carbohydrates contributed 77.0% and 64.4% of the total energy in the basic meal and the three-dish meal, respectively, whereas lipid contributed only 8.3% and 11.6%. The energy from lipid did not exceed 12% of the total energy supply until a five-dish meal, which was considered a luxury, was analyzed.

The traditional Korean meal was estimated to be able to supply from 2000 to 2500 calories and from 80 to 90 g of protein per day. The energy constituents were 73% to 77% carbohydrates, 15% to 18% proteins, and 10% to 12% lipids. Animal protein was 20% to 30% of the total protein. The contribution of lipid energy in total calorie intake did not significantly change by increasing the number of side dishes to five, but that of protein did increase. It appears that the Korean traditional meal is well balanced in the view of modern nutritional science. It could supply sufficient protein, minerals, and vitamins to nourish an adult male if the energy intake exceeded 2000 calories per day (10).

High amounts of carbohydrates, which are mainly supplied by cereals and vegetables, and low amounts of animal meat and fat are characteristics of a traditional Korean diet. In the traditional Korean culture, food was considered to be the fundamental source of health, and it was believed that all diseases could be cured by the control of food intake. Without knowledge of the chemical composition of foods, Koreans have developed a well-balanced diet by using their *yin* and *yang* and Five Phases theories. Whereas nutritional science in Western society is based on analytical techniques confirmed with animal experiments, the Eastern concept of food and nutrition is developed for the harmonization of human beings and nature through long experience with human trials.

On the basis of the health and nutritional concepts of Korea, Hong Seon Pyo (13) proposed dietary guidelines in his *Book of Korean Cookery*, published in 1940, as follows:

1. Eat only when hungry.
2. Eat hard materials, with adequate mastication.
3. Stop eating before achieving satisfaction.
4. Eat raw food wherever possible.

He suggested using certain principles in selecting ingredients for the preparation of healthy food:

1. Fresh
2. Raw
3. Natural
4. Long-lived plants and animals
5. Dense texture
6. Young plants and animals
7. Materials produced nearby
8. Nonstimulating foods

He also recommended the reduction of salt and refined sugar intake. His dietary guidelines and principles of selecting food materials are widely accepted today.

Considering food to be medicine, practitioners of traditional medicine studied each food ingredient for its property, taste, and medicinal effects. Their knowledge has been compiled in numerous medicinal books in China and Korea for thousands of years and has been practiced in everyday life at the household level as part of Korean dietary custom.

Food preparation was likened to prescription of medicine for the individuals in a household. The word *yaknyum,* the general term for "seasoning", means "thought of medicine." This mentality refuses to accept processed food made in a mass production system. The enormous size of the health food market today in Korea and neighboring countries reflects the tradition of "food as medicine."

A 1996 survey of consumer attitudes toward health food and their perceptions on health and food habits in Korea revealed that the people considered their food habits as the most important factor in the maintenance of health, followed by physical exercise. More than 90% of the people believed that food habits were the most important factor determining the condition of health of human beings, and that diseases could be cured by adjusting food habits (14). One-half of the subjects had the experience of using health foods, and 68% believed in their effectiveness (15).

IV. HARMONIZATION FOR PARADIGM SHIFT

Food technology and modern nutritional science have improved the well-being and health status of humankind tremendously for the last century. The chemical analysis of food components makes us evaluate not only the nutritional value of food, but also physiological function at the microgram level of minor components. Food is no longer merely an energy and essential nutrient supplier; it is now proved by modern analytical tools to be a disease-preventing and even a disease-curing agent. The border between food and medicine is obscure, as shown in Fig. 2 (16).

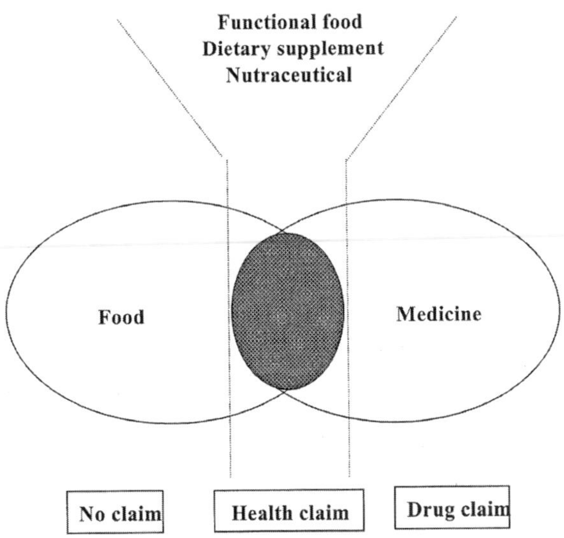

Figure 2 Definition and regulatory status of health foods.

The regulations adopted from Western society strictly distinguish food from medicine, creating severe conflict with the general concept of food and medicine in traditional societies. The range of health claim allowances for food varies from country to country for this reason. The demand for health claims in food will increase as modern analytical techniques explore more of the physiologically effective substances in food.

It appears that the old Eastern concept of food as medicine is now being proved by Western analytical methods. However, the approaches taken to utilize natural products for health benefits of humankind are different in East and West. Western analytical science continues to identify the effective components in food materials and purify them and make into pills, as they have for vitamins. It is the main driving force of research today to produce profitable nutraceuticals and drugs from natural products. Biotechnology is being used to maximize the production of useful components in the cell and facilitate their separation. This trend of product development will probably cause the same quagmires we have experienced in Western society, such as removal of fiber from food materials, causing constipation and colon cancer, and consumption of high-protein and high-fat diet, aggravating cancer, heart disease, and osteoporosis.

The Eastern approach is blending various components to maximize beneficial effects and minimize side effects. Eastern people know that no single component can provide absolute benefit to the body, which is made of various organs and tissues that have different functions and chemical reactions. Many of the health and medicinal effects of natural products are derived not from a single component but from the integrated action of numerous components. Emphasizing only a known beneficial effect by concentrating the component responsible may potentially create numerous unknown adverse effects. This caution makes the Eastern medical practitioners use blending of herbs and other natural products in medicine. Some of the ingredients are used only for the reduction of an adverse effect of the major ingredients. The same principles have been applied for selection of food ingredients, as represented by the *yin–yang* and Five Phases theories mentioned.

Now the answer is rather clear about the direction the paradigm shift should follow. The analytical approach of Western science and technology has gone too far in the "differential" region and is likely to overlook the whole picture. It is necessary to consider matters with an integral view, considering harmony of human and universe, sparing natural resources, and practicing ethical behavior with our fellow creatures. We often discuss "sustainable development" and "environmentally friendly production systems". These tendencies are all in the same direction. Modern biotechnology should pay great attention to this proposition, so that it will not be hindered by social resistance and rejection.

REFERENCES

1. J Needham. Science and Civilisation in China, Vol. 1. Introductory Orientations. Cambridge: The University Press, 1954, p.18.
2. J Robbins. Diet for a New America. Tiburon, CA: H J Kramer Book, 1987, p.xiv.
3. US Senates, Selected Committee on Nutrition and Human Needs. Dietary goals for the United States, 2nd ed., U. S. Government Printing Office, Washington, D. C., 1977.
4. E Frazao. High cost of poor eating patterns in the United States. In: America Eating Habits. USDA-ERS, AIB-750, 2000.
5. CY Lee. Future food, Harmonization of Eastern and Western food systems. Abstracts of the 11th World Congress of Food Science and Technology, April 22–27, COEX, Seoul, Korea, 2001, p.74.
6. CH Lee. Country paper, Republic of Korea, In International Trade and Food Security in Asia. Asian Productivity Organization, Tokyo, 2000, pp.195–208.
7. KHIDI (Korea Health Industry Development Institute). Report on 1998 National Health and Nutrition Survey, 1999.
8. CH Lee. Changes in the dietary patterns, health, and nutritional status of Koreans during the last century. Korean and Korean American Studies Bulletin, 6(1), 32–47, 1995.
9. SW Lee. *Hankook Sikpum Sahoesa* (History of Korean Food and Society). Kyomunsa, Seoul, 1984, pp. 168.
10. CH Lee. Fermentation Technology in Korea. Korea University Press, Seoul, 2001, p.1–22.
11. MH Kim. *Nature, Man and Traditional Korean Medicine*. Yoksa Bipyungsa, Seoul, 1995, p. 187.
12. CH Lee, SS Ryu. Nutritional evaluation of Korea traditional diet. *Korean Journal of Dietary Culture*, 3(3), 275–280, 1988.
13. SP Hong. *Chosun Yorihak* (Book of Korean Cookery). Korea, 1940.
14. EJ Lee, SO Ro, CH Lee. A survey of consumer attitude toward health food in Korea. 1. Consumer perceptions of health and food habits. *Korean Journal of Dietary Coulture*, 11(4), 475–485. 1996.
15. EJ Lee, SO Ro, CH Lee. A survey of consumer attitude toward health food in Korea. 2. Consumer perceptions of health food. *Korean Journal of Dietary Coulture*, 11(4), 487–495, 1996.
16. CH Lee. Improvement of health claim regulation. Proceedings of Health Food Symposium, Seoul: Korean Institute of Food Science and Technology, 1998, pp.73–86.
17. WHO, World Health Statistics, 1997–1999.
18. Korea National Statistical Office, Statistics and Causes of Death, 1998–1999.

21

Consumer Attitudes Toward Biotechnology: Implications for Functional Foods

Christine M. Bruhn
University of California, Davis, Davis, California, U.S.A.

Concern about biotechnology continues to be low on consumers' list of concerns. When asked to volunteer food-related concerns, only 2% express concerns about the safety of foods modified by biotechnology. People support applications that benefit the environment; modifications that provide direct consumer benefits, such as increased nutritional value or better taste, are endorsed by slightly fewer people. Most consumer research has focused on plant applications of biotechnology; modification of animals may be more emotionally charged. Few European consumers consider themselves knowledgeable about biotechnology. Knowledge of basic biology is lacking, putting people at risk for misinformation. A sizable percentage of Europeans consider genetically modified plants very different from traditional plants and believe their personal genetic material will change if they consume genetically modified food. Among applications of biotechnology, more Europeans consider medical applications useful and appropriate for support when compared to other applications. Consumers are interested in information on the relationship between health and diet. Health-related applications may be well received with the appropriate communication program. Communication with Europeans intended to correct misinformation and describe potential benefits of biotechnology is challenging since few consumers trust government and industry information sources. Experience in the United States indicates that communication can change attitudes. Frequent and effective communication that highlights potential benefits and addresses public concerns is a prerequisite for increasing public acceptance.

I. INTRODUCTION

Modern biotechnology can be applied in a broad range of areas, offering advantages in food production and processing, environment stewardship, and human health (1). Crop production can be modified to provide increased yields on the same acreage, lower production costs, and use less of pesticides and herbicides. Plant breeders have used recombinant deoxyribonucleic acid (rDNA) technology to develop corn that is nutritionally more dense and easier for animals to digest. Human food can be modified to provide improved nutritional characteristics, lower levels of natural toxins, and increased quality. In the future, proteins that trigger allergic reactions may be removed, so people with allergies can enjoy previously prohibited foods. Although the range of promising applications is broad, not all consumers have responded positively to them.

II. ATTITUDES TOWARD BIOTECHNOLOGY

A. U.S. Consumers

Consumer surveys provide general information about consumer attitudes toward product innovations. Most U.S. consumers have a positive attitude toward biotechnology, and 61% believe biotech will benefit them and their family within the next 5 years (2). Furthermore, in a 1000-person nationwide U.S. telephone survey conducted in 2001, 65% of consumers indicated that they would purchase produce modified by biotechnology to reduce pesticide use, and 52% said that they would purchase products modified for better taste (2). When told that biotechnology-modified plants could be used to produce cooking oil with less saturated fat, only 14% indicated that the use of biotechnology would have a negative effect on their likelihood of purchase.

Few U.S. consumers perceive modifications by biotechnology as risky, and labeling food as genetically modified is not a priority. When asked in an opened-ended question about food safety concerns, only 2% volunteered concerns about genetically modified food (2). In contrast, concerns about food borne disease and safe handling were mentioned by 30% and 32%, respectively. When asked whether there was information not currently on a food label they would like to see added, only 1% asked for information indicating whether the food was genetically altered (2). When asked by another investigator to select one item from a list of potential label additions, 17% chose labeling that indicated whether the product was genetically altered, 33% selected information whether pesticides were used in production, 8% whether the product was imported; 16% responded that they needed no additional information; and 15% said they did not know (3).

Attitudes toward biotechnology can be influenced by perception of the technology's impact beyond the safety or quality of food. Surveys reveal that the public considers potential environmental risks caused by genetic modification important. When respondents were asked whether a series of potential risks are very, somewhat, or not at all important, the potential for contamination of plant species by genetic transfer was considered a very important risk by 64% of consumers (4). Other potential risks and the percentage of consumers considering the risk very important include the potential to create superweeds, 57%; to develop pesticide-resistant insects, 57%; to reduce genetic diversity, 49%; and to produce modified plants that could harm others, 48%.

The percentage of the public having a positive view of biotechnology has decreased over time. In 2001, 61% of consumers said they expect to receive personal benefits from biotechnology, whereas in 1997, 78% held that view (2). Similarly in 2001, 65% of consumers indicated they would purchase products modified to reduce pesticide use, whereas in 1997, 71% said they would buy modified products offering this benefit. This attitude change could be related to negative media coverage and the perception that potential risks were not under control. Media coverage of biotechnology increased from less than 1% of articles in 1997 to 6% in 1999 and 12% in 2001 (5). A content analysis of articles in 1999 found that claims of harm were expressed in 70% of the articles, whereas discussions of benefits were included in only 30% of the stories (6). Discussion of harm focused on environmental or human health, whereas benefits were generally limited to increased production. This is not likely to be an important benefit to consumers in areas where food is abundant. In 2001, articles on Starlink corn dominated media coverage at 73%, while the ability to detect biotechnology components was the focus of 11% of the articles, and labeling of biotechnology foods was discussed in 10% of the articles (5). Negative comments exceeded benefits by an 8 to 1 ratio. Although Starlink corn led to no known human illness, the potential for an allergic response was frequently mentioned. Furthermore, the incidence illustrated that modified plant products could unexpectedly be present in a wide variety of foods.

Many consumers value potential benefits made possible through biotechnology. When specific benefits are identified, 74% rated cleaning toxic pollutants as very important (4). Other potential benefits and the percentage of consumers considering the benefit very important include reducing soil erosion, 73%; using less fertilizer, 72%; developing drought-resistant plants, 68%; developing disease-resistant trees, 67%; and using less pesticide, 61%.

Few consumer studies have focused on animal applications of biotechnology. Early work found fewer than 40% of consumers indicate support for traditional cross-breeding practices (7). In contrast, when asked about the techniques of biotechnology in conjunction with specific benefits, more than 40% expressed support for using biotechnology to produce leaner meat, and more than 50% supported use of biotech to enhance animal disease resistance.

B. European Consumers

Few Europeans (11%) consider themselves informed about biotechnology (8). Questions related to facts of biology indicate basic knowledge is lacking among the general population. In 1999, only 34% of European consumers correctly responded that genetically modified animals are not always larger than conventional animals. Only 35% correctly responded that the following statement is false: "Ordinary tomatoes do not contain genes while genetically modified tomatoes do." Furthermore, only 42% recognized that eating genetically modified fruit does not change personal genes, 24% believed human genes would be changed, and the remaining were uncertain. A comparison of responses in 1996 and 1999 indicates that there has been little increase in public knowledge. In fact, fewer consumers responded correctly in 1999 to the statement that eating genetically modified fruit changes human genes than in 1996: 42% correct responses in 1999 compared to 48% in 1996.

Only about half of European consumers were aware of many of the applications of biotechnology (8). Slightly over half, 56%, were aware that genetic modification could be used to make plants resistant to insect attack. About half were aware that these tools could be used to detect inherited diseases (53%), prepare human medicine or vaccine from animals (52%), and clone human cells to replace those that are not functioning (49%). Slightly fewer (44%) were aware that tools of biotechnology could be used to introduce human genes into bacteria to make medicine such as insulin for diabetics. Only 28% knew biotechnology could be used to clean toxic spills. Similarly, a national survey in the United Kingdom found 22% said that there were no benefits of genetically modified food and 31% indicated they did not know about benefits (9). Furthermore, 69% of UK consumers indicated they did not know enough about genetically modified foods to made a decision about their use.

When basic knowledge about biology is lacking, the belief persists that consumption of modified food can change human genetic material, and awareness of benefits is limited, it is not unexpected that European consumers are concerned about potential risks associated with eating genetically modified food.

When respondents were asked to rate whether a biotechnology application is useful, is risky, or should be encouraged, medical applications were considered most useful, including using tools of biotechnology to detect hereditary disease, using human genes in bacteria to produce medicine, and replacing nonfunctional human tissue (8). Environmental remediation, described as developing genetically modified (GM) bacteria to clean toxic spills, was also more highly regarded.The usefulness rating for modifying food to obtain high protein, longer shelf life, or changed taste was lower than for the other applications explored: a usefulness rating at the midpoint. The support for environmental applications is consistent with a national survey in France. A majority of consumers (63%) indicated that it was acceptable to grow genetically modified crops in France if doing so allowed the development of more environmentally friendly practices (9).

Europeans considered all applications of genetically engineering somewhat risky (8). The applications considered most risky were food production and cloning of animals. Using tools of biotechnology to detect hereditary disease received the lowest risky rating but was still at the midpoint of risk.

When respondents were asked to rate the moral acceptability of various biotechnology applications, developing GM bacteria for environmental remediation, using biotechnology to detect hereditary disease, and cloning animals to produce medicine or vaccine were rated highest in acceptability (8). Two applications, making plants resistant to insects and modifying food to increase protein, shelf life, or taste, were below the midpoint of moral acceptability.

Those applications European consumers felt should be encouraged corresponded to perceptions of usefulness and moral acceptability (8). Medical and environmental applications received the highest ratings; production of insect-resistant plants and food-related modifications received the lowest. Since consumers respond differently to various applications of biotechnology, it is not the technology itself that is acceptable or unacceptable, but the way the technology is used. In an analysis of European attitudes, Gaskell (2000) observed that as the perceived usefulness of applications declines, there are an increase in perceived risk and a decline in moral acceptability and support (10). Perception of the value or worth of potential benefits is key to acceptance of the application of biotechnology and the products produced.

Communication about potential benefits and risks is critical to acceptance. When asked who is doing a good job in the area of genetic modification, more European consumers supported the work of consumer organizations, 70%, followed by newspapers, 59%, and environmental organizations, 58%

(8). The government and the food industry were perceived as doing a good job by only 45% and 30% of Europeans, respectively. Similarly, few European consumers, 3%–4%, indicated they trusted international or national public authorities in regard to work with genetic modifications; greatest trust was placed in consumer organizations, 26%, and medical professionals, 24%.

C. Attitude Change

People can change their attitudes in response to information from a credible source. Many groups are involved in education in the United States. The biotechnology industry funds television advertisements that describe the benefits of this technology. Professional societies, such as the American Dietetic Association and the Institute of Food Technologists, have prepared material for their members, the media, regulators, and the public. Universities and colleges also have outreach programs to keep the public informed about new developments.

A university-sponsored program delivered in California and Indiana provided information about biotechnology for community groups, such as Kiwanis, Lions, Rotary, and Soroptomist (11). A 14-minute videotape highlighting current applications of biotechnology was shown during the organization's normal meeting time, followed by a discussion with members as to potential risks and benefits of the technology. Tracking of attitudes of participants before and after the program revealed that the percentage of consumers who believe biotechnology offers society benefits increased from 68% before the program to 96% afterward. Similarly those who believe biotechnology presents society with risks increased from 46% to 65%. In an assessment of the overall impact of biotechnology on human health, those believing the impact would be positive increased from 71% initially to 90% after the program. Those believing the overall impact of biotechnology on the environment would be positive increased from 65% initially to 83% after the program.

Similarly, consumer surveys after television advertisements describing potential benefits of biotechnology show increased public recognition that biotechnology can be used to reduce pesticides, produce healthier foods, produce hardier corps, and develop new medicines (12). Those agreeing that biotechnology can be used to develop new medicines and health care techniques increased from 68% in March 2000 to 78% in July 2002. Similarly, those believing biotechnology develops healthier foods such as food that is higher in nutrients increased from 45% in March 2000 to 56% in July 2002. Support for biotechnology continues to be higher among people who have heard more about it. In July 2002, 67% of those who had heard some or a lot about biotechnology supported developing new varieties of crops, compared to 51% who had heard little or nothing about biotechnology.

III. ATTITUDES TOWARD HEALTH-PROMOTING FOODS

Consumer interest in diet and health is high. Most consumers check nutritional labels when first purchasing items (13–15). Consumers in the United States indicated an interest in receiving information about the health benefits of food. In a nationwide survey, 78% of consumers said they were interested in receiving information about food that boosts the immune system, 77% were interested in information about food that reduces the risk of disease, and 53% were interested in information about active cultures in yogurt (16). A focus group study conducted in California investigating consumer response to probiotic bacteria in dairy products found strong interest in product benefits (17). Consumers indicated their belief in the validity of health effects was increased when health claims were reviewed by the U.S. Food and Drug Administration and additional endorsement was provided by recognized health organizations (17). A survey of food manufacturers, retailers, and ingredient producers projected a 60% growth in functional foods in Europe and a 78% growth in the United States between 2000 and 2015 (18). Although consumer interest is high, national surveys find that the majority of Americans support increased government regulation to ensure that claims about health benefits are true (19).

IV. CONCLUSIONS

The key to consumer acceptance of biotechnology is perceived benefit. Consumers expect human, animal, and environmental safety to be protected. Consumer attitudes are more likely to be positive when people understand why animals or plants are modified, they view the potential benefit as important, and neither the animal nor the environment is harmed. Support of medical application of biotechnology is strong. Consumer attitudes toward modification of functional foods for health benefits may be aligned with medical applications when appropriately presented.

Messages about potential benefits should be presented many times, using a variety of media. Although consumers in the United States have positive attitudes, few are aware of all the potential applications under development. European consumers are less aware, and many do not trust regulators or the food industry. Messages about biotechnology that highlight potential benefits, address consumer concerns, and are delivered by trusted knowledge sources are critical to long term-acceptance.

REFERENCES

1. Institute of Food Technologists. IFT Expert Report on Biotechnology and Foods. 2000, Chicago, pp. 1–56.
2. W Wirthlin. More US consumers expect biotech benefits. 2001, International Food Information Council. http://ific.org. Last accessed July, 2002.
3. Bruskin. National opinion polls on labeling of genetically modified foods. 2001, Center for Science and the Public Interest. http://www.cspinet.org/new/poll_gefoods.html. Last accessed May, 2002.
4. Pew Initiative. Environmental savior or saboteur? Debating the impacts of genetic engineerng. 2002, Pew Charitable Trust. http://www.pewtrust.com. Last accessed June 2002.
5. International Food Information Council. Food for thought IV, Reporting on diet, nutrition and food safety 2001 vs 1999 vs 1997 vs 1995. 2002, Washington DC, pp.1– 7.
6. International Food Information Council. Food for thought III, Reporting on diet, nutrition and food safety 1999 vs 1997 vs 1995. 2000, Washington DC, pp. 1–7.
7. T Hoban, P Kendall. Consumer attitudes about food biotechnology, Project Report. 1993, North Carolina State University Cooperative Extension Service: Raleigh, NC, pp. 1–36.
8. INRA-Europe ECOSA. Eurobarometer 52.1 The Europeans and biotechnology. 2002, http://europa.eu.int/comm/public_opinion/archives/eb/ebs_134_en.pdf. Last accessed June 2002.
9. ABE. Public Attitudes to agricultural biotechnology, in Issue Paper 2. 2002, p. 9. http://www.ABEurope.info. Last accessed April 2002.
10. G Gaskell. Agricultural biotechnology and public attitudes in the European Union. 2000, http://www.agbioforum.org/. Last accessed May 2002.
11. C Bruhn, A Mason. Community leader response to educational information about biotechnology. J Food Science 67(1):399–403, 2002.
12. Council for Biotechnology Information. US Tracking Results, July. 2002, KRC Research & Consulting: Washington DC, pp. 13.
13. MM Bender, BM Derby. Prevalence of reading nutritional information and ingredient information on food labels among adult Americans: 1982–1988. J Nutr Ed 24(6):292–297, 1992.
14. MN Rodolfo, D Lipinski N Savur. Consumers' use of nutritional labels while food shopping and at home. J Consumer Affairs 32(1):106–120, 1998.
15. J Buzby, R Ready. Do consumers trust food safety information? Food Review 19(1):46–49, 1996.
16. Health Focus International, Health focus trend report. 2001: Atlanta, Georgia, p. 250.

17. CM Bruhn, JC Bruhn, A Cotter, C Garrett, M Klenk, C Powell, G Stanford, Y Steinbring, E West. Consumer attitudes toward use of probiotic cultures. J Food Science 67(5):1969–1972, 2002.
18. R Fonden, G Mogensen, R Tanaka, S Salminen. Trends in attitudes towards probiotics. In Effect of Culture-Containing Dairy Products in Health, Supplement F-Doc 294. International Dairy Federation, Athens, Greece, 1999, pp 42–46.
19. RJ Blendon, CM DesRoches, JM Bennsion, M Brodie, DE Altman. American's views on the use and regulation of dietary supplements. Arch Intern Med 161(6):805–810, 2001.

Index